Wastewater Treatments Based on Adsorption, Catalysis, Biodegradation, and Beyond

Wastewater Treatments Based on Adsorption, Catalysis, Biodegradation, and Beyond

Guest Editors

**Yongchang Sun
Dimitrios Giannakoudakis**

Basel • Beijing • Wuhan • Barcelona • Belgrade • Novi Sad • Cluj • Manchester

Guest Editors

Yongchang Sun
School of Water and Environment
Chang'an University
Xi'an
China

Dimitrios Giannakoudakis
Faculty of Chemistry
Maria Curie-Skłodowska University
Lublin
Poland

Editorial Office
MDPI AG
Grosspeteranlage 5
4052 Basel, Switzerland

This is a reprint of the Special Issue, published open access by the journal *Molecules* (ISSN 1420-3049), freely accessible at: www.mdpi.com/journal/molecules/special_issues/49Y2WXE772.

For citation purposes, cite each article independently as indicated on the article page online and using the guide below:

Lastname, A.A.; Lastname, B.B. Article Title. *Journal Name* **Year**, *Volume Number*, Page Range.

ISBN 978-3-7258-2752-7 (Hbk)
ISBN 978-3-7258-2751-0 (PDF)
https://doi.org/10.3390/books978-3-7258-2751-0

Cover image courtesy of Dimitrios Giannakoudakis

© 2024 by the authors. Articles in this book are Open Access and distributed under the Creative Commons Attribution (CC BY) license. The book as a whole is distributed by MDPI under the terms and conditions of the Creative Commons Attribution-NonCommercial-NoDerivs (CC BY-NC-ND) license (https://creativecommons.org/licenses/by-nc-nd/4.0/).

Contents

About the Editors vii

Preface ix

Dimitrios A. Giannakoudakis and Yongchang Sun
Wastewater Treatments Based on Adsorption, Catalysis, Biodegradation, and Beyond
Reprinted from: *Molecules* 2024, 29, 5470, https://doi.org/10.3390/molecules29225470 1

Jie Zhang, Hao Ji, Zepeng Liu, Liping Zhang, Zihao Wang and Ying Guan et al.
3D Porous Structure-Inspired Lignocellulosic Biosorbent of *Medulla tetrapanacis* for Efficient Adsorption of Cationic Dyes
Reprinted from: *Molecules* 2022, 27, 6228, https://doi.org/10.3390/molecules27196228 5

Chiara Mongioví, Maélys Jaillet, Dario Lacalamita, Nadia Morin-Crini, Michael Lecourt and Sandra Tapin-Lingua et al.
A Strategy to Valorize a By-Product of Pine Wood (*Pinus pinaster*) for Copper Removal from Aqueous Solutions
Reprinted from: *Molecules* 2023, 28, 6436, https://doi.org/10.3390/molecules28186436 22

Lei Zhang, Heng Liu, Jiaqi Zhu, Xueling Liu, Likun Li and Yanjun Huang et al.
Effective Removal of Fe (III) from Strongly Acidic Wastewater by Pyridine-Modified Chitosan: Synthesis, Efficiency, and Mechanism
Reprinted from: *Molecules* 2023, 28, 3445, https://doi.org/10.3390/molecules28083445 41

Maria Philippou, Ioannis Pashalidis and Charis R. Theocharis
Uranium Isotope (U-232) Removal from Waters by Biochar Fibers: An Adsorption Study in the Sub-Picomolar Concentration Range
Reprinted from: *Molecules* 2022, 27, 6765, https://doi.org/10.3390/molecules27196765 59

Heng Liu, Jiaqi Zhu, Qimei Li, Likun Li, Yanjun Huang and Yi Wang et al.
Adsorption Performance of Methylene Blue by $KOH/FeCl_3$ Modified Biochar/Alginate Composite Beads Derived from Agricultural Waste
Reprinted from: *Molecules* 2023, 28, 2507, https://doi.org/10.3390/molecules28062507 68

Tushar Kanti Sen
Agricultural Solid Wastes Based Adsorbent Materials in the Remediation of Heavy Metal Ions from Water and Wastewater by Adsorption: A Review
Reprinted from: *Molecules* 2023, 28, 5575, https://doi.org/10.3390/molecules28145575 82

Risheng Li, Wen Sun, Longfei Xia, Zia U, Xubo Sun and Zhao Wang et al.
Adsorption of Toxic Tetracycline, Thiamphenicol and Sulfamethoxazole by a Granular Activated Carbon (GAC) under Different Conditions
Reprinted from: *Molecules* 2022, 27, 7980, https://doi.org/10.3390/molecules27227980 104

Marina Gutiérrez, Paola Verlicchi and Dragana Mutavdžić Pavlović
Study of the Influence of the Wastewater Matrix in the Adsorption of Three Pharmaceuticals by Powdered Activated Carbon
Reprinted from: *Molecules* 2023, 28, 2098, https://doi.org/10.3390/molecules28052098 119

Yuan Zhou, Yingying He, Ruixue Wang, Yongwei Mao, Jun Bai and Yan Dou
Modification of Multiwalled Carbon Nanotubes and Their Mechanism of Demanganization
Reprinted from: *Molecules* 2023, 28, 1870, https://doi.org/10.3390/molecules28041870 140

Maria Sadia, Izaz Ahmad, Zain Ul-Saleheen, Muhammad Zubair, Muhammad Zahoor and Riaz Ullah et al.
Synthesis and Characterization of MIPs for Selective Removal of Textile Dye Acid Black-234 from Wastewater Sample
Reprinted from: *Molecules* **2023**, *28*, 1555, https://doi.org/10.3390/molecules28041555 162

Mohammed F. Hamza, Eric Guibal, Khalid Althumayri, Thierry Vincent, Xiangbiao Yin and Yuezhou Wei et al.
New Process for the Sulfonation of Algal/PEI Biosorbent for Enhancing Sr(II) Removal from Aqueous Solutions—Application to Seawater
Reprinted from: *Molecules* **2022**, *27*, 7128, https://doi.org/10.3390/molecules27207128 179

Nur Anis Liyana Kamaruddin, Mohd Faisal Taha and Cecilia Devi Wilfred
Synthesis and Characterization of Novel Thiosalicylate-based Solid-Supported Ionic Liquid for Removal of Pb(II) Ions from Aqueous Solution
Reprinted from: *Molecules* **2023**, *28*, 830, https://doi.org/10.3390/molecules28020830 220

Thi Quynh Hoa Kieu, Thi Yen Nguyen and Chi Linh Do
Treatment of Organic and Sulfate/Sulfide Contaminated Wastewater and Bioelectricity Generation by Sulfate-Reducing Bioreactor Coupling with Sulfide-Oxidizing Fuel Cell
Reprinted from: *Molecules* **2023**, *28*, 6197, https://doi.org/10.3390/molecules28176197 237

Basma G. Alhogbi and Ghadeer S. Al Balawi
An Investigation of a Natural Biosorbent for Removing Methylene Blue Dye from Aqueous Solution
Reprinted from: *Molecules* **2023**, *28*, 2785, https://doi.org/10.3390/molecules28062785 253

Qiongyao Wang, Yongchang Sun, Mingge Hao, Fangxin Yu and Juanni He
Hydrothermal Synthesis of a Technical Lignin-Based Nanotube for the Efficient and Selective Removal of Cr(VI) from Aqueous Solution
Reprinted from: *Molecules* **2023**, *28*, 5789, https://doi.org/10.3390/molecules28155789 269

Soumya K. Manikandan, Pratyasha Pallavi, Krishan Shetty, Debalina Bhattacharjee, Dimitrios A. Giannakoudakis and Ioannis A. Katsoyiannis et al.
Effective Usage of Biochar and Microorganisms for the Removal of Heavy Metal Ions and Pesticides
Reprinted from: *Molecules* **2023**, *28*, 719, https://doi.org/10.3390/molecules28020719 287

About the Editors

Yongchang Sun

Yongchang Sun is a professor of Environmental Science and Technology in the School of Water and Environment, Chang'an University, Xi'an, China, where he has been actively involved in teaching and research since 2014. In 2008, he received his bachelor's degree in chemistry, and he received his Ph. D. in forest product chemical engineering from Beijing Forestry University, Beijing, China, in 2014. He completed his postdoctoral studies at the Department of Environmental Science and Technology, Chang'an University in Xi'an in 2015–2017. He is a member of the American Chemical Society and the Chinese Chemical Society. Professor Sun is a Qinling Mountain Ecological Protection Young Scholar in China and an outstanding scholar in the environmental science and technology research area. He also received the Chang'an Young Scholars award from Chang'an University. He conducts fundamental and applied research in water and soil environment protection and environmental chemistry, with a specific focus on wastewater treatment, groundwater pollution remedy, pollution control (heavy metals, organics, emerging contaminants, etc.), soil pollution control, (nano)material and biochar preparation and utilization, biomass fractionation, and environmental catalysis.

To date, Professor Sun is the author of nearly 60 peer-reviewed scientific papers, published in the journals of SCI, and he currently serves as a Topical Advisory Panel Member of *Molecules* (2023–). Recently, he became one of the editors for the special issue for the *Journal of Molecular Catalysis, Water, and Molecules*, etc. He earned the Excellent Guest Editor award from MDPI in 2024. He has already guided more than 20 graduate students (including international students).

Dimitrios Giannakoudakis

Dr. Dimitrios A. Giannakoudakis graduated as a chemist from the Aristotle University of Thessaloniki (AUTh) in Greece, where he also obtained two MSc degrees (one in physical and electro-chemistry and another in new educational technologies). He received his PhD at the City University of New York (CUNY) and was a sub-disciple of "Nanotechnology and Materials Chemistry" in 2017, with full scholarship from the CUNY from the Onassis Foundation. He continued as a postdoctoral researcher and adjunct lecturer at the City College of New York (CCNY), Aristotle University of Thessaloniki (AUTh), and Institute of Physical Chemistry in Warsaw (IChF). Afterwards, he served at IChF as an adjunct associate professor and lecturer of "Modern Topics in Physical Chemistry". During the COVID pandemic, he moved to AUTh as a research associate. Nowadays, he serves as an assistant professor and as a PI of a SONATA grant in the Faculty of Chemistry, Maria Curie-Sklodowska University (UMCS), Poland.

His research focuses on photo/thermo/sono-catalysis (solely or in combination) for biomass valorization; (micro)plastic upcycling, environmental, and energy applications; and the design, functionalization, and characterization of nano-engineered materials. DG is included in the top 2% of the world's most influential scientists (single years 2021, 2022, and 2023, Stanford list, Elsevier). He was elected as vice-chair at the European Chemistry Society (EuChemS), division of Physical Chemistry (2023–25). To date, he has co-authored >125 articles in leading journals (average IF as first or last author above 11) that are cited more than 4500 times, one monograph, six edited books, seventeen book chapters, and four invention patents, with his work being presented more than 100 times in conferences. He is on the editorial boards of the journals *Molecular Catalysis* (Elsevier), *Carbon Research* (Springer), and *Sustainable Chemistry* (MDPI). His experience in teaching includes more than 15 semesters in USA, Poland, and Greece for graduate and undergraduate levels.

Preface

The continuous widespread technological progress and industrial expansion over the last several decades has been accompanied with the serious environmental problem of water pollution. The pollution of water is related to the uncontrollable contamination of water bodies such as lakes, rivers, oceans, and groundwater by a plethora of substances/pollutants that can be harmful for humans, as well as for flora and fauna, even at very low concentrations. For example, heavy metals/metalloids can accumulate in living organisms, causing permanent cell damage and disorders leading to disease and even death. The unmanageable disposal of pharmaceuticals and personal care products (PPCPs) in water bodies can cause serious ecotoxicological problems and pose extraordinary threats to ecosystems or organisms. Microplastics are a class of emerging pollutants that generate severe environmental issues because of their small size, unique morphology, and enhanced chemical heterogeneity but above all due to their stability and ability to act as pollutant carriers.

Environmental protection is regarded as key in the design and development of a sustainable future and hence, the attention of researchers and the public is concentrated on novel remediation approaches. Intense emphasis is placed on the removal of heavy metals, PPCPs, MPs, and other pollutants from water and wastewater. The diverse composition of polluted water bodies and of industrial wastewater requires a variety of treatment methods. Heavy metal ions are most often removed by the precipitation of their hardly soluble compounds. Ion exchange and sorption methods are also widely used. Sorption is an effective method for the removal of emerging contaminants and heavy metals from water and wastewater. Sorbents derived from discarded biomass, wastes, and other feedstocks are widely studied for the treatment of contaminated water, since this material development strategy is within the framework of the sustainable (bio)economy. Additionally, biodegradation and catalytic degradation methods are receiving attention for the removal of PPCPs and MPs. Therefore, the research and development of novel and efficient materials for environmental remediation applications and especially for the removal of pollutants from water bodies remains an active field of research.

This Special Issue features 14 original research articles and two reviews that explore recent advancements in sorption, biodegradation, catalytic degradation, and beyond, using new materials for effective (waste)water treatment and purification. We extend our gratitude to the 86 authors from countries including China, France, Vietnam, Cyprus, Saudi Arabia, Italy, India, Egypt, Pakistan, Croatia, Malaysia, and Greece. We also greatly appreciate the dedicated efforts of the reviewers!

Yongchang Sun and Dimitrios Giannakoudakis
Guest Editors

Editorial

Wastewater Treatments Based on Adsorption, Catalysis, Biodegradation, and Beyond

Dimitrios A. Giannakoudakis [1,2,*] and Yongchang Sun [3,*]

1. Institute of Chemical Sciences, Faculty of Chemistry, Maria Curie-Sklodowska University, Maria Curie Sklodowska Sq. 3, 20-031 Lublin, Poland
2. School of Chemistry, Aristotle University of Thessaloniki, 54124 Thessaloniki, Greece
3. School of Water and Environment, Chang'an University, Xi'an 710054, China
* Correspondence: dagchem@gmail.com (D.A.G.); ycsun@chd.edu.cn (Y.S.)

Citation: Giannakoudakis, D.A.; Sun, Y. Wastewater Treatments Based on Adsorption, Catalysis, Biodegradation, and Beyond. *Molecules* **2024**, *29*, 5470. https://doi.org/10.3390/molecules29225470

Received: 25 October 2024
Accepted: 11 November 2024
Published: 20 November 2024

Copyright: © 2024 by the authors. Licensee MDPI, Basel, Switzerland. This article is an open access article distributed under the terms and conditions of the Creative Commons Attribution (CC BY) license (https://creativecommons.org/licenses/by/4.0/).

The ongoing technological advancements and industrial growth over the past few decades have resulted in significant environmental challenges, with one of the most notable being water pollution caused by the improper disposal of organic and inorganic pollutants. Protecting the environment is an essential component of a sustainable future, leading researchers and the public to focus on innovative remediation strategies. As a result, the development of new "green"-oriented, low-cost, and efficient materials for environmental remediation, especially those which facilitate the removal of pollutants from water bodies, remains a vibrant area of research. This Special Issue encompasses 14 original research articles and 2 review papers which aim to explore new methods outside of sorption, biodegradation, and catalytic degradation, while also introducing new materials and composites for effective (waste) water treatment and purification.

In their study, Gao et al. used a biomass, Medulla tetrapanacis (MT), as a sorbent. Known as "da-tong-cao" in China, where it is also famously used as a traditional medicine [1], MT is a well-developed porous 3D structure with an ultra-thin cell wall, predominantly consisting of holocellulose (~82 wt.%) and ash/minerals (~11 wt.%). MT showed high remediation efficiency against methylene blue (MB) and crystal violet (CV), achieving 411 and 553 mg/g adsorption, respectively. Interestingly, the depleted samples were pyrolyzed to biochars, achieving adsorption performances of 320 mg/g for Cu^{2+} and 840 mg/g for Pb2. Crini et al. utilized individualized fibers of a pine wood by-product (Pinus pinaster) were utilized as a sorbent for copper in poly-contaminated aqueous solutions [2]. Upon activation with sodium carbonate, the pine fibers successfully removed 2.5 mg/g copper regardless of the changing pH levels, which ranged from 3 to 5. The presence of the Na^+ cation at concentrations of 0.1 M did not affect the performance of the material. The adsorption process was rapid, as most of the copper was absorbed within the first 10 min of exposure. The underlying mechanism is considerably more complex due to physisorption, chemisorption, and/or diffusion phenomena. Wang and coworkers followed a multistep protocol to synthesize chitosan modified with pyridine and crosslinked with glutaraldehyde (PYCS) material that achieved 66.2 mg/g of Fe(III) adsorption at pH = 2.5 [3]. The surface pyridine groups were responsible for the formation of stable chelate with Fe(III) ions. PYCS successfully maintained Fe sorption efficiency, since the removal performance only decreased by 29% after six regeneration cycles. Another lignocellulosic biomass composed of Luffa cylindrica fibers was used by Theocharis et al. to synthesize biochar (LCC) after thermal treatment in a nitrogen atmosphere (at 650 °C for 1 h), which was further oxidized (LCC_ox) using HNO_3 [4]. The materials showed elevated adsorption of U-232 radionuclide at sub-picomolar initial concentrations. Oxidation of the biochar had a positive effect on uranium removal, with the adsorption process being affected by the dispersion pH. The interactions were found to be entropy-driven (ΔH° and ΔS° > 0) based on the creation of inner-sphere complexes. Zhang et al. used recycled agricultural waste from corncobs and KOH/FeCl$_3$ to prepare a biochar/alginate composite bead (MCB/ALG) adsorbent via high-temperature pyrolysis (at 800 °C for 2 h) [5]. Although MCB/ALG did not exhibit high porosity (129 m^2/g), it achieved 1373 mg/g MB removal, and the process was found to be endothermic and spontaneous. This performance was linked to the filling of the pores, electrostatic interactions, and hydrogen bonding. After five cycles of regeneration, the MB removal remained high (85% compared to the first cycle). Tushar Kanti Sen delivered a comprehensive review based on an up-to-date literature overview, focusing on studies that utilized a wide range of solid waste agricultural biomass-based adsorbents (raw, modified, and treated) to purify water that had been polluted with various metal ions [6]. Various important influential physicochemical process parameters, like metal concentration, adsorbent dose, the initial pH of the solution, and the temperature were considered as potential perspectives and conclusions for future work.

Deng and coworkers applied activated carbon in granular form (GAC) for water purification upon the co-presence of three antibiotics, specifically thiamphenicol (THI), sulfamethoxazole (SMZ), and tetracycline (TC) [7]. The carbon granules were obtained from corn stover biomass that was dried, chopped into small pieces, and heated at 500 °C for 2 h and then at 700 °C for 2 h under a N$_2$ atmosphere, then underwent further activation via a superheated steam (600 °C, 2.0 MPa, 2 h). GAC had a surface area of 1059 m^2/g and was predominately microporous (Vmic = 0.488 cm^3/g). The maximum capacities were around 27, 17, and 30 mg/g for SMZ, TC, and THI, respectively; the sorption was exothermic and spontaneous. The Weber–Morris intraparticle diffusion model and the Boyd kinetic model demonstrated that diffusion across the boundary layer was the primary determinant of the adsorption process. Activated porous carbons (PACs) were studied in another work by Pavlovic et al. as adsorbents of three pharmaceuticals: sulfamethoxazole, trimethoprim, and diclofenac. The removal efficiency was studied using ultra-pure water, humic acid solution, or liquor from real samples collected from wastewater treatment plants [8]. The best removal results were recorded for trimethoprim, followed by diclofenac and sulfamethoxazole. The most notable differentiating factors were linked to the charge and hydrophobicity of the pharmaceuticals at a specific pH. The maximum capacities varied depending on the water matrix; the highest capacity was achieved for sulfamethoxazole and diclofenac in humic acid solution. Multiwalled carbon nanotubes (MWCNTs) were modified through oxidation and acidification [9] using concentrated HNO$_3$ and H$_2$SO$_4$, respectively, and were studied as Mn(II) adsorbents by Dou et al. The chemical treatment had a positive impact on the Mn remediation efficiency, which was almost four times higher than that of the pristine nanotubes. The reliability of the experimental results was thoroughly validated by the PSO-BP simulation, and the findings can serve as a basis for subsequent simulations.

A molecularly imprinted polymer (MIP) was synthesized by Zekker and coworkers through bulk polymerization and utilized in wastewater treatment to enhance the adsorption of specific template molecules [10]. This process utilized ethylene glycol dimethacrylate (EGDMA) as the crosslinker, methacrylic acid (MAA) as the functional monomer, acid black-234 (AB-234) as the template, 2,2$'$-azobisisobutyronitrile (AIBN) as the initiator, and methanol as the porogenic solvent. The adsorption capacity of the MIP for AB-234 was significantly greater (94%) than that of the NIP (31%) at pH 5, with an estimated maximum capacity of 83 mg/g at 298 K based on the Langmuir model. The MIP exhibited an imprinted factor (IF) of 5.13 and a Kd value of 0.53. Li and coworkers designed a new sulfonation process for an algal/polyethyleneimine (PEI) composite [11]. The functionalized sorbent showed a triple sorption capacity for Sr(II) at pH 4 compared to its non-sulfonated counterpart. The sulfonate groups played a key role in adsorption (as revealed by IR), and sulfonation had a positive impact on thermal stability and porosity. The co-presence of NaCl had a negligible influence on sorption, explaining the good Sr(II) remediation in seawater samples, which was reproducible for at least five cycles. Wilfed and coworkers tested solid-supported ionic liquid consisting of activated silica gel combined with 1-methyl-3-(3-trimethoxysilylpropyl) imidazolium thiosalicylate-based ionic liquid as an extractant of Pb(II) ions from aqueous solutions [12]. The formation of covalent bonds was confirmed by solid-state NMR. The maximum Pb(II) removal capacity was recorded as ~9 mg/g, with the kinetics to be better described by a pseudo-second order model.

Linh Do et al. developed a system that utilizes sulfate-reducing and sulfide-oxidizing processes to treat organic wastewater with high sulfate and sulfide levels [13]. The effects of the COD/SO_4^{2-} ratio and hydraulic retention time (HRT) on the removal efficiencies of sulfate, COD, sulfide, and electricity generation were examined. The results indicated that the removal efficiencies for COD and sulfate were stable, reaching ~95 and 93%, respectively, throughout the operation. A power density of 18.0 ± 1.6 mW/m^2 was recorded alongside a sulfide removal efficiency of 93%. However, both sulfide removal efficiency and power density gradually declined after 45 days. Scanning electron microscopy combined with energy-dispersive X-ray analysis revealed that sulfur accumulated on the anode, leading to decreases in sulfide oxidation and electrical generation. This study presents a promising treatment system that could be scaled up for practical applications related to the management of this type of wastewater.

Alhogbi and Balawi presented composite adsorbents consisting of kaolin-derived zeolite and different mass ratios of palm tree fibers (Zeo-FPT) [14]. Mixing the two counterparts resulted in a composite that achieved faster and more efficient removal of methylene blue. The remediation efficiency of the material, which consisted of a one-to-one mass ratio of fibers and zeolite was strong; it achieved >99% removal of a low initial MB concentration in bottled, tap, and well water. He and co-authors prepared a photo-active composite using aminated lignin (AL) and titanate nanotubes [15]. AL was prepared according to the Mannich reaction by modifying technical lignin (TL), while the composite (AL-TiNTs) was formed via a hydrothermal synthesis approach. AL-TiNTs had a specific surface area of 189 m^2/g, indicating the formation of titanate nanotubes. AL-TiNTs showed an ability to reduce Cr(VI) to Cr(III) and Cr under exposure to visible light, with maximum chromium removal of 90 mg/g. AL-TiNTs also achieved high adsorption performances against Zn^{2+} (64 mg/g), Cd^{2+} (59 mg/g), and Cu^{2+} (66 mg/g), with high efficiency in simulated wastewater for up to four cycles.

Nair and co-authors reviewed the application of biochar-based materials as adsorbents and support for microorganisms to achieve efficient bioremediation of various heavy metal ions and/or pesticides, emphasizing and summarizing the predominant interaction mechanisms and how the immobilized bacteria on biochar contribute to the improvement of bioremediation strategies [16]. In conclusion, this paper outlines the future scopes of this field based on the reviewed and discussed research.

Author Contributions: D.A.G. and Y.S.: Writing-original draft preparation and review and editing. All authors have read and agreed to the published version of the manuscript.

Acknowledgments: The Guest Editors would like to express their gratitude to all the authors for their contributions to this Special Issue, to the reviewers for their efforts in assessing the submitted articles, and to the editorial team of *Molecules* for their valuable support.

Conflicts of Interest: The authors declare no conflicts of interest.

References

1. Zhang, J.; Ji, H.; Liu, Z.; Zhang, L.; Wang, Z.; Guan, Y.; Gao, H. 3D Porous Structure-Inspired Lignocellulosic Biosorbent of *Medulla tetrapanacis* for Efficient Adsorption of Cationic Dyes. *Molecules* **2022**, *27*, 6228. [CrossRef]
2. Mongiovi, C.; Jaillet, M.; Lacalamita, D.; Morin-Crini, N.; Lecourt, M.; Tapin-Lingua, S.; Crini, G. A Strategy to Valorize a By-Product of Pine Wood (*Pinus pinaster*) for Copper Removal from Aqueous Solutions. *Molecules* **2023**, *28*, 6436. [CrossRef]
3. Zhang, L.; Liu, H.; Zhu, J.; Liu, X.; Li, L.; Huang, Y.; Fu, B.; Fan, G.; Wang, Y. Effective Removal of Fe (III) from Strongly Acidic Wastewater by Pyridine-Modified Chitosan: Synthesis, Efficiency, and Mechanism. *Molecules* **2023**, *28*, 3445. [CrossRef] [PubMed]
4. Philippou, M.; Pashalidis, I.; Theocharis, C.R. Uranium Isotope (U-232) Removal from Waters by Biochar Fibers: An Adsorption Study in the Sub-Picomolar Concentration Range. *Molecules* **2022**, *27*, 6765. [CrossRef] [PubMed]
5. Liu, H.; Zhu, J.; Li, Q.; Li, L.; Huang, Y.; Wang, Y.; Fan, G.; Zhang, L. Adsorption Performance of Methylene Blue by KOH/FeCl$_3$ Modified Biochar/Alginate Composite Beads Derived from Agricultural Waste. *Molecules* **2023**, *28*, 2507. [CrossRef] [PubMed]

6. Sen, T.K. Agricultural Solid Wastes Based Adsorbent Materials in the Remediation of Heavy Metal Ions from Water and Wastewater by Adsorption: A Review. *Molecules* **2023**, *28*, 5575. [CrossRef] [PubMed]
7. Li, R.; Sun, W.; Xia, L.; Zia, U.; Sun, X.; Wang, Z.; Wang, Y.; Deng, X. Adsorption of Toxic Tetracycline, Thiamphenicol and Sulfamethoxazole by a Granular Activated Carbon (GAC) under Different Conditions. *Molecules* **2022**, *27*, 7980. [CrossRef] [PubMed]
8. Gutiérrez, M.; Verlicchi, P.; Pavlović, D.M. Study of the Influence of the Wastewater Matrix in the Adsorption of Three Pharmaceuticals by Powdered Activated Carbon. *Molecules* **2023**, *28*, 2098. [CrossRef] [PubMed]
9. Zhou, Y.; He, Y.; Wang, R.; Mao, Y.; Bai, J.; Dou, Y. Modification of Multiwalled Carbon Nanotubes and Their Mechanism of Demanganization. *Molecules* **2023**, *28*, 1870. [CrossRef] [PubMed]
10. Sadia, M.; Ahmad, I.; Ul-Saleheen, Z.; Zubair, M.; Zahoor, M.; Ullah, R.; Bari, A.; Zekker, I. Synthesis and Characterization of MIPs for Selective Removal of Textile Dye Acid Black-234 from Wastewater Sample. *Molecules* **2023**, *28*, 1555. [CrossRef] [PubMed]
11. Hamza, M.F.; Guibal, E.; Althumayri, K.; Vincent, T.; Yin, X.; Wei, Y.; Li, W. New Process for the Sulfonation of Algal/PEI Biosorbent for Enhancing Sr(II) Removal from Aqueous Solutions—Application to Seawater. *Molecules* **2022**, *27*, 7128. [CrossRef] [PubMed]
12. Kamaruddin, N.A.L.; Taha, M.F.; Wilfred, C.D. Synthesis and Characterization of Novel Thiosalicylate-based Solid-Supported Ionic Liquid for Removal of Pb(II) Ions from Aqueous Solution. *Molecules* **2023**, *28*, 830. [CrossRef] [PubMed]
13. Kieu, T.Q.H.; Nguyen, T.Y.; Do, C.L. Treatment of Organic and Sulfate/Sulfide Contaminated Wastewater and Bioelectricity Generation by Sulfate-Reducing Bioreactor Coupling with Sulfide-Oxidizing Fuel Cell. *Molecules* **2023**, *28*, 6197. [CrossRef] [PubMed]
14. Alhogbi, B.G.; Al Balawi, G.S. An Investigation of a Natural Biosorbent for Removing Methylene Blue Dye from Aqueous Solution. *Molecules* **2023**, *28*, 2785. [CrossRef] [PubMed]
15. Wang, Q.; Sun, Y.; Hao, M.; Yu, F.; He, J. Hydrothermal Synthesis of a Technical Lignin-Based Nanotube for the Efficient and Selective Removal of Cr(VI) from Aqueous Solution. *Molecules* **2023**, *28*, 5789. [CrossRef] [PubMed]
16. Manikandan, S.K.; Pallavi, P.; Shetty, K.; Bhattacharjee, D.; Giannakoudakis, D.A.; Katsoyiannis, I.A.; Nair, V. Effective Usage of Biochar and Microorganisms for the Removal of Heavy Metal Ions and Pesticides. *Molecules* **2023**, *28*, 719. [CrossRef] [PubMed]

Disclaimer/Publisher's Note: The statements, opinions and data contained in all publications are solely those of the individual author(s) and contributor(s) and not of MDPI and/or the editor(s). MDPI and/or the editor(s) disclaim responsibility for any injury to people or property resulting from any ideas, methods, instructions or products referred to in the content.

Article

3D Porous Structure-Inspired Lignocellulosic Biosorbent of *Medulla tetrapanacis* for Efficient Adsorption of Cationic Dyes

Jie Zhang, Hao Ji, Zepeng Liu, Liping Zhang, Zihao Wang, Ying Guan and Hui Gao *

School of Forestry and Landscape Architecture, Anhui Agricultural University, Hefei 230036, China
* Correspondence: huigaozh@ahau.edu.cn

Abstract: The focus of this work was on developing a green, low-cost, and efficient biosorbent based on the biological structure and properties of MT and applying it to the remediation of cationic dyes in dye wastewater. The adsorption performance and mechanism of MT on methylene blue (MB) and crystal violet (CV) were investigated by batch adsorption experiments. The results demonstrated that the highest adsorption values of MT for MB (411 mg/g) and CV (553 mg/g) were greatly higher than the reported values of other biosorbents. In addition, the adsorption behaviors of methylene blue (MB) and crystal violet (CV) by MT were spontaneous exothermic reactions and closely followed the pseudo-second-order (PSO) kinetics and Langmuir isotherm. Further, the depleted MT was regenerated using pyrolysis mode to convert depleted MT into MT-biochar (MBC). The maximum adsorption of Cu^{2+} and Pb^{2+} by MBC was up to 320 mg/g and 840 mg/g, respectively. In conclusion, this work presented a new option for the adsorption of cationic dyes in wastewater and a new perspective for the treatment of depleted biosorbents.

Keywords: *Medulla tetrapanacis*; cationic dye; adsorption; biosorbent

1. Introduction

The treatment of dye wastewater has always been one of the challenges of water pollution. Cationic dyes have been employed in a wide range of fields, such as wool, printing and dyeing, and pharmaceuticals, etc. The presence of cationic dyes in wastewater has been reported to disrupt ecological balance [1] and posed certain risks to human health [2]. Therefore, wastewater would be purified to meet discharge standards. Various modalities have been developed to eliminate heavy metal ions and dye molecules from wastewater, including flotation [3–5], chemical precipitation, coagulation, bio-oxidation, ion exchange, membrane filtration, and adsorption, etc. [6]. The adsorption method has the characteristics of simplicity, high efficiency, and wide application, and is considered a feasible method for treating dye wastewater. Currently, various types of adsorbents have been designed to treat dye wastewater, including organic–inorganic composites [7], activated carbon [1], hydrogel [8], biochar [9], and other adsorption materials. However, adsorption materials are faced with problems such as unsatisfactory adsorption performance, high cost, and secondary pollution. Therefore, new adsorbents with high adsorption capacity, sustainability, green attributes, and non-toxicity are urgently needed to solve the problem of dye-polluted wastewater [10].

Tetrapanax papyriferus (*Araliaceae*) is mainly distributed in southwest China and is characterized by fast growth and high reproductive capacity. MT is a columnar, white, lightweight material derived from the pith of *Tetrapanax papyriferus*. The diameter and length of MT range as 5–30 mm and 80–100 cm, respectively. MT is very famous as traditional medicine in China, which is usually cut into slices for sale. The chemical composition of MT was determined in the previous study [11], and holocellulose reached 82.03%, lignin content was only 5.37%, and ash content and ethanol/benzene extractive content were 11.39% and 1.21%, respectively.

The primary chemical constituents of natural plants are cellulose, hemicellulose, lignin, and some minor components such as ethanol/benzene extractive, etc. The conversion of natural biomass into biosorbents with high adsorption properties has proven to be a promising research direction [12]. The feasibility of plant or industrial plant waste as potential biosorbents for cationic dyes has been demonstrated by different studies, such as with fallen leaves [13], *Pine elliottii* waste [14], walnut shells [15], straw and sugarcane waste [12], and tree bark [16]. These biosorbents have captured interest because of their extensive feedstock, environmentally friendly preparation processes, non-toxicity, and low cost. Compared to traditional biomass materials, MT possesses a higher holocellulose content and a special 3D porous microstructure. The high cellulose content means that the adsorbent surface has more electronegative oxygen-containing functional groups, which provide attachment sites for the adsorption of cationic dyes, while the regular porous structure provides wide diffusion channels for the dye molecules [17]. Therefore, MT is expected to be a biosorbent with excellent performance for cationic dyes. According to the discovery of Lawal et al. [18], the high content of cellulose contributed substantially to evolving mesopores, and biochar produced by high-cellulose biomass material was more effective in the treatment of dye wastewater. The study of Lawal provided a new idea for the treatment of depleted biosorbent. Additionally, citric acid was selected as a chemically modified reagent to treat MT. CA is an edible acid, which has been widely applied to modify various waste biomass materials to adsorb dyes. One of the advantages of this modification method is the use of non-toxic reactants [19]. A previous study reported that citric acid was used for kenaf fiber modification, and the adsorption efficiency for MB (131.6 mg/g) was significantly enhanced [20]. Similarly, pine sawdust modified with CA also improved the removal capability of MB (111.46 mg/g) [21]. It is known that related studies prepared biochar from MT as a substrate for supercapacitors [22] and as an adsorbent to adsorb heavy metal ions [11]. However, the excellent performance of MT itself has not been noticed. The studies related to the application of MT as a biosorbent for the removal of cationic dyes have not been reported.

In this study, the adsorption performance of MT for cationic dyes was evaluated using two typical cationic dyes (MB, CV) as model dyes. The adsorption mechanism of MT on cationic dyes was investigated. In addition, CA was applied as a modifier to improve the adsorption properties of MT. Finally, MBC was prepared using depleted biosorbents to avoid secondary pollution to the environment, and the adsorption performance of biochar on Cu^{2+}, Pb^{2+}, MB, and Congo red (CR) was detected.

2. Results and Discussion

2.1. Characterization of MT

MT was considered to be an excellent adsorbent material because of its well-developed pore structure, ultra-thin cell wall, and high holocellulose content. The SEM images of MT sections are shown in Figure 1a–c. As shown in Figure 1a, the cross-section of MT showed a regular honeycomb-like structure with an average pore size of about 200 μm. As can be seen in Figure 1b,c, both the radial and tangential sections of MT were observed to exhibit a honeycomb-like shape, which was obviously different from the other natural plant cell wall structures. Therefore, it could be inferred that the MT was a naturally occurring 3D penetrating porous structure. The 3D simulation of the MT spatial structure was shown in Figure 1f. As shown in Figure 1d,e, the MT pore walls became thicker and smoother after the adsorption of dyes. The phenomenon indicated that the originally white surface of MT was also changed to the corresponding color due to adsorption of dyes. These results demonstrated that MT had good adsorption capacity for cationic dyes.

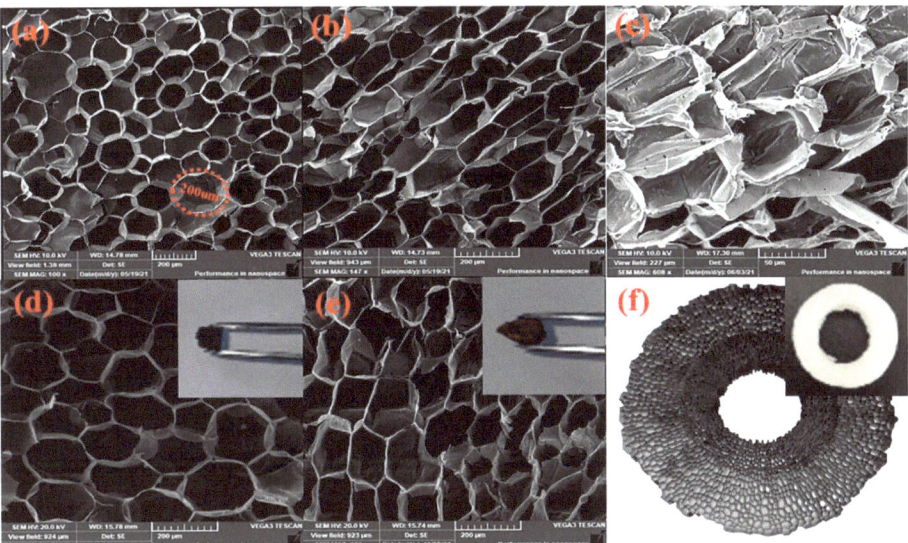

Figure 1. SEM images of MT (**a**) cross-section, (**b**) radial-section, and (**c**) tangential-section. SEM images of MT cross-section after adsorption of (**d**) MB, (**e**) CV. (**f**) The 3D simulation of the MT spatial structure.

The FTIR spectra of MT (a), MT-MB (b), and MT-CV (c) are presented in Figure 2. The stretching vibrations of O-H and C-H bonds on biomass materials correspond to the peaks at 3340 cm^{-1} and 2920 cm^{-1}, respectively. The peak at 1725 cm^{-1} was reported to be associated with -COO$^-$ [23]. It was evident that the intensities of the absorption peaks of MT at 1335 cm^{-1} and 886 cm^{-1} were increased after the adsorption of MB. The bands at 1335 cm^{-1} and 886 cm^{-1} were attributed to the C-N bond in the MB molecules [24] and the C-H bond in the aromatic ring [25], respectively. A noticeable enhancement of the absorption peak at 1174 cm^{-1} was observed after the adsorption of CV, which was attributed to the C=N on the CV molecule [24]. In conclusion, the electronegative oxygen-containing functional groups on the surface of MT could be served as effective adsorption points for binding to cationic dyes.

The pore structure characteristics of different biosorbents and activated carbon are listed in Table 1. As seen in Table 1, the pore characteristics of MT were noticed to be similar to other biosorbents, but was significantly different from activated carbon. The specific surface area of MT (2.37 m^2/g) was greatly lower than activated carbon (249 m^2/g), indicating that the adsorption of MT on dyes was mainly identified to be related to the active groups on the surface of the biosorbent rather than physical filling [26]. The well-developed macropore structure of MT was considered to facilitate the swelling and promote the binding of dye molecules to the chemical groups on the MT surface [24].

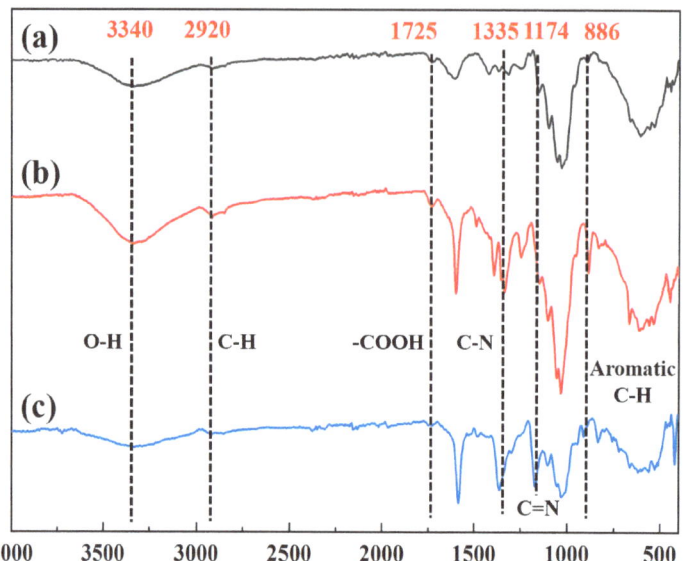

Figure 2. FTIR spectra of (**a**) MT, (**b**) MT-MB, and (**c**) MT-CV.

Table 1. Pore characteristics of different adsorbents.

Adsorbents	Surface Area (m^2/g)	Pore Volume (m^3/g)	Pore Diameter (nm)	Sources
Straw	0.37	–	–	[26]
Eucalyptus leaves	7.40	0.0086	–	[27]
Leaves of Magnoliaceae	7.85	0.0088	7.05	[13]
Sunflower pith	64.49	0.0500	1.34	[23]
Activated carbon	249	0.1330	3.09	[28]
MT	2.37	0.0026	4.41	This study

2.2. Adsorption Study

2.2.1. Effect of pH and MT Dosage on MB and CV Adsorption

The adsorption capacity of the biosorbent was significantly influenced by the pH of solution mainly in two aspects: (1) The degrees of ionization of solute molecules were controlled by the pH of the solution. (2) The pH value of the solution had an effect on the structural stability and surface chemical properties of biosorbent [29]. The adsorption mechanism of MT for cationic dyes can be further investigated by analyzing the influence of solution pH. According to the preliminary experiment, the adsorbance of the MB solution was observed to stabilize under acidic and slightly alkaline conditions (pH < 9), followed by a drastic decline with increasing pH of the solution (10 < pH < 13). According to Lin et al. [30], MB molecules had the property of converting into complex compounds under alkaline conditions. The solution of CV was observed to exhibit a blue color under acidic conditions (pH < 2) and gradually changed to green as the pH of the solution decreased. To exclude the interference of pH on the adsorbance of the dye solution, the initial pH of the dye solution was set to 3.0–9.0 in this study. The impact of pH on the adsorption capacity is presented in Figure 3a. According to Figure 3a, the adsorption capacity of MT for both MB and CV showed an increasing trend as the pH of the solution was increased. In an acidic medium, the surface of MT was rich in positive charge, and part of the active sites were occupied by protons. In an alkaline medium, the MT surface was deprotonated, and more active sites were exposed, which was responsible for the significant increase in MT adsorption capacity. Therefore, the initial pH of the dye solution was adapted to 9.0

for further experiments. Based on the above analysis, electron donor–acceptor may be an important mechanism for the adsorption of cationic dyes by MT, which was consistent with the study of Dallel et al. [31]. It was also reasonable to infer that MT loaded with cationic dyes could be desorbed and regenerated by competitive adsorption of H^+ and MB^+ under acidic conditions to achieve the purpose of recycling. This result had also been verified in subsequent experiments.

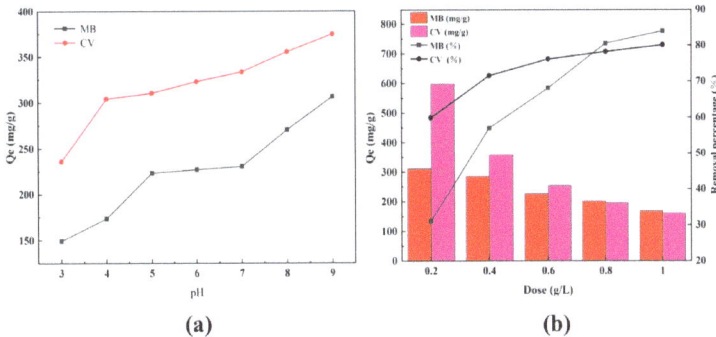

Figure 3. The effect of (**a**) pH and (**b**) biosorbent dosage on MT adsorption capacity.

The influence of adsorbent dosage and removal rate is shown in Figure 3b. According to Figure 3b, the removal rate showed an increasing trend as the adsorbent dosage increased, which could be explained by the fact that more adsorbents provided more adsorption sites [32]. However, the number of dye molecules in a fixed volume of solution were limited. The excess of adsorption sites over the saturation adsorption requirement indicated that a lot of effective adsorption points were underutilized [33]. Obviously, the adsorption capacity of MT for MB and CV was significantly reduced when the amount of adsorbent addition was increased. Finally, the adsorbent dosage of 0.4 g/L was chosen to be applied in the following experiments.

2.2.2. Effect of Adsorption Time on MB (CV) Adsorption and Kinetics Study

The influence of adsorption time on the adsorption capacity of MB and CV is displayed in Figure 4a. The adsorption equilibrium time of MT for MB was 960 min, and that of CV was 1500 min. The kinetic parameters of the MB and CV adsorption are presented in Figure 4b–d and Table 2. The PSO kinetic model had a better fit than the PFO dynamics model. Meanwhile, the adsorption capacity obtained by PSO kinetic fitting was closely in accordance with the experimental adsorption capacity. The adsorption data closely followed the PSO model, indicating that the adsorption for MB and CV by MT was dominated by chemisorption, including electronic exchange or sharing [34]. The fitting curves of the intra-particle diffusion model are presented in Figure 4c,d. The intra-particle diffusion model for MB and CV adsorption by MT could be divided into three stages, which was consistent with the study of Wang et al. [1]. In the first stage, the faster rate of adsorption was attributed to the large difference between the internal and external concentration of the adsorbent. In addition, the C value representing the boundary layer thickness tended toward 0, indicating that there was a favorable surface compatibility between the MB (or CV) molecules and MT. Therefore, intra-particle diffusion was identified as the main rate-controlling step. In the second stage, part of the efficient sites on the biosorbent were occupied, which led to a reduction in the adsorption rate. Adsorption to equilibrium was observed in the third stage, where the efficient sites of biosorbent were completely occupied, and the adsorption and desorption of dye molecules on MT reached a dynamic equilibrium.

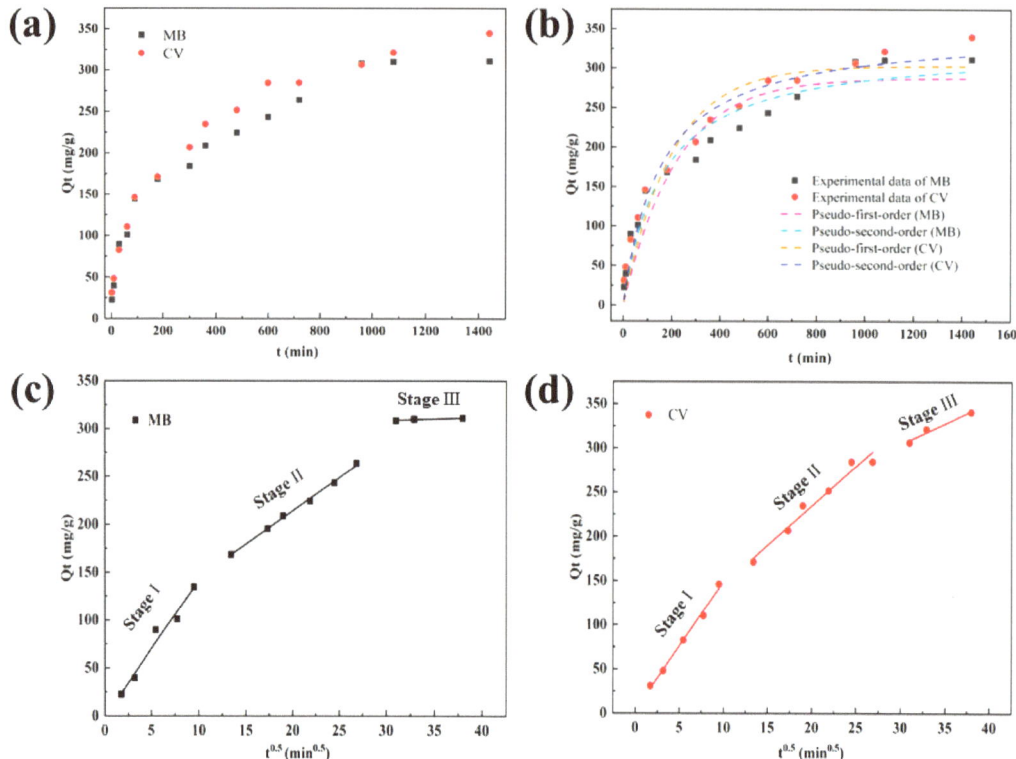

Figure 4. (a) The effect of contact time on the performance for the adsorption of MB and CV by MT. Adsorption kinetics for the adsorption of MB and CV by MT: (b) pseudo-first-order model, pseudo-second-order model, (c,d) intra-particle diffusion model.

Table 2. Adsorption kinetics and intra-particle diffusion model parameters for the adsorption of MB and CV by MT.

Model	Parameters	MB	CV
Pseudo-first-order	Q_e (mg/g) K_1 (min^{-1}) R^2	287.70 0.0046 0.89	303.09 0.0050 0.94
Pseudo-second-order	Q_e (mg/g) K_2 (g/mg·min) R^2	328.99 0.000019 0.95	348.97 0.000018 0.97
Intra-particle diffusion	K_{id1} (mg·g^{-1}·min$^{-0.5}$) C_1 (mg/g) R^2	14.19 0.91 0.97	14.52 3.54 0.99
	K_{id2} (mg·g^{-1}·min$^{-0.5}$) C_2 (mg/g) R^2	7.20 69.23 0.99	8.93 53.96 0.96
	K_{id3} (mg·g^{-1}·min$^{-0.5}$) C_3 (mg/g) R^2	0.34 298.12 0.89	2.67 227.92 0.78

2.2.3. Effect of Initial Concentration on MB (CV) Adsorption and Isotherm Study

The influence of the initial concentration of the dye for the adsorption of MB and CV by MT was determined, as presented in Figure 5a. According to Figure 5a, the adsorption ability of MB and CV showed an increasing trend as the initial concentration of dye increased, and then tended toward equilibrium. This phenomenon ascribed to the presence of a concentration grading between the adsorbent and the bulk of the solution, and the larger molecular mass transfer force, was provided by the solution with higher initial concentration. When the concentration of dye further increased, the effective sites of the adsorbent were completely consumed, and the adsorption capacity tended to balance [35]. Adsorption isotherm is usually adopted to evaluate the complex interplay between adsorbent and adsorbate, which is another important method to explore the mechanism of adsorption [36]. Langmuir, Freundlich, and Temkin adsorption isotherms were employed in this study. According to Figure 5b,c and Table 3, the Langmuir isotherm provided high determination coefficients for the adsorption of MB (0.99) and CV (0.98). The maximum adsorption capacity (Q_{max}) predicted by the Langmuir model was correlated closely with the experimental result, indicating that the adsorption process of MT for dye molecules was monolayer. Moreover, the maximum R_L values of MB (0.48) and CV (0.71) were calculated to be lower than 1, which was due to the adsorption procedure being favorable. The adsorption capacity of various materials for MB and CV are compared in Table 4. As evidenced by the table, MT has a good adsorption capacity for MB and CV.

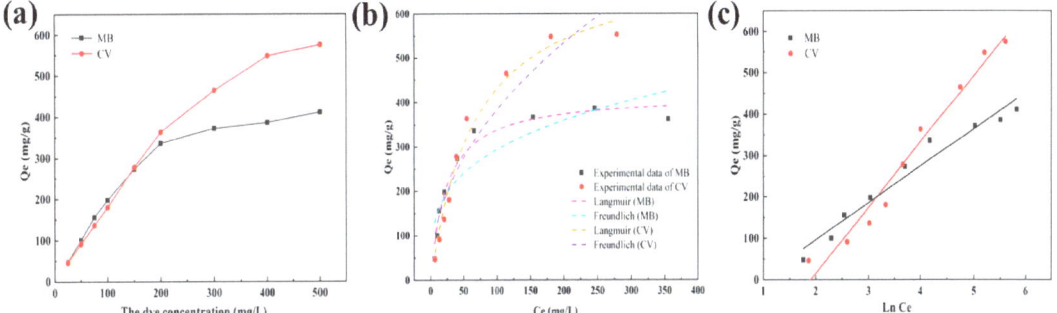

Figure 5. (a) The effect of initial concentration on the performance for the adsorption of MB and CV by MT. Adsorption isotherms for the adsorption of MB and CV by MT: (b) Langmuir model, Freundlich model, and (c) Temkin model.

Table 3. Adsorption isotherm parameters for the adsorption of MB and CV by MT.

Dyes	Langmuir				Freundlich			Temkin		
	Q_{max} (mg/g)	K_L (L/mg)	R^2	R_L	K_F (mg·g^{-1}·mg$^{1/n}$·L$^{-1/n}$)	n	R^2	K_T (L/g)	b_T (J/mol)	R^2
MB	400	0.043	0.99	0.044–0.48	38.28	2.2	0.83	0.40	28.23	0.96
CV	588	0.016	0.98	0.11–0.71	16.98	1.5	0.94	0.15	15.83	0.97

Table 4. Comparison of adsorption capacity of different adsorbents for MB and CV.

Adsorbents	Adsorption Capacity (mg/g)	Dyes	Sources
CMC/PAA/GO composite	138.4	MB	[37]
Rapanea ferruginea	106	CV	[38]
Tea waste/Fe_3O_4 magnetic composite	333.3	CV	[39]
Hickory chip biochars	310	MB	[40]
PVDF/PDA/PPy composite	370.4	MB	[41]
CNC/MnO_2/SA composite	114.5	MB	[8]
Biochar from crisp persimmon peel	59.7	MB	[42]
ZnO-Chitosan Nanocomposites	97.9	MB	[43]
MT	400	MB	This study
MT	553	CV	This study

2.2.4. Thermodynamic Study

The thermodynamic parameters are shown in Figure 6 and Table 5. The spontaneous and exothermic characteristics of MT for MB and CV adsorption could be explained by the negative Gibbs free energy (ΔG^o) and the negative enthalpy (ΔH^o). This result indicated that the adsorption of MT for cationic dyes proceed spontaneously, which provided a good premise for the practical application of MT. Enthalpy change represented the energy that was changed during the adsorption process. When the adsorption process with lower energy change (absolute value of ΔH^o is 0–20 KJ/mol) was described as physical adsorption, and with higher energy change (absolute value of ΔH^o is 80–400 KJ/mol) it was described as chemical adsorption [44]. As presented in Table 5, the absolute values of the enthalpy of adsorption for MB and CV by MT were 25.75 and 14.98 KJ/mol, respectively. Analysis of the data showed that the adsorption of MT for cationic dyes was a complex procedure, which involved both physical and chemical adsorption. The negative entropy (ΔS^o) for the adsorption of MB and CV by MT illustrated a promoted reordering with reduced randomness of biosorbent [10].

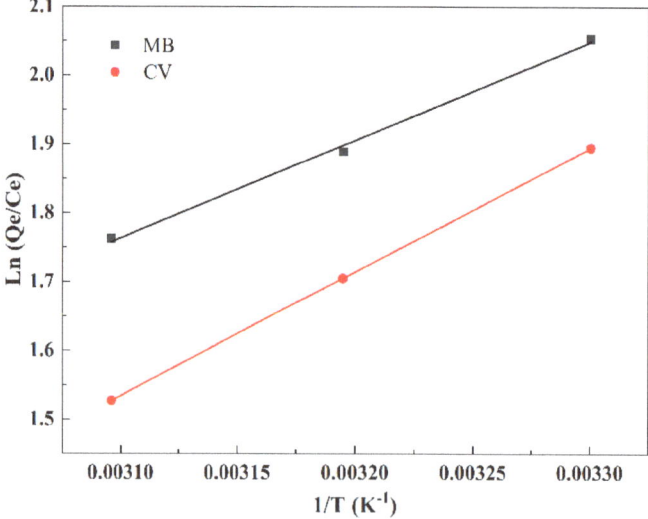

Figure 6. The van't Hoff plot for the adsorption of MB and CV by MT.

Table 5. Adsorption thermodynamic parameters for the adsorption of MB and CV by MT.

Dyes	$\Delta S°$ (J/mol × K)	$\Delta H°$ (KJ/mol)	$\Delta G°$ (kJ/mol)			R^2
			303K	313K	323K	
MB	−72.50	−25.75	−3.79	−3.06	−2.34	0.99
CV	−33.68	−14.98	−4.77	−4.43	−4.10	0.99

2.2.5. Adsorption Mechanism

The adsorption of MT on cationic dyes was mainly influenced by the electronegative oxygen-containing functional groups (the hydroxyl and the carboxyl group) of cellulose and hemicellulose. The important role of oxygen-containing groups in the adsorption process was further validated by FTIR analysis. Cationic dyes can be immobilized by electronegative functional groups on MT through electron donor–acceptor. The adsorption mechanism was further discussed through the analysis of the adsorption model. The adsorption procedure closely followed the Langmuir isotherm and PSO kinetic model, suggesting that the adsorption procedure was monolayer adsorption and chemisorption, respectively. Meanwhile, intra-particle diffusion was the primary rate-controlling step in the adsorption of dye molecules by MT in the early stages of the adsorption procedure. It was noteworthy that the adsorption process may involve both chemical and physical processes by thermodynamic analysis, which suggested that hydrogen bonding and electrostatic interaction were equally important for the adsorption procedure [11]. The mechanism of MT adsorption of dyes is demonstrated by Figure 7. The unique 3D penetrating porous structure of MT was thought to facilitate swelling and provided channels for mass transfer of dye molecules.

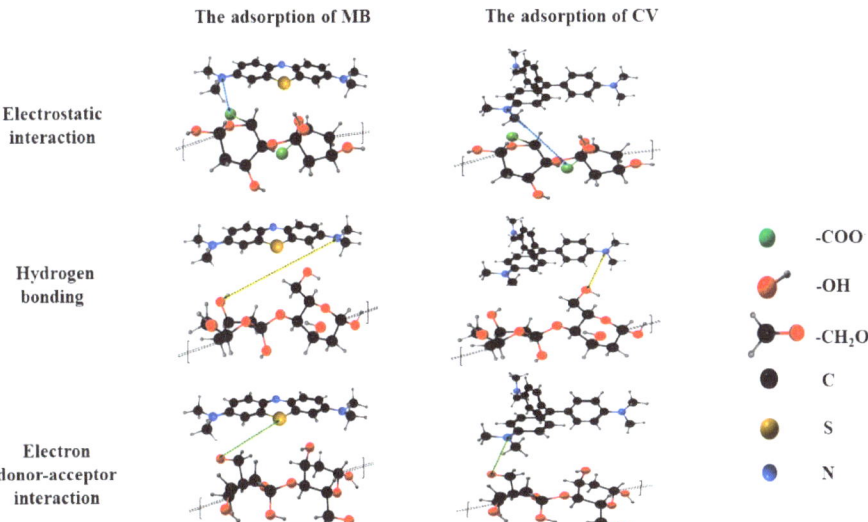

Figure 7. Possible adsorption mechanism.

2.2.6. Reusability of MT

The regeneration performance is an important criterion for evaluating the quality of adsorbent. In this study, HCl (0.1 mol/L) was employed to remove the loaded dye molecules on MT. As seen in Figure 8, the adsorption capacity and removal rate of dyes by MT were observed to decrease slightly after three adsorption–desorption experiments, but MT still exhibited high adsorption capacity. However, the structure of the cell wall was destroyed by acids acting on glycosidic bonds hydrolyzing the matrix polysaccharides (cellulose and hemicellulose) [45]. In practice, the loose structure of MT after several cycles of desorption was not conducive to recovery. In this regard, MBC was prepared by pyrolysis of the depleted MT in nitrogen atmosphere. The adsorption properties of MBC for various contaminants are discussed in the next section.

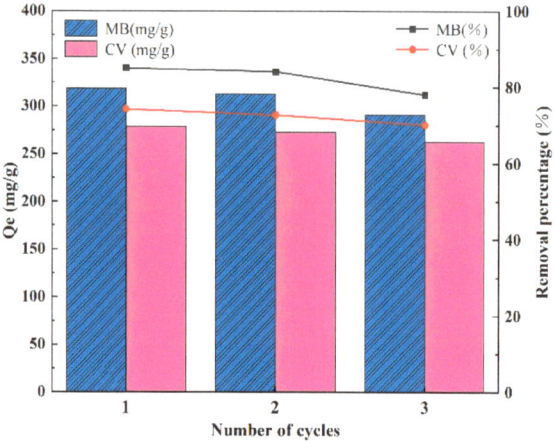

Figure 8. Reusability of MT for the adsorption of MB and CV.

2.2.7. Study on Adsorption Capacity of MBC

The depleted adsorbent will cause secondary pollution to the environment if it is not properly treated. We innovatively recovered depleted MT, and the biochar was prepared under a high-temperature nitrogen atmosphere using the depleted MT as raw material. The results of exploration on the adsorption performance of MBC for MB, CR, Cu^{2+}, and Pb^{2+} are presented in Figure 9. The adsorption capacity of MBC for Cu^{2+} and Pb^{2+} reached 320 mg/g and 840 mg/g, respectively, which was in accordance with the previous conclusion that MBC had good affinity for heavy metal ions. The adsorption capacity of MBC on CR and MB were 125 mg/g and 80 mg/g, respectively. Apparently, the adsorption performance of MBC on dyes did not meet our expectation. It will be our future work to improve the adsorption capacity of MBC on dyes.

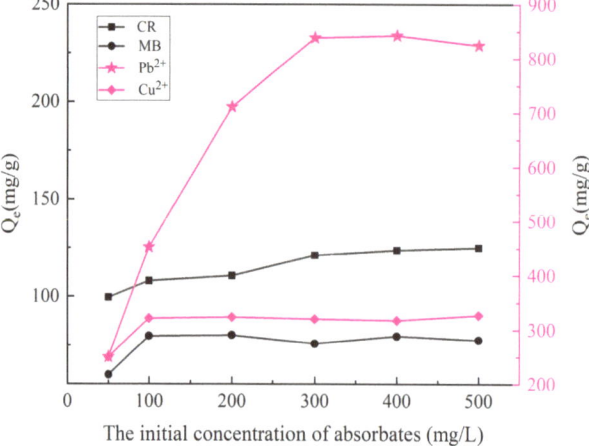

Figure 9. The adsorption capacity of MBC for Cu^{2+}, Pb^{2+}, MB, and CR.

2.2.8. Performance Study of Citric Acid Modified MT

CA was employed as a modifier for the chemical modification of MT in this study. The modification effect of CA on MT was in two aspects. Firstly, the internal structure of MT was changed after the reaction with the CA solution. Secondly, the surface of MT introduced carboxyl groups through CA modification. The various properties of CAMT are presented in Figure 10. As evidenced by Figure 10a, the pore wall structure of MT was retained after CA modification, but the pore walls became thinner. It was noteworthy that inter-connected cellulose fiber networks were formed inside the CAMT. These fiber bundles were derived from the cell wall. During chemical treatment, part of the cellulose was detached from the cell wall and freeze-dried to form an interconnected network structure, which provided a better prerequisite for adsorption [46]. The change of the main peak of the MT and CAMT occurred at 1725 cm^{-1} in Figure 10b, which was caused by the introduction of carboxyl groups (COO-) after MT modification [47]. As seen in Figure 10c, the maximum adsorption of CAMT for MB enhanced from 400 mg/g to 526 mg/g, while the maximum adsorption of CV enhanced from 553 mg/g to 627 mg/g. After CA modification, the maximum adsorption capacities of CAMT for MB and CV were increased by 31.5% and 13.4%, respectively. From Figure 10d, it could be observed that the adsorption equilibrium times of MT for MB and CV were 960 min and 1500 min, and that of CAMT were 360 min and 720 min for MB and CV, respectively. The time consumed to achieve kinetic equilibrium was critical to evaluate the efficiency and feasibility of adsorbents for water pollution control. The adsorption equilibrium times of CAMT for MB and CV were reduced by 62.5% and 52%, respectively. The above data analysis demonstrated that CA modification can effectively reduce the time spent on the adsorption process and improve adsorption efficiency.

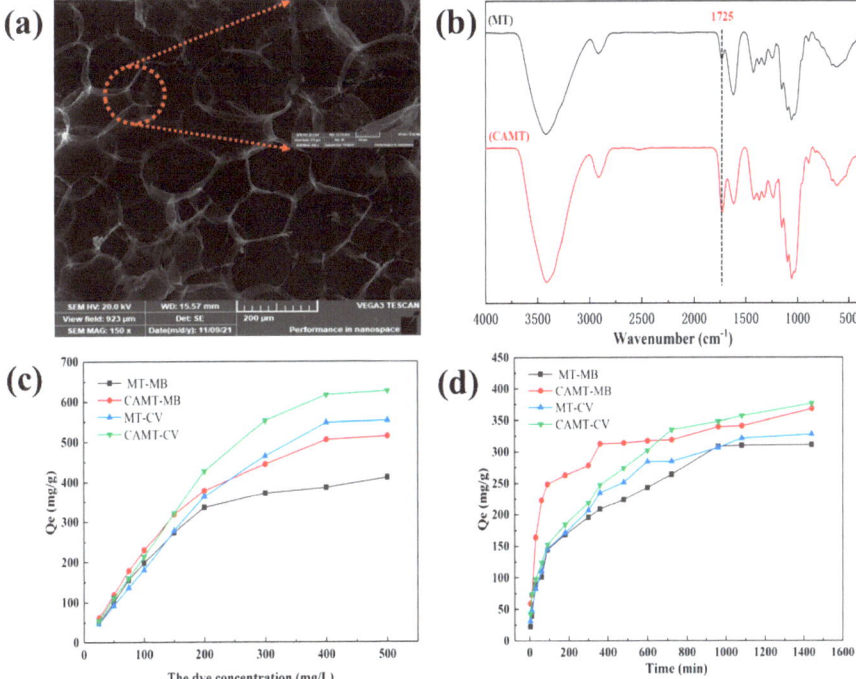

Figure 10. (**a**) SEM image of CAMT. (**b**) FTIR spectra of MT and CAMT. (**c**) Effect of initial concentration of dye on the performance of MT and CAMT. (**d**) Effect of adsorption time on the performance of MT and CAMT.

3. Materials and Methods

3.1. Materials

MT was obtained in Bozhou (Anhui province, China). Sodium bicarbonate ($NaHCO_3$), copper nitrate ($Cu(NO_3)_2$), MB, sodium hydroxide (NaOH), CR, citric acid monohydrate (CA), CV, hydrochloric acid (HCl), and lead nitrate ($Pb(NO_3)_2$) were acquired from Xilong Scientific Co., Ltd. (Shantou, China). All other chemicals were used without further refinement. All experiments used deionized water.

3.2. Preparation of CAMT and MBC

MT was cut into the size of 5 × 5 mm and then placed in a beaker with 0.4 mol/L citric acid solution (w/w 1:50). The mixture was activated at 80 °C for 2 h, and then transferred to an autoclave lined with Teflon, and subsequently placed in an oven at 120 °C for 0.5 h for the esterification reaction. After completion of the reaction, MT was washed with $NaHCO_3$ (0.1 mol/L) three times, and then deionized water was used to wash MT several times to remove residual $NaHCO_3$. Finally, the modified biosorbent was freeze-dried, and named CAMT. The depleted biosorbent was dried at room temperature. The dried samples were placed in a vacuum high-temperature tubular sintering furnace (BTF-1200C-II, BEQ, Hefei, China) and pyrolyzed under nitrogen atmosphere. The heating rate was controlled at 5 °C/min, and the temperature was held at 700 °C for 2 h. After pyrolysis, the biochar samples were first washed with hydrochloric acid (10 wt %) to remove ash, and then with deionized water to neutral. After drying, the biochar was obtained and named MBC.

3.3. Characterization

Scanning electron microscope (VEGA3, TESCAN, Brno, Czech Republic) was applied to determine the physical structure of the samples. Before measurement, the samples were placed on conductive adhesive for vacuum gold spraying. Fourier-transform infrared spectrometer (Bruker Tensor II, Bruker, Karlsruhe, Germany) was applied to identify the functional groups of the samples. The pore structure of the samples was determined with the surface area and porosity analyzer (ASAP2460, Micromeritics, Norcross, GA, USA). The contents of MB and CV in aqueous solutions were measured at 665 nm and 590 nm by UV-vis spectrophotometer (TU-1810PC, PERSEE, Beijing, China), respectively. The concentrations of Cu^{2+} and Pb^{2+} in the samples were detected by atomic absorption spectrophotometer (TAS-990, PERSEE, Beijing, China).

3.4. Batch Adsorption Experiments

In batch adsorption experiments, 20 mg of biosorbent was added to a triangular flask containing 50 mL and 200 mg/L of MB (or CV) solution, the pH of the suspension was set to 9, and the suspension was subjected to a water bath oscillator at 303 K for 24 h at 120 rpm. The pH of the suspension was adjusted to 3.0–9.0 using 0.1 mol/L HCl or NaOH to explore the influence of pH on the adsorption capacity. The dosage of biosorbent was set at 10–50 mg to evaluate the influence of biosorbent dosage on the adsorption ability. The exploration process of MBC for MB, CR, Cu^{2+}, and Pb^{2+} adsorption capacity was similar to the above procedure. All experiments were set up with three parallel groups of samples, and the relative error of the experimental data was within 5%. The calculation formulas of adsorption capacity and removal rate are as follows:

$$Q_e = \frac{(C_0 - C_e)V}{m} \qquad (1)$$

$$Q_t = \frac{(C_0 - C_t)V}{m} \qquad (2)$$

$$R = \frac{(C_0 - C_e)}{C_0} \times 100\% \qquad (3)$$

where Q_e and Q_t (mg/g) represent the adsorption capacities at adsorption equilibrium and at time t, respectively. C_0, C_t, and C_e (mg/L) represent the adsorbate concentrations in the solution at the initial time, the fixed time t, and adsorption equilibrium, respectively. V (L) represents the volume of solution. R (%) represents the removal rate. m (g) represents the weight of adsorbent used.

3.5. Adsorption Kinetics Study

The concentration of MB (or CV) in suspension at different time points within 24 h was determined to investigate the impact of adsorption time on MB (or CV) adsorption ability. To elaborate adsorption mechanism precisely, pseudo-first-order (PFO) model, PSO model, and intra-particle diffusion model are adopted to describe the adsorption process and the equations can be expressed as [19]:

the PFO model:
$$Q_t = Q_e \left(1 - e^{-K_1 t}\right) \tag{4}$$

the PSO model:
$$Q_t = \frac{Q_e^2 K_1 \times t}{Q_e K_1 \times t + 1} \tag{5}$$

the intra-particle diffusion model:
$$Q_t = K_{id} t^{\frac{1}{2}} + C \tag{6}$$

where Q_e and Q_t (mg/g) represent the adsorption capacity at adsorption equilibrium and at time t, respectively. t (min) represents the adsorption time. C (mg/g) represents the intercept. K_1 (min^{-1}), K_2 (g/mg × min), and K_{id} (g/mg × min) are the constant of the PFO rate, the PSO rate, and the intra-particle diffusion rate, respectively.

3.6. Adsorption ISOTHERM Study

The starting concentration of MB (or CV) was specified as 25–500 mg/L to evaluate the impact of the starting concentration on the adsorption ability. Adsorption isotherms are another critical methodology to study the adsorption mechanism and are crucial for the optimization and use of adsorbents [36]. Langmuir, Freundlich, and Temkin isotherm models were adopted to describe the equilibrium data. The equations are as follows [19]:

Langmuir:
$$Q_e = \frac{K_L C_e Q_{max}}{1 + K_L C_e} \tag{7}$$

$$R_L = \frac{1}{1 + K_L C_0} \tag{8}$$

Freundlich:
$$Q_e = K_F C_e^{\frac{1}{n}} \tag{9}$$

Temkin:
$$Q_e = \frac{RT}{b_T} \ln K_T + \frac{RT}{b_T} \ln C_e \tag{10}$$

where C_0 and C_e (mg/L) represent the starting and at-adsorption equilibrium concentration of MB (CV), respectively. Q_e and Q_{max} (mg/g) represent the maximum adsorption value at equilibrium and theoretical maximum adsorption value, respectively. K_L (L/mg), K_F (mg × g^{-1} × mg$^{1/n}$ × L$^{-1/n}$), and K_T (L/mg) represent the Langmuir, Freundlich, and Temkin adsorption constant, respectively. n is the adsorption constant. R_L is a dimensionless separation coefficient for determining whether the adsorption process is favorable. T (K) is the absolute temperature. b_T (KJ/mol) represents the characteristic constant of the Temkin model. R (8.314 J/mol × K) represents the universal gas constant.

3.7. Adsorption Thermodynamic Study

The temperature range was set to 303–323 K to explore the influence of different temperatures on the adsorption ability of MT. The adsorption mechanism and the driving force of adsorption can be further explained through the analysis of adsorption thermodynamics. Gibbs free energy (ΔG°, KJ/mol), enthalpy (ΔH°, KJ/mol), and entropy (ΔS°, J/mol·K) are calculated as follows [6]:

$$K_d = \frac{Q_e}{C_e} \quad (11)$$

$$\ln K_d = \frac{\Delta S^\circ}{R} - \frac{\Delta H^\circ}{RT} \quad (12)$$

$$\Delta G^\circ = \Delta H^\circ - T\Delta S^\circ \quad (13)$$

where Q_e (mg/g) represents the adsorption capacity. C_e (mg/L) represents the MB (or CV) concentrations. T (K) represents the temperature in Kelvin. R (8.314 J/mol × K) represents the universal gas constant. K_d is the thermodynamic equilibrium constant.

3.8. Regeneration Experiment

Desorption cycle experiment was used to assess the reusability of MT. An amount of 20 mg MT was added to a conical flask with 50 mL MB (or CV) solution (150 mg/L). The MT was collected after equilibrium adsorption and desorbed in HCl solution (0.1 mol/L) for 24 h. After desorption, MT was washed to neutral and then freeze-dried. The cycle experiment was repeated 3 times, and the adsorption capacity of MT in each cycle was measured.

4. Conclusions

In this study, MB and CV were used as model dyes, and the adsorption potential of MT for cationic dyes was evaluated. Further, the mechanism of MT adsorption for cationic dyes was investigated. The main findings are summarized as follows:

(1) The maximum adsorption capacities of MT for MB and CV reached 411 mg/g and 553 mg/g, respectively. In addition, the adsorption efficiency of MT for dyes could be effectively improved by CA modification. Meanwhile, MT still possessed good adsorption performance after three adsorption–desorption experiments. The above data proved that MT was a promising biosorbent for the treatment of cationic dye contamination.

(2) The thermodynamic study demonstrated that the adsorption process was spontaneous and exothermic. The adsorption procedure closely followed PSO kinetic, and the intraparticle diffusion was the main rate-controlling step. The 3D pore structure of MT was mainly used for mass transfer of dye molecules rather than pore filling.

(3) The maximum adsorption capacities of MBC for Cu^{2+} and Pb^{2+} were 320 mg/g and 840 mg/g, respectively, which indicated that the preparation of biochar for adsorption of heavy metal ions using depleted MT as a carbon source is a feasible method to treat depleted MT.

Author Contributions: All authors contributed to the study conception and design. Methodology, J.Z.; writing—original draft, H.J.; formal analysis, Z.L.; data curation, L.Z.; supervision, Z.W.; writing—review and editing, Y.G. and H.G. All authors have read and agreed to the published version of the manuscript.

Funding: This research was funded by Natural Science Foundation of Anhui Province (2008085MC99); Anhui Provincial Training Program of Innovation and Entrepreneurship for Undergraduates (S202110364179); Anhui Agricultural University Training Program of Innovation and Entrepreneurship (XJDC2021169) and (XJDC 2021433).

Institutional Review Board Statement: Not applicable.

Informed Consent Statement: Not applicable.

Data Availability Statement: The data supporting this study are available when reasonably requested from the corresponding author.

Acknowledgments: This work was supported by Anhui Agricultural University.

Conflicts of Interest: The authors declare that there are no conflict of interest associated with this manuscript.

Sample Availability: Samples of the compounds are available from the authors.

References

1. Wang, B.; Zhai, Y.B.; Wang, T.F.; Li, S.H.; Peng, C.; Wang, Z.X.; Li, C.T.; Xu, B.B. Fabrication of bean dreg-derived carbon with high adsorption for methylene blue: Effect of hydrothermal pretreatment and pyrolysis process. *Bioresour. Technol.* **2019**, *274*, 525–532. [CrossRef] [PubMed]
2. Santoso, E.; Ediati, R.; Kusumawati, Y.; Bahruji, H.; Sulistiono, D.O.; Prasetyoko, D. Review on recent advances of carbon based adsorbent for methylene blue removal from waste water. *Mater. Today Chem.* **2020**, *16*, 100233. [CrossRef]
3. Zhao, W.J.; Wang, M.L.; Yang, B.; Feng, Q.C.; Liu, D.W. Enhanced sulfidization flotation mechanism of smithsonite in the synergistic activation system of copper–ammonium species. *Miner. Eng.* **2022**, *187*, 107796. [CrossRef]
4. Han, G.; Wen, S.M.; Wang, H.; Feng, Q.C. Sulfidization regulation of cuprite by pre-oxidation using sodium hypochlorite as an oxidant. *Int. J. Min. Sci. Technol.* **2021**, *31*, 1117–1128. [CrossRef]
5. Wang, H.; Wen, S.M.; Han, G.; He, Y.X.; Feng, Q.C. Adsorption behavior and mechanism of copper ions in the sulfidization flotation of malachite. *Int. J. Min. Sci. Technol.* **2022**, *32*, 897–906. [CrossRef]
6. Li, Q.; Li, Y.H.; Ma, X.M.; Du, Q.J.; Sui, K.Y.; Wang, D.C.; Wang, C.P.; Li, H.L.; Xia, Y.Z. Filtration and adsorption properties of porous calcium alginate membrane for methylene blue removal from water. *Chem. Eng. J.* **2017**, *316*, 623–630. [CrossRef]
7. Dai, H.J.; Huang, Y.; Zhang, Y.H.; Zhang, H.; Huang, H.H. Green and facile fabrication of pineapple peel cellulose/magnetic diatomite hydrogels in ionic liquid for methylene blue adsorption. *Cellulose*. **2019**, *26*, 3825–3844. [CrossRef]
8. Li, J.L.; Zhou, L.J.; Song, Y.K.; Yu, X.; Li, X.L.; Liu, Y.X.; Zhang, Z.R.; Yuan, Y.; Yan, S.; Zhang, J.M. Green fabrication of porous microspheres containing cellulose nanocrystal/MnO$_2$ nanohybrid for efficient dye removal. *Carbohydr. Polym.* **2021**, *270*, 118340. [CrossRef]
9. Liu, L.H.; Yue, T.T.; Liu, R.; Lin, H.; Wang, D.Q.; Li, B.X. Efficient absorptive removal of Cd(II) in aqueous solution by biochar derived from sewage sludge and calcium sulfate. *Bioresour. Technol.* **2021**, *336*, 12533. [CrossRef]
10. Jiang, L.; Wen, Y.Y.; Zhu, Z.J.; Liu, X.F.; Shao, W. A Double cross-linked strategy to construct graphene aerogels with highly efficient methylene blue adsorption performance. *Chemosphere* **2021**, *265*, 129169. [CrossRef]
11. Zhang, L.P.; Li, W.Q.; Cao, H.S.; Hu, D.; Chen, X.; Guan, Y.; Tang, J.; Gao, H. Ultra-efficient sorption of Cu^{2+} and Pb^{2+} ions by light biochar derived from *Medulla tetrapanacis*. *Bioresour. Technol.* **2019**, *291*, 121818. [CrossRef] [PubMed]
12. Halysh, V.; Sevastyanova, O.; Pikus, S.; Dobele, G.; Pasalskiy, B.; Gun'ko, V.M.; Kartel, M. Sugarcane bagasse and straw as low-cost lignocellulosic sorbents for the removal of dyes and metal ions from water. *Cellulose* **2020**, *27*, 8181–8197. [CrossRef]
13. Guo, D.; Li, Y.X.; Cui, B.H.; Hu, M.; Luo, S.Y.; Ji, B.; Liu, Y. Natural adsorption of methylene blue by waste fallen leaves of *Magnoliaceae* and its repeated thermal regeneration for reuse. *J. Clean. Prod.* **2020**, *267*, 121903. [CrossRef]
14. Bortoluz, J.; Ferrarini, F.; Bonetto, R.L.; Crespo, J.D.S.; Giovanela, M. Use of low-cost natural waste from the furniture industry for the removal of methylene blue by adsorption: Isotherms, kinetics and thermodynamics. *Cellulose* **2020**, *27*, 6445–6466. [CrossRef]
15. Halysh, V.; Sevastyanova, O.; Riazanova, A.V.; Budnyak, B.P.; Lindstrom, M.E.; Kartel, M. Walnut shells as a potential low-cost lignocellulosic sorbent for dyes and metal ions. *Cellulose* **2018**, *25*, 4729–4742. [CrossRef]
16. Hernandes, P.T.; Oliveira, M.L.S.; Georgin, J.; Franco, D.S.P.; Allasia, D.; Dotto, G. Adsorptive decontamination of wastewater containing methylene blue dye using golden trumpet tree bark (*Handroanthus albus*). *Environ. Sci. Pollut. Res.* **2019**, *26*, 31924–31933. [CrossRef]
17. Wang, Q.; Luo, C.M.; Lai, Z.Y.; Chen, S.Q.; He, D.W.; Mu, J. Honeycomb-like cork activated carbon with ultra-high adsorption capacity for anionic, cationic and mixed dye: Preparation, performance and mechanism. *Bioresour. Technol.* **2022**, *357*, 127636. [CrossRef]
18. Lawal, A.A.; Hassan, M.A.; Zakaria, M.R.; Yusoff, Z.M.; Norrrahim, M.N.F.; Mokhtar, M.N.; Shirai, Y. Effect of oil palm biomass cellulosic content on nanopore structure and adsorption capacity of biochar. *Bioresour. Technol.* **2021**, *332*, 125070. [CrossRef]
19. Sun, L.; Chen, D.M.; Wan, S.G.; Yu, Z.B. Performance, kinetics, and equilibrium of methylene blue adsorption on biochar derived from eucalyptus saw dust modified with citric, tartaric, and acetic acids. *Bioresour. Technol.* **2015**, *198*, 300–308. [CrossRef]
20. Sajab, M.S.; Chia, C.H.; Zakaria, S.; Jani, S.M.; Ayob, M.K.; Chee, K.L.; Khiew, P.S.; Chiu, W.S. Citric acid modified kenaf core fibres for removal of methylene blue from aqueous solution. *Bioresour. Technol.* **2011**, *102*, 7237–7243. [CrossRef]
21. Zou, W.H.; Bai, H.J.; Gao, S.P.; Li, K. Characterization of modified sawdust, kinetic and equilibrium study about methylene blue adsorption in batch mode. *Korean J. Chem. Eng.* **2013**, *30*, 111–122. [CrossRef]
22. Liu, B.; Yang, M.; Yang, D.G.; Chen, H.B.; Li, H.M. Medulla tetrapanacis-derived O/N co-doped porous carbon materials for efficient oxygen reduction electrocatalysts and high-rate supercapacitors. *Electrochim. Acta.* **2018**, *272*, 88–96. [CrossRef]

23. Ma, X.T.; Liu, Y.R.; Zhang, Q.B.; Sun, S.L.; Zhou, X.; Xu, Y. A novel natural lignocellulosic biosorbent of sunflower stem-pith for textile cationic dyes adsorption. *J. Clean. Prod.* **2021**, *331*, 129878. [CrossRef]
24. Pang, X.N.; Sellaoui, L.; Franco, D. Adsorption of crystal violet on biomasses from pecan nutshell, para chestnut husk, araucaria bark and palm cactus: Experimental study and theoretical modeling via monolayer and double layer statistical physics models. *Chem. Eng. J.* **2019**, *378*, 122101. [CrossRef]
25. Feng, Y.F.; Dionysiou, D.D.; Wu, Y.H.; Zhou, H.; Xue, L.H.; He, S.Y.; Yang, L.Z. Adsorption of dyestuff from aqueous solutions through oxalic acid-modified swede rape straw: Adsorption process and disposal methodology of depleted bio-adsorbents. *Bioresour. Technol.* **2013**, *138*, 191–197. [CrossRef]
26. Liu, Q.M.; Li, Y.Y.; Chen, H.F.; Lu, J. Superior adsorption capacity of functionalised straw adsorbent for dyes and heavy-metal ions. *J. Hazard. Mater.* **2019**, *382*, 121040. [CrossRef]
27. Ghosh, K.; Bar, N.; Biswas, B.A.; Das, S.K. Elimination of crystal violet from synthetic medium by adsorption using unmodified and acid-modified eucalyptus leaves with MPR and GA application. *Sustain. Chem. Pharm.* **2021**, *19*, 100370. [CrossRef]
28. Sahu, S.; Pahi, S.; Sahu, K.J.; Sahu, U.K.; Patel, R.K. Kendu (*Diospyros melanoxylon Roxb*) fruit peel activated carbon—An efficient bioadsorbent for methylene blue dye: Equilibrium, kinetic, and thermodynamic study. *Environ. Sci. Pollut. Res.* **2020**, *27*, 22579–22592. [CrossRef]
29. Deng, W.J.; Tang, S.W.; Zhou, X.; Liu, Y.; Liu, S.J.; Luo, J.W. Honeycomb-like structure-tunable chitosan-based porous carbon microspheres for methylene blue efficient removal. *Carbohydr. Polym.* **2020**, *247*, 116736. [CrossRef]
30. Lin, D.C.; Shi, M.; Zhang, Y.M.; Wang, D.; Cao, J.F.; Yang, J.W.; Peng, C.S. 3D Crateriform and Honeycomb Polymer Capsule with Nano Re-entrant and Screen Mesh Structures for the Removal of Multi-component Cationic Dyes from Water. *Chem. Eng. J.* **2019**, *375*, 121911. [CrossRef]
31. Dallel, R.; Kesraoui, A.; Seffen, M. Biosorption of cationic dye onto "*Phragmites australis*" fibers: Characterization and mechanism. *J. Environ. Chem. Eng.* **2018**, *6*, 7247–7256. [CrossRef]
32. Dawood, S.; Sen, T.K. Removal of anionic dye Congo red from aqueous solution by raw pine and acid-treated pine cone powder as adsorbent: Equilibrium, thermodynamic, kinetics, mechanism and process design. *Water Res.* **2012**, *46*, 1933–1946. [CrossRef] [PubMed]
33. Yang, X.X.; Li, Y.H.; Du, Q.J.; Sun, J.K.; Chen, L.; Hu, S.; Wang, Z.H.; Xia, Y.Z.; Xia, L.H. Highly effective removal of basic fuchsin from aqueous solutions by anionic polyacrylamide/graphene oxide aerogels. *J. Colloid Interface Sci.* **2015**, *453*, 107–114. [CrossRef] [PubMed]
34. Mayakaduwa, S.S.; Kumarathilaka, P.; Herath, I.; Ahmad, M.; Al-Wabel, M.; Ok, Y.S.; Usman, A.; Abduljabbar, A.; Vithanage, M. Equilibrium and kinetic mechanisms of woody biochar on aqueous glyphosate removal. *Chemosphere* **2016**, *144*, 2516–2521. [CrossRef] [PubMed]
35. Mahdi, Z.; Yu, Q.J.; Hanandeh, A.E. Investigation of the kinetics and mechanisms of nickel and copper ions adsorption from aqueous solutions by date seed derived biochar. *J. Environ. Chem. Eng.* **2018**, *6*, 1171–1181. [CrossRef]
36. Xu, Y.; Liu, S.B.; Tan, X.F.; Zeng, G.M.; Zeng, W.; Ding, Y.; Cao, W.C.; Zheng, B.H. Enhanced adsorption of methylene blue by citric acid modification of biochar derived from water hyacinth (*Eichornia crassipes*). *Environ. Sci. Pollut. Res.* **2016**, *23*, 23606–23618. [CrossRef]
37. Hosseini, H.; Zirakjou, A.; McClements, D.J.; Goodarzi, V.; Chen, W.H. Removal of methylene blue from wastewater using ternary nanocomposite aerogel systems: Carboxymethyl cellulose grafted by polyacrylic acid and decorated with graphene oxide. *J. Hazard. Mater.* **2022**, *421*, 126752. [CrossRef]
38. Chahm, T.; Martins, B.A.; Rodrigues, C.A. Adsorption of methylene blue and crystal violet on low-cost adsorbent: Waste fruits of *Rapanea ferruginea* (ethanol-treated and H_2SO_4-treated). *Environ. Earth Sci.* **2018**, *77*, 508. [CrossRef]
39. Kumbhar, P.; Narale, D.; Bhsale, R. Synthesis of tea waste/Fe_3O_4 magnetic composite (TWMC) for efficient adsorption of crystal violet dye: Isotherm, kinetic and thermodynamic studies. *J. Environ. Chem. Eng.* **2022**, *10*, 107893. [CrossRef]
40. Zhang, Y.; Zheng, Y.L.; Yang, Y.C.; Huang, J.S.; Zimmerman, A.R.; Chen, H.; Hu, X.; Gao, B. Mechanisms and adsorption capacities of hydrogen peroxide modified ball milled biochar for the removal of methylene blue from aqueous solutions. *Bioresour. Technol.* **2021**, *337*, 125432. [CrossRef]
41. Ma, F.F.; Zhang, D.; Zhang, N.; Huang, T.; Wang, Y. Polydopamine-assisted deposition of polypyrrole on electrospun poly (vinylidene fluoride) nanofibers for bidirectional removal of cation and anion dyes. *Chem. Eng. J.* **2018**, *354*, 432–444. [CrossRef]
42. Xie, L.Q.; Jiang, X.Y.; Yu, J.G. A Novel Low-Cost Bio-Sorbent Prepared from Crisp Persimmon Peel by Low-Temperature Pyrolysis for Adsorption of Organic Dyes. *Molecules* **2022**, *27*, 5160. [CrossRef] [PubMed]
43. Zango, Z.U.; Dennis, J.O.; Aljameel, A.I.; Usman, F.; Ali, M.K.M.; Abdulkadir, B.A.; Algessair, S.; Aldaghri, O.A.; Ibnaouf, K.H. Effective Removal of Methylene Blue from Simulated Wastewater Using ZnO-Chitosan Nanocomposites: Optimization, Kinetics, and Isotherm Studies. *Molecules* **2022**, *27*, 4746. [CrossRef] [PubMed]
44. Zhang, H.Q.; Khanal, S.K.; Jia, Y.Y.; Song, S.L.; Lu, H. Fundamental insights into ciprofloxacin adsorption by sulfate-reducing bacteria sludge: Mechanisms and thermodynamics. *Chem. Eng. J.* **2019**, *378*, 122103. [CrossRef]
45. Shah, R.; Huang, S.X.; Pingali, S.V.; Sawada, D.; Pu, Y.Q.; Rodriguez, M., Jr.; Ragauskas, A.J.; Kim, S.H.; Evans, B.R.; Davison, B.H.; et al. Hemicellulose–Cellulose Composites Reveal Differences in Cellulose Organization after Dilute Acid Pretreatment. *Biomacromolecules* **2018**, *20*, 893–903. [CrossRef]

46. Chen, C.J.; Song, J.W.; Cheng, J. Highly Elastic Hydrated Cellulosic Materials with Durable Compressibility and Tunable Conductivity. *ACS Nano* **2020**, *14*, 16723–16734. [CrossRef]
47. Feng, Y.F.; Zhou, H.; Liu, G.H.; Qiao, J.; Wang, J.H.; Lu, H.Y.; Yang, L.Z.; Wu, Y.H. Methylene blue adsorption onto swede rape straw (*Brassica napus* L.) modified by tartaric acid: Equilibrium, kinetic and adsorption mechanisms. *Bioresour. Technol.* **2012**, *125*, 138–144. [CrossRef]

Article

A Strategy to Valorize a By-Product of Pine Wood (*Pinus pinaster*) for Copper Removal from Aqueous Solutions

Chiara Mongiovi [1], Maélys Jaillet [1], Dario Lacalamita [1], Nadia Morin-Crini [1], Michael Lecourt [2], Sandra Tapin-Lingua [2] and Grégorio Crini [1,*]

[1] Chrono-Environnement, Université de Franche-Comté, CNRS, Faculté des Sciences, 25000 Besançon, France; chiara.mongiovi@univ-fcomte.fr (C.M.); maelys.mydh@gmail.com (M.J.); dario.lacalamita@univ-fcomte.fr (D.L.); nadia.crini@univ-fcomte.fr (N.M.-C.)

[2] Institut FCBA, Institut Technologique Forêt Cellulose Bois-Construction Ameublement, Domaine Universitaire, CS 90251, cedex 9, 38044 Grenoble, France; michael.lecourt@fcba.fr (M.L.); sandra.tapin-lingua@fcba.fr (S.T.-L.)

* Correspondence: gregorio.crini@univ-fcomte.fr

Citation: Mongioví, C.; Jaillet, M.; Lacalamita, D.; Morin-Crini, N.; Lecourt, M.; Tapin-Lingua, S.; Crini, G. A Strategy to Valorize a By-Product of Pine Wood (*Pinus pinaster*) for Copper Removal from Aqueous Solutions. *Molecules* 2023, 28, 6436. https://doi.org/10.3390/molecules28186436

Academic Editors: Yongchang Sun and Dimitrios Giannakoudakis

Received: 12 July 2023
Revised: 28 August 2023
Accepted: 1 September 2023
Published: 5 September 2023

Copyright: © 2023 by the authors. Licensee MDPI, Basel, Switzerland. This article is an open access article distributed under the terms and conditions of the Creative Commons Attribution (CC BY) license (https://creativecommons.org/licenses/by/4.0/).

Abstract: This study describes the valorization of a pine wood by-product (*Pinus pinaster*) in the form of individualized fibers to a complex copper or more broadly metals present in an aqueous solution using a batch process. The adsorption results show that pine fibres activated by sodium carbonate are effective in recovering copper ions from monocontaminated or polycontaminated solutions of varying concentrations in a few minutes. One gram of material captures 2.5 mg of copper present in 100 mL of solution at pH 5 in less than 10 min. The results are perfectly reproducible and independent of pH between 3 and 5. The presence of the Na^+ cation at concentrations of 0.1 M has no impact on material performance, unlike that of Ca^{2+} ions, which competes with Cu^{2+} ions for active sites. The adsorption process can be considered as rapid, as most of the copper is adsorbed within the first 10 min of exposure. Investigation of modeling possibilities shows some limitations. Indeed, the Weber and Morris and Elovich models show poor possibilities to describe all the kinetic data for copper adsorption on fibres. This may prove that the mechanism is far more complex than simple physisorption, chemisorption and/or diffusion. Complexation by wood fibers can be extended to solutions containing several types of metals. The results of this study show that the field of selective metal recovery could be a new way of valorizing by-products from the wood industry.

Keywords: wood; copper; adsorption; water decontamination

---

1. Introduction

In Europe, under the 2000 European Water Directive (WFD), companies that use large volumes of water are classified for environmental protection (ICPE), and their substance discharges are controlled to comply with regulatory values known as emission limit values (ELVs). ICPE industries have to consider improving the treatment of their discharge in order to improve chemical and ecological quality of water [1,2]. In France, the Région Bourgogne Franche-Comté is particularly concerned with the problem of metal emissions. It is a region of excellence, with a large number of companies in the surface treatment (ST) sector of various industries, from automotive to aeronautics, electricity, metallurgy, tanning and luxury goods (watchmaking, fashion industry, leather goods) [3].

The ST sector uses large volumes of water and chemical substances, and consequently produces significant quantities of wastewater contaminated by a mixture of substances with variable pH levels and high salinity. Among these substances, copper occupies an important place. This metallic element is considered a priority substance by the French Water Agency for ecological water quality. In addition, along with seven other metals (cadmium, mercury, arsenic, lead, nickel, chromium and zinc), copper is used to calculate an index called METOX (article R. 213-48-3 of the French Environment Code) to establish a toxicity threshold linked to the importance of metals present in the aquatic environment. Copper, as a trace element and an industrial substance, is indispensable; it is found in many products (food supplements, for example) and used in many sectors (agriculture, viticulture, livestock farming, pharmaceuticals, cosmetology, etc.). However, it can also be toxic if present in high concentrations. This is why this metal is the subject of special attention [4–7].

In general, the technique used to decontaminate metal-contaminated water is physico-chemical, which involves precipitating the metal elements by adding an alkalizing agent such as lime or soda and separating the clarified water from the sludge formed by filtration or sedimentation processes [8]. If the discharge complies with current regulations (for copper, the ELV is 2 mg/L), it is released back to the environment. The current context is increasingly focused on preserving the quality of water resources; consequently, manufacturers are being asked to implement techniques capable of achieving zero discharge of pollutants, or at least of significantly reducing the ELV down to 2 mg/L for copper. However, this is not an easy task, since metals are present in trace amounts and are accompanied by the organic load in complex polycontaminated and highly saline mixtures. In theory, there are several techniques available to achieve a higher level of decontamination, complementary to chemical insolubilization, the physico-chemical technique traditionally used by the ST sector to remove metals from their wastewater. Coupling physico-chemistry with complementary treatments such as membrane filtration (nanofiltration, reverse osmosis), adsorption, ion exchange or evaporation makes it possible to achieve the objective of zero pollutant discharge. Each of these methods has its advantages and disadvantages [9].

In recent years, a new challenge has emerged: recovering metals present in wastewater (diluted effluents, concentrated baths) or in treated water (discharges) in trace amounts and recycling them [5,8–11]. The aim in this case is to recover metals instead of discharging them into the environment. This challenge is relevant considering the current crisis situation (rising raw material prices, financial speculation, shortages, environmentally unfriendly extraction of metals, etc.). One of the techniques capable of retaining and complexing metals is ion exchange. This treatment method is already used by ST manufacturers to produce demineralized water, protect treatment baths and maintain their quality, recycle rinses, or reduce the ELVs of the main metals (Cu, Ni, Cr, Al, etc.) present in water treated by physical chemistry [9]. The ion exchange technology is also mainly used to recover high-value metals (Au, Ag, Pd, Pt) present in low-concentration water. However, the ion exchange technology is highly sensitive to the presence of suspended solids (SS), organic load (characterized by the values of chemical oxygen demand, COD), salts, and pH. This is why, in wastewater treatment plants (WWTPs), ion exchange is coupled with preliminary stages of filtration to remove SS and activated carbon adsorption to remove COD. However, for small- and medium-sized businesses, operating and maintenance costs can be significant. In addition, it is difficult to effectively and selectively remove metals from complex mixtures, especially because of the competition between different chemical species and the specific parameters (highly variable pH, high salinity) of each type of effluent. In addition, carbon reactors and resin filters are known to quickly become saturated. They must then either be replaced (in the case of coals), or regenerated (resins), which generates additional costs. Therefore, effective, simple, viable and environmentally friendly techniques for recovering metals are to be identified [7,9,11–25].

The adsorption of pollutants on inexpensive lignocellulosic materials is well known [26–33]. These media are used as biosorbents in ion exchange, chelation or adsorption processes to complex metals, dyes, hydrocarbons, bacteria and viruses present in an aqueous solution or in leachates. Numerous studies have been published, and the materials proposed take the form of sawdust, bark and hurds (shives), for example, from wood, hemp and flax, either raw or chemically modified. Lignocellulosic materials modified by grafting carboxylic functions onto their surface have been proposed as adsorbents because of the ion exchange properties of these functions. This grafting strongly modifies their affinities for dyes or their selectivity towards hard or soft metals, as, for example, in [7,12,14,15,26,27,30,34–45]. Fewer studies have been carried out on fibers extracted directly from wood, without chemical modification or grafting [14,15,44].

Maritime pine (*Pinus pinaster*) is a cultivated species in southern Europe. It is used in a variety of applications: as lumber in sawmills for applications in construction, furniture and processing industries such as particleboard or pulp for paper. Controlled mechanical defibration of wood produces isolated fibers or fiber bundles that can be recovered in bulk or in panels by dedicated plants. New ways of adding value to the wood industry by-products need to be identified to strengthen the sector. This study is part of an industrial project to develop low-cost, effective technologies for treating wastewater contaminated with metals such as copper from the surface treatment process. Two industrial partners are heavily involved: The first wishes to explore new ways of adding value to wood, and more specifically maritime pine fibers, and the second wishes to selectively recover the copper present in concentrated baths, finding methods of treating polycontaminated wastewater capable of reducing the copper emission limit value set at 2 mg/L for its discharges. To our knowledge, this is the first time that pine material has been used for these purposes.

In this study, wood fibers, activated by sodium carbonate Na_2CO_3, are used as an adsorbent material of complex copper ions present in synthetic solutions reproducing real-life conditions of industrial baths and discharges. The complexation properties of the materials, obtained by the batch method and expressed in terms of adsorption capacity or % reduction, are described. The influence of several batch variables such as initial copper concentration, material dose, contact time, initial solution pH, as well as the presence of salts (NaCl, $CaCl_2$) in the solution to be treated is evaluated and discussed. Experimental data are modeled using equations (adsorption isotherms and kinetic models) commonly used in the literature.

2. Results and Discussion

The activation of materials with sodium carbonate is a process proposed by the industrial partner to activate carboxylic functions. Preliminary studies have shown that this type of activation leads to changes in both the structure of materials and their chemical composition, in line with numerous published studies on the subject [13–15,44]. SEM images and EDS spectra of maritime pine materials in raw (Pin-R), washed (Pin-W) and sodium carbonate-activated (Pin-C) forms are shown in Figure 1. In the spectrum of the Pin-R sample, the presence of the calcium signal can be observed, which disappears after a simple water wash. For the Na_2CO_3-activated sample (Pin-C), a new sodium signal is observed. SEM images also show a smoother, softer surface for Pin-W and Pin-C materials. Some fine elements observed onto the fibers generate during mechanical processing are washed away.

Figure 2 compares the copper elimination obtained on washed (Pin-W) and activated (Pin-C) pine materials, for three initial copper concentrations present in an aqueous solution with an initial pH close to 5 (pH generally encountered in industrial water from the surface treatment process). The copper elimination is expressed in % of removed pollutant (abatements). Whatever the concentration, the abatements measured for sodium carbonate-activated pine are always higher than those for washed pine. These eliminations are high: 1 g of the activated material is capable of complexing all of the 2.5 g of copper introduced in a 100 mL volume, the reduction being close to 100%, whereas for washed pine, the value is close to 60%. This result demonstrates the interest of activating the adsorbent material with sodium carbonate, as carboxylic acids are converted into carboxylate functions, which then enable the metal cations to be complexed. This interpretation was confirmed by two phenomena: the first concerned the systematic increase in the final pH value, indicating the presence of chemical interactions; the second concerned the increase in sodium concentrations found in the supernatants after adsorption, confirming interactions by ion exchange, with the sodium cation carried by the active sites of the fibers being replaced by the metal cation in the aqueous solution. Similar interpretations have been reported in the literature [13,40,41,44]. Finally, the results in Figure 2 were repeated five times under identical batch conditions, and the standard deviation values showed the reproducibility of the data.

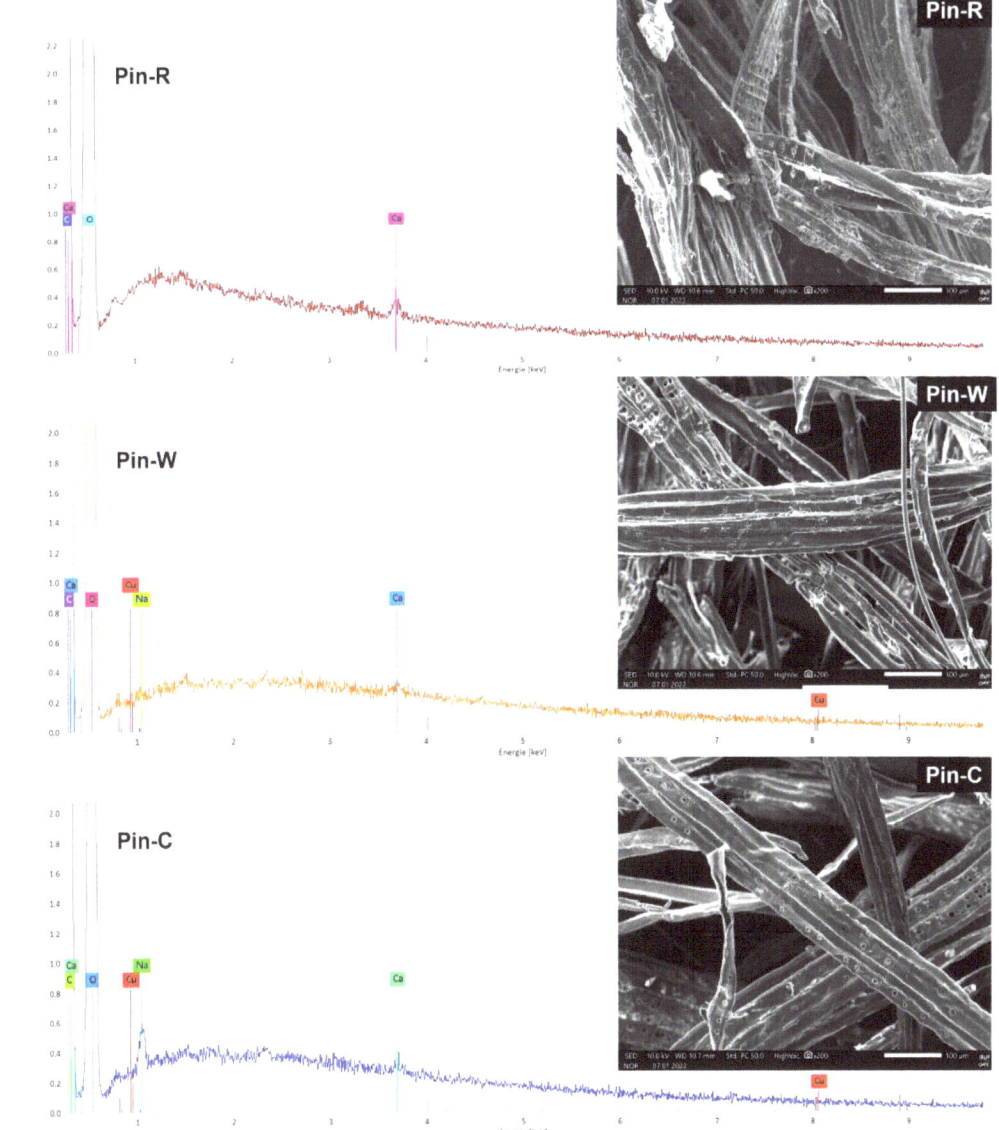

Figure 1. Comparison of scanning electron microscopy images and elemental analysis spectra by energy-dispersive spectroscopy of raw pine (Pin-R), washed pine (Pin-W) and activated pine (Pin-C) materials.

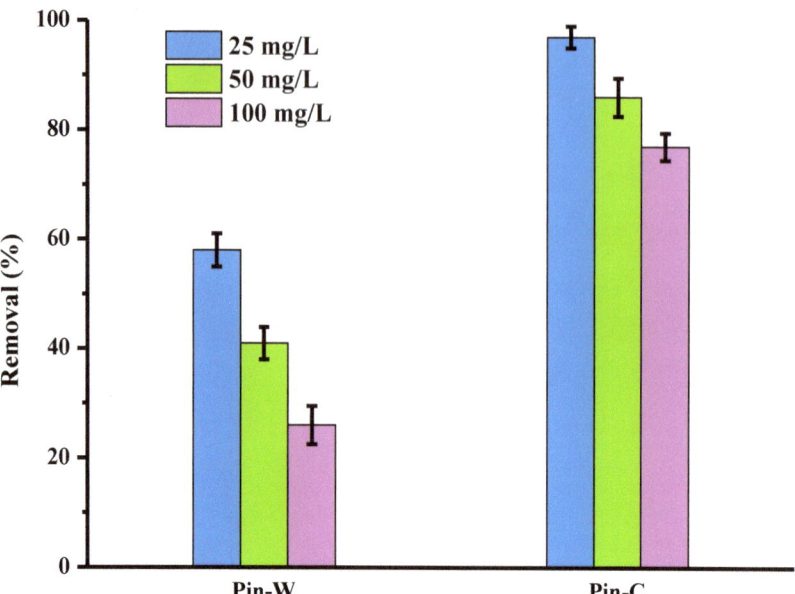

Figure 2. Comparison of copper abatements (expressed in % of removed pollutant) obtained on two pine materials, washed (Pin-W) and activated (Pin-C), at three initial concentrations of copper (other conditions: volume of solution = 100 mL; mass of material = 1 g; contact time = 120 min; T = 22 ± 1 °C; pH_i = 5.1–5.4 ± 0.1; agitation = 250 rpm; n = 5).

The results in Figure 2 were obtained with a contact time of 120 min, the same as that used by the manufacturers in their batch tests. Nevertheless, the effect of contact time on copper removal by both materials was investigated (Figure 3). The study was conducted over a time range from 5 to 240 min and at two different copper concentrations, 25 and 50 mg/L, with the other batch conditions held constant. The experimental results were represented by plotting the amount of adsorbed copper (q_t in mg/g) at time t as a function of the time variable (t in min). The data in Figure 3 show that q_t values increase with time and initial copper concentration until equilibrium is reached. Irrespective of concentration, higher adsorbed quantities were obtained for the activated material compared to the washed pine. At the start of the adsorption process, the presence of a large number of accessible open active sites on the surface favored rapid copper adsorption: q_t values increased very rapidly. Then, as contact time was extended, ions moved through the pores of the adsorbent (internal diffusion). The number of free sites and their accessibility then began to decrease, and reaction equilibrium was reached. The data clearly indicated that the adsorption process can be considered rapid, since the greatest quantity of copper was adsorbed by the fibers within the first 10 min of adsorption. This result can be explained by the presence of strong interactions between the materials and the metal ions via surface adsorption, diffusion into the wood material network and chemisorption.

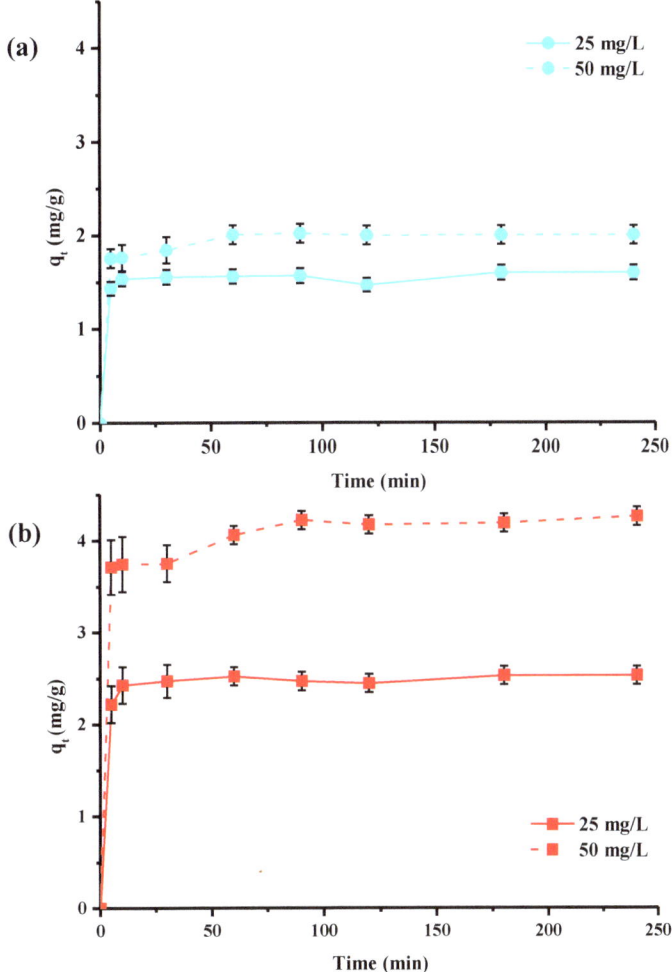

Figure 3. Kinetics of adsorption of copper ions by washed Pin-W (**a**) and activated Pin-C (**b**) pine materials at two different concentrations, 25 and 50 mg/L (other conditions: solution volume = 100 mL; mass of material = 1 g; T = 22 ± 1 °C; pH_i = 5.2–5.4 ± 0.1; stirring = 250 rpm; n = 3).

Experimental kinetic data were modeled using various empirical models commonly used in the literature [46–53]. The graphical representations of these models are shown in Figure 4, and the calculated parameters are presented in Table 1. The first model used is the pseudo-first-order model established by Lagergren [54], which considers that the occupancy rate of adsorption sites is proportional to the number of unoccupied sites. This model is expressed by Equation (3), in which q_e and q_t, respectively, represent the quantity of metal adsorbed at equilibrium and time t (mg/g), and k_1 is the pseudo-first-order rate constant (min^{-1}). Interpretation of these results showed that Lagergren's model did not apply to our data set (Figure 4; Table 1), and that copper was therefore not adsorbed by maritime pine fibers by occupying a localized adsorption site.

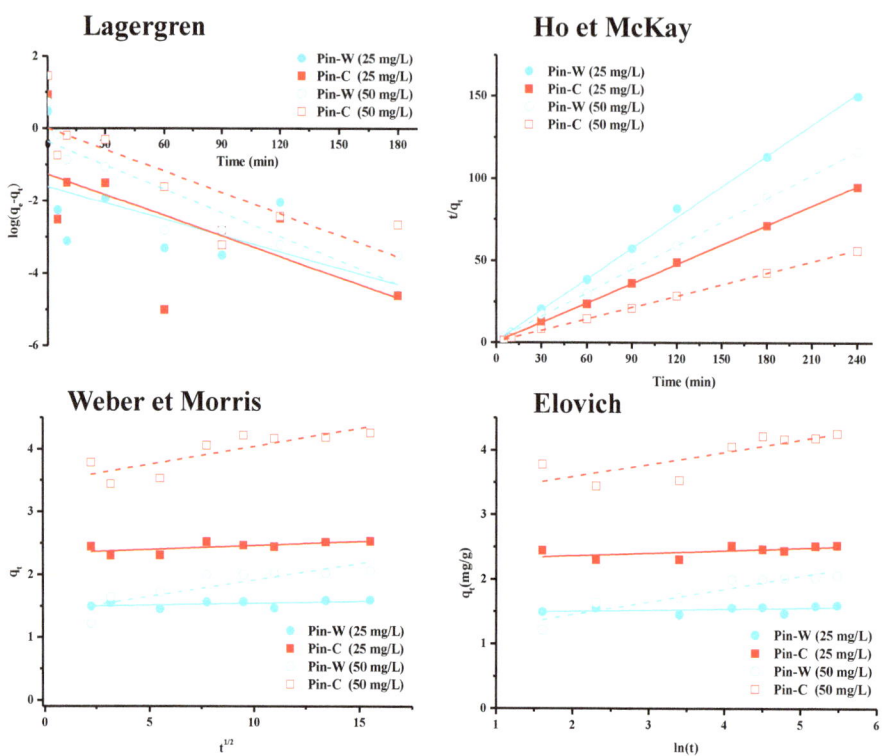

Figure 4. Graphical representations of the kinetic models of Lagergren, Ho and McKay, Weber and Morris, and Elovich for Pin-W washed pine and Pin-C activated pine materials at two different concentrations, 25 and 50 mg/L.

Table 1. Theoretical parameters for the kinetic models of Lagergren, Ho and McKay, Weber and Morris, and Elovich (units: C_i in mg/L; $q_{e,exp}$ et $q_{e,cal}$ in mg/g; k_1 in min^{-1}; k_2 in g/mg min et K_p in mg/g min$^{1/2}$; α in g/mg et β en mg/g).

				Lagergren			Ho et McKay			Weber et Morris			Elovich		
		C_i	$q_{e,exp}$	$q_{e,cal}$	k_1	R^2	$q_{e,cal}$	k_2	R^2	K_p	C	R^2	α	β	R^2
Pin-W		25	1.60	0.19	0.014	0.3087	1.59	0.303	0.9980	0.006	1.48	0.1448	6×10^{35}	58	0.0632
		50	2.06	0.28	0.018	0.3261	2.53	0.250	0.9996	0.013	2.33	0.3825	7×10^{22}	24	0.2835
Pin-C		25	2.53	0.69	0.020	0.7598	2.08	0.122	0.9997	0.052	1.38	0.6661	4×10^{1}	5	0.8622
		50	4.26	0.101	0.020	0.6089	4.35	0.073	0.9994	0.058	3.47	0.6495	3×10^{6}	5	0.6049

The second model is that of Ho and McKay, or the pseudo-second-order model [55], which is widely used in the literature due to its simplicity. It is expressed by Equation (4) in which q_e and q_t represent, respectively, the amount of metal adsorbed at equilibrium and time t (mg/g) and k_2 is the pseudo-second-order rate constant (g/mg min). Plotting $1/q_t$ as a function of time t systematically yields a linear plot with regression coefficients close to one (Table 1). The application of the Ho and McKay model enabled us to fit our experimental data to those obtained by calculation (Figure 4, Table 1). Indeed, the q_e values found theoretically by this model were closer to those found experimentally than those calculated by the pseudo-first-order model. However, this model, being empirical and without scientific basis, provided no information on the adsorption mechanism. Like the Lagergren model, the Ho and McKay model simply suggested that adsorption depends on the adsorbent–adsorbent couple. In the literature, other simplified models such as those of Weber and Morris (intraparticle diffusion model) [56] or Elovich (chemisorption or ion exchange model) [57] have been used to obtain information on these mechanisms.

The intraparticle diffusion model, widely used in the literature despite numerous recent criticisms, is based on a concept of diffusion in the pores of the material. In addition to the models of Lagergren and Ho and McKay, the applicability of the Weber–Morris model makes it possible to hypothesize the interactions (physisorption, chemisorption and, above all, diffusion) involved in the adsorption process, and thus to define the limit stage of the kinetics [56]. The graphical representation of this model (Equation (5)) offers a constant for the rate of intraparticle diffusion (mg/g min$^{1/2}$) and an index for the thickness of the boundary layer (the higher the index, the thicker the boundary layer). In general, three lines were observed: the first, and clearest, corresponded to the external diffusion stage (film diffusion), the second represented the internal or intraparticle diffusion stage, and the third was the final equilibrium stage (adsorption reaction). Analysis of kinetic data over the whole range from 0 to 240 min contact time showed the non-adjustment of experimental data with those calculated from the Weber and Morris intraparticle diffusion model (Table 1). The values of the correlation coefficients were indeed low for the entire time range studied. Consequently, for Pin-W and Pin-C materials and for both copper concentrations, the data set did not follow the Weber and Morris model, so that internal or external diffusion was not the main mechanism controlling adsorption kinetics.

The final model is that of Elovich [57]. In general, this chemisorption model is used to describe adsorption processes characterized by the presence of a heterogeneous surface and to assess the chemical nature of the adsorption mechanism. However, the R^2 values obtained for the two pine materials showed that this model did not apply to the experimental data (Table 1), regardless of the concentration used.

The structure of the fibers, made up of tangles of cellulose microfibrils coated with hemicellulose and lignin in the outer layer, and the anatomy of the fibers, with a more or less accessible lumen, justify such an approach to diffusion and chemisorption. However, a full analysis of the modeling via these Weber and Morris and Elovich models has shown that neither of these two models is entirely satisfactory in describing all the kinetic data for copper adsorption on maritime pine. The mechanism is far more complex than simple physisorption, chemisorption and/or diffusion [46–53]. Indeed, lignocellulosic materials such as those in this work are known to have a complex structure, with fibers consisting of tangles of cellulose microfibrils covered with hemicellulose and lignin in the outer layer, as well as fiber anatomy with a more or less accessible lumen.

The results in Figure 5 describe the impact of initial copper concentration on the purification performance of washed and activated pine materials. For this study, a series of experiments was carried out with initial concentrations ranging from 10 to 200 mg/L, while all other batch conditions were constant. For the same quantity of wood fiber, increasing the concentration lead to a decrease in the abatement rate through progressive saturation of the active sites. Activated pine once again demonstrated high performance even at high copper concentrations, confirming the interest of the chemical activation of the material. Indeed, for an initial concentration of 100 mg/L, the abatement rate of activated pine was of the order of 80%, and could therefore enable the recovery of copper present in industrial baths. For concentrations of industrial interest (below 10 mg/L), Pin-C was capable of complexing all the copper present in the solution, and thus of achieving the decontamination objective (reduction of ELV < 2 mg/L). Finally, however, it should be noted that the abatement rate observed for washed pine did not follow the same trend as that for activated pine.

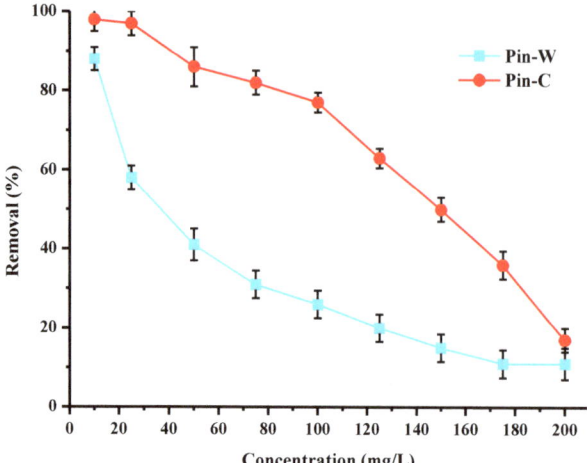

Figure 5. Effect of the initial copper concentration on the purification performance (expressed in % of removed copper) of washed (Pin-W) and activated (Pin-C) pine materials (other conditions: volume of solution = 100 mL; mass of material = 1 g; contact time = 120 min; T = 22 ± 1 °C; pH_i = 4.7–5.6 ± 0.1; stirring = 250 rpm; n = 3).

To determine the theoretical isotherm followed by copper adsorption by maritime pine fibers and predict, by calculation, the maximum adsorption capacity (q_m) of this material, three empirical equations were applied: Langmuir [58], Freundlich [59], and Temkin [60]. The simplest model used in the literature to describe adsorption properties is that proposed by Langmuir. This simplified model is given by Equation (7), where q_e represents the equilibrium adsorption capacity (mg/g), Ce is the equilibrium concentration (mg/L) and K is the Langmuir constant (L/g). Its graphical representation enables us to estimate the maximum adsorption capacity of a given material (q_m in mg/g), corresponding to adsorbent saturation [58]. According to the assumptions of this model, the following information can be deduced: uniform material surface, localized adsorption on a fixed number of sites, instantaneous and reversible reaction, identical active sites capable of binding only one pollutant, no interaction between pollutants, constant adsorption energy and monolayer adsorption with the existence of a saturation plateau value [3,48,50].

The Freundlich equation is also widely used. According to this model, the material surface is heterogeneous and interactions are possible between the adsorbates forming a multilayer [59]. In addition, this empirical equation assumes that there is a large number of active sites available on the material surface, acting simultaneously, each with a different adsorption free energy. Adsorption is reversible and infinite. In Equation (8), K_F is the Freundlich constant ($mg^{1-(1/n)} L^{1/n}/g$) and n_F is a heterogeneity coefficient. The value of $1/n_F$ is between zero and one and indicates the degree of non-linearity between solution concentration and adsorption. If the value of $1/n_F$ is equal to unity, adsorption is linear; if the value is less than unity, this implies that the adsorption process is favorable; and if the value is greater than unity, the adsorption is an unfavorable process. The more heterogeneous the surface, the closer the value of $1/n_F$ to zero.

Temkin and Pyzhev studied the effects of certain indirect interactions between adsorbate and adsorbate on adsorption isotherms [60]. They suggested that, due to these interactions and ignoring concentration values, the heat of adsorption of all adsorbates in the layer would decrease linearly with coverage. This concept is expressed by the so-called Temkin Equation (9), in which K_T is the equilibrium binding constant (L/g) corresponding to the maximum binding energy and B is another constant (J/mol) related to the adsorption energy [46,49].

The results described in Figure 6 and Table 2, respectively, show graphical representations of the experimental data modeled by these three equations and the corresponding parameters calculated. Interpretation of these data showed a better match with the experimental data by applying the Langmuir model. The Langmuir equation can also be used to calculate q_m: for Pin-W and Pin-C, values of 2.2 and 3.9 mg of copper per gram of material, respectively, were obtained. Similar values were recently reported in the literature [7]. These values are of the same order of magnitude as those described in the literature for other lignocellulosic materials. The data in Table 2 show $1/n_F$ values below one, indicating a favorable adsorption process. Finally, the Temkin and Pyzhev model is not applicable to our data set.

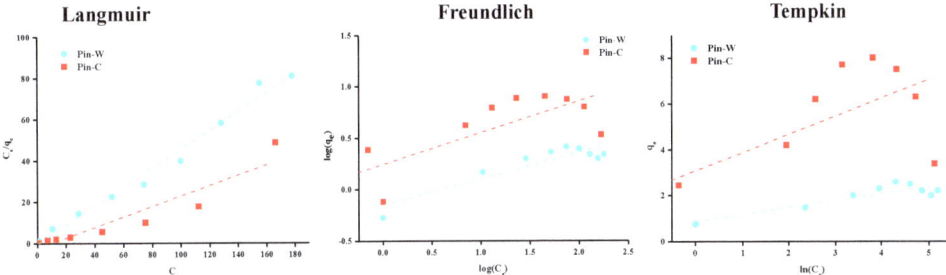

Figure 6. Representations of adsorption isotherms of experimental results modelled by the Langmuir, Freundlich and Temkin equations.

Table 2. Theoretical parameters calculated from the Langmuir, Freundlich and Temkin equations (units: q_m in mg/g; K in L/mg; K_F in mg$^{1-(1/n)}$ L$^{1/n}$/g; K_T in L/g; B in J/mol).

	Langmuir			Freundlich			Temkin		
	q_m	K	R^2	$1/n_F$	K_F	R^2	B	K_T	R^2
Pin-W	2.2	1.01	0.9813	0.26	0.70	0.8621	0.30	7.9	0.5659
Pin-C	3.9	0.10	0.8681	0.31	1.76	0.6002	0.79	27.9	0.4377

The data in Figure 7 describe the effect of the initial dose of washed and activated materials on the adsorption of copper present at concentrations of 25 and 50 mg/L. These results show that, for the same initial concentration, the increase in the concentration of maritime pine fibres in the water leads to an increase in the abatement rate due to the increase in the number of active sites available for the adsorption of metal cations. For activated pine, a dose of 1 g is sufficient to obtain the highest purifying percentages at the concentration of 25 mg/L. For 50 mg/L of copper, abatement rates are lower. However, to increase these levels, it is enough to increase the amount of fiber present in the water to be treated. The replacement of a conventional adsorbent by a bioadsorbent implies that the latter has a similar or even better performance at the same or lower cost, which is the case, our material being an abundant and cheap co-product.

One of the most important parameters in an adsorption process is pH because it modifies both the surface charge of the adsorbents and the chemistry of the elements to be complexed. Figure 8 describes the effect of the initial pH of the solution to be treated on the reduction in copper by the two types of fibers at two different doses of copper. This study was carried out for pH values between two and five because at values above five, copper begins to precipitate [7]. Analysis of the data showed that the optimal pH range for the highest performance is between four and five, especially for activated pine. At the two doses studied, the trends were identical for both materials. The lower the pH, the lower the abatement values. Especially at pH 2, copper abatement was very low, because at this pH, cupric ions compete with H+ ions for active sites [7].

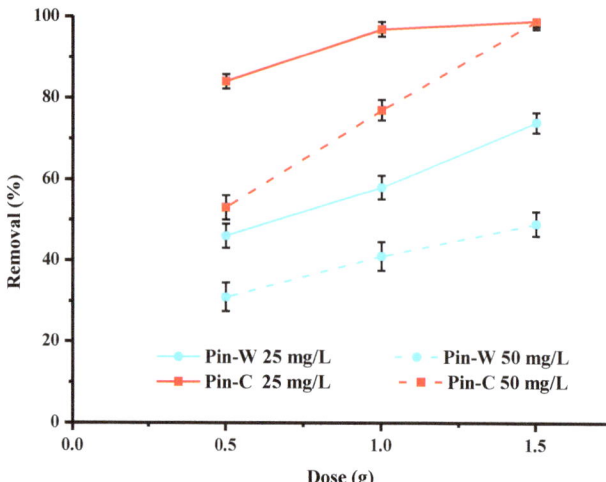

Figure 7. Effect of the mass of adsorbent used on copper abatement (expressed in % of removed pollutant) by washed (Pin-W) and activated (Pin-C) pine materials at two different concentrations, 25 and 50 mg/L (other conditions: volume of solution = 100 mL; contact time = 120 min; T = 22 ± 1 °C; pH_i = 5.2–5.4 ± 0.1; agitation = 250 rpm; n = 3).

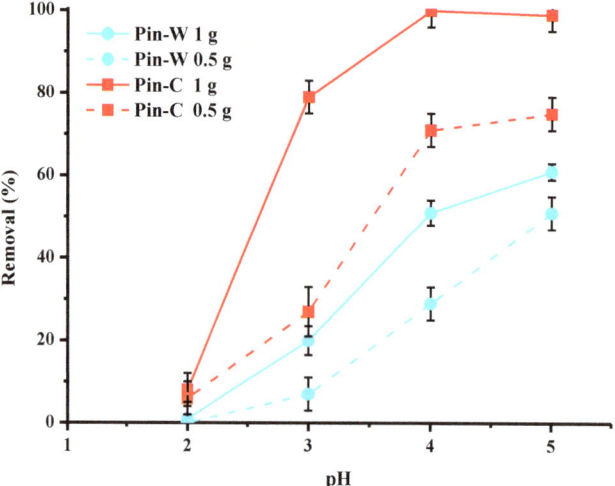

Figure 8. Effect of the initial pH of the solution on copper abatement (expressed in % of removed pollutant) by washed (Pin-W) and activated (Pin-C) pine materials at two different masses, 0.5 and 1 g (other conditions: volume of solution = 100 mL; [Cu] = 25 mg/L; contact time = 120 min; T = 22 ± 1 °C; stirring = 250 rpm; n = 3).

In order to study the impact of ionic strength on the purification performance of materials, different concentrations of NaCl and $CaCl_2$ salts were explored, as these salts are widely used in surface treatment. Three molar concentrations were studied: 0.1, 0.5 and 1 mol/L, which correspond to 5.8, 29 and 58 g/L for NaCl and 11, 55 and 110 g/L for $CaCl_2$. Values between 0.1 and 0.5 M are representative of those found in effluents and industrial discharges. The results in Figure 9 show a significant drop in the purification performance of materials with the increase in salt concentration. Indeed, the adsorption rates for washed pine decreased from 58% to 10% for Na+ and from 58% to 5% for Ca^{2+}, respectively, in the absence of salt and at a concentration of 1 M of Na^+ or Ca^{2+}. For activated pine, the abatements were reduced from 97% to 30% for Na^+ and 9% for Ca^{2+}. These results can be explained by the suppression of electrostatic interactions. Indeed, when the amount of salt increased, i.e., an increase in the ionic strength of the solution occurred, the amount of Ca^{2+} or Na^+ ions increased, which masked the active sites of pine fibers for Cu^{2+} ions. Nevertheless, at the concentration of 25 mg/L of copper and in the case of the addition of NaCl, activated pine was able to remove about 30% of copper even at the concentration of 1 M of Na^+, which demonstrates a very important selectivity. Another interesting observation was that the presence of the Ca^{2+} cation had more effect on copper adsorption than the presence of the Na^+ cation, which can be explained by the fact that Ca^{2+} ions are more similar to Cu^{2+} ions in terms of their physical and chemical properties (ionic radius, charge, etc.). There is, therefore, competition between Ca^{2+} and Cu^{2+} ions for active sites.

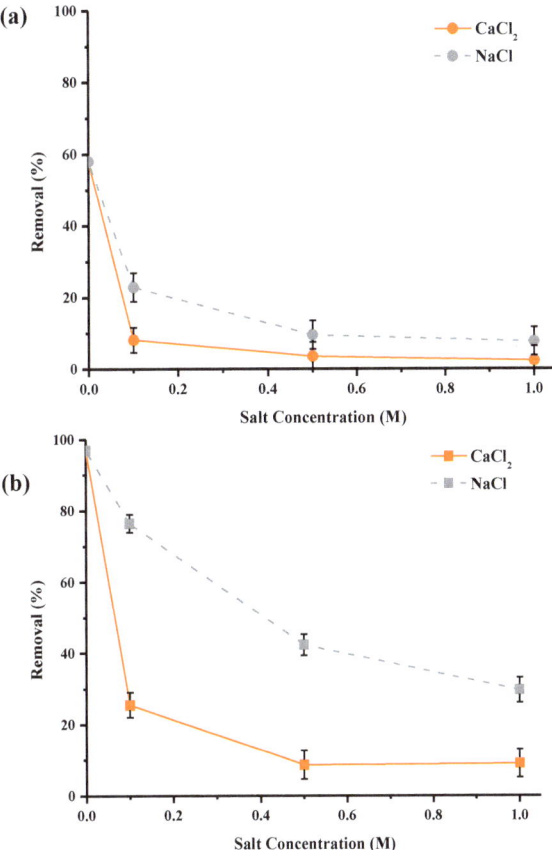

Figure 9. Effect of presence of NaCl and $CaCl_2$ in the solution on copper abatement (expressed in % of removed pollutant) by Pin-W (**a**) and activated Pin-C (**b**) pine materials (other conditions: volume of solution = 100 mL; mass of material = 1 g; [Cu] = 25 mg/L; $[Na^+]$ = $[Ca^{2+}]$ = 0.1, 0.5 and 1 mol/L; contact time = 120 min; T = 22 ± 1 °C; stirring = 250 rpm; pH = 5.4 ± 0.1; n = 3).

Some industrial baths resulting from the processes are generally characterized by high concentrations of copper (of the order of g/L) but relatively insignificantly contaminated by other chemical species. On the contrary, in industrial discharges that are sent into aquatic environments, copper can be found, but also other cations such as Ni, Mn, Zn, Na and Ca. However, the concentrations are lower, of the order of a few tens of mg/L. The results of Figure 10 describe the purification performance of activated pine material towards several cations present in a monocontaminated solution at a concentration of 25 mg/L. Other batch conditions are constant, as in the case of monocontaminated copper solutions. For this concentration, the following order of affinity was obtained: Cu > Ca > Mn > Ni ~ Zn >>> Na, with abatements of 100%, 96%, 90%, 76%, 71% and 0%. The sodium was not adsorbed at all by the treated maritime pine and was even released into solution in all experiments, which confirms the exchange of ions. On the contrary, calcium was largely adsorbed by activated pine, which showed a great affinity of this material vis-à-vis the mineral. The difference in the degree of biosorption can be attributed to the physical and chemical characteristics of each element, namely ionic radius, molar mass and electronegativity [61].

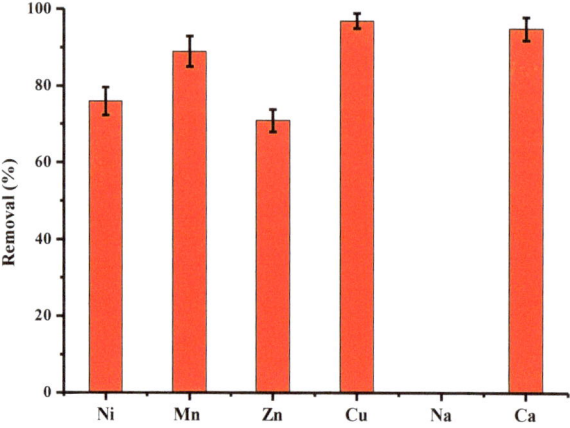

Figure 10. Comparison between adsorption capacities (expressed in % of removed pollutant) of activated pine material towards several cations present in monocontaminated solution at a concentration of 25 m/L (other conditions: volume of solution = 100 mL; mass of material = 1 g; contact time = 120 min; T = 22 ± 1 °C; stirring = 250 rpm; pH$_i$ = 5.4–5.9 ± 0.1; n = 5).

Adsorption experiments in solutions polycontaminated with four metals (Ni, Mn, Zn and Cu) were performed by selecting the initial concentration of each metal of 25 mg/L or 100 mg/L in total metals. The influence of the presence of Na$^+$ and Ca^{2+} ions was studied by adding to the initial solutions a quantity of salt at the concentration of 0.1 mol/L, i.e., 2.3 g of Na$^+$ and 4 g of Ca^{2+}. The results are shown in Figure 11. In the mixture, regardless of the experience, the order of affinity was as follows: Cu >> Zn > Ni > Mn. This order was independent of the presence of salt. Of the 10 mg of metals present in the 100 mL of the initial solution, 1 g of activated pine adsorbed 4.12 mg of the four metals. In the presence of NaCl or CaCl$_2$, this value increased to three, 21 and 0.92 mg, respectively. Activated pine, therefore, has an excellent biosorption capacity for the four metals present in the polycontaminated solution, in particular for the 2.5 mg copper present in the 100 mL of the batch solution. In addition, it can be seen that this selectivity for copper remains important despite the presence of Na$^+$ and Ca^{2+} ions.

Characterization analyses by EDS spectroscopy were performed for the fibers of Pin-C after copper adsorption (Pin-CCu). On the spectrum obtained after adsorption, a new peak due to copper could be observed (Figure 12).

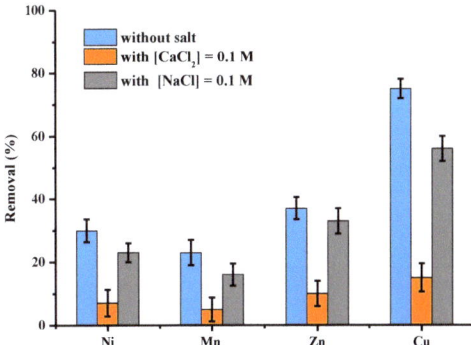

Figure 11. Comparison between adsorption capacities (expressed in % of removed pollutant) of activated pine material towards 4 cations (Ni, Mn, Zn and Cu) present in polycontaminated solution at a total concentration of 100 mg/L in the presence or without $CaCl_2$ and NaCl (other conditions: volume of solution = 100 mL; mass of material = 1 g; individual concentration of each cation = 25 mg/L; $[Na^+] = [Ca^{2+}] = 0.1$ mol/L contact time = 120 min; T = 22 ± 1 °C; stirring = 250 rpm; pH = 5.3 ± 0.1; n = 3).

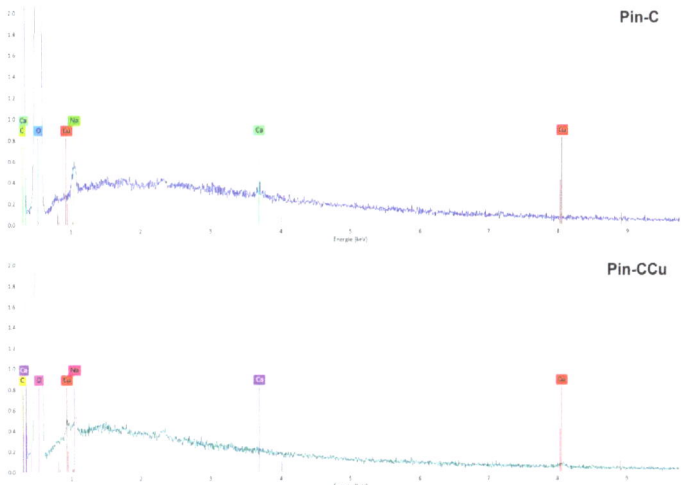

Figure 12. EDS spectra of Na_2CO_3 activated maritime pine samples before (Pin-C) and after adsorption (Pin-CCu).

3. Materials and Methods

3.1. Materials

Maritime pine fibers derived from a mechanical defibration process were supplied by the FCBA Institute (France). These fibers were treated by simple washing with water and activation with sodium carbonate Na_2CO_3, according to a protocol developed by the industrial partner. Table 3 describes the chemical composition of the washed and carbonate-treated fibers. Extractive content was measured using the acetone-water extraction sequence on a Soxtec Automatic extractor apparatus. Klason lignin content was performed according to Tappi T-222 om-02 standard methods.

Table 3. Chemical composition of raw (Pin-R), washed (Pin-W), and activated (Pin-C) pine materials.

	Pin-R	Pin-W	Pin-C
Klason lignin (%)	30.7	25.4	26.5
Lignin soluble (%)	0.3	0.3	0.3
Total lignin (%)	31.0	25.8	26.7
Extractive soluble water (%)	4.8	4.4	0.9
Acetone soluble extractives (%)	1.7	1.6	0.1
Total mining and quarrying (%)	6.5	6.0	1.0
Total cellulose (%)	41.9	41.9	48.2
Total hemicelluloses (%)	20.0	18.6	21.1
Ashes at 550 °C (%)	0.	0.4	0.6

3.2. Synthetic Solutions

The studied metals were copper, zinc, manganese and nickel, the cationic species found in ST effluents. Sulfate salts of these metals (purity greater than 99.9%), supplied by the industrial partner, were dissolved at a concentration of 200 mg/L (stock solution) in demineralized water. Monocontaminated solutions of the desired concentration were obtained by simple dilution. Similarly, from these solutions, water polycontaminated with 4 metals (Cu, Zn, Mn and Ni) was prepared at a concentration of 25 mg/L for each metal. This total concentration of 100 mg/L was selected as it is representative of the total metal load found in the industrial partner's water. The calcium and sodium salts were in chloride form and were also supplied by the industrial partner (purity greater than 99.9%). The exact concentration of each initial solution was measured by inductively coupled plasma atomic emission spectrometry (ICP-AES), as was the pH before and after adsorption.

3.3. Experimental Protocol

The analytical technique used to determine the purification performance of wood samples is the so-called batch technique [46], which involves bringing a given mass of material into contact with a volume of solution to be treated under pre-defined experimental conditions. The mixture is then stirred for a certain period of time and filtered. The supernatant is then analyzed. The batch method is widely used in the literature and in the industry, as it is simple to implement, it is fast, and it requires standard equipment. It is also reproducible, making it easy to model results. The procedure used in this study is as follows: 1 g of sample is added to 100 mL of solution to be treated, of known initial concentration and pH; the solution is then stirred at 250 rpm and room temperature for 120 min; at the end of the experiment, the solution is filtered through a 0.45 µm filter; the supernatant is recovered and analyzed by ICP; the pH of the final solution is also systematically measured. This batch protocol is similar to that used by the industrial partner in the surface treatment sector. Several parameters were studied in order to measure their impact on the purification performance of the materials: the initial copper concentration (from 10 to 200 mg/L), the adsorbent dose (from 0.5 to 1.5 g), the contact time (from 5 to 240 min), the pH (from 2 to 5) and the ionic strength; this was achieved by adding two salts (NaCl and $CaCl_2$) at concentrations of 0.1, 0.5 and 1 M. All adsorption experiments were repeated (n = 3–5) under identical conditions.

3.4. Chemical Analysis

The cation determination was carried out by ICP-AES spectrometry by the PEA^2t platform of Chrono-environment (Besançon, France). Based on these analyses, the percentage abatement of the cation concerned was calculated according to Equation (1), where C_i and C_f represent the initial and final concentrations, respectively, expressed in mg/L. The results were also expressed using the amount of metal adsorbed per gram of adsorbent material (q_t in mg/g), given by Equation (2), where V represents the volume of solution (L) and m is the mass of material (g).

$$R = \frac{C_i - C_f}{C_i} * 100, \quad (1)$$

$$q_t = \frac{V(C_i - C_f)}{m}. \quad (2)$$

3.5. Modeling

The experimental results for kinetics and adsorption, in a ternary adsorbent/adsorbent/water system, can be modeled by simple, empirical, two-, three- or n-dimensional equations [46–53]. For kinetic results, the models of Lagergren (Equation (3)) [54], Ho and McKay (Equation (4)) [55], Weber and Morris (Equation (5)) [56], and Elovich (Equation (6)) [57] were chosen, and for adsorption data, the adsorption isotherms of Langmuir (Equation (7)) [58], Freundlich (Equation (8)) [59], and Temkin (Equation (9)) [60] were selected. The graphical representation of the various linear equations was used to calculate the maximum quantities of adsorbed copper and the model constants and parameters. The validity of a model was expressed by the value of the regression coefficient of the line and by the accuracy of the calculated parameters with those obtained by experiment [46–53].

$$\ln(q_e - q_t) = \ln q_e - k_1 t, \quad (3)$$

$$\frac{t}{q_t} = \frac{1}{k_2 q_e^2} + \frac{1}{q_e} t, \quad (4)$$

$$q_t = k_p t^{1/2} + C, \quad (5)$$

$$q_t = \frac{\ln(\alpha\beta)}{\beta} + \frac{1}{\beta} \ln t, \quad (6)$$

$$\frac{C_e}{q_e} = \frac{1}{K q_m} + \frac{C_e}{q_m}, \quad (7)$$

$$\ln(q_e) = \ln(K_F) + \frac{1}{n_F}, \quad (8)$$

$$q_e = B \ln(K_T) + B \ln(C_e). \quad (9)$$

3.6. Microscopic and Spectroscopic Characterization of Materials

Scanning electron microscopy (SEM) analyses were carried out after metallization of the samples by carbon deposition. A field-effect scanning electron microscope (JEOL-JSM_IT500HR, Tokyo, Japan) was used simultaneously for qualitative energy-dispersive X-ray microanalysis (EDS, Brucker, Karlsruhe, Germany). SEM-EDS analysis enables high-resolution image acquisition and chemical analysis of the samples. Images were acquired using secondary electrons at an accelerating voltage of 10 kV and a working distance of 5 mm.

4. Conclusions

The results of this study highlighted the value of using a material derived from chemically activated maritime pine fibers to capture copper present in both monocontaminated and polycontaminated metal solutions, whichever concentrations considered. This co-product of the wood industry is interesting not only for its adsorption properties, but also because it is a renewable, abundant and cheap resource. In addition, the proposed chemical modification is relatively simple to implement, free of solvent or other harmful chemicals.

Abatements measured for sodium carbonate-activated pine are always higher than those measured for washed pine, proving the efficiency of the activation. These eliminations are high, the reduction being up to 100%. The adsorption process can be considered rapid, as most of the copper was adsorbed within the first 10 min of exposure. Investigating modeling possibilities showed some limitations. Indeed, Weber and Morris and Elovich models showed poor possibilities to describe all the kinetic data for copper adsorption on fibres. This may prove that mechanism is far more complex than simple physisorption, chemisorption and/or diffusion. Lower pH levels were beneficial for performances, whereas a drop in the purification performance of materials with the increase in salt concentration was measured. The proposed material could thus find a new outlet for selective metal recovery. The next step will be to extrapolate these results to other effluents (rinses and baths) and discharges in order to evaluate the impact of the chemical variability of these waters (due to industrial processes) on the purification performance of activated pine fibers. As a consequence, semi-industrial tests on larger volumes are planned. Further spectroscopic characterization of materials, in particular by Raman and XPS measurements as suggested by one of the three reviewers, before and after adsorption, will validate the hypotheses about adsorption mechanisms. Finally, we are currently studying the recovery of metal-laden materials by combustion and their recovery in jewelry (confidential results).

Author Contributions: Conceptualization: N.M.-C., M.L. and G.C.; methodology: C.M., M.J., D.L. and G.C.; software: M.J., C.M., D.L. and S.T.-L.; validation: N.M.-C., M.L. and G.C.; investigation: C.M., M.J., D.L., N.M.-C., M.L., S.T.-L. and G.C.; resources: N.M.-C., M.L. and G.C.; data curation: N.M.-C. and G.C.; writing—original draft preparation: C.M., M.J., D.L. and N.M.-C.; writing—review and editing: C.M., M.J., D.L., N.M.-C., M.L., S.T.-L. and G.C.; supervision, G.C.; project administration, G.C.; funding acquisition, N.M.-C., M.L. and G.C. All authors have read and agreed to the published version of the manuscript.

Funding: This research was supported by the Université de Franche-Comté, the Région Bourgogne Franche-Comté, and the Institut FCBA Grenoble (CHRYSALIDE 2022-2023 and FINEAU 2020-2024 projects: Plant-Based Cellulosic Materials for Wastewater Treatment_2020-0058).

Institutional Review Board Statement: Not applicable.

Informed Consent Statement: Not applicable.

Data Availability Statement: Not applicable.

Acknowledgments: The authors thank the Université de Franche-Comté, the Région Bourgogne Franche-Comté, and the Institut FCBA Grenoble, and the two industrial partners of the project for financing this study through the CHRYSALIDE 2022-2023 and FINEAU 2020-2024 projects, which made it possible to recruit Chiara Mongioví, Maélys Jaillet and Dario Lacalamita. The authors also thank the PEA^{2}t analytical platform that manages the analytical equipment used in this study.

Conflicts of Interest: The authors declare no conflict of interest.

References

1. Morin-Crini, N.; Lichtfouse, E.; Crini, G. *Emerging Contaminants—Occurrence and Impact. Environmental Chemistry for a Sustainable World*; Springer Nature: Cham, Switzerland, 2021; ISBN 978-3-030-69078-6.
2. Druart, C.; Morin-Crini, N.; Euvrard, E.; Crini, G. Chemical and ecotoxicological monitoring of discharge water from a metal-finishing factory. *Environ. Processes* **2016**, *3*, 59–72. [CrossRef]
3. Morin-Crini, N.; Crini, G. *Eaux Industrielles Contaminées—Réglementation, Paramètres Chimiques et Biologiques & Procédés d'Épuration Innovants*; Presses Universitaires de Franche-Comté: Besançon, France, 2017; ISBN 978-2-84867-589-3.
4. Chowdhury, S.; Mazumder, M.A.J.; Al-Attas, O.; Husain, T. Heavy metals in drinking water: Occurrences, implications, and future needs in developing countries. *Sci. Total Environ.* **2016**, *569*, 476–488. [CrossRef] [PubMed]
5. Al-Saydeh, S.A.; El-Naas, M.H.; Zaidi, S.J. Copper removal from industrial wastewater: A comprehensive review. *J. Ind. Eng. Chem.* **2017**, *56*, 35–44. [CrossRef]
6. Rehman, M.; Liu, L.; Wang, Q.; Saleem, M.H.; Bashir, S.; Ullah, S.; Peng, D.X. Copper environmental toxicology, recent advances, and future outlook: A review. *Environ. Sci. Pollut. Res.* **2019**, *26*, 18003–18016. [CrossRef]
7. Orozco, C.I.; Freire, S.; Gómez-Díaz, D.; González-Álvarez, J. Removal of copper from aqueous solution by biosorption onto pine sawdust. *Sustain. Chem. Pharm.* **2023**, *32*, 101016. [CrossRef]

8. Qasem, N.A.A.; Mohammed, R.H.; Lawal, D.U. Removal of heavy metal ions from wastewater: A comprehensive and critical review. *NPJ Clean Water* **2021**, *4*, 36. [CrossRef]
9. Crini, G.; Lichtfouse, E. Advantages and disadvantages of techniques used for wastewater treatment. *Environ. Chem. Lett.* **2019**, *17*, 145–155. [CrossRef]
10. Crini, G.; Lichtfouse, E. *Green Adsorbents for Pollutant Removal. Innovative Materials. Environmental Chemistry for a Sustainable World*; Springer Nature: Cham, Switzerland, 2018; ISBN 978-3-319-92162-4.
11. Bashir, M.; Tyagi, S.; Annachhatre, A.P. Adsorption copper from aqueous solution onto agricultural adsorbent: Kinetics and isotherm studies. *Mater. Today Proc.* **2020**, *28*, 1833–1840. [CrossRef]
12. Nagy, B.; Mânzatu, C.; Török, A.; Indolean, C.; Măicăneanu, A.; Tonk, S.; Majdik, C. Isotherm and thermodynamic studies of Cd(II) removal process using chemically modified lignocellulosic adsorbent. *Rev. Roum. Chim.* **2015**, *60*, 257–264.
13. Emenike, P.C.; Omole, D.O.; Ngene, B.U.; Tenebe, I.T. Potentiality of agricultural adsorbent for the sequestering of metal ions from wastewater. *Glob. J. Environ. Sci. Manag.* **2016**, *2*, 411–442.
14. Kumar, R.; Sharma, R.K.; Singh, A.P. Cellulose based grafted biosorbents—Journey from lignocellulose biomass to toxic metal ions sorption applications—A review. *J. Mol. Liq.* **2017**, *232*, 62–93. [CrossRef]
15. Malik, D.S.; Jain, C.K.; Yadav, A.K. Removal of heavy metals from emerging cellulosic low-cost biosorbents: A review. *Appl. Water Sci.* **2017**, *7*, 2113–2136. [CrossRef]
16. Afroze, S.; Sen, T.K. A review on heavy metal ions and dye adsorption from water by agricultural solid waste adsorbents. *Water Air Soil Pollut.* **2018**, *229*, 225. [CrossRef]
17. Joseph, L.; Jun, B.M.; Flora, J.R.V.; Park, C.M.; Yoon, Y. Removal of heavy metals from water sources in the developing world using low-cost materials: A review. *Chemosphere* **2019**, *229*, 142–159. [CrossRef]
18. Omer, A.M.; Dey, R.; Eltaweil, A.S.; Abd El-Monaem, E.M.; Ziora, Z.M. Insights into recent advances of chitosan-based adsorbents for sustainable removal of heavy metals and anions. *Arab. J. Chem.* **2022**, *15*, 103543. [CrossRef]
19. Wang, S.; Liu, Y.; Yang, A.; Zhu, Q.; Sun, H.; Sun, P.; Yao, B.; Zang, Y.X.; Du, X.H.; Dong, L.M. Xanthate-modified magnetic $Fe_3O_4@SiO_2$-based polyvinyl alcohol/chitosan composite material for efficient removal of heavy metal ions from water. *Polymers* **2022**, *14*, 1107. [CrossRef]
20. Dong, L.M.; Shan, C.Y.; Liu, Y.; Sun, H.; Yao, B.; Gong, G.Z.; Jin, X.D.; Wang, S.F. Characterization and mechanistic study of heavy metal adsorption by facile synthesized magnetic xanthate-modified chitosan/polyacrylic acid hydrogels. *Int. J. Environ. Res. Public Health* **2022**, *19*, 11123. [CrossRef]
21. Surgutskaia, N.S.; Di Martino, A.; Zednik, J.; Ozaltin, K.; Lovecka, L.; Bergerova, E.V.; Kimmer, D.; Svoboda, J.; Sedlarik, V. Efficient Cu^{2+}, Pb^{2+} and Ni^{2+} ion removal from wastewater using electrospun DTPA-modified chitosan/polyethylene oxide nanofibers. *Sep. Purif. Technol.* **2020**, *247*, 116914. [CrossRef]
22. Li, S.J.; Cai, M.J.; Wang, C.C.; Liu, Y.P. Ta3N5/CdS Core-shell S-scheme heterojunction nanofibers for efficient photocatalytic removal of antibiotic tetracycline and Cr(VI): Performance and mechanism insights. *Adv. Fiber Mater.* **2023**, *5*, 994–1007. [CrossRef]
23. Li, X.L.; Liu, T.; Zhang, Y.; Cai, J.F.; He, M.Q.; Li, M.Q.; Chen, Z.G.; Zhang, L. Growth of BiOBr/ZIF-67 nanocomposites on carbon fiber cloth as filter-membrane-shaped photocatalyst for degrading pollutants in flowing wastewater. *Adv. Fiber Mater.* **2022**, *4*, 1620–1631. [CrossRef]
24. Liu, X.G.; Zhang, Y.; Guo, X.T.; Pang, H. Electrospun metal-organic framework nanofiber membranes for energy storage and environmental protection. *Adv. Fiber Mater.* **2022**, *4*, 1463–1485. [CrossRef]
25. Han, J.; Xing, W.Q.; Yan, J.; Wen, J.; Liu, Y.T.; Wang, Y.Q.; Wu, Z.F.; Tang, L.C.; Gao, J.F. Stretchable and superhydrophilic polyaniline/halloysite decorated nanofiber composite evaporator for high efficiency seawater desalination. *Adv. Fiber Mater.* **2022**, *4*, 1233–1245. [CrossRef]
26. de Quadros Melo, D.; de Oliveira Sousa Neto, V.; de Freitas Barros, F.C.; Cabral Raulino, G.S.; Bastos Vidal, C.; do Nascimento, R.F. Chemical modifications of lignocellulosic materials and their application for removal of cations and anions from aqueous solutions. *J. Appl. Polym. Sci.* **2016**, *133*, 43286. [CrossRef]
27. Litefti, K.; Freire, M.S.; Stitou, M.; González-Álvarez, J. Adsorption of an anionic dye (Congo red) from aqueous solutions by pine bark. *Sci. Rep.* **2019**, *9*, 16530. [CrossRef] [PubMed]
28. Bakar, N.A.; Othman, N.; Yunus, Z.M.; Hamood Altowayti, W.A.; Tahir, M.; Fitriani, N.; Mohd-Salleh, S.N.A. An insight review of lignocellulosic materials as activated carbon precursor for textile wastewater treatment. *Environ. Technol. Innov.* **2021**, *22*, 101445. [CrossRef]
29. Roa, K.; Oyarce, E.; Boulett, A.; AlSamman, M.; Oyarzun, D.; Del, C.; Pizarro, G.; Sánchez, J. Lignocellulose-based materials and their application in the removal of dyes from water: A review. *Sustain. Mater. Technol.* **2021**, *29*, e00320. [CrossRef]
30. Alonso-Esteban, J.I.; Carocho, M.; Barros, D.; Velho, M.V.; Heleno, S.; Barros, L. Chemical composition and industrial applications of maritime pine (*Pinus pinaster* Ait.) bark and other non-wood parts. *Rev. Environ. Sci. Biotechnol.* **2022**, *21*, 583–633. [CrossRef]
31. Chakhtouna, H.; Benzeid, H.; Zari, N.; Qaiss, A.E.K.; Bouhfid, R. Recent advances in eco-friendly composites derived from lignocellulosic biomass for wastewater treatment. *Biomass. Convers. Biorefin.* **2022**. [CrossRef]
32. Nayak, A.K.; Naik, K.R.; Pal, A. Lignocellulosic-based sorbents. In *Wastewater Treatment—Recycling, Management, and Valorization of Industrial Solid Wastes*; Fahim, I.S., Said, L., Eds.; CRC Press: Boca Raton, FL, USA, 2023; Chapter 9; p. 74.
33. Vasić, V.; Kukić, D.; Šćiban, M.; Đurišić-Mladenović, N.; Velić, N.; Pajin, B.; Crespo, J.; Farre, M.; Šereš, Z. Lignocellulose-based biosorbents for the removal of contaminants of emerging concern (CECs) from water: A review. *Water* **2023**, *15*, 1853. [CrossRef]

34. Vázquez, G.; Antorrena, G.; González, J.; Doval, M.D. Adsorption of heavy metal ions by chemically modified Pinus pinaster bark. *Bioresour. Technol.* **1994**, *48*, 251–255. [CrossRef]
35. Vázquez, G.; González-Álvarez, J.; Freire, S.; López-Lorenzo, M.; Antorrena, G. Removal of cadmium and mercury ions from aqueous solution by sorption on treated *Pinus pinaster* bark: Kinetics and isotherms. *Bioresour. Technol.* **2002**, *82*, 247–251. [CrossRef] [PubMed]
36. Vázquez, G.; Alonso, R.; Freire, S.; González-Álvarez, J.; Antorrena, G. Uptake of phenol from aqueous solutions by adsorption in a *Pinus pinaster* bark packed bed. *J. Hazard. Mater. B* **2006**, *133*, 61–67. [CrossRef]
37. Vázquez, G.; González-Álvarez, J.; García, A.I.; Freire, S.; Antorrena, G. Adsorption of phenol on formaldehyde-pretreated in a *Pinus pinaster* bark: Equilibrium and kinetics. *Bioresour. Technol.* **2007**, *98*, 1535–1540. [CrossRef]
38. Brás, I.P.; Santos, L.; Alves, A. Organochlorine pesticides removal by pinus bark sorption. *Environ. Sci. Technol.* **1999**, *33*, 631–634. [CrossRef]
39. Haussard, M.; Gaballah, I.; Donato, P.D.; Barrès, O.; Mourey, A. Removal of hydrocarbons from wastewater using treated bark. *J. Air Waste Manag. Assoc.* **2001**, *51*, 1351–1358. [CrossRef] [PubMed]
40. Basso, M.C.; Cerrella, E.G.; Cukierman, A.L. Lignocellulosic materials as potential biosorbents of trace toxic metals from wastewater. *Ind. Eng. Chem. Res.* **2002**, *41*, 3580–3585. [CrossRef]
41. Shin, E.W.; Karthikeyan, K.G.; Tshabalala, M.A. Adsorption mechanism of cadmium on juniper bark and wood. *Bioresour. Technol.* **2007**, *98*, 588–594. [CrossRef]
42. Sen, T.K.; Afroze, S.; Ang, H.M. Equilibrium, kinetics and mechanism of removal of methylene blue from aqueous solution by adsorption onto pine cone biomass of *Pinus radiata*. *Water Air Soil Pollut.* **2011**, *218*, 499–515. [CrossRef]
43. Sousa, S.; Jiménez-Guerrero, P.; Ruiz, A.; Ratola, N.; Alves, A. Organochlorine pesticides removal from wastewater by pine bark adsorption after activated sludge treatment. *Environ. Technol.* **2011**, *32*, 673–683. [CrossRef] [PubMed]
44. Abdolali, A.; Guo, W.S.; Ngo, H.H.; Chen, S.S.; Nguyen, N.C.; Tung, K.L. Typical lignocellulosic wastes and by-products for biosorption process in water and wastewater treatment: A critical review. *Bioresour. Technol.* **2014**, *160*, 57–66. [CrossRef]
45. Arim, A.L.; Cecílio, D.F.; Quina, M.J.; Gando-Ferreira, L.M. Development and characterization of pine bark with enhanced capacity for uptaking Cr (III) from aqueous solutions. *Can. J. Chem. Eng.* **2018**, *96*, 855–864. [CrossRef]
46. Mongioví, C.; Crini, G. Copper recovery from aqueous solutions by hemp shives: Adsorption studies and modeling. *Processes* **2023**, *11*, 191. [CrossRef]
47. Foo, K.Y.; Hameed, B.H. Insights into the modeling of adsorption isotherm systems. *Chem. Eng. J.* **2010**, *156*, 2–10. [CrossRef]
48. Hamdaoui, O.; Naffrechoux, E. Modeling of adsorption isotherms of phenol and chlorophenols onto granular activated carbon: Part II. Models with more than two parameters. *J. Hazard. Mater.* **2007**, *147*, 401–411. [CrossRef]
49. Saadi, R.; Saadi, Z.; Fazaeli, R.; Fard, N.E. Monolayer and multilayer adsorption isotherm models for sorption from aqueous media. *Korean J. Chem. Eng.* **2015**, *32*, 787–799. [CrossRef]
50. Ayawei, N.; Ebelegi, A.N.; Wankasi, D. Modelling and interpretation of adsorption isotherms. *J. Chem.* **2017**, *2017*, 3039817. [CrossRef]
51. Tran, H.N.; You, S.J.; Hosseini-Bandegharaei, A.; Chao, H.P. Mistakes and inconsistencies regarding adsorption of contaminants from aqueous solutions: A critical review. *Water Res.* **2017**, *120*, 88–116. [CrossRef] [PubMed]
52. Al-Ghouti, M.A.; Da'ana, D.A. Guidelines for the use and interpretation of adsorption isotherms models: A review. *J. Hazard. Mater.* **2020**, *393*, 122383. [CrossRef]
53. González-López, M.E.; Laureano-Anzaldo, C.M.; Pérez-Fonseca, A.A.; Arellano, M.; Robledo-Ortíz, J.R. A critical overview of adsorption models linearization: Methodological and statistical inconsistencies. *Sep. Purif. Rev.* **2021**, *51*, 358–372. [CrossRef]
54. Lagergren, S. Zur theorie der sogenannten adsorption gelster stoffe. Kungliga Svenska Vetenskapsakademiens Handlingar Band. *Handlingar* **1898**, *24*, 1–39.
55. Ho, Y.S.; McKay, G. Kinetic models for the sorption of dye from aqueous solution by wood. *Process Saf. Environ. Prot.* **1998**, *76*, 183–191. [CrossRef]
56. Weber, W.J.; Morris, J.C. Kinetics of adsorption of carbon from solution. *J. Sanit. Eng. Div.* **1963**, *89*, 31–59. [CrossRef]
57. Elovich, S.Y.; Larinov, O.G. Theory of adsorption from solutions of non electrolytes on solid (I) equation adsorption from solutions and the analysis of its simplest form, (II) verification of the equation of adsorption isotherm from solutions. Izvestiya Akademii Nauk. *SSSR Otd. Khimicheskikh Nauk.* **1962**, *2*, 209–216.
58. Langmuir, I. The constitution and fundamental properties of solids and liquids. *J. Am. Chem. Soc.* **1916**, *38*, 2221–2295. [CrossRef]
59. Freundlich, H.M.F. Uber die adsorption in lösungen. *Z. Für Phys. Chem.* **1906**, *57*, 385–471. [CrossRef]
60. Temkin, M.J.; Pyzhev, V. Kinetics of ammonia synthesis on promoted iron catalysts. *Acta Physicochim. URSS* **1940**, *12*, 217–256.
61. Mongioví, C.; Morin-Crini, N.; Lacalamita, D.; Bradu, C.; Raschetti, M.; Placet, V.; Ribeiro, A.R.L.; Ivanovska, A.; Kostić, M.; Crini, G. Biosorbents from plant fibers of hemp and flax for metal removal: Comparison of their biosorption properties. *Molecules* **2021**, *26*, 4199. [CrossRef]

Disclaimer/Publisher's Note: The statements, opinions and data contained in all publications are solely those of the individual author(s) and contributor(s) and not of MDPI and/or the editor(s). MDPI and/or the editor(s) disclaim responsibility for any injury to people or property resulting from any ideas, methods, instructions or products referred to in the content.

Article

Effective Removal of Fe (III) from Strongly Acidic Wastewater by Pyridine-Modified Chitosan: Synthesis, Efficiency, and Mechanism

Lei Zhang [1,*], Heng Liu [1], Jiaqi Zhu [1], Xueling Liu [1], Likun Li [2,*], Yanjun Huang [1], Benquan Fu [3], Guozhi Fan [1] and Yi Wang [1]

[1] School of Chemistry and Environmental Engineering, Wuhan Polytechnic University, Wuhan 430023, China; wangyi2020@whpu.edu.cn (Y.W.)
[2] China-Ukraine Institute of Welding, Guangdong Academy of Sciences, Guangzhou 510650, China
[3] R & D Center of Wuhan Iron and Steel Company, Wuhan 430080, China
* Correspondence: zhanglei@whpu.edu.cn (L.Z.); lilk@gwi.gd.cn (L.L.)

Citation: Zhang, L.; Liu, H.; Zhu, J.; Liu, X.; Li, L.; Huang, Y.; Fu, B.; Fan, G.; Wang, Y. Effective Removal of Fe (III) from Strongly Acidic Wastewater by Pyridine-Modified Chitosan: Synthesis, Efficiency, and Mechanism. *Molecules* **2023**, *28*, 3445. https://doi.org/10.3390/molecules28083445

Academic Editors: Yongchang Sun and Dimitrios Giannakoudakis

Received: 13 March 2023
Revised: 11 April 2023
Accepted: 12 April 2023
Published: 13 April 2023

Copyright: © 2023 by the authors. Licensee MDPI, Basel, Switzerland. This article is an open access article distributed under the terms and conditions of the Creative Commons Attribution (CC BY) license (https:// creativecommons.org/licenses/by/ 4.0/).

Abstract: A novel pyridine-modified chitosan (PYCS) adsorbent was prepared in a multistep procedure including the successive grafting of 2-(chloromethyl) pyridine hydrochloride and crosslinking with glutaraldehyde. Then, the as-prepared materials were used as adsorbents for the removal of metal ions from acidic wastewater. Batch adsorption experiments were carried out to study the impact of various factors such as solution pH value, contact time, temperature, and Fe (III) concentration. The results showed that the absorbent exhibited a high capacity of Fe (III) and the maximum adsorption capacity was up to 66.20 mg/g under optimal experimental conditions (the adsorption time = 12 h, pH = 2.5, and T = 303 K). Adsorption kinetics and isotherm data were accurately described by the pseudo-second-order kinetic model and Sips model, respectively. Thermodynamic studies confirmed that the adsorption was a spontaneous endothermic process. Moreover, the adsorption mechanism was investigated using Fourier transform infrared spectroscopy (FTIR) and X-ray photoelectron spectroscopy (XPS). The results revealed the pyridine group forms a stable chelate with iron (III) ions. Therefore, this acid-resistant adsorbent exhibited excellent adsorption performance for heavy metal ions from acidic wastewater compared to the conventional adsorbents, helping realize direct decontamination and secondary utilization.

Keywords: acidic wastewater; heavy metal ions; pyridine-modified chitosan adsorbent; adsorption mechanism

1. Introduction

A large amount of strongly acidic wastewater is produced during the flue gas desulfurization process in the coking industry. This type of wastewater has a high concentration of sulfuric acid (1–3wt%) and contains high concentrations of iron ions and traces of arsenic, lead, and mercury ions [1,2]. It is reported that about 5.0–7.4 million tons of strongly acidic wastewater are generated from the coking industry per year in China [3]. The most common practice is to add an alkali for neutralization and formation of hydroxide precipitants, which can be separated by coagulation/flocculation, precipitation, adsorption, and filtration methods [4]. However, this process needs further improvements due to generating large amounts of solid hazardous waste, which poses serious environmental risks. More recently, the recycling of acidic wastewater as dilute sulfuric acid for chemical production of, for example, $(NH_4)_2SO_4$ is considered a reliable method [5,6]. Because the high concentration of heavy metals will affect the quality of downstream products, heavy metal ions in acidic wastewater should be effectively removed.

In recent years, various technologies have been employed to remove heavy metal ions from acidic wastewater, such as membrane filtration, adsorption, solvent extraction, and ion exchange [7–10]. Compared with other techniques, adsorption is considered an effective and promising method in terms of low cost, easy operation, and high efficiency [11]. Recently, the development and utilization of inexpensive and eco-friendly biopolymers have attracted great attention. Chitosan is a linear polysaccharide extracted from chitin, which is the second most abundant natural polymer; it can be found in the crustacean shells of shrimps, crabs, and lobsters [12,13]. The hydroxyl and amino functional groups on the backbone of chitosan can serve as coordination sites to form complexes with various heavy metal ions [14]. Nonetheless, water solubility under acidic conditions and low adsorption capacity restricts chitosan as an efficient adsorbent [15]. Therefore, the modification of chitosan is of great significance to enhance its adsorption performance.

According to the reports, chitosan is modified by various functional groups for heavy metals removal, such as ethylenediamine, polyaniline, α-ketoglutaric acid, ionic liquid [16–19], etc. However, the removal efficiency of most chitosan-based adsorbents decreases significantly at lower pH, which limits their application in strong acidic wastewater. Pyridine is an excellent basic ligand with nitrogen donor atoms. Due to the electron absorption effect of aromatic group, the nitrogen atom of pyridine group exhibits a low pKa, which makes it difficult to protonate. Therefore, pyridine groups can combine with heavy metal ions and form complexes at very low pH value [20,21]. The introduction of pyridine groups into chitosan structure will enhance its adsorption performance in strongly acidic solutions.

In this work, a pyridine-modified and glutaraldehyde-crosslinked chitosan (PYCS) was prepared and employed for the removal of Fe (III) from strongly acidic wastewater. The obtained adsorbent PYCS was characterized using Fourier transform infrared spectroscopy (FTIR), scanning electron microscopy (SEM), thermogravimetric analysis (TGA), and an X-ray diffractometer (XRD). The effects of parameters such as pH value, contact time, initial Fe (III) concentration, and temperature on adsorption behaviors were investigated. Isotherm, kinetic, thermodynamics, and inter-particle diffusion models were used to evaluate the adsorption process. Furthermore, the reusability of PYCS was studied in an adsorption–desorption experiment. Finally, the adsorption mechanism of Fe (III) on PYCS was clarified using FTIR and XPS.

2. Results and Discussion

2.1. Characterization

Figure 1 shows the SEM images of chitosan (CS) and pyridine-modified chitosan (PYCS). The pristine CS exhibited a smooth surface and layered structure (Figure 1a) while the CS modified with pyridine displayed some irregular aggregates with a loose honeycomb structure (Figure 1b). The flake-like particles were formed by glutaraldehyde crosslinking chitosan. The surface with plenty of irregular pores increased the surface area, indicating abundant binding sites for metal ion adsorption.

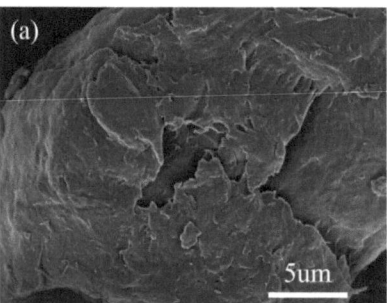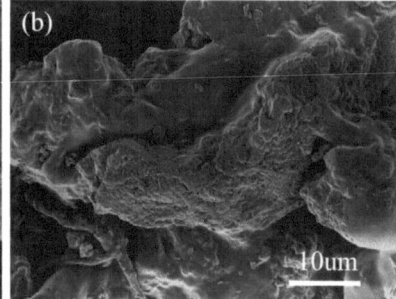

Figure 1. SEM images ((**a**): CS, (**b**): PYCS).

In order to detect the chemical bonds of modified chitosan, FTIR spectroscopy of CS and PYCS materials were detected at 4000 cm^{-1}~400 cm^{-1}, and the results are shown in Figure 2a. The strong and broad peaks at 3420 cm^{-1} and 3442 cm^{-1} represent the absorption peaks of −OH and −NH$_2$, which overlap to produce multiple broadened absorption peaks [22]. The CS had a strong absorption peak at 1600 cm^{-1}, which was caused by the N−H in-plane bending vibration of the primary amine, and the vibration peak disappeared after modification [23]. After modification with pyridine, a new absorption peak near 1660 cm^{-1} was the characteristic C=N peak on the pyridine ring and C=C. In addition, the new absorption peak of skeletal vibration of pyridine occurred at 1590 cm^{-1} and 1450 cm^{-1}, and the adsorption feature near 765 cm^{-1} was caused by the =C−H bending vibration of pyridine ring. The above analysis showed that pyridine group was successfully grafted onto CS. The characteristic absorption peak of β-D pyranoside near 900 cm^{-1} still existed, indicating that the grafting reaction did not destroy the pyran ring of chitosan.

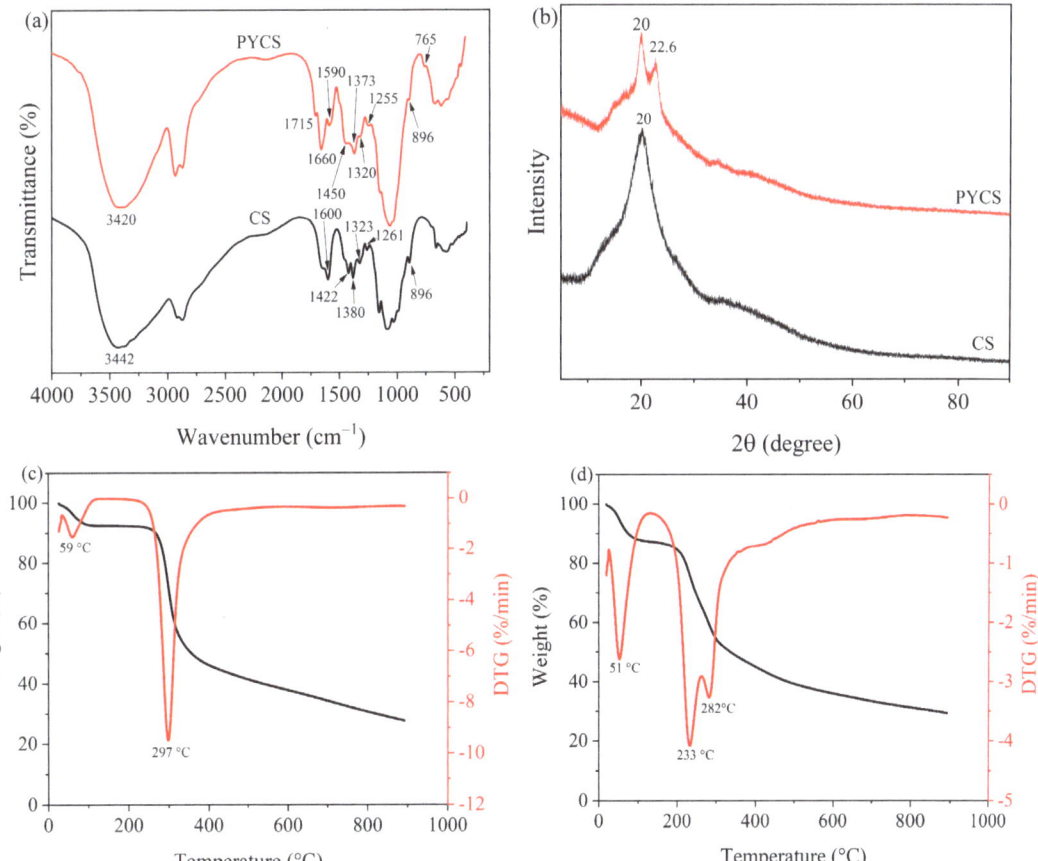

Figure 2. (**a**) FTIR spectra, (**b**) XRD spectra, and TGA–DTG curves ((**c**): CS, (**d**): PYCS).

Figure 2b presents the XRD spectra of CS and PYCS. The CS showed a broad peak at 20.0°, indicating the presence of amorphous regions and regular crystalline regions [24]. When pyridine was grafted into chitosan via the substitution reaction, the intensity of the diffraction peak at 20.0° decreased, indicating the decrease in crystallinity. The crystallinity of PYCS decreased due to the destruction of the sequence of chitosan molecular chains by the crosslinking reaction. According to Thien et al. [25], the typical functional groups of modified CS would become more active and flexible in the amorphous state, which would enable PYCS to effectively form a complex with many metal ions.

The TGA–DTG curves of CS and PYCS are shown in Figure 2c,d. It can be seen from the figure that the TG curves of CS and PYCS are similar, indicating that they have similar skeleton structures. The thermal profile of CS was divided into two stages: the first peak of CS at 59 °C, where the weight loss rate was approximately 7.50%, which was due to water loss [26]. The second stage was the rapid weight loss degradation stage, which occurred at 215–550 °C, and the peak weight loss of CS in the second stage was at 297 °C. The weight loss rate was 52.04%, which was caused by the breakage of glycosidic bonds and the degradation of the structure of chitosan itself [27]. However, the thermal weight loss of PYCS occurred at three unique peaks, specifically, 51 °C, 233 °C, and 282 °C with varied percentage weight reductions of 11.96%, 27.47%, and 28.38%, respectively. The first stage of mass loss was due to the release of typically strong hydrogen-bonded water [28]. The thermal degradation at 130–260 °C observed in the PYCS may have been due to the elimination of the crosslinker by the breakage of the ionic interaction between the CS and the glutaraldehyde. Regarding the thermal degradation observed in cases at 260–550 °C, this stage was the result of glycosidic bond cleavage, main chain degradation, and concomitant graft decomposition [29]. These results indicated that the thermal stability of PYCS was lower than that of CS, which may be caused by the destruction of intramolecular and intermolecular hydrogen bonds of chitosan by the newly introduced side chains [30].

2.2. Effect of pH on Adsorption

The pH value of solution is one of the most important factors for adsorption of PYCS, because it influences the ionization state of functional groups [31,32]. The pH should be set below 3 because of the metal precipitation at higher pH levels is not in line with the real wastewater quality. The Fe (III) adsorption experiments were performed at different pH values (from 0.5 to 3). It can be seen from Figure 3a that with the increase in pH value, the removal efficiency of Fe (III) increased significantly. The overall trend was increasing, but it was not obvious when the acid was strong. The reason was that the N atom on the pyridine ring was protonated and positively charged in extremely acidic solution [33], which weakened its chelating ability with Fe (III). The lower the pH, the more evident the protonation, thus leading the removal efficiency to rise slowly [34]. When the pH was 3, the removal efficiency was significantly improved, mainly due to part of Fe (III) precipitation. Additionally, the pH_{PZC} value of PYCS adsorbent was examined to be 4.3 using the pH drift method (Figure 3b). When the solution pH value was less than pH_{PZC}, the surface of the PYCS adsorbent was positively charged, which was not conducive to Fe (III) adsorption. Based on the precipitation of Fe (III) at a pH greater than 2.7, the solution pH was chosen to be 2.5 in the subsequent experiments.

2.3. Adsorption Kinetics

As plotted in Figure 4a, the uptake capacity of Fe (III) increased as the contact time continued, and gradually reached adsorption equilibrium. PYCS had many free active sites, and the driving force for adsorption was very strong initially. The equilibrium concentration of Fe (III) decreased with time as the number of free active sites adsorbed declined, then the adsorption rate gradually decreased, and finally tended to saturate. The data first demonstrated that the time varied greatly for the adsorption of different Fe (III) concentrations to reach equilibrium, with values of 400 and 720 min for 100 and 200 mg/L, respectively. In order to reach the adsorption equilibrium sufficiently, the contact time of 12 h was selected in the following isotherm experiments.

To investigate the rate-limiting step in the adsorption mechanisms, the kinetic experimental data were fitted with Lagergren's pseudo-first-order, Ho's pseudo-second-order, and intra-particle diffusion kinetic models [35–37]. The nonlinear form equations for the three models were expressed as follows in Equations (1)–(3).

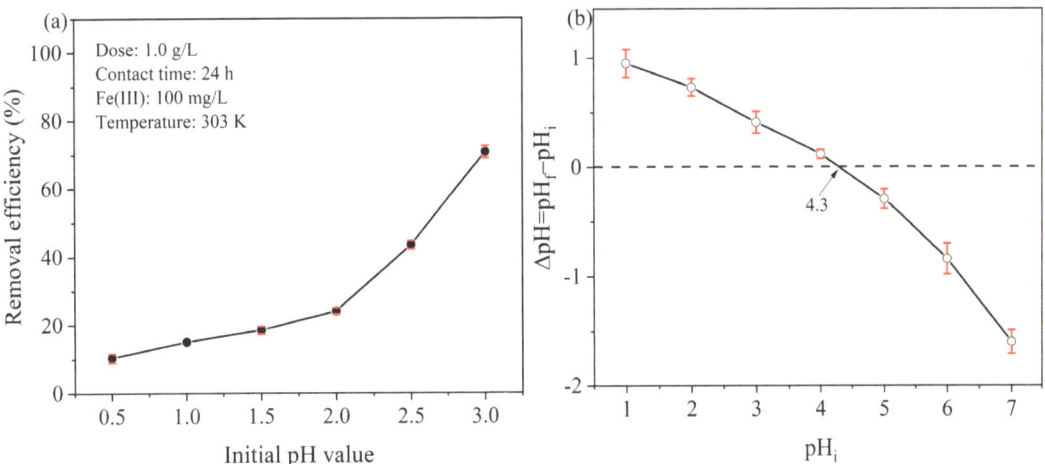

Figure 3. Effect of solution pH value on Fe (III) (**a**) removal efficiency and (**b**) the point of zero charge of PYCS.

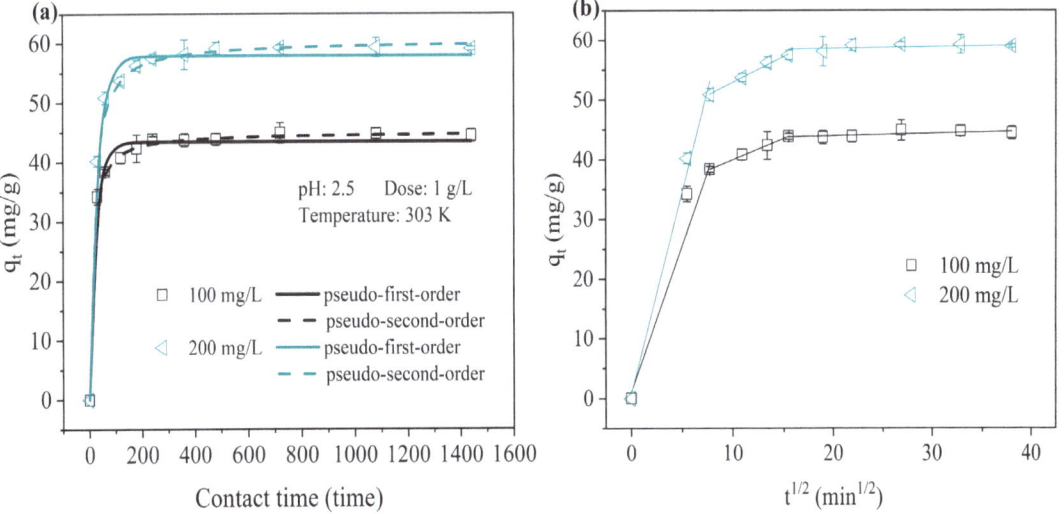

Figure 4. Kinetic profiles of Fe (III) adsorption by PYCS: (**a**) The pseudo-first- and pseudo-second-order kinetic plots. (**b**) The intra-particle diffusion kinetic model.

The pseudo-first-order kinetic model is expressed by

$$q_t = q_e[1 - e^{-k_1 t}] \tag{1}$$

The pseudo-second-order kinetic model is expressed by

$$q_t = \frac{q_e^2 k_2 t}{1 + k_2 q_e t} \tag{2}$$

The intra-particle diffusion kinetic model is expressed by

$$q_t = k_i t^{1/2} + C \tag{3}$$

where q_t and q_e were the adsorption capacities (mg/g) of adsorbents at contact time t (min) and equilibrium, respectively. Whereas, k_1 (min^{-1}) and k_2 (g/mg·min) were the rate constants of the pseudo-first-order and pseudo-second-order kinetic models, respectively. k_i (mg/g·min$^{1/2}$) was the rate constant for the intra-particle diffusion model, and C was the thickness of boundary layer [38,39].

In this work, the root mean square error (RMSE) was used to measure the value deviation between the experimental data and the predicted values. In principle, the lower the RMSE value, the more suitable the model fits. RMSE could be calculated using the following equation:

$$RMSE = \sqrt{\frac{1}{n}\sum_{i=1}^{n}(q_{cali} - q_{\exp i})} \qquad (4)$$

where q_{cali} and $q_{\exp i}$ were the predicted and measured values of the adsorption capacity at time t, respectively, and n was the number of experimental data.

The experimental data for the adsorption kinetics were fitted using the non-linear form of the pseudo-first-order and pseudo-second-order models, as shown in Figure 4a, and the values of kinetic parameters and their RMSE values are summarized in Table 1. Much better fitting results were obtained when the pseudo-second-order model was adopted for the adsorption kinetics of the different Fe (III) concentrations by PYCS. In addition, the theoretical adsorption capacity predicted by the pseudo-second-order model was much closer to the experimental value. These results showed that the pseudo-second-order model was more suitable for describing the adsorption process than the pseudo-first-order model, which indicated that the rate-determining step of PYCS may be a chemisorption process.

Table 1. Kinetic parameters and RMSE values for adsorption of Fe (III) on PYCS.

Kinetic Models	Parameters	Concentration (mg/L)	
		100	200
Pseudo-first-order	$q_{e,exp}$ (mg/g)	44.88	59.29
	$q_{e,cal}$ (mg/g)	43.43	57.87
	k_1 (min^{-1})	0.047	0.037
	RMSE	1.36	1.47
Pseudo-second-order	$q_{e,cal}$ (mg/g)	44.96	60.38
	k_2 (g/mg·min)	0.0022	0.0012
	RMSE	0.38	0.72
Intra-particle diffusion	$k_{i,1}$ (mg/g·min$^{1/2}$)	4.90	6.75
	$k_{i,2}$ (mg/g·min$^{1/2}$)	0.71	0.86
	$k_{i,3}$ (mg/g·min$^{1/2}$)	0.04	0.03
	$C_{i,1}$ (mg/g)	1.10	0.86
	$C_{i,2}$ (mg/g)	32.89	44.30
	$C_{i,3}$ (mg/g)	43.19	58.03

The intra-particle diffusion model is further applied to determine the rate-controlling step of the adsorption processes. According to Figure 4b, it can be observed that the three adsorption curves were divided into three linear parts and did not pass through the origin. The modeling results indicated that the adsorption process was mainly controlled by intra-particle diffusion, as well as other mechanisms including surface chemisorption process. The first stage of the straight line represented the boundary layer diffusion process of Fe (III) adsorption by PYCS. The next stage represented the diffusion of Fe (III) into PYCS channels and the process of gradual adsorption (intra-particle diffusion). The slope of the third stage was almost zero, indicating the equilibrium stage of adsorption and desorption [40–42]. Simultaneously, the diffusion rate constants followed the order of $k_{i,1} > k_{i,2} > k_{i,3}$ which may be affected by the change in the number of PYCS diffusion sites. The rate constant in the first stage was much higher than that in other stages, which meant that the film diffusion was the main rate-limiting step in the whole process, but the adsorption rate was also affected and controlled by the external diffusion step and the adsorbent surface diffusion.

2.4. Adsorption Isotherm

Adsorption isotherm is usually used to study the interaction between the adsorbent and pollutant in equilibrium, and can also estimate the maximum adsorption capacity of the adsorbent for further comparison with adsorbents reported in other studies. As can be seen from Figure 5, with the increase in the initial Fe (III) concentration, the adsorption of Fe (III) by PYCS gradually increased and reached surface saturation at high concentration. This was because at a lower initial Fe (III) concentration, PYCS adsorption active sites were enough to hold the adsorbate, and the interaction sites were not saturated. As the concentration of Fe (III) increased, almost all active sites on the surface of the adsorbent were occupied, leaving the Fe (III) uptake unchanged, then the adsorption capacity for Fe (III) would reach the maximum. In this study, the maximum adsorption capacity of PYCS was 66.20 mg/g.

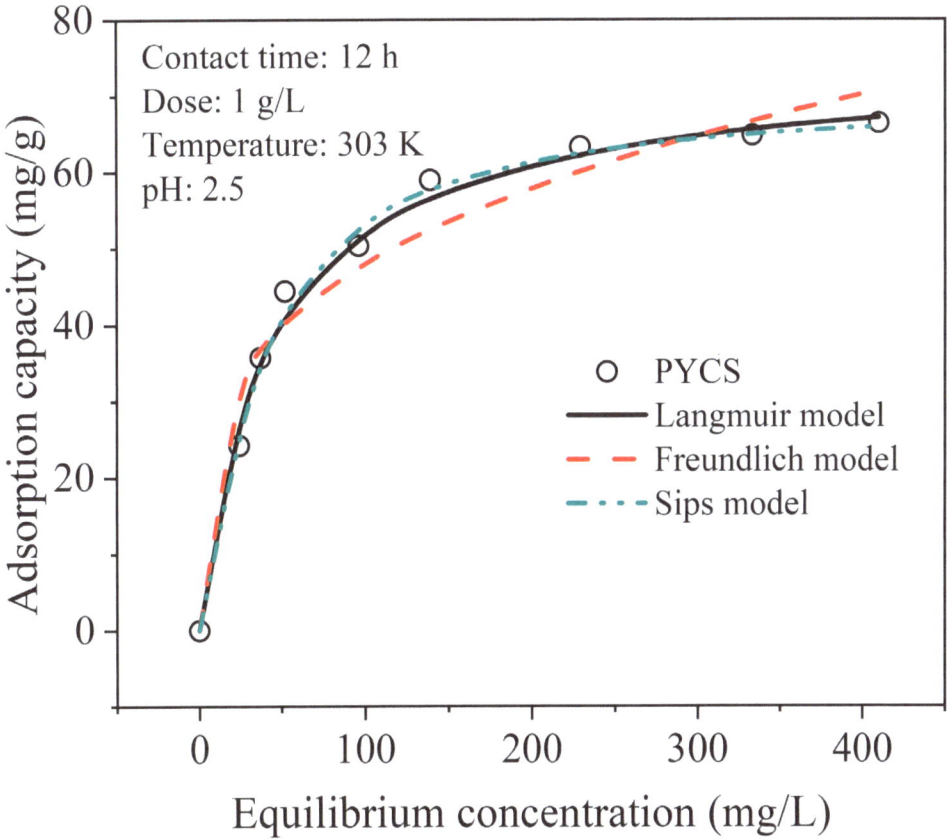

Figure 5. Adsorption isotherms of Fe (III) on PYCS.

To investigate the adsorption behavior and mechanism of PYCS, the three common adsorption isotherm models were employed to fit the experimental data, and their nonlinear equations were given as follows.

The Langmuir isotherm model assumes that monolayer adsorption occurs on the adsorbent surface with equivalent adsorption sites, and there is no interaction between the adsorbed pollutants [43,44].

$$q_e = \frac{q_m K_L C_e}{1 + K_L C_e} \tag{5}$$

The Freundlich equation is one of the earliest empirical equations used to describe equilibrium data and adsorption characteristics for a heterogeneous surface [45].

$$q_e = K_F C_e^{1/n_F} \tag{6}$$

The Sips model is a hybrid of the Langmuir and the Freundlich isotherms [46].

$$q_e = \frac{q_m (K_S C_e)^{n_S}}{1 + (K_S C_e)^{n_S}} \tag{7}$$

where q_e is the amount of Fe (III) adsorbed at equilibrium (mg/g), and q_m is the maximum adsorption capacity of Fe (III) on PYCS (mg/g); C_e represents the equilibrium concentration of Fe (III) (mg/L); K_L is the Langmuir isothermal constant which is related to the affinity of binding sites (L/mg). K_F is the Freundlich isothermal constant (mg$^{(1-n)}$·L^n/g), and n_F is the heterogeneity factors. K_S is the Langmuir equilibrium constant (L/mg), and n_S is comparable to the Freundlich heterogeneity factor ($n_S = 1/n_F$).

The adsorption isotherm parameters simulated from the three isotherm models are summarized in Table 2. Based on the value of the RMSE, the adsorption isotherms of Fe (III) were fitted better by the Sips model. In addition, the heterogeneity factor n_S values shown in Table 2 were more than unit, and the experimental ($q_{m,\,exp}$) value was in good agreement with calculated ($q_{m,\,cal}$) value, indicating that the adsorption of Fe (III) ions onto the prepared adsorbent was a heterogeneous process [40–42]. Table 3 lists the maximum adsorption capacities (q_m) of Fe (III) on PYCS with various adsorbents. It can be seen that the maximum adsorption capacity of PYCS was higher than the other adsorbents listed in Table 4. Nevertheless, the maximum adsorption capacity of PYCS was measured at pH 2.5, while the maximum adsorption capacities of other adsorbents were measured at pH greater than 3. This showed that PYCS displayed high adsorption capacity, which is beneficial to remove Fe (III) from strong acid solution.

Table 2. Isothermal model constants, RMSE values and adsorption capacities of PYCS.

Isothermal Models	Parameters	
Langmuir	$q_{m,exp}$ (mg/g)	66.20
	$q_{m,cal}$ (mg/g)	73.83
	K_L (L/mg)	0.024
	RMSE	1.76
Freundlich	n_F	3.70
	K_F (mg$^{(1-n)}$ L^n/g)	13.93
	RMSE	4.33
Sips	$q_{e,cal}$ (mg/g)	68.93
	n_S	1.27
	K_S (L/mg)	0.027
	RMSE	1.41

Table 3. Comparison of maximum adsorption capacities of Fe (III) on PYCS with other adsorbents.

Adsorbent	Conditions		Adsorption Capacity (mg/g)	Reference
	pH	T (K)		
Chitosan/attapulgite	3	308	47.17	[47]
Chitosan/MMT	5.5	298	7.03	[48]
Chitosan films	4.5	298	299.04	[49]
Chitosan/PVA	3	298	136	[50]
Carboxymethylated chitosan	4.7	298	18.5	[51]
Chitosan-EGDE	3	298	46.30	[52]
Amine-modified chitosan resins	2.5	298	109.61	[53]
PYCS	2.5	303	66.20	This work

Table 4. Thermodynamic parameters for Fe (III) adsorption on PYCS.

Sample	ΔS (J/mol·K)	ΔH (kJ/mol)	ΔG (kJ/mol)			
			303 K	313 K	323 K	333 K
PYCS	36.49	8.14	−2.91	−3.27	−3.64	−4.00

2.5. Adsorption Thermodynamics

The temperature is an essential factor affecting the adsorption effect. Gibbs free energy (ΔG), enthalpy (ΔH), and entropy (ΔS) are important parameters reflecting the thermodynamic reaction. The thermodynamic relations are depicted in the following equations [54,55].

$$K_d = q_e/C_e \tag{8}$$

$$\Delta G = \Delta H - T\Delta S \tag{9}$$

$$\ln K_d = \Delta S/R - \Delta H/RT \tag{10}$$

where R (8.314 J/(mol·K)) is the ideal gas constant; K_d is the adsorption equilibrium constant; T (K) is the experimental temperature; ΔS (J/(mol·K)), ΔH (kJ/mol), and ΔG (kJ/mol) refer to entropy change, enthalpy change, and Gibbs free energy change, respectively.

The calculation results of thermodynamic equilibrium coefficient and thermodynamic parameters of PYCS adsorption process are shown in Table 4. The ΔH values were positive, which indicated that the adsorption process was endothermic. Meanwhile, the ΔG values were negative, thus indicating that the adsorption process was spontaneous. ΔS values were positive which reveals that the reaction system was chaotic and that Fe (III) had good contact with PYCS. Therefore, the reaction between PYCS and Fe (III) was a spontaneous endothermic process, and increasing the temperature can make this reaction proceed more efficiently.

2.6. Regeneration and Reusability

As the reusability of the adsorbent was very important for its industrial application, the adsorption–desorption cycles were conducted six times to further evaluate the reusability of the as-prepared PYCS. As observed from Figure 6, the adsorption ability slightly decreased as the number of cycles increased. After six cycles, the adsorption capacity of PYCS was maintained at 71% of that of the first cycle. The decrease of adsorption capacity may be the loss or degradation of adsorbent [39]. In addition, the incomplete removal of Fe (III) ions in the adsorbent by the eluent may be another reason for the reduction of adsorption capacity. Overall, PYCS showed a good reusability and had great potential as an efficient adsorbent for the removal of Fe (III) ions from polluted water.

2.7. Application in Real Acidic Wastewater

The prepared materials were used for the adsorption of heavy metals in real acidic wastewater to evaluate the practical applicability. The results are in Figure 7. It can be found that the removal rates of Fe (III), Pb (II), and As (III) were 98.9%, 34.5%, and 28.6%, respectively. The results proved PYCS as a potential material for heavy metal removal in water treatment with high efficiency.

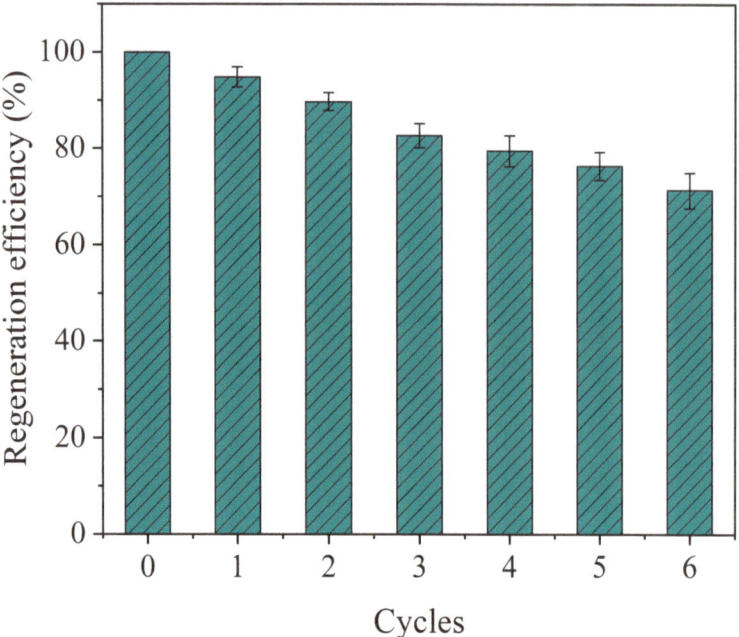

Figure 6. Reuse of PYCS in removal of Fe (III) from acidic wastewater.

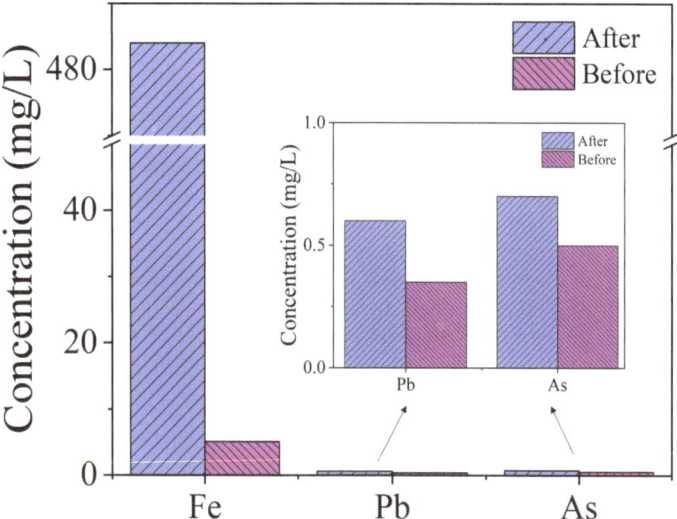

Figure 7. The removal of metal ions from real wastewater.

2.8. Adsorption Mechanism

The adsorption mechanism of PYCS was investigated by examining the XPS and FT-IR data. The FT-IR spectra of the PYCS before and after loading with Fe (III) are shown in Figure 8. For PYCS, the adsorption band near 1450 cm^{-1} corresponded to the stretching vibrations of C−N, and the adsorption feature near 1590 cm^{-1} was caused by skeletal vibration of pyridine. After adsorption, the strength of the C=N (1660 cm^{-1}) bonds reduced and moved at a minor wave length, confirming that pyridines were the functional groups of PYCS, forming coordinate bonds with Fe (III) [56]. The stretching vibration of pyridine at 1590 cm^{-1} and that of C−N at 1450 cm^{-1} were significantly weakened and almost disappeared after adsorption, indicating that pyridine was involved in the chelating interaction with Fe (III). The appearance of new peaks at 1110 cm^{-1} and 618 cm^{-1} corresponded to the stretching vibration of SO_4^{2-} ions after adsorption [20].

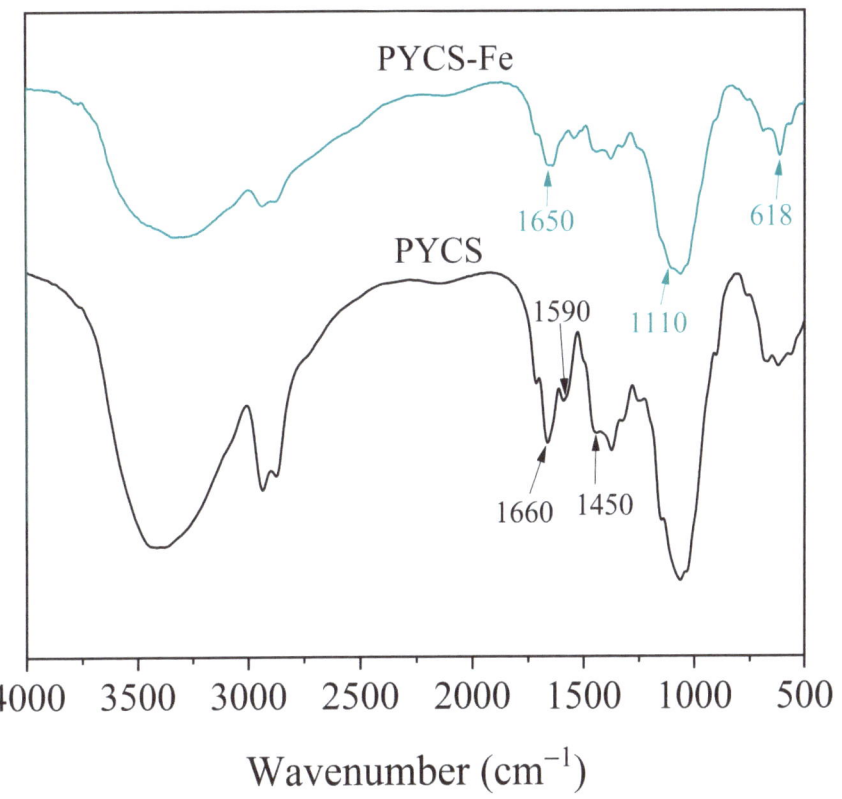

Figure 8. FTIR spectra of PYCS before and after adsorbing Fe (III).

To further distinguish the species of functional groups and confirm the coordination interaction between nitrogen atoms and Fe (III), XPS characterization of PYCS before and after Fe trapping were performed. The XPS wide scans and N1s spectra of PYCS are exhibited in Figure 9. For the raw PYCS, the N1s core-level XPS spectra were deconvoluted into 4 peaks at 397.66, 400.40, 398.48, and 404.99 eV, which corresponded to the nitrogen atoms in the neutral amine (−NH or C−N), protonated amine (−NH$^+$), pyridine (C=N), and nitrate ions (NO$_3^-$), respectively [57–60]. Meanwhile, the existence of nitrate ions was due to the synthetic PYCS cleaned by HNO$_3$, which was consistent with the results of FT-IR. After the adsorption of Fe (III) on PYCS at pH 2.5, the minor shift for −NH or C−N, and the obvious shift for C=N for PYCS before and after the adsorption of Fe (III) may be interpreted such that the pyridine nitrogen (C=N) atoms in the organic functional groups on the surface of PYCS had coordinative chelation with Fe (III) during the adsorption process [61]. In addition, the peak value of nitrate ions decreased significantly at 404.99 eV, which can prove that nitrate ions were replaced by sulfate ions.

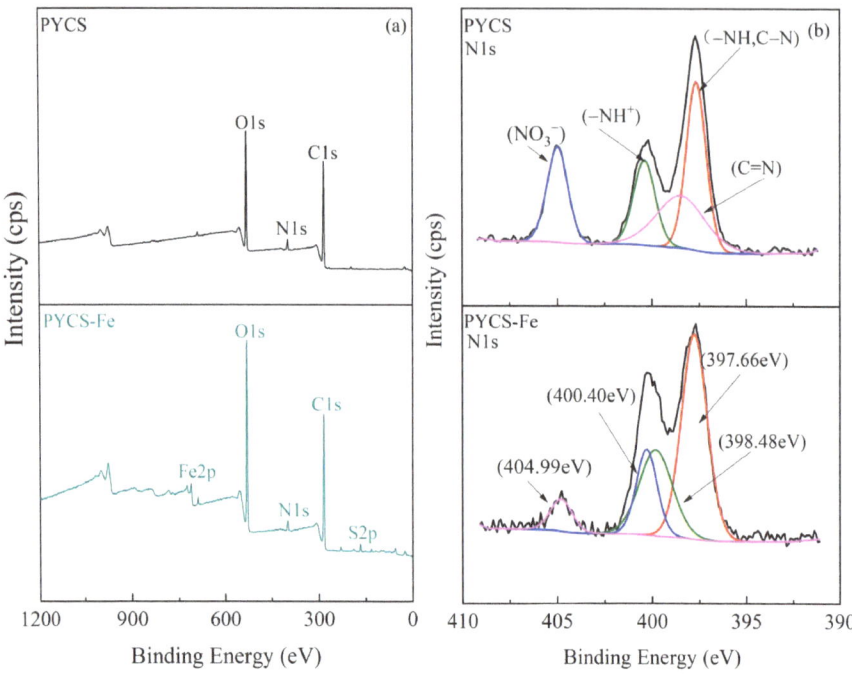

Figure 9. Characterization of PYCS before and after Fe (III) adsorption: (**a**) XPS wide scan spectra. (**b**) XPS spectra of N1s.

The possible mechanism is shown in Figure 10. The adsorbent PYCS was capable of capturing Fe (III) through nitrogen atoms of pyridine that participated in the chelating effect at pH 2.5. According to literature reports, the polyamine adsorbent lost its adsorption ability below pH 3 [62], but pyridine-modified adsorbent could adsorb Fe (III). Owing to the weak alkalinity of the aliphatic amines and pyridine ring, competitive adsorption would occur between Fe (III) ions and H^+ under low pH. Hence, the active adsorption sites of the nitrogen atoms would be partially protonated, resulting in lower adsorption capacity.

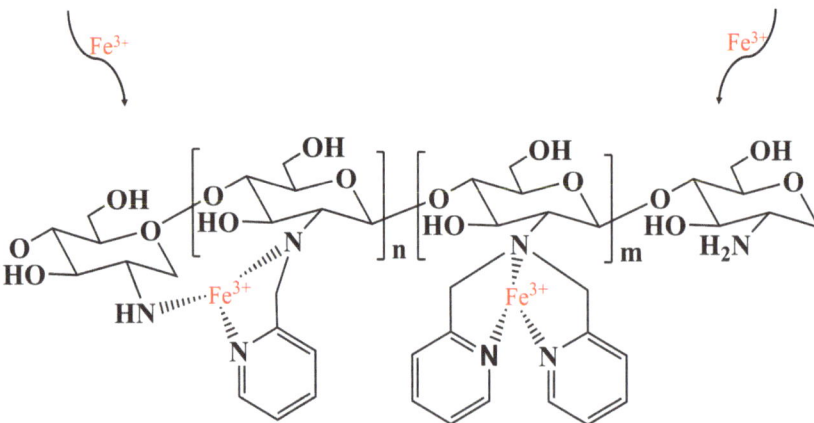

Figure 10. Possible mechanism of the adsorption process.

3. Materials and Methods

3.1. Chemical and Materials

Chitosan (>95% deacetylated, 100–200 mPa s, M.wt%. 6.9×10^5) was supplied from the Aladdin Industrial Corporation (Shanghai, China). Acetic acid (CH_3COOH, AR), sodium hydroxide (NaOH, AR), and 2-Chloromethylpyridine hydrochloride ($C_6H_6ClN \cdot HCl$, AR) were all purchased from Macklin Biochemical Co., Ltd. (Shanghai, China). Ethanol (CH_3CH_2OH, AR), glutaraldehyde, sodium carbonate (Na_2CO_3, AR), nitric acid (HNO_3, AR), ferric sulfate ($Fe_2(SO_4)_3$, AR), sulfuric acid (H_2SO_4, AR), and hydrochloric acid (HCl, AR) were supplied by Sinopharm Chemical Reagent Co., Ltd. (Shanghai, China), all reagents of which were analytical grade and used directly without further purification. All water mentioned in this study was deionized water (18.2 MΩ cm). The metal concentrations in as-pretreated real acidic wastewater from an iron and steel industry in Central China, were determined, as listed in Table 5.

Table 5. Metal concentration (mg/L) in strongly acidic wastewater from coking plant.

Fe (III)	As (III)	Pb (II)	SO_4^{2-}	pH
484 ± 10	0.7 ± 0.1	0.6 ± 0.1	1260	0.5

3.2. Preparation of Pyridine-Modified Chitosan

The preparation procedures of pyridine-modified chitosan (PYCS) adsorbent were as follows: Firstly, 1 g of chitosan was dissolved in 50 mL of acetic acid (2% v/v). Then, 250 mL of NaOH (0.5 mol/L) was slowly added dropwise to form chitosan beads. The mixture was stirred for 3 h. After separation, the particles were washed to neutrality with deionized water and ethanol.

Then, the as-obtained chitosan beads, 3.06 g 2-(chloromethyl) pyridine hydrochloride (dissolved in 50 mL ethanol), and 0.83 g sodium carbonate were introduced into a 250 mL round-bottom flask. The reaction was carried out through reflux condensing at 363 K for 24 h. After filtration, the product was immersed into a mixture solution containing 2.5wt% glutaraldehyde for 12 h at 303 K with continuously shaking, then rinsed with ethanol and deionized water.

After drying in vacuum at 313 K for 12 h, the adsorbent PYCS was obtained. The synthetic procedure of PYCS is shown in Figure 11.

3.3. Characterization

The chemical constituents of samples were recorded using the Fourier transform infrared spectrometer (FTIR, Nicolet iS10, ThermoFisher Scientific, Waltham, MA, USA) in a range of 4000 to 400 cm^{-1}. The surface morphology of the samples was characterized using a scanning electron microscope (SEM, S4800, Hitachi, Tokyo, Japan). Thermal decomposition characteristic analysis was carried out on a thermogravimetric analyzer (TGA, STA 7300, Hitachi, Tokyo, Japan) at a heating rate 10 °C/min from room temperature to 900 °C in a N_2 flow. The X-ray photoelectron spectroscopy (XPS, K-Alpha, Thermo Scientific, Waltham, MA, USA) was used to investigated the surface element component of the samples. The crystallinity of the samples was characterized on the X-ray diffractometer (XRD, XRD-7000, Shimadzu, Kyoto, Japan) over the 2 thetas range of 5°–90°. The concentration of each heavy metal ion was determined using inductively coupled plasma mass spectrometry (ICP, iCAP Q, Thermo Fisher Scientific, Waltham, MA, USA). The pH of the zero charge (pH_{pzc}) of the prepared PYCS was determined using the batch equilibrium method [63].

3.4. Adsorption Experiments

Batch adsorption tests were undertaken using a centrifuge tube at 303 K. In general, 0.02 g of PYCS and 20 mL aqueous solution containing Fe (III) were added to the centrifuge tube and agitated at 170 rpm on a shaker. To examine the effect of pH on Fe (III) removal, the adsorption was conducted at the Fe (III) concentration of 100 mg/L in the pH range of

0.5–3.0 for 24 h. The initial pH of Fe (III) aqueous solution was measured by pH meter and was adjusted by using 0.1 mol/L H_2SO_4 and NaOH (aq).

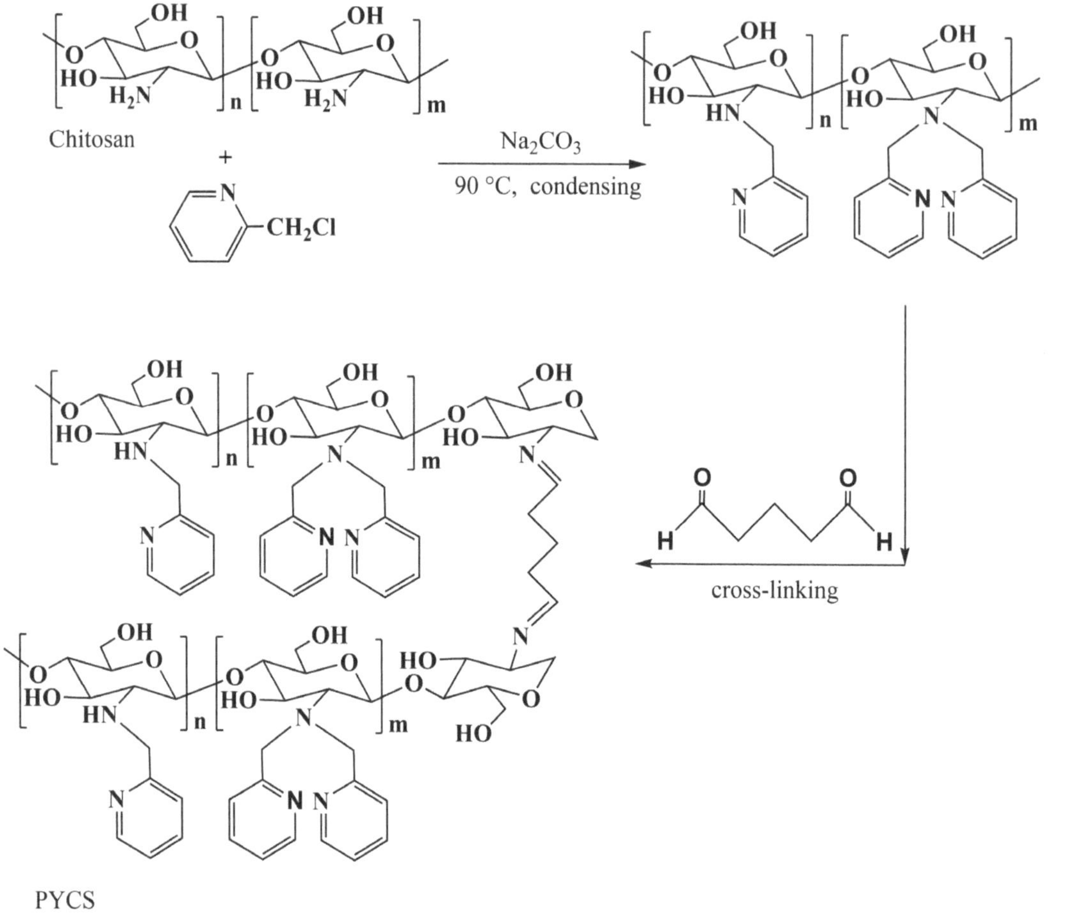

Figure 11. Synthetic procedure of PYCS.

For adsorption kinetic experiments, 0.02 g of PYCS was added into 20 mL of an aqueous solution containing Fe (III). The initial pH and Fe (III) concentration value of the solution were 2.5, 100 mg/L, respectively. The adsorption isotherm was investigated at 303 K, and the initial concentration of the Fe (III) varied from 20 to 500 mg/L at pH 2.5. To evaluate adsorption thermodynamics, 303 K, 313 K, 323 K, and 333 K were selected for the adsorption. After adsorption, the reaction solution was sampled and immediately filtered through a 0.45 μm membrane filter. Finally, the residual Fe (III) concentration in the wastewater was determined using ICP-MS. The adsorption capacity of adsorbent was calculated using the following Equation.

$$q_e = \frac{(C_0 - C_e)V}{m} \quad (11)$$

where q_e (mg/g) represents the amount of Fe (III) adsorbed onto the adsorbent at equilibrium, C_0 and C_e (mg/L) are the initial and equilibrium concentration of Fe (III) ions, respectively, V (L) represents the volume of solution, and m (g) represents dosage of adsorbent.

3.5. Regeneration Studies

In order to evaluate the regeneration and reuse of PYCS, 0.02 g PYCS was first added to 20 mL of a solution containing 100 mg/L of Fe (III) ions, wherein the solution pH was 2.5, and then was agitated at 170 rpm for 24 h at 303 K. After the adsorption process, the absorbent saturated with Fe (III) was filtered and washed with water to remove the un-adsorbed Fe (III), then agitated with 20 mL of elution solution (3 mol/L HCl) for another 24 h. The regenerated PYCS was filtered and washed with water several times and then was stored for the next experiment. The adsorption regeneration experiment was repeated 6 times.

3.6. Application in Real Acidic Wastewater

In this work, PYCS was used to treat real acidic wastewater from a coking plant in Wuhan. The detailed metal concentration of the real acidic wastewater is listed in Table 5. 1 g PYCS was added into 20 mL of real acidic wastewater and agitated at 170 rpm for 24 h at 303 K. Afterwards, the solution samples were filtrated with 0.45 μm filters, and the residual Fe (III) concentration in the solution was analyzed using ICP-MS.

4. Conclusions

A pyridine-modified chitosan (PYCS) was successfully prepared by modifying pyridine onto chitosan, which was then used to remove Fe (III) from acidic wastewater. The adsorbent has a distinguished adsorption effect for Fe (III), and the removal rates of Fe (III), Pb (II), and As (III) in real acidic wastewater were 98.9%, 34.5%, and 28.6%, respectively. The maximum adsorption capacity was 66.20 mg/g, which was better than most reported adsorbents. Comprehensive adsorption kinetics and isotherm results showed that adsorption was heterogeneous chemisorption process. Fe (III) adsorption on PYCS was endothermic and spontaneous in nature. In addition, after 6 cycles of adsorption, the adsorption capacity of the material could still reach 71%, which indicated that it had important practical application value. In short, the material may be used as a promising, efficient, and environmentally friendly adsorbent for the remediation of heavy metal pollution.

Author Contributions: L.Z.: Conceptualization, Writing—original draft, and acquisition of funding. H.L.: Writing—original draft, Investigation, Validation, and Formal analysis; J.Z.: Formal analysis and Investigation; X.L.: Investigation; L.L.: Visualization, Investigation, and Methodology; Y.H.: Investigation and Formal analysis; B.F.: Formal analysis; G.F.: Investigation; Y.W.: Formal analysis, Investigation, Methodology, and Project administration. All authors have read and agreed to the published version of the manuscript.

Funding: This work was supported by the Research and Innovation Initiatives of WHPU (2022Y16) and the Scientific Research Foundation of Wuhan Polytechnic University (118-53210052171, 118-53210052144 and 118-53210052136).

Institutional Review Board Statement: Not applicable.

Informed Consent Statement: Not applicable.

Data Availability Statement: Compound data sets are publicly available. Samples are available from the authors.

Acknowledgments: The authors would like to thank Chengzhang Liu from Shiyanjia Lab (www.shiyanjia.com (accessed on 15 February 2023)) for the XRD and XPS analysis.

Conflicts of Interest: The authors declare no conflict of interest.

Sample Availability: Not available.

References

1. Zhang, W.; Oswal, H.; Renew, J.; Ellison, K.; Huang, C.H. Removal of heavy metals by aged zero-valent iron from flue-gas-desulfurization brine under high salt and temperature conditions. *J. Hazard. Mater.* **2019**, *373*, 572–579. [CrossRef] [PubMed]
2. Solinska, A.; Bajda, T. Modified zeolite as a sorbent for removal of contaminants from wet flue gas desulphurization wastewater. *Chemosphere* **2022**, *286*, 131772. [CrossRef] [PubMed]
3. Hu, X.; Zhu, F.; Kong, L.; Peng, X. Sulfate radical-based removal of chloride ion from strongly acidic wastewater: Kinetics and mechanism. *J. Hazard. Mater.* **2021**, *410*, 124540. [CrossRef] [PubMed]
4. You, S.; Lu, J.; Tang, C.Y.; Wang, X. Rejection of heavy metals in acidic wastewater by a novel thin-film inorganic forward osmosis membrane. *Chem. Eng. J.* **2017**, *320*, 532–538. [CrossRef]
5. Pi, X.; Sun, F.; Qu, Z.; Li, Y.; Gao, J. Hierarchical pore configuration in activated coke boosting direct desorption of desulfurization product H_2SO_4: A combined experimental and computational investigation. *Fuel* **2021**, *298*, 120697. [CrossRef]
6. Guo, H.; Yuan, P.; Pavlovic, V.; Barber, J.; Kim, Y. Ammonium sulfate production from wastewater and low-grade sulfuric acid using bipolar- and cation-exchange membranes. *J. Clean. Prod.* **2021**, *285*, 124888. [CrossRef]
7. You, X.; Chen, J.; Pan, S.; Lu, G.; Teng, L.; Lin, X.; Zhao, S.; Lin, J. Piperazine-functionalized porous anion exchange membranes for efficient acid recovery by diffusion dialysis. *J. Membr. Sci.* **2022**, *654*, 120560. [CrossRef]
8. Akoto, J.D.; Chai, F.; Repo, E.; Yang, Z.; Wang, D.; Zhao, F.; Liao, Q.; Chai, L. Polyethyleneimine stabilized nanoscale zero-valent iron-magnetite (Fe_3O_4@nZVI-PEI) for the enhanced removal of arsenic from acidic aqueous solution: Performance and mechanisms. *J. Environ. Chem. Eng.* **2022**, *10*, 108589. [CrossRef]
9. Silva, J.E.; Paiva, A.P.; Soares, D.; Labrincha, A.; Castro, F. Solvent extraction applied to the recovery of heavy metals from galvanic sludge. *J. Hazard. Mater.* **2005**, *120*, 113–118. [CrossRef]
10. Nenov, V.; Dimitrova, N.; Dobrevsky, I. Recovery of sulphuric acid from waste aqueous solutions containing arsenic by ion exchange. *Hydrometallurgy* **1997**, *44*, 43–52. [CrossRef]
11. Wang, P.; Tang, Y.; Liu, Y.; Wang, T.; Wu, P.; Lu, X.Y. Halloysite nanotube@carbon with rich carboxyl groups as a multifunctional adsorbent for the efficient removal of cationic Pb(II), anionic Cr(VI) and methylene blue (MB). *Environ. Sci. Nano* **2018**, *5*, 2257–2268. [CrossRef]
12. Huang, Y.; Zheng, H.; Hu, X.; Wu, Y.; Tang, X.; He, Q.; Peng, S. Enhanced selective adsorption of lead(II) from complex wastewater by DTPA functionalized chitosan-coated magnetic silica nanoparticles based on anion-synergism. *J. Hazard. Mater.* **2022**, *422*, 126856. [CrossRef] [PubMed]
13. Wang, J.; Zhuang, S. Chitosan-based materials: Preparation, modification and application. *J. Clean. Prod.* **2022**, *355*, 131825. [CrossRef]
14. Saleh, A.S.; Ibrahim, A.G.; Elsharma, E.M.; Metwally, E.; Siyam, T. Radiation grafting of acrylamide and maleic acid on chitosan and effective application for removal of Co(II) from aqueous solutions. *Radiat. Phys. Chem.* **2018**, *144*, 116–124. [CrossRef]
15. Zhang, C.; Wen, H.; Huang, Y.; Shi, W. Adsorption of anionic surfactants from aqueous solution by high content of primary amino crosslinked chitosan microspheres. *Int. J. Biol. Macromol.* **2017**, *97*, 635–641. [CrossRef]
16. Hu, X.J.; Wang, J.S.; Liu, Y.G.; Li, X.; Zeng, G.M.; Bao, Z.L.; Zeng, X.X.; Chen, A.W.; Long, F. Adsorption of chromium(VI) by ethylenediamine-modified cross-linked magnetic chitosan resin: Isotherms, kinetics and thermodynamics. *J. Hazard. Mater.* **2011**, *185*, 306–314. [CrossRef]
17. Karthik, R.; Meenakshi, S. Removal of Pb(II) and Cd(II) ions from aqueous solution using polyaniline grafted chitosan. *Chem. Eng. J.* **2015**, *263*, 168–177. [CrossRef]
18. Zhao, Z.; Huang, Y.; Wu, Y.; Li, S.; Yin, H.; Wang, J. α-ketoglutaric acid modified chitosan/polyacrylamide semi-interpenetrating polymer network hydrogel for removal of heavy metal ions. *Colloids Surf. A Physicochem. Eng. Asp.* **2021**, *628*, 127262. [CrossRef]
19. Rosli, N.; Yahya, W.Z.N.; Wirzal, M.D.H. Crosslinked chitosan/poly(vinyl alcohol) nanofibers functionalized by ionic liquid for heavy metal ions removal. *Int. J. Biol. Macromol.* **2022**, *195*, 132–141. [CrossRef]
20. Zong, L.; Liu, F.; Chen, D.; Zhang, X.; Ling, C.; Li, A. A novel pyridine based polymer for highly efficient separation of nickel from high-acidity and high-concentration cobalt solutions. *Chem. Eng. J.* **2018**, *334*, 995–1005. [CrossRef]
21. Zou, B.; Zhang, S.; Sun, P.; Zhao, Q.; Zhang, W.; Zhang, X.; Ran, L.; Zhou, L.; Ye, Z. Synthesis of a novel Poly-chloromethyl styrene chelating resin containing Tri-pyridine aniline groups and its efficient adsorption of heavy metal ions and catalytic degradation of bisphenol A. *Sep. Purif. Technol.* **2021**, *275*, 119234. [CrossRef]
22. Zahedifar, M.; Es-haghi, A.; Zhiani, R.; Sadeghzadeh, S.M. Synthesis of benzimidazolones by immobilized gold nanoparticles on chitosan extracted from shrimp shells supported on fibrous phosphosilicate. *RSC Adv.* **2019**, *9*, 6494–6501. [CrossRef] [PubMed]
23. Ibrahim, A.G.; Fouda, A.; Elgammal, W.E.; Eid, A.M.; Elsenety, M.M.; Mohamed, A.E.; Hassan, S.M. New thiadiazole modified chitosan derivative to control the growth of human pathogenic microbes and cancer cell lines. *Sci. Rep.* **2022**, *12*, 21423. [CrossRef] [PubMed]
24. Samuels, R.J. Solid state characterization of the structure of chitosan films. *J. Polym. Sci. Polym. Phys. Ed.* **1981**, *19*, 1081–1105. [CrossRef]
25. Thien, D.; An, N.; Hoa, N. Preparation of Fully Deacetylated Chitosan for Adsorption of Hg(II) Ion from Aqueous Solution. *Chem. Sci.* **2015**, *6*, 1.
26. Ali, N.; Khan, A.; Bilal, M.; Malik, S.; Badshah, S.; Iqbal, H.M.N. Chitosan-Based Bio-Composite Modified with Thiocarbamate Moiety for Decontamination of Cations from the Aqueous Media. *Molecules* **2020**, *25*, 226. [CrossRef]

27. Yu, C.; Liu, X.; Pei, J.; Wang, Y. Grafting of laccase-catalysed oxidation of butyl paraben and p-coumaric acid onto chitosan to improve its antioxidant and antibacterial activities. *React. Funct. Polym.* **2020**, *149*, 104511. [CrossRef]
28. Zawadzki, J.; Kaczmarek, H. Thermal treatment of chitosan in various conditions. *Carbohydr. Polym.* **2010**, *80*, 394–400. [CrossRef]
29. Chung, Y.C.; Tsai, C.F.; Li, C.F. Preparation and characterization of water-soluble chitosan produced by Maillard reaction. *Fish Sci.* **2006**, *72*, 1096–1103. [CrossRef]
30. Chen, L.; Tang, J.; Wu, S.; Wang, S.; Ren, Z. Selective removal of Au(III) from wastewater by pyridine-modified chitosan. *Carbohydr. Polym.* **2022**, *286*, 119307. [CrossRef]
31. Zhan, W.; Xu, C.; Qian, G.; Huang, G.; Tang, X.; Lin, B. Adsorption of Cu(II), Zn(II), and Pb(II) from aqueous single and binary metal solutions by regenerated cellulose and sodium alginate chemically modified with polyethyleneimine. *RSC Adv.* **2018**, *8*, 18723–18733. [CrossRef] [PubMed]
32. Xie, L.; Yu, Z.; Islam, S.M.; Shi, K.; Cheng, Y.; Yuan, M.; Zhao, J.; Sun, G.; Li, H.; Ma, S.; et al. Remarkable acid stability of polypyrrole-MoS$_4$: A highly selective and efficient scavenger of heavy metals over a wide pH range. *Adv. Funct. Mater.* **2018**, *28*, 1800502. [CrossRef]
33. Zou, B.; Zhang, S.; Sun, P.; Ye, Z.; Zhao, Q.; Zhang, W.; Zhou, L. Preparation of a novel Poly-chloromethyl styrene chelating resin containing heterofluorenone pendant groups for the removal of Cu(II), Pb(II), and Ni(II) from wastewaters. *Colloids Interface Sci. Commun.* **2021**, *40*, 100349. [CrossRef]
34. Ma, J.; Shen, J.; Wang, C.; Wei, Y. Preparation of dual-function chelating resin with high capacity and adjustable adsorption selectivity to variety of heavy metal ions. *J. Taiwan Inst. Chem. Eng.* **2018**, *91*, 532–538. [CrossRef]
35. Lagergren, S. About the theory of so-called adsorption of solution substances, Kung. *Sven. Veten. Hand.* **1898**, *24*, 1–39.
36. Ho, Y.S.; McKay, G. Sorption of dye from aqueous solution by peat. *Chem. Eng. J.* **1998**, *70*, 115–124. [CrossRef]
37. Weber Walter, J.; Morris, J.C. Kinetics of Adsorption on Carbon from Solution. *J. Sanit. Eng. Div. Am. Soc. Civ. Eng.* **1963**, *89*, 31–59. [CrossRef]
38. Liu, L.; Hu, S.; Shen, G.; Farooq, U.; Zhang, W.; Lin, S.; Lin, K. Adsorption dynamics and mechanism of aqueous sulfachloropyridazine and analogues using the root powder of recyclable long-root Eichhornia crassipes. *Chemosphere* **2018**, *196*, 409–417. [CrossRef]
39. Huang, Y.; Keller, A.A. EDTA functionalized magnetic nanoparticle sorbents for cadmium and lead contaminated water treatment. *Water Res.* **2015**, *80*, 159–168. [CrossRef]
40. Repo, E.; Warchol, J.K.; Kurniawan, T.A.; Sillanpää, M.E.T. Adsorption of Co(II) and Ni(II) by EDTA- and/or DTPA-modified chitosan: Kinetic and equilibrium modeling. *Chem. Eng. J.* **2010**, *161*, 73–82. [CrossRef]
41. Repo, E.; Warchoł, J.K.; Bhatnagar, A.; Sillanpää, M. Heavy metals adsorption by novel EDTA-modified chitosan–silica hybrid materials. *J. Colloid Interface Sci.* **2011**, *358*, 261–267. [CrossRef] [PubMed]
42. Zhao, F.; Repo, E.; Yin, D.; Sillanpää, M.E.T. Adsorption of Cd(II) and Pb(II) by a novel EGTA-modified chitosan material: Kinetics and isotherms. *J. Colloid Interface Sci.* **2013**, *409*, 174–182. [CrossRef] [PubMed]
43. Foo, K.Y.; Hameed, B.H. Insights into the modeling of adsorption isotherm systems. *Chem. Eng. J.* **2010**, *156*, 2–10. [CrossRef]
44. Langmuir, I. The constitution and fundamental properties of solids and liquids. *J. Am. Chem. Soc.* **1917**, *183*, 102–105. [CrossRef]
45. Freundlich, H. Über die Adsorption in Lösungen. *Z. Phys. Chem.* **1907**, *57U*, 385–470. [CrossRef]
46. Zhao, F.; Repo, E.; Yin, D.; Meng, Y.; Jafari, S.; Sillanpää, M. EDTA-Cross-Linked β-Cyclodextrin: An Environmentally Friendly Bifunctional Adsorbent for Simultaneous Adsorption of Metals and Cationic Dyes. *Environ. Sci. Technol.* **2015**, *49*, 10570–10580. [CrossRef] [PubMed]
47. Zou, X.; Pan, J.; Ou, H.; Wang, X.; Guan, W.; Li, C.; Yan, Y.; Duan, Y. Adsorptive removal of Cr(III) and Fe(III) from aqueous solution by chitosan/attapulgite composites: Equilibrium, thermodynamics and kinetics. *Chem. Eng. J.* **2011**, *167*, 112–121. [CrossRef]
48. Shehap, A.M.; Nasr, R.A.; Mahfouz, M.A.; Ismail, A.M. Preparation and characterizations of high doping chitosan/MMT nanocomposites films for removing iron from ground water. *J. Environ. Chem. Eng.* **2021**, *9*, 104700. [CrossRef]
49. Marques, J.L.; Lütke, S.F.; Frantz, T.S.; Espinelli, J.B.S.; Carapelli, R.; Pinto, L.A.A.; Cadaval, T.R.S. Removal of Al(III) and Fe(III) from binary system and industrial effluent using chitosan films. *Int. J. Biol. Macromol.* **2018**, *120*, 1667–1673. [CrossRef]
50. Habiba, U.; Siddique, T.A.; Talebian, S.; Lee, J.J.L.; Salleh, A.; Ang, B.C.; Afifi, A.M. Effect of deacetylation on property of electrospun chitosan/PVA nanofibrous membrane and removal of methyl orange, Fe(III) and Cr(VI) ions. *Carbohydr. Polym.* **2017**, *177*, 32–39. [CrossRef]
51. Wang, M.; Xu, L.; Zhai, M.; Peng, J.; Li, J.; Wei, G. γ-ray radiation-induced synthesis and Fe(III) ion adsorption of carboxymethylated chitosan hydrogels. *Carbohydr. Polym.* **2008**, *74*, 498–503. [CrossRef]
52. Ngah, W.S.W.; Ab Ghani, S.; Kamari, A. Adsorption behaviour of Fe(II) and Fe(III) ions in aqueous solution on chitosan and cross-linked chitosan beads. *Bioresour. Technol.* **2005**, *96*, 443–450. [CrossRef] [PubMed]
53. Khalil, M.M.H.; Al-Wakeel, K.Z.; Rehim, S.S.A.E.; Monem, H.A.E. Efficient removal of ferric ions from aqueous medium by amine modified chitosan resins. *J. Environ. Chem. Eng.* **2013**, *1*, 566–573. [CrossRef]
54. Li, W.; Wei, H.; Liu, Y.; Li, S.; Wang, G.; Han, H. Fabrication of novel starch-based composite hydrogel microspheres combining Diels-Alder reaction with spray drying for MB adsorption. *J. Environ. Chem. Eng.* **2021**, *9*, 105929. [CrossRef]
55. Sultan, M.; Mansor, E.S.; Nagieb, Z.A.; Elsayed, H. Fabrication of highly efficient nano-composite films based on ZnO-g-C$_3$N$_4$ @ PAA-g-(HEC/PVA)-Fe^{3+} for removal of methylene blue dye from water. *J. Water Process. Eng.* **2021**, *42*, 102184. [CrossRef]

56. Hu, H.; Gao, H.; Chen, D.; Li, G.; Tan, Y.; Liang, G.; Zhu, F.; Wu, Q. Ligand-Directed Regioselectivity in Amine–Imine Nickel-Catalyzed 1-Hexene Polymerization. *ACS Catal.* **2015**, *5*, 122–128. [CrossRef]
57. Bertóti, I.; Mohai, M.; László, K. Surface modification of graphene and graphite by nitrogen plasma: Determination of chemical state alterations and assignments by quantitative X-ray photoelectron spectroscopy. *Carbon* **2015**, *84*, 185–196. [CrossRef]
58. Outirite, M.; Lagrenée, M.; Lebrini, M.; Traisnel, M.; Jama, C.; Vezin, H.; Bentiss, F. Ac impedance X-ray photoelectron spectroscopy and density functional theory studies of 3,5-bis(n-pyridyl)-1,2,4-oxadiazoles as efficient corrosion inhibitors for carbon steel surface in hydrochloric acid solution. *Electrochim. Acta* **2010**, *55*, 1670–1681. [CrossRef]
59. Lebrini, M.; Lagrenée, M.; Traisnel, M.; Gengembre, L.; Vezin, H.; Bentiss, F. Enhanced corrosion resistance of mild steel in normal sulfuric acid medium by 2,5-bis(n-thienyl)-1,3,4-thiadiazoles: Electrochemical, X-ray photoelectron spectroscopy and theoretical studies. *Appl. Surf. Sci.* **2007**, *253*, 9267–9276. [CrossRef]
60. Latha, G.; Rajendran, N.; Rajeswari, S. Influence of alloying elements on the corrosion performance of alloy 33 and alloy 24 in seawater. *J. Mater. Eng. Perform.* **1997**, *6*, 743–748. [CrossRef]
61. Wang, L.L.; Ling, C.; Li, B.S.; Zhang, D.S.; Li, C.; Zhang, X.P.; Shi, Z.F. Highly efficient removal of Cu(II) by novel dendritic polyamine-pyridine-grafted chitosan beads from complicated salty and acidic wastewaters. *RSC Adv.* **2020**, *10*, 19943–19951. [CrossRef] [PubMed]
62. Zhang, X.P.; Liu, F.Q.; Zhu, C.Q.; Xu, C.; Chen, D.; Wei, M.-M.; Liu, J.; Li, C.H.; Ling, C.; Li, A.M.; et al. A novel tetraethylenepentamine functionalized polymeric adsorbent for enhanced removal and selective recovery of heavy metal ions from saline solutions. *RSC Adv.* **2015**, *5*, 75985–75997. [CrossRef]
63. Babić, B.M.; Milonjić, S.K.; Polovina, M.J.; Kaludierović, B.V. Point of zero charge and intrinsic equilibrium constants of activated carbon cloth. *Carbon* **1999**, *37*, 477–481. [CrossRef]

Disclaimer/Publisher's Note: The statements, opinions and data contained in all publications are solely those of the individual author(s) and contributor(s) and not of MDPI and/or the editor(s). MDPI and/or the editor(s) disclaim responsibility for any injury to people or property resulting from any ideas, methods, instructions or products referred to in the content.

Article

Uranium Isotope (U-232) Removal from Waters by Biochar Fibers: An Adsorption Study in the Sub-Picomolar Concentration Range

Maria Philippou, Ioannis Pashalidis * and Charis R. Theocharis

Department of Chemistry, University of Cyprus, P.O. Box 20537, Nicosia 1678, Cyprus
* Correspondence: pspasch@ucy.ac.cy; Tel.: +357-22892785

Abstract: The adsorption of the U-232 radionuclide by biochar fibers in the sub-picomolar concentration range has been investigated in laboratory aqueous solutions and seawater samples. The adsorption efficiency (K_d values and % relative removal) of untreated and oxidized biochar samples towards U-232 has been investigated as a function of pH, adsorbent mass, ionic strength and temperature by means of batch-type experiments. According to the experimental data, the solution pH determines to a large degree the adsorption efficiency, and adsorbent mass and surface oxidation lead to significantly higher K_d values. The ionic strength and temperature effect indicate that the adsorption is based on the formation of inner-sphere complexes, and is an endothermic and entropy-driven process ($\Delta H°$ and $\Delta S° > 0$), respectively. Regarding the sorption kinetics, the diffusion of U-232 from the solution to the biochar surface seems to be the rate-determining step. The application of biochar-based adsorbents to treat radioactively (U-232) contaminated waters reveals that these materials are very effective adsorbents, even in the sub-picomolar concentration range.

Keywords: U-232; sub-picomolar range; modified biochar fibers; seawater; adsorption; K_d values

1. Introduction

The interaction of uranium with biochar-based adsorbents has been widely studied using different biomass types as starting materials [1–4], including plant fibers, which in addition to their abundance and low value present desirable adsorbent properties such as a tubular structure and large external surface, allowing for fast material exchange and increased sorption capacities [5–9]. These physical properties remain almost intact after carbonization and chemical oxidation by nitric acid. The latter, which is a chemical modification, results in adsorbent materials with enormous mechanical and chemical resistance, and specificity towards cationic species/metal ions due to the formation of surface carboxylic groups [6,7,10]. Moreover, the derivatization of surface moieties results in selectivity towards uranyl cations or other metal ions [11,12].

Uranium is a natural, ubiquitous element, the concentration of which varies depending on the geological background, and it is almost constant in oceans (~3.3 ppb) [13]. However, various activities including the nuclear fuel cycle and the use of depleted uranium have resulted in the production of huge amounts of uranium-containing solid and liquid waste, and in local environmental contamination by depleted uranium [14]. The uranium sorption by biochar fibers has been investigated with respect to uranium removal from contaminated waters using adsorption-based technologies, as well as for the recovery of uranium from industrial process and waste waters, to secure a long-term supply of uranium as fuel for nuclear power reactors [15].

The investigations related to uranium sorption by biochar-based materials have been performed in the micromole to millimole concentration range, which is relevant for studies that aim to determine the sorption capacity and carry out spectroscopic studies (e.g., Fourier-transform infrared (FTIR), Raman, X-ray photoelectron (XPS), etc.) for surface species characterization [6,7,10–12]. However, there are almost no studies in the nanomole to picomole range, which is for radionuclides of particular interest, because these can be hazardous to living organisms and human beings even at ultratrace levels. On the other hand, it is of particular interest to compare the sorption behavior in a wide concentration scale, including the effect of various parameters affecting the sorption efficiency (e.g., pH, temperature (T), ionic strength (I), adsorbent mass (m), etc.).

The present paper deals with the interaction of two different types of biochar fibers with uranium (e.g., U-232) in laboratory and seawater solutions at ultratrace levels (picomole range). The adsorption efficiency was expressed by the partition coefficient, K_d, and investigated as a function of pH, adsorbent mass, ionic strength (I) and temperature.

2. Results and Discussion

2.1. The Effect of Contact Time on the U-232 Adsorption by LCC and LCC_ox

The effect of contact time was studied in order to evaluate the time needed to reach equilibrium in the system. The corresponding kinetic data are summarized in Figure 1 and indicate that equilibrium was reached after almost two days. Hence, the following experiments were performed after three days of contact time to assure equilibrium conditions. Compared to studies using higher uranium concentrations [6,8], in which equilibrium conditions were reached after a few hours of contact time, equilibrium in the sub-picomolar concentration range is achieved at significantly higher contact times (>50 h), because at such low concentrations, the diffusion of the uranium cations from the solution to the biochar surface is the adsorption rate-limiting step.

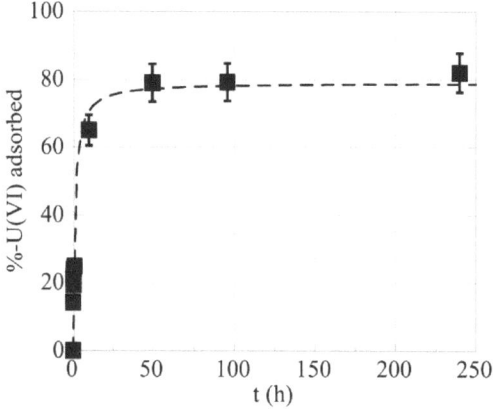

Figure 1. The %-relative amount of U(VI) adsorbed by oxidized biochar fibers (LCC_ox) as a function of time. [U-232] = 8.6 × 10^{-14} mol/L, biochar mass = 0.01 g, pH 4 and ambient conditions.

2.2. The Effect of pH on the U-232 Adsorption by LCC and LCC_ox

The solution pH was expected to affect to sorption efficiency (K_d values) because the proton concentration determines the U(VI) speciation in solution and the surface charge of the biochar materials. The latter is associated with π-proton interaction [16] and carboxylic dissociation [2] for the LCC and LCC_ox, respectively. Regarding U(VI), the predominant species in solution at pH 2 and pH 4, pH 7, and pH 9 were expected to be UO_2^{2+}, UO_2CO_3, and $UO_2(CO_3)_3^{4-}$, respectively [15]. On the other hand, at pH 2, the biochar surface carboxylic moieties are extensively protonated (pKa < 4) and partly deprotonated at pH 4. Only at pH 7 and pH 9 are the carboxylic groups extensively deprotonated and the surface negatively charged [6].

According to Figure 2, the lowest sorption efficiency ($\log_{10}K_d$ (LCC) = 0.7 ± 0.3 and $\log_{10}K_d$ (LCC_ox) = 4.1 ± 0.3) is observed at pH 2, which is related to the competitive interaction of protons with the π-system of LCC and the partial deprotonation of the carboxylic moieties on the LCC_ox surface, respectively. The highest K_d values are observed at pH 4 ($\log_{10}K_d$(LCC) = 3.8 ± 0.3 and $\log_{10}$K (LCC_ox) = 5.2 ± 0.3) because of the partial deprotonation of the surface moieties, which attract the uranyl cation (UO_2^{2+}) to form the π–uranyl cation and U(VI)-carboxylate surface species. Furthermore, at pH 7 ($\log_{10}K_d$(LCC) = 3.0 ± 0.3 and $\log_{10}$K (LCC_ox) = 4.9 ± 0.3) and pH 9 ($\log_{10}K_d$(LCC) = 2.5 ± 0.3 and $\log_{10}$K (LCC_ox) = 4.7 ± 0.3) the adsorption efficiency declines gradually due to the stabilization of U(VI) in solution in the form of the stable U(VI)-carbonato species (UO_2CO_3 and $UO_2(CO_3)_3^{4-}$), and the repulsive forces between the negatively charged U(VI) species and the biochar surface, particularly in the case of LCC_ox.

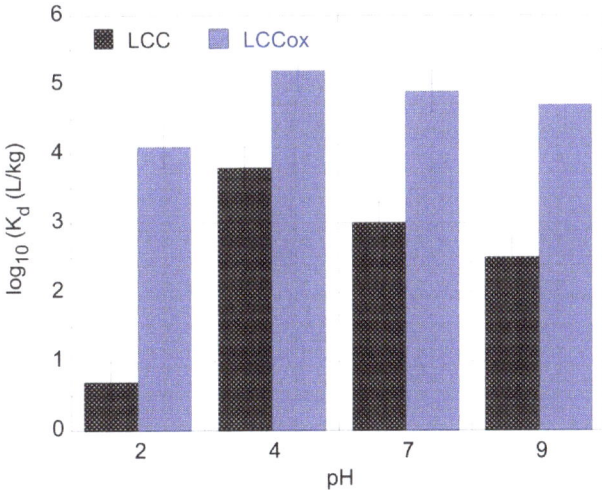

Figure 2. The K_d values for the U(VI) adsorption by two different types of biochar fibers as a function of pH. [U-232] = 8.6 × 10^{-14} mol/L, biochar mass = 0.01 g and ambient conditions.

By a comparison of the pH effect observed in this study with previous studies performed at significantly higher concentration (six to nine orders of magnitude) [6–9], it is obvious that the effect of pH is similar, assuming that the sorption chemistry is governed in both cases by the surface characteristics of the adsorbent and the U(VI) speciation in solution.

2.3. The Effect of Adsorbent Mass on the U-232 Adsorption by LCC and LCC_ox

In order to evaluate the effect of the adsorbent mass on the sorption efficiency, adsorption experiments were performed using similar solutions ([U-232] = 8.6 × 10^{-14} mol/L, pH 2) and varying adsorbent masses. These experiments were performed at a pH of 2 in order to obtain, in all cases, a relative removal below 100% for a better comparison.

The activity concentration of the remaining U-232 in solution was determined by alpha-spectroscopy and the corresponding spectra are summarized in Figure 3, indicating the effect of the biochar mass on the sorption efficiency, particularly in the case of LCC_ox.

The alpha-spectroscopic data were evaluated to calculate the relative U(VI) removal and the corresponding data are graphically presented in Figure 4. It is obvious that the sorption efficiency increased exponentially with the adsorption mass and the highest relative removal values are observed for LCC_ox. This is in agreement with previous studies performed at elevated concentrations, indicating the affinity of biochar materials for U(VI) and the importance of surface modification (e.g., surface oxidation) to improving the affinity and adsorption capacity of the biochar materials towards metal/metalloid species (e.g., uranium) in aqueous solutions [6–9].

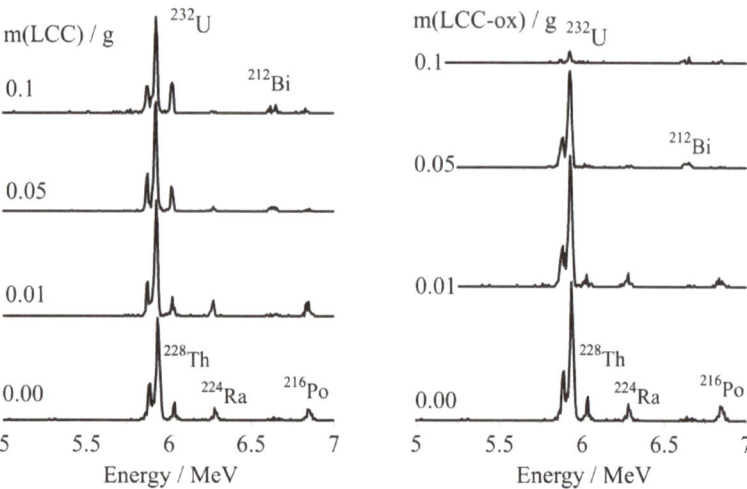

Figure 3. The alpha spectra of the U-232-traced aqueous solutions after treatment with different masses of unmodified (m(LCC) in g) and oxidized biochar (m(LCC_ox) in g). [U-232] = 8.6×10^{-14} mol/L, pH 4 and ambient conditions. It has to be noted that Bi-212, Ra-224, Th-228 and Po-216 are daughter nuclides of U-232 and are therefore present in the alpha spectra.

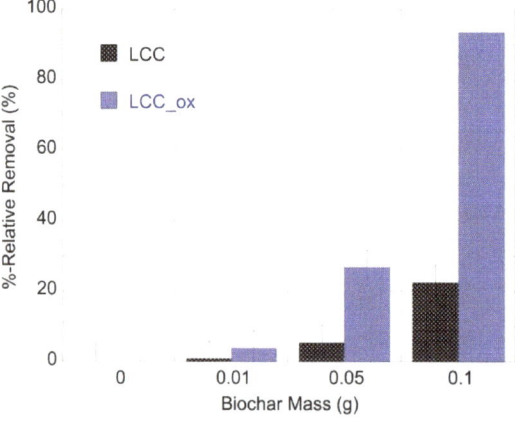

Figure 4. The %-relative removal efficiency of U-232-traced aqueous solutions after treatment with different masses of unmodified (LCC) and oxidized biochar (LCC_ox). [U-232] = 8.6×10^{-14} mol/L, pH 4 and ambient conditions.

2.4. The Effect of Ionic Strength on the U-232 Adsorption by LCC and LCC_ox

At increased metal ion concentrations, spectroscopic methods (such as FTIR, Raman and XPS) may be used to evaluate the adsorption mechanism at the molecular level [5–9,12,16–18]. However, this was impossible at the ultratrace levels used in the present study and therefore the effect of the ionic strength on the adsorption efficiency was employed to indicate the type of surface complexes formed (e.g., inner- or outer-sphere complexes). Generally, the decline of the adsorption efficiency with increasing ionic strength indicates non-specific/electrostatic interactions and the formation of outer-sphere complexes. On the other hand, insignificant changes in the adsorption efficiency with increasing ionic strength are associated with the formation of specific interactions and the formation of inner-sphere complexes [6,7].

According to the data in Figure 5, which show the adsorption efficiency ($\log_{10} K_d$ values) as a function of the ionic strength, there was almost no effect of the ionic strength on adsorption efficiency, assuming specific interactions and formation of inner-sphere complexes between the uranyl cation and biochar surface moieties, which are carboxylic/carboxylate groups particularly on the surface of LCC_ox. It has to be noted that oxygen-containing moieties were also present (to a lesser extent) on the surface of untreated biochar contributing significantly to U(VI) adsorption by LCC, particularly at pH 4.

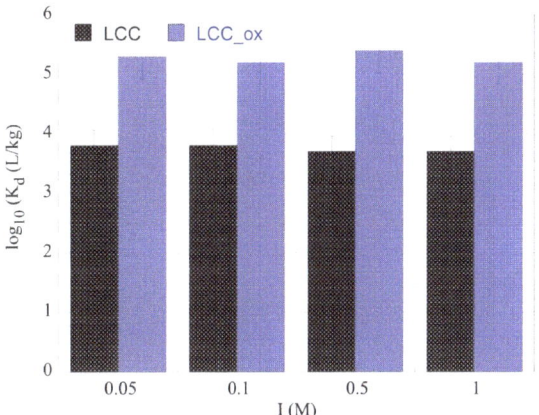

Figure 5. The K_d values for the U(VI) sorption by two different types of biochar fibers (LCC and LCC_ox) as a function of ionic strength. [U-232] = 8.6 × 10^{-14} mol/L, biochar mass = 0.01 g, pH 4 and ambient conditions.

2.5. The Effect of Temperature on the U-232 Adsorption by LCC and LCC_ox

The evaluation of the thermodynamic parameters (ΔH^0; and ΔS^0) for the U(VI) adsorption by the two different biochar materials was performed by determining the corresponding K_d values at three different temperatures and plotting $\ln K_d$ versus $1/T$ according to the Van 't Hoff equation:

$$\ln K_d = -\frac{\Delta H^0}{R \cdot T} + \frac{\Delta S^0}{R} \qquad (1)$$

The $\ln K_d - 1/T$ plot is shown in Figure 6; the thermodynamic parameters were calculated using the slope and intercept obtained from the linear regression of the associated experiment, and were found to amount to ΔH^0 = 1.4 kJ mol^{-1} and ΔS^0 = 5.4 J K^{-1} mol^{-1}, and ΔH^0 = 1.5 kJ mol^{-1} and ΔS^0 = 6.1 J K^{-1} mol^{-1}, for the adsorption of U(VI) by LCC and LCC_ox, respectively. These data clearly indicate that the adsorption of U(VI) by both biochar materials is an endothermic and entropy-driven process, and are in agreement with previous studies. However, the values obtained are up to two orders lower than the corresponding values obtained from experiments performed using higher uranium concentrations [6,7]. This is attributed to the fact that at higher concentrations, the assumption of a large excess of binding sites compared to the metal species is not viable, resulting in those significant differences.

2.6. The Removal of U-232 from Seawater by LCC and LCC_ox

The applicability of the biochar materials regarding the removal of ultratrace amounts of the radionuclide from natural waters was tested using U-232-contaminated seawater solutions and varying biochar amounts (e.g., 0.01, 0.05 and 0.1 g). After 24 h of contact, the remaining radionuclide concentration in solution was determined by alpha-spectroscopy and the representative spectra are shown in Figure 7. The relative U-232 removal from the seawater solution (20 mL) using various amounts of biochar is graphically summarized in Figure 8. The data (Figure 7) clearly show that increasing biochar mass resulted in increasing removal efficiency and that the oxidized counterpart (LCC-ox) presented a higher removal efficiency, which was almost 100% for the seawater sample treated with 0.1 g.

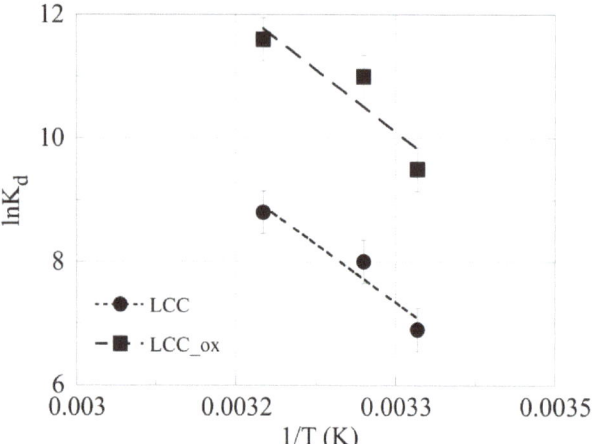

Figure 6. The $\ln K_d$ as a function of $1/T$ for the sorption of U(VI) by LCC and LCC_ox, at an initial uranium concentration of 8.6×10^{-14} mol/L, biochar mass = 0.01 g, pH 4 and 3 days of contact time.

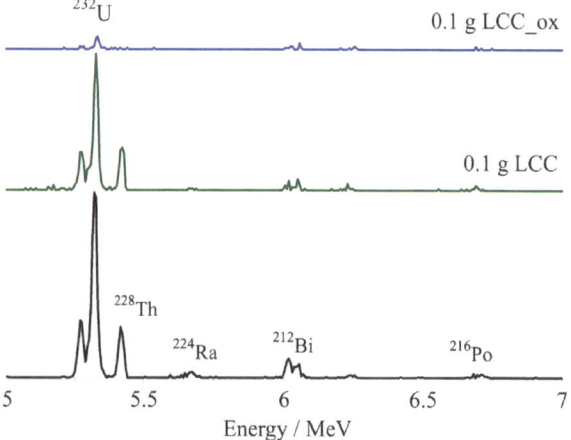

Figure 7. The alpha spectra of U-232-traced seawater solutions treated with 0.1 g of unmodified (LCC) and oxidized biochar (LCC_ox). [U-232] = 8.6×10^{-14} mol/L, pH 8.3 and ambient conditions.

Compared to the corresponding data obtained from laboratory solutions (pH 2), it is obvious that the removal efficiency in seawater solutions is generally higher for both biochar materials (LCC and LCC_ox) despite the presence of competing cations (e.g., Ca^{2+}, Fe^{3+}) and complexing ligands (e.g., CO_3^{2-}) that disfavor U(VI) surface complexation [6,7]. This is ascribed to the higher pH of seawater solutions compared to the laboratory solutions (pH 2), and the associated higher affinity of the surface moieties for U(VI) at pH 8.3 compared to pH 2. Nevertheless, at lower adsorbent amounts, the removal efficiency decreases dramatically, because the surface active sites are quantitatively occupied by competing cations (e.g., Ca^{2+}, Fe^{3+}), indicating that the materials are non-selective adsorbents and can bind polyvalent metal ions that can specifically interact and form complexes with the carboxylic moieties, present particularly on the LCC_ox surface.

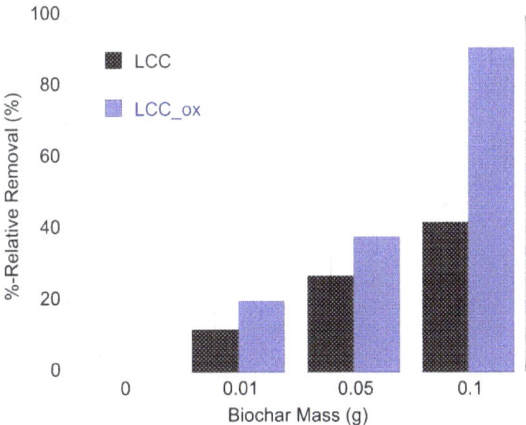

Figure 8. The %-relative removal efficiency of U-232-traced seawater solutions after treatment with different masses of unmodified (LCC) and oxidized biochar (LCC_ox). [U-232] = 8.6×10^{-14} mol/L, pH 8.3 and ambient conditions.

3. Materials and Methods

All experiments were carried out in 30-mL polyethylene (PE) screw-capped bottles under ambient conditions (23 ± 2 °C). The uranium isotope, U-232 ($t_{1/2}$ = 68.9 years), was used for the studies. Reference and test solutions had the same initial activity concentration (0.5 mBq/mL) and were prepared from a standard tracer solution (4.923 kBq/g) purchased from NPL (United Kingdom). The studied biochar samples included biochar prepared from *Luffa cylindrica* fibers (LCC) and its surface-oxidized counterpart (LCC_ox). The preparation and characterization of unmodified and oxidized biochar fibers is extensively investigated and described elsewhere [6,7]. In contrast to LCC, which is characterized by a graphite-like surface, the carboxylic moieties govern the surface charge and chemistry of LCC_ox. The experiments were performed in laboratory solutions using deionized water of different pH (i.e., 2, 4, 7 and 9) and in seawater collected from a local coastal area. The composition of the seawater samples is given elsewhere [19]. The pH was adjusted using HCl and NaOH solutions, and measured using a combined glass electrode and a pH meter (Hanna Instruments, Woonsocket, RI, USA), which was calibrated prior to each experiment with buffer solutions (Scharlau, Barcelona, Spain). The radiometric analysis of the uranium isotope (U-232) was performed by means of an alpha-spectrometer (Canberra France, Loches, France), as described elsewhere [20]. In addition, the reference and control samples from the U-232 analysis were obtained by liquid scintillation counting (LSC, Triathler, Hidex, Turku, Finland). Alpha-spectrometry measurements were carried out in duplicate and LSC measurements were also performed in parallel to compare the data obtained from both radiometric methods. The detection limits were evaluated to be 0.05 mBq and 0.03 mBq for the LSC and alpha-spectrometric measurements, respectively.

The adsorption studies were carried out by contacting 0.01 g of the biochar with 20 mL of the U-232 solution at an activity concentration of 25 Bq/L ([U-232]= 8.6×10^{-14} mol/L) in 30-mL screw-capped PE bottles. Specifically, the effect of contact time was explored at a pH of 4 and under the previously mentioned conditions. Similarly, the pH effect was investigated for the pH of 2, 4, 7 and 9. The effect of the biochar mass was investigated in laboratory solutions (pH 2) and seawater solutions (pH 8.3) by adding 0.01, 0.05 and 0.1 g of biochar and the reference solution (without biochar). The effect of ionic strength (I) was considered using aqueous $NaClO_4$ solutions of varying concentrations (0.05, 0.1, 0.5 and 1M). The effect of temperature was investigated at 25, 40 and 60 °C at a pH of 4. The suspensions were agitated on a rocking shaker (SK-R1807, DLAB, Beijing China) at an agitation rate of 65 min^{-1}. For the uranium analysis, aliquots of 200 µL were used to determine the radionuclide concentration in solution using an alpha-spectrometer or a liquid scintillation counter, which were previously calibrated using standard reference solutions and sources.

The partition coefficient, K_d, was used to evaluate the sorption efficiency, because of the sub-picomolar uranium concentrations used and the great excess of surface binding sites (B). The partition coefficient, K_d, is defined as:

$$K_d = [U(VI)]_{ads}/[U(VI)]_{aq} \; (L/kg) \qquad (2)$$

where $[U(VI)]_{ads}$ (Bq/g) is the activity of the U-232 adsorbed by the biochar and $[U(VI)]_{aq}$ (Bq/L) is the U-232 activity concentration in solution at equilibrium. The amount of U-232 adsorbed by biochar (dry mass) was calculated from the total activity of U-232 adsorbed minus the activity of U-232 adsorbed by the plastic bottle walls, which was not negligible and had to be taken into account. In addition, under ambient conditions, the U(VI) was expected to be the predominant oxidation state of uranium in solution [21].

In addition, the sorption efficiency is expressed as %-relative removal and is calculated by:

$$\%\text{-relative removal} = 100 \cdot ([U(VI)]_R - [U(VI)]_{aq})/[U(VI)]_R \qquad (3)$$

where $[U(VI)]_R$ is the U-232 concentration in the reference solution.

The experiments were performed in triplicate, and the mean values and uncertainties were used for the data evaluation and graphical presentations.

4. Conclusions

Biochar materials have the ability to adsorb radionuclides (U-232) even in the sub-picomolar concentration range. The adsorption efficiency strongly depends on the solution pH that governs surface charge and U(VI) speciation in solution. Surface modification, such as the oxidation of the biochar, results in a significantly higher sorption capacity and affinity of the adsorbent for U(VI) due to the formation of surface carboxylic moieties. Variation of the ionic strength does not have any significant effect on the sorption efficiency, indicating the formation of inner-sphere surface complexes, and the evaluation of the data obtained from the experiments performed at different temperatures revealed endothermic and entropy-driven adsorption processes. In seawater, that has a pH~8.3 and contains anions (e.g., Ca^{2+}, Fe^{3+}) which can compete with U(VI) and occupy sorption sites on the MPs surface, and anions (CO_3^{2-}, SO_4^{2-}) which can act as ligands and stabilize U(VI) in the aqueous phase, the sorption efficiency decreases dramatically. Nevertheless, increasing the biomass amount, particularly in the case of LCC_ox, can result in the almost quantitative removal of the radionuclide even from seawater solutions.

Future studies could include experiments related to the adsorption efficiency of biochars for other radionuclides (e.g., Ra, Am, Pu) and surface derivatization to enhance selectivity towards certain radionuclides of particular interest, especially for use in nuclear medicine, as well as experiments to study the competitive adsorption in radionuclide mixtures.

Author Contributions: Conceptualization, C.R.T. and I.P.; methodology, I.P.; validation, C.R.T. and I.P.; investigation, M.P.; data curation, M.P.; writing—original draft preparation, I.P.; writing—draft preparation, M.P.; writing—review and editing, C.R.T. and I.P. All authors have read and agreed to the published version of the manuscript.

Funding: This research received no external funding.

Institutional Review Board Statement: Not applicable.

Informed Consent Statement: Not applicable.

Data Availability Statement: Not applicable.

Conflicts of Interest: The authors declare no conflict of interest.

References

1. Philippou, K.; Pashalidis, I. Chapter 11—Polyvalent metal ion adsorption by chemically modified biochar fibers. In *Biomass-Derived Materials for Environmental Applications*; Anastopoulos, I., Lima, E., Meili, L., Giannakoudakis, D., Eds.; Elsevier: Amsterdam, The Netherlands, 2022; pp. 267–286.
2. Hopkins, D.; Hawboldt, K. Biochar for the removal of metals from solution: A review of lignocellulosic and novel marine feedstocks. *J. Environ. Chem. Eng.* **2020**, *8*, 103975. [CrossRef]
3. Duwiejuah, A.B.; Abubakari, A.H.; Quainoo, A.K.; Amadu, Y. Review of Biochar Properties and Remediation of Metal Pollution of Water and Soil. *J. Health Pollut.* **2020**, *10*, 200902. [CrossRef] [PubMed]
4. Inyang, M.I.; Gao, B.; Yao, Y.; Xue, Y.; Zimmerman, A.; Mosa, A.; Pullammanappallil, P.; Ok, Y.S.; Cao, X. A review of biochar as a low-cost adsorbent for aqueous heavy metal removal. *Crit. Rev. Environ. Sci. Technol.* **2016**, *46*, 406–433. [CrossRef]
5. Tomlinson, J.B.; Theocharis, C.R. Studies of Steam-Activated Viscose Rayon Chars. *Carbon* **1992**, *30*, 907–911. [CrossRef]
6. Hadjittofi, L.; Pashalidis, I. Uranium sorption from aqueous solutions by activated biochar fibres investigated by FTIR spectroscopy and batch experiments. *J. Radioanal. Nucl. Chem.* **2015**, *304*, 897–904. [CrossRef]
7. Liatsou, I.; Michael, G.; Demetriou, M.; Pashalidis, I. Uranium binding by biochar fibres derived from Luffa cylindrical after controlled surface oxidation. *J. Radional. Nucl. Chem.* **2017**, *311*, 871–875. [CrossRef]
8. Philippou, K.; Savva, I.; Pashalidis, I. Uranium(VI) binding by pine needles prior and after chemical modification. *J. Radional. Nucl. Chem.* **2018**, *318*, 2205–2211. [CrossRef]
9. Stasi, C.; Georgiou, E.; Ioannidis, I.; Pashalidis, I. Uranium removal from laboratory and environmental waters by oxidised biochar prepared from palm tree fibres. *J. Radioanal. Nucl. Chem.* **2022**, *331*, 375–381. [CrossRef]
10. Hadjittofi, L.; Prodromou, M.; Pashalidis, I. Activated biochar derived from cactus fibres—Preparation, characterization and application on Cu(II) removal from aqueous solutions. *Bioresour. Technol.* **2014**, *159*, 460–464. [CrossRef] [PubMed]
11. Liatsou, I.; Pashalidis, I.; Nicolaides, A. Triggering selective uranium separation from aqueous solutions by using salophen-modified biochar fibers. *J. Radioanal. Nucl. Chem.* **2018**, *318*, 2199–2203. [CrossRef]
12. Liatsou, I.; Pashalidis, I.; Dosche, C. Cu(II) adsorption on 2-thiouracil-modified Luffa cylindrica biochar fibres from artificial and real samples, and competition reactions with U(VI). *J. Hazard. Mater.* **2020**, *383*, 120950. [CrossRef] [PubMed]
13. Tissot, F.L.H.; Dauphas, N. Uranium isotopic compositions of the crust and ocean: Age corrections, U budget and global extent of modern anoxia. *Geochim. Cosmochim. Acta* **2015**, *167*, 113–143. [CrossRef]
14. Bleise, A.; Danesi, P.R.; Burkart, W. Properties, use and health effects of depleted uranium (DU): A general overview. *J. Environ. Radioact.* **2003**, *64*, 93–112. [CrossRef]
15. Haneklaus, N.; Sun, Y.; Bol, R.; Lottermoser, B.; Schnug, E. To Extract, or not to Extract Uranium from Phosphate Rock, that is the Question. *Environ. Sci. Technol.* **2017**, *51*, 753–754. [CrossRef] [PubMed]
16. Reddy, A.S.; Sastry, G.N. Cation [M = H^+, Li^+, Na^+, K^+, Ca^{2+}, Mg^{2+}, NH_4^+, and NMe_4^+] interactions with the aromatic motifs of naturally occurring amino acids: A theoretical study. *J. Phys. Chem. A* **2005**, *109*, 8893–8903. [CrossRef] [PubMed]
17. Noli, F.; Kapashi, E.; Pashalidis, I.; Margellou, A.; Karfaridis, D. The effect of chemical and thermal modifications on the biosorption of uranium in aqueous solutions using winery wastes. *J. Mol. Liq.* **2022**, *351*, 118665. [CrossRef]
18. Philippou, K.; Anastopoulos, I.; Dosche, C.; Pashalidis, I. Synthesis and characterization of a novel Fe_3O_4-loaded oxidized biochar from pine needles and its application for uranium removal. Kinetic, thermodynamic, and mechanistic analysis. *J. Environ. Manag.* **2019**, *252*, 109677. [CrossRef] [PubMed]
19. Georgiou, E.; Raptopoulos, G.; Papastergiou, M.; Paraskevopoulou, P.; Pashalidis, I. Extremely Efficient Uranium Removal from Aqueous Environments with Polyurea-Cross-Linked Alginate Aerogel Beads. *ACS Appl. Polym. Mater.* **2022**, *4*, 920–928. [CrossRef]
20. Kiliari, T.; Pashalidis, I. Simplified alpha-spectroscopic analysis of uranium in natural waters after its separation by cation-exchange. *Radiat. Meas.* **2010**, *45*, 966–968. [CrossRef]
21. Konstantinou, M.; Pashalidis, I. Speciation and Spectrophotometric Determination of Uranium in Seawater. *Mediterr. Mar. Sci.* **2004**, *5*, 55–60. [CrossRef]

Article

Adsorption Performance of Methylene Blue by KOH/FeCl₃ Modified Biochar/Alginate Composite Beads Derived from Agricultural Waste

Heng Liu [1], Jiaqi Zhu [1], Qimei Li [1], Likun Li [2,*], Yanjun Huang [1], Yi Wang [1], Guozhi Fan [1] and Lei Zhang [1,*]

[1] School of Chemistry and Environmental Engineering, Wuhan Polytechnic University, Wuhan 430023, China
[2] China-Ukraine Institute of Welding, Guangdong Academy of Sciences, Guangzhou 510650, China
* Correspondence: lilk@gwi.gd.cn (L.L.); zhanglei@whpu.edu.cn (L.Z.)

Abstract: In this study, high-performance modified biochar/alginate composite bead (MCB/ALG) adsorbents were prepared from recycled agricultural waste corncobs by a high-temperature pyrolysis and KOH/FeCl₃ activation process. The prepared MCB/ALG beads were tested for the adsorption of methylene blue (MB) dye from wastewater. A variety of analytical methods, such as SEM, BET, FTIR and XRD, were used to investigate the structure and properties of the as-prepared adsorbents. The effects of solution pH, time, initial MB concentration and adsorption temperature on the adsorption performance of MCB/ALG beads were discussed in detail. The results showed that the adsorption equilibrium of MB dye was consistent with the Langmuir isothermal model and the pseudo-second-order kinetic model. The maximum adsorption capacity of MCB/ALG−1 could reach 1373.49 mg/g at 303 K. The thermodynamic studies implied endothermic and spontaneous properties of the adsorption system. This high adsorption performance of MCB/ALG was mainly attributed to pore filling, hydrogen bonding and electrostatic interactions. The regeneration experiments showed that the removal rate of MB could still reach 85% even after five cycles of experiments, indicating that MCB/ALG had good reusability and stability. These results suggested that a win-win strategy of applying agricultural waste to water remediation was feasible.

Keywords: corncob biochar; KOH/FeCl₃ modified; methylene blue; adsorption performance; adsorption mechanism

Citation: Liu, H.; Zhu, J.; Li, Q.; Li, L.; Huang, Y.; Wang, Y.; Fan, G.; Zhang, L. Adsorption Performance of Methylene Blue by KOH/FeCl₃ Modified Biochar/Alginate Composite Beads Derived from Agricultural Waste. *Molecules* **2023**, *28*, 2507. https://doi.org/10.3390/molecules28062507

Academic Editors: Yongchang Sun and Dimitrios Giannakoudakis

Received: 19 February 2023
Revised: 6 March 2023
Accepted: 8 March 2023
Published: 9 March 2023

Copyright: © 2023 by the authors. Licensee MDPI, Basel, Switzerland. This article is an open access article distributed under the terms and conditions of the Creative Commons Attribution (CC BY) license (https://creativecommons.org/licenses/by/4.0/).

1. Introduction

In recent decades, chemical dyes have been widely used in textile, leather, paint and other industries, resulting in a large amount of dye wastewater [1]. Due to toxicity and poor biodegradability, it is essential that the dye wastewater is properly treated before being discharged into aquatic systems [2]. As a common cationic dye, methylene blue (MB) is widely used in print, leather and textile industries [3–5]. However, it has been proven to cause various health problems, such as vertigo, retching and eye burns [6]. Therefore, in recent years, it has been of great significance to remove MB from industrial wastewater [7].

Conventional treatment methods for MB include membrane separation, adsorption, electrochemical oxidation and the Fenton method [8–11]. Among various treatment technologies, the adsorption method is considered as the most promising method because of its good stability and efficiency [12]. Numerous adsorbents, such as graphene oxide, chitosan, activated carbon and bentonite, have been used to remove MB from aqueous solution [13–16]. However, the application of these adsorbents is limited by a high cost and restricted regeneration performance. Developing alternative adsorbents with the merits of low cost and eco-friendliness has always been pursued by scholars all over the world.

Biochar (BC) has attracted considerable attention as economical absorbents because of its large pore volume and facile modification [17–19]. Biochar is derived from agricultural and forestry wastes (e.g., peanut shells, corncob, rice husk, walnut shell, etc.) and prepared by pyrolysis under high temperature and under oxygen-free conditions [20–22]. In China, most corncobs are directly discarded or burned, resulting in resource waste and serious air pollution [23]. Corncob is a good raw material for the production of biochar to solve problems such as effective utilization of agricultural and forestry by-product wastes [24]. However, the biochar obtained by one-step direct pyrolysis usually has small adsorption capacity. In order to enhance its adsorption capacity, additional chemical modification is typically undertaken. KOH modification could improve the surface groups of biochar, thus improving the adsorption capacity of biochar [25]. Ma et al. [26] reported that the adsorption capacity of KOH modified corncob biochar for pollutants is six times higher than that of unmodified biochar. Furthermore, the modification of ferric chloride ($FeCl_3$) can increase carbon yield and increase the number of functional groups of biochar, thus enhancing its adsorption performance [27]. However, the difficulties in cleaning and separating powdered biochar restrict its further industry application. Thus, the preparation of solid biochar is the key to the industrial application of biochar.

Sodium alginate (SA) is a kind of natural polysaccharide obtained mainly from marine brown algae, with rich hydroxyl and carboxyl functional groups in its polymer chain. It is widely used as the supporting material of various powder adsorbents in various methods, and shows a good application prospect in wastewater pollutant removal [28]. However, the prepared adsorbent has poor mechanical strength. To enhance the mechanical strength, glutaraldehyde is selected as the crosslinking agent. Sodium alginate composite corncob biochar pellets in the form of beads are deliberated in this research, which may overcome these disadvantages and are easy to reuse in batch studies. It is believed that the framework of $KOH/FeCl_3$ modified corncob biochar loaded alginate matrix beads might show high selectivity towards the removal of MB from water.

Based on the hypothesis, the current work had the aim of fabricating facile, eco-friendly and biocompatible $KOH/FeCl_3$ modified biochar/alginate composite beads for application in the adsorption of MB from water. The adsorbent was characterized by SEM, BET, XRD and FTIR. The effects of solution pH, initial concentration and contact time, adsorption temperature on the removal MB process were optimized. At last, the adsorption kinetics, isotherms and thermodynamics of adsorption of MB from water carried out to explore adsorption behavior.

2. Results and Discussion

2.1. Characterization

SEM images of the corncob biochar (CB), modified corncob biochar (MCB) and modified biochar/alginate composite beads (MCB/ALG) are shown in Figure 1. The surface of CB (Figure 1a) is relatively smooth, while the surface of MCB (Figure 1b) becomes rough, with more pore structures and more cracks. This is because the activation reaction between KOH and carbonaceous structure promotes the development of porous structure of porous carbon [29]. Obviously, this rich porosity may be conducive to improving the absorption performance of biochar. Although MCB/ALG (Figure 1c,e,g) has a spherical shape, its surface is relatively rough due to the irregular accumulation of biochar. It is worth noting that this phenomenon becomes more pronounced with the increase in MCB content in MCB/ALG. Due to the addition of large amounts of MCB, inhomogeneities can be observed on the surface of MCB/ALG−3.

To gain insight into the porous structure of the samples, N_2 adsorption–desorption tests are performed using the isotherms and BJH pore size distribution curves shown in Figure 2. The porous structure parameters are summarized in Table 1. According to the IUPAC classification, the adsorption–desorption isotherms (Figure 2a) of three samples are considered to be type IV with H3 type hysteresis loop, implying the presence of porous structures in these samples. Compared to CB, the specific surface area of MCB is increased by more than twenty times. However, the poor pore structure of MCB/ALG compared to MCB suggests that some micropores and mesopores in MCB particles may be blocked after the composite process. Figure 2b shows the pore diameter distribution curves of the samples. It is observed that the pore diameters of three samples are mainly distributed around 8 nm, indicating that they have a mesoporous structure [30]. Table 1 shows that the modification leads to an increase in specific surface area and pore volume, and a decrease in pore diameter, indicating that the modification can induce a small porous structure in CB. In contrast, after the recombination process, the pore diameter of MCB/ALG increases slightly. In conclusion, the relatively high specific surface area and large pore diameter of MCB/ALG may be more conducive to the diffusion of contaminants into the internal pores and subsequent removal by adsorption.

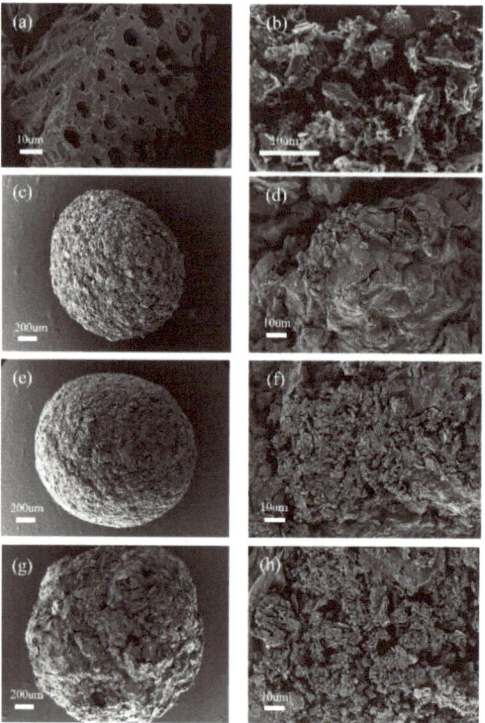

Figure 1. SEM images of (**a**) CB, (**b**) MCB, (**c**,**d**) MCB/ALG−1, (**e**,**f**) MCB/ALG−2, (**g**,**h**) MCB/ALG−3.

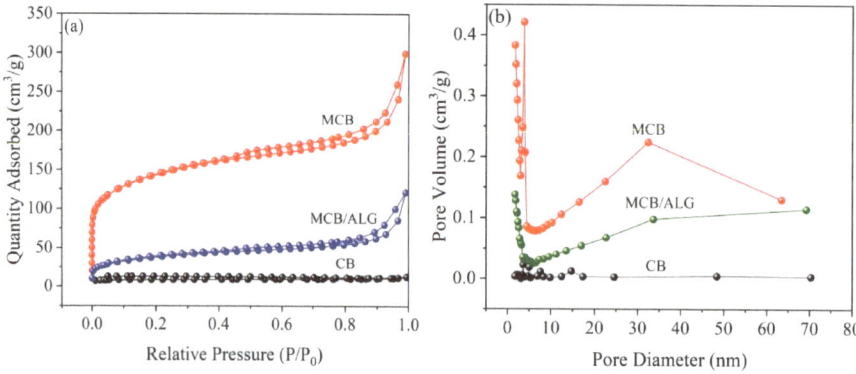

Figure 2. (**a**) N_2 adsorption–desorption isotherms and (**b**) the pore size distribution of BC samples.

Table 1. BET parameters of CB, MCB and MCB/ALG.

Sample	BET Surface Area (m²/g)	Pore Volume (cm³/g)	Pore Diameter (nm)
CB	25.9	0.006	7.6
MCB	468.4	0.3	6.0
MCB/ALG	128.8	0.1	8.1

Figure 3a shows the FTIR spectra of the samples to determine the various functional groups. The surface functional groups of corncob biochar were changed during the chemical modification of KOH and FeCl$_3$ and high temperature carbonization process. All samples exhibit broad peaks around 3420 cm^{-1}, indicating the presence of hydroxyl (O–H) stretching vibrations [31]. The increase in peak intensity of –OH after alkali modification compared to unmodified indicates an increase in oxygen-containing functional groups on the surface of biochar. These oxygen-containing functional groups enhance the adsorption of MB by providing active sites [32]. The peak around 1590 cm^{-1} and 1400 cm^{-1} are the aromatic ring (C=C) and C–N stretching vibrations, respectively. The peaks at 880 cm^{-1} and 810 cm^{-1} could be attributed to the C–H bending vibration outside the aromatic plane. It has been reported that the occurrence of aromatization processes during the modification process may lead to the enhancement of the intensity of the above-mentioned peaks. It can be seen that the prepared MCB/ALG inherits the functional groups of MCB and SA. It proves that the composite beads were successfully synthesized. The XRD spectra of the samples are illustrated in Figure 3b. For the four samples in the figure, the broad diffraction peaks near 2θ = 22° are characteristic peaks for cellulose and hemicellulose [33]. After modification, FeCl$_3$ significantly changed the XRD spectrum of CB. Both MCB and MCB/ALG−1 samples show a wide diffraction peak at a 2θ value of 43.2°, belonging to the plane of graphite carbon. The diffraction peaks of Fe$_3$O$_4$ (2θ = 30.2°, 43.2°, 57.2°, 62.7°) and γ-Fe$_2$O$_3$ (2θ = 35.5°) can be clearly observed on MCB and MCB/ALG [34]. This is mainly due to the addition of FeCl$_3$ in the modification process, which introduces a small amount of Fe. The results demonstrate the presence of graphene structures in MCB and MCB/ALG, which facilitates the adsorption of MB [35].

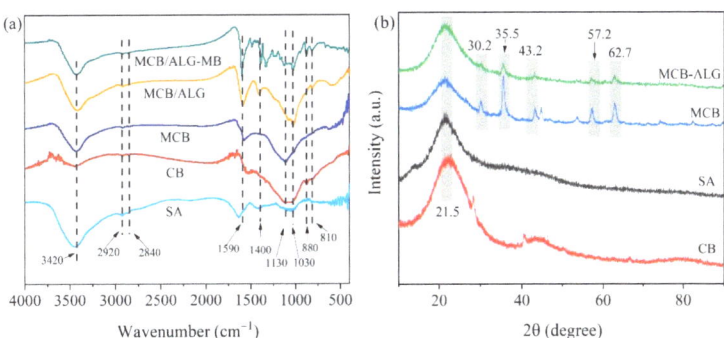

Figure 3. (a) FTIR and (b) XRD spectra of samples.

2.2. Effect of pH on Adsorption

As an important factor in the adsorption process, the initial pH value can affect the charge change on the adsorbent surface, thus affecting the adsorption efficiency of MB [36]. The effect of pH on MB adsorption from water on adsorbents is presented in Figure 4a. From Figure 4a, it can be seen that the removal efficiencies of MB have changed slightly in the pH value range of 3.0–10.0. This suggests that the adsorption process is less affected by the electrostatic force between MB and adsorbent under the solution pH value of 3.0–7.0, and the adsorption process may be dominated by pore filling and hydrogen bonding.

As the solution pH value enhances, the removal efficiency of MB increased significantly. Therefore, it is not conducive for the adsorption of MB on MCB/ALG samples under acidic conditions. Compared with three MCB/ALG adsorbents, MCB/ALG−1 shows a better adsorption performance. The affinity of biochar to MB gradually increases with increasing pH. The solution has a strong electrostatic attraction between the positively charged biochar adsorbent and the negatively charged MB. As shown in Figure 4b, the adsorbent surface shows a more negative charge density at higher pH values. The surface of biochar contains chemical groups, such as –OH and –COOH, and the pH of the dye changes its charge. At acidic conditions, the capacity of the binding sites on biochar is limited. At low pH conditions, MB is a positively charged cationic dye, which leads to electrostatic repulsion between MB and biochar, thereby reducing the removal rate. Through later analysis, the study found that electrostatic attraction is not the only adsorption method [37]. In addition to the adsorption of surface functional groups, it also includes intra-particle diffusion. More adsorption mechanisms have been explored later. Hence, increasing the pH of the solution after reaching the adsorption equilibrium does not significantly reduce the removal rate. When the pH value increased to 10, the absolute value of ΔpH of biochar decreased, showing a slight decrease in removal rate. So, exploring suitable adsorption conditions is conducive to the maximization of adsorption efficiency.

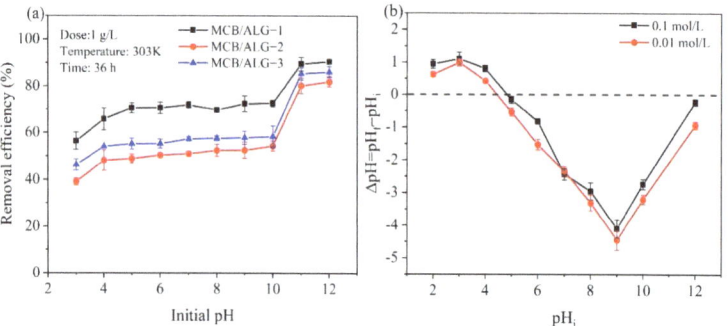

Figure 4. Effect of pH on MB by MCB/ALG−1, MCB/ALG−2 and MCB/ALG−3: (**a**) removal efficiency; (**b**) determination of pH_{PZC} of MCB/ALG−1 in NaCl solutions by batch equilibrium.

2.3. Adsorption Kinetics

Figure 5a shows the time evolution of MB adsorption on MCB/ALG−1 and MCB. In order to evaluate the adsorption kinetic behavior of MB on adsorbent, pseudo-first-order, pseudo-second-order and intra-particle diffusion models were used to fit the experimental data [38].

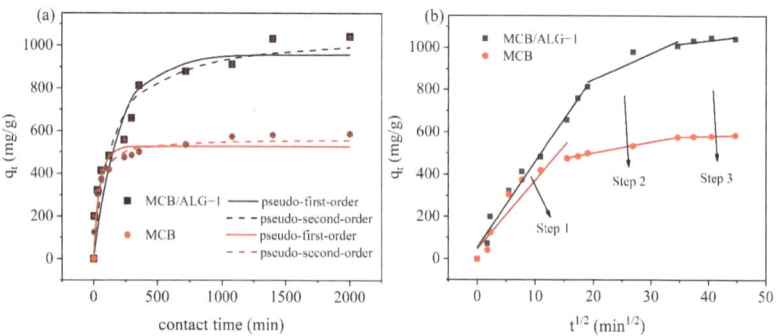

Figure 5. (**a**) Kinetics fitting results of MB adsorption on MCB/ALG−1 and MCB; (**b**) intra-particle diffusion model for MB uptake onto MCB/ALG−1 and MCB.

The pseudo-first-order kinetic model:

$$q_t = q_e[1 - e^{-k_1 t}] \quad (1)$$

The pseudo-second-order kinetic model:

$$q_t = \frac{q_e^2 k_2 t}{1 + k_2 q_e t} \quad (2)$$

The intra-particle diffusion kinetic model:

$$q_t = k_i t^{1/2} + C \quad (3)$$

where q_t (mg/g) represents the amount of MB adsorbed by a mass unit of adsorbent at a predetermined time t (min), whereas k_1 (/min) and k_2 (g/mg·min) are the rate constants of the pseudo-first-order and pseudo-second-order kinetic models, respectively. k_i (mg/g·min$^{1/2}$) is the rate constant for the intra-particle diffusion model, and C (mg/g) is a constant that represents the boundary layer thickness.

In this work, the kinetic parameters were obtained by using non-linear fitting method, and the applicability of adsorption kinetic model was determined by the root mean squared error (*RMSE*). In principle, the lower the *RMSE* value, the more suitable the model fits. *RMSE* could be calculated using the following Equation:

$$RMSE = \sqrt{\frac{1}{n}\sum_{i=1}^{n}(q_{cali} - q_{expi})} \quad (4)$$

where q_{cali} and q_{expi} are the predicted and measured values of the adsorption capacity at time t, respectively, and n is the number of experimental data.

As shown in Figure 5a, the adsorption process can be divided into three stages: the fast adsorption stage, the slow adsorption stage, and the stabilization stage. During the fast adsorption stage, about 90% of the adsorption occurred within 250 min, which may be because the high concentration of MB at the interface between the adsorbents and the solution promoted a large mass transfer driving force, resulting in MB rapidly occupying the adsorption site. In addition, the values of kinetic parameters and their *RMSE* values are summarized in Table 2. As a key parameter, the root mean square error of the pseudo-second-order model is lower than that of the pseudo-first-order model. In contrast, the fitted curves of the pseudo-second-order model are always closer to the experimental data points, indicating that the pseudo-second-order model can better describe the adsorption kinetics. These results show that the pseudo-second-order model is more suitable for describing the adsorption process than the pseudo-first-order model, which indicates that the rate–determining step of MCB/ALG−1 may be a chemisorption process.

Table 2. Kinetic parameters for the adsorption of MB on MCB/ALG−1 and MCB.

Sample	$q_{e,exp}$ (mg/L)	Pseudo-First-Order Model			Pseudo-Second-Order Model		
		$q_{e,cal}$ (mg/g)	k_1 (/min)	RMSE	$q_{e,cal}$ (mg/g)	k_2 (g/mg·min)	RMSE
MCB/ALG−1	1040.7	957.3	0.005	105.1	1055.2	7.31×10^{-6}	81.4
MCB	585.92	525.8	0.02	46.0	562.7	6.09×10^{-5}	25.1

The intra-particle diffusion model is further applied to determine the rate-controlling step of the adsorption processes. According to Figure 5b, the plots of q_t versus $t^{0.5}$ are comprised of three linear segments and do not pass through the origin. It is proved that the adsorption process of MB on MCB/ALG−1 and MCB is multi-step, and intra-particle diffusion is not the only rate-limiting step [39]. In the first stage, the fastest adsorption rate (maximum slope) is associated with film diffusion, where MB molecules migrate from solution to the outer surface of MCB/ALG−1 and MCB, as the large concentration gradient provides sufficient driving force. The second stage has a relatively high adsorption rate and shows a progressive adsorption phase corresponding to the intra-particle diffusion of MB molecules through the internal pores and cavities of the adsorbent. In the third stage, the diffusion rate in the pores is further reduced and the adsorption gradually reaches equilibrium. Compared to the rate constants of the stages, the rate constants of the first stage are much higher than those of the other stages, which implies that the film diffusion is the dominant rate limiting step in the whole process; while the linear part near the platform in the late stage, indicating that intra-particle diffusion is not the only rate-limiting factor, the adsorption rate is also affected and controlled by the external diffusion step and the surface diffusion of the adsorbent [36,40].

2.4. Adsorption Isotherm

The adsorption isotherm curve is helpful to analyze the interaction between the adsorbent and the adsorbate and the characteristics of the adsorption layer. Langmuir and Freundlich isotherm models are used to describe the adsorption data of MB on MCB/ALG−1 samples at 303, 313 and 323 K. The Langmuir and Freundlich isotherm models can be expressed as follows [41].

The Langmuir isothermal model:

$$q_e = \frac{q_m K_L C_e}{1 + K_L C_e} \tag{5}$$

The Freundlich isothermal model:

$$q_e = K_F C_e^{1/n_F} \tag{6}$$

where q_e and q_m (mg/g) represent the equilibrium adsorption amount and the maximum adsorption amount, respectively; C_e (mg/L) represents the equilibrium adsorption concentration; K_L is the Langmuir constant which is related to the affinity of binding sites (L/mg). K_F is the Freundlich isothermal constant and n_F is the heterogeneity factors.

The characteristic constants of Langmuir and Freundlich isotherm models are shown in Table 3, and the curves are shown in Figure 6. The Langmuir isothermal model assumes that monolayer adsorption occurs on the adsorbent surface with equivalent adsorption sites, while the Freundlich isothermal model is used to describe equilibrium data and adsorption characteristics for a heterogeneous surface. The Langmuir model has a lower root mean square error compared to the Freundlich model, indicating that the Langmuir isotherm model is a better fit than the Freundlich isotherm model for the adsorption data in the current experiments, in agreement with a previous report [42]. Thus, the adsorption of MB on MCB/ALG−1 samples is consistent with monolayer adsorption. In addition, an increasing trend of MB adsorption is observed in the temperature range of 303–323 K, clearly indicating its endothermic nature. Meanwhile, the parameter n_F of the Freundlich model is greater than 1, indicating a strong affinity between the MCB/ALG−1 samples and the MB molecule, which is a favorable adsorption process. The maximum adsorption capacities (q_m) of MCB/ALG−1 calculated according to the Langmuir model were 1373.49, 1457.28 and 1485.03 mg/g at 303, 313 and 323 K, respectively. Table 4 lists the maximum adsorption capacities (q_m) for MB on biochar derived from various waste reported in previous studies. It is noted that the adsorption capacity of MCB/ALG−1 is superior or comparable to that of most adsorbents reported in the literature.

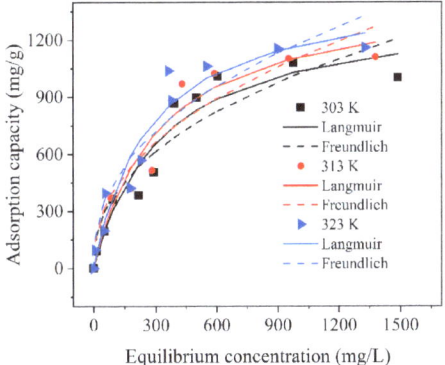

Figure 6. Adsorption isotherms of MB on MCB/ALG−1 sample at different temperature.

Table 3. Adsorption parameters obtained from Langmuir and Freundlich isotherm models.

T (K)	Langmuir			Freundlich		
	q_m (mg/g)	K_L (L/mg)	RMSE	K_F (L/mg)	n_F	RMSE
303	1373.49	0.0030	105.05	55.93	2.37	127.88
313	1457.28	0.0031	94.47	59.26	2.3	116.74
323	1485.03	0.0037	106.91	70.73	2.45	127.25

Table 4. Comparison of various biochar materials for MB removal.

Raw	Pyrolysis Temperature (°C)	q_m (mg/g)	Reference
Tamarind seed	500	102.77	[43]
Sodium carboxymethyl cellulose	900	249.6	[44]
Bamboo	600	286.1	[45]
Alfalfa	600	326.90	[35]
Rattan stalks	600	359	[46]
Coffee grounds	600	367	[47]
Waste tea	450	683.6	[48]
Corncobs	800	1373.49	This work
Corncob-to-xylose residue	850	1563.9	[49]

2.5. Adsorption Thermodynamics

Temperature is an important parameter to control species adsorption in the system. The Gibbs free energy change (ΔG), enthalpy change (ΔH) and entropy change (ΔS) play a key role in determining the heat exchange and spontaneity of the adsorption process. Application of the flow equation to calculate thermodynamic parameters [50].

$$\Delta G = -RT \ln K_d \tag{7}$$

$$\Delta G = \Delta H - T\Delta S \tag{8}$$

$$\ln K_d = \frac{\Delta S}{R} - \frac{\Delta H}{RT} \tag{9}$$

where K_d is the thermodynamic equilibrium constant; R (8.314 J/(mol·K)) stands for the gas constant; T (K) represents the experimental temperature; ΔS (J/(mol·K)) represents the entropy change of the system; ΔH (kJ/mol) represents the enthalpy change and ΔG (kJ/mol) represents the Gibbs free energy.

The calculation results of thermodynamic equilibrium coefficient are shown in Table 5. The enthalpy of the adsorption process become positive ($\Delta H > 0$) at different temperatures, indicating that the adsorption of MB on MCB/ALG−1 sample is a heat absorption reaction and the increase in temperature favors the adsorption process, which is consistent with the isothermal model analysis. The change in entropy is also positive ($\Delta S > 0$), proving that the adsorption process is irreversible and proceeds along the direction of increasing system disorder. Meanwhile, the ΔG values are negative, thus indicating that the adsorption process was spontaneous. Hence, the MB adsorption on MCB/ALG−1 is endothermic and spontaneous.

Table 5. Thermodynamic parameters for adsorption of MB on MCB/ALG−1.

T (K)	ΔG (kJ·mol^{-1})	ΔH (kJ·mol^{-1})	ΔS (kJ·mol^{-1} K^{-1})
303	−0.2551		
313	−0.3726	9.4163	0.0754
323	−0.6369		

2.6. Regeneration and Reusability

In practical applications, it is necessary to perform adsorption–desorption tests on the adsorbent, since the reusability of the adsorbent is a key factor for economic results [51]. From Figure 7, it can be found that the MCB/ALG−1 sample can still reach 85% MB removal after five adsorption–desorption cycles. Therefore, the MCB/ALG−1 sample is a promising adsorbent with good reusability.

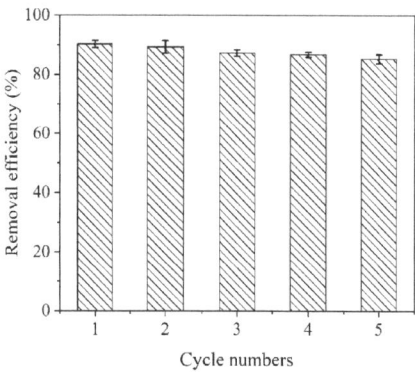

Figure 7. Removal efficiency of the regeneration cycle.

3. Materials and Methods

3.1. Materials

The corncobs were obtained from the countryside of Wuhan, Hubei province, China. Methylene blue, ferric chloride hexahydrate (FeCl$_3$·6H$_2$O, AR, 98%), potassium hydroxide (KOH, GR, 85%), hydrochloric acid (HCl, 36%), glutaraldehyde (25%), calcium chloride (CaCl$_2$, AR, 96%) and sodium alginate (200 ± 20 mPa s), sodium chloride (NaCl, AR, 99.5%), nitric acid (HNO$_3$), sodium hydroxide (NaOH, AR, 95%) were supplied from Sinopharm Chemical Reagent Co., Ltd (Shanghai, China). All reagents are analytical grade and used directly. All water mentioned in this paper was deionized water (18.2 MΩ cm).

3.2. Preparation of KOH/FeCl$_3$ Modified Corncob Biochar (MCB)

To prepare KOH/FeCl$_3$ modified corncob biochar (MCB), the corncob was washed with deionized water 3 times to remove impurities on its surface. After that, the corncob powder and KOH (mass ratio of corncob powder: KOH = 1:1.87) were mixed with 100 mL 2.5 mol/L FeCl$_3$·6H$_2$O for modification [35]. The mixture was stirred in a 60 °C water bath for 2 h and maintained overnight in an oven at 60 °C. Subsequently, the corncob powder was pyrolyzed in a tubular furnace at 800 °C for 2 h at a heating rate of 10 °C/min [51]. During the pyrolysis process, N_2 was used as a protective gas at a flow rate of 200 mL/min. Finally, after cooling to room temperature, the biochar was washed with 1 mol/L HCl, and then with distilled water at room temperature up to neutral pH. The final sample was dried in an oven at 60 °C. The resulting modified biochar was denoted by MCB, and the unmodified biochar was denoted as CB.

3.3. Preparation of MCB/ALG Composite Beads

Generally, 0.2 g SA was dissolved in 10 mL deionized water. MCB (0.2, 0.4, 0.6 g) was added, and the mixture was stirred to obtain a homogeneous system. Then, the above mixture was poured slowly into a syringe (1 mL) and was injected into 1% $CaCl_2$ solution at a uniform rate for crosslinking of 12 h. The beads were rinsed repeatedly with deionized water to remove Ca^{2+} and residual MCB particles. Subsequently, the obtained beads were immersed into a mixture solution containing 1 wt% glutaraldehyde for 12 h with continuously shaking and were rinsed 5–6 times with deionized water. The resulting sample was denoted MCB/ALG−1, MCB/ALG−2 and MCB/ALG−3, according to the mass ratio of MCB and SA in the modified biochar/alginate composite beads. Finally, the prepared MCB/ALG were freeze-dried. Figure 8 illustrates the process of MCB/ALG preparation.

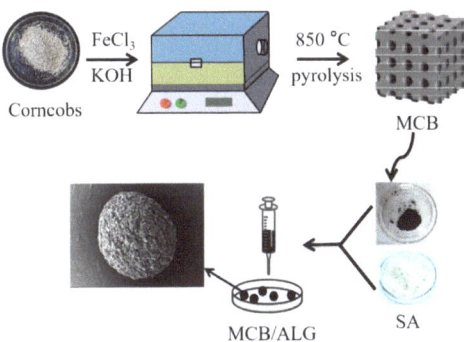

Figure 8. Schematic illustration of the formation of MCB/ALG.

3.4. Characteristics of Samples

The surface morphology and particle size of adsorbents were studied by scanning electron microscopy (SEM, Hitachi, S-3000N, Tokyo, Japan). Fourier transform infrared spectroscopy (FTIR, Thermo Scientific, Nicolet IS10, Waltham, MA, USA) recorded chemical constituents of adsorbents in the range from 4000 to 400 cm^{-1}. The specific surface area, pore size and pore volume of samples were measured by the Brunauer–Emmet–Teller (BET, Micromeritics, ASAP2460, Atlanta, GA, USA). The crystalline state of the adsorbents was analyzed by X-ray diffraction (XRD, Shimadzu, XRD-7000, Kyoto, Japan). The point of zero charge (pH_{PZC}) for MCB-ALG beads was measured by the batch equilibrium method. The pH values of 0.1 and 0.01 mol/L NaCl solutions were adjusted from 3.0 to 12.0 by adding HNO_3 or NaOH solution. The solution was puffed with nitrogen at room temperature to remove dissolved carbon dioxide until the initial pH (pH_i) was stable. Then, 0.1 g of MCB/ALG−1 was introduced into 50 mL, respectively, and the suspension was shaken for 24 h. The final pH values (pH_f) of supernatant were recorded. The difference between the initial and final pH values ($\Delta pH = pH_f - pH_i$) was plotted against the pH_i. The point of intersection of the resulting curve at which $\Delta pH = 0$ was the pH_{PZC}.

3.5. Batch Adsorption Experiments

The stock solution of MB dye with a concentration of 10,000 mg/L was prepared and was diluted to obtain the MB solution required for the experiment. Each sample was tested three times under the same conditions, and the average value of the results was taken.

Ten milligrams of adsorbent (MCB, MCB/ALG) was placed into a centrifuge tube containing 10 mL MB solution and agitated at 170 rpm on a shaker. After stirring at 25 °C for 24 h, the supernatant was separated by a 0.45 μm membrane filter. The concentration of MB in the supernatant was measured by UV–visible spectrophotometer (Beijing general analytical Instrument, T6, China) at a wavelength of 664 nm. The adsorption capacity (q_e, mg/g) and removal efficiency (R, %) were calculated according to Equations (10) and (11).

$$q_e = \frac{(C_0 - C_e)V}{m} \tag{10}$$

$$R = \frac{(C_0 - C_e)}{C_0} \times 100\% \tag{11}$$

where C_0 and C_e are the initial and equilibrium concentrations of MB, respectively, m (g) represents the dosage of adsorbent, V (L) is the MB solution volume.

In order to study the effect of pH value on the removal efficiency of MB, the pH value of solution was changed from 3 to 12. The initial pH value of the MB solution measured by a pH meter and adjusted by using 0.1 mol/L HNO_3 and NaOH (aq). For adsorption kinetics experiments, the MB solution with a concentration of 1200 mg/L was shaken at 25 °C for a range of 10 to 2000 min. Adsorption isotherm experiments was tested at the MB solution concentration of ranging from 25 to 4000 mg/L at 303, 313 and 323 K. The thermodynamic studies were conducted at different temperatures under the same experimental conditions.

3.6. Regeneration Experiment

For desorption and regeneration studies, experiments were performed in 10 mL of MB solution (1200 mol/L) with 10 mg MCB/ALG−1 and shaken at 25 °C for 24 h. After saturation, the spent MCB/ALG−1 beads were washed with 50 mL solution containing 1 mol/L HNO_3 for 12 h. Then, MCB/ALG−1 beads were washed thoroughly with deionized water and dried at 40 °C. The adsorption–desorption experiment was repeated five times.

4. Conclusions

In this study, biochar/alginate composite beads from waste corncobs were successfully fabricated by integrating a pyrolysis process and $KOH/FeCl_3$ modification. By comparing the properties and adsorption performance of MCB/ALG with different KOH modified ratios, it was found that the MCB/ALG−1 sample had the most graphitized structure and the largest adsorption capacity, with maximum adsorption capacities of 1373.49 mg/g at 303 K. The adsorption followed the pseudo-second-order kinetic model and the Langmuir isotherm model. Both the intra-particle diffusion model and the liquid film diffusion were involved in the adsorption process, and the film diffusion was the main rate-limiting step. The negative values of ΔG and the positive value of ΔH confirmed the spontaneous and endothermic nature of MB adsorption on MCB/ALG−1. In addition, MCB/ALG−1 showed the excellent adsorption capacities and considerable reusability. Therefore, MCB/ALG−1 can be used as a promising adsorbent material towards the treatment of MB dye-polluted water or wastewater.

Author Contributions: H.L.: Writing—original draft, Investigation, Validation and Formal analysis; J.Z.: Investigation, Methodology, Funding acquisition. Q.L.: Methodology, Funding acquisition; L.L.: Visualization, Investigation and Methodology; Y.H.: Investigation and Formal analysis; G.F.: Investigation; Y.W.: Formal analysis, Investigation, Methodology, Funding acquisition, Project administration; L.Z.: Conceptualization, Writing—original draft, and acquisition of funding. All authors have read and agreed to the published version of the manuscript.

Funding: This work was supported by the Research and Innovation Initiatives of WHPU (2022Y16), the Scientific Research Foundation of Wuhan Polytechnic University (118-53210052171, 118-53210052144 and 118-53210052136) and the National Key Research and Development Program of China (2020YFE0205300).

Institutional Review Board Statement: Not applicable.

Informed Consent Statement: Not applicable.

Data Availability Statement: Compound data sets are publicly available. Samples are available from the authors.

Acknowledgments: The authors would like to thank Chengzhang Liu from Shiyanjia Lab (www.shiyanjia.com (accessed on 8 February 2022)) for the XRD and BET analysis.

Conflicts of Interest: The authors declare they have no competing interests.

References

1. Xu, H.; Yang, B.; Liu, Y.; Li, F.; Shen, C.; Ma, C.; Tian, Q.; Song, X.; Sand, W. Recent advances in anaerobic biological processes for textile printing and dyeing wastewater treatment: A mini-review. *World J. Microbiol. Biotechnol.* **2018**, *34*, 165. [CrossRef]
2. Ulu, A.; Alpaslan, M.; Gultek, A.; Ates, B. Eco-friendly chitosan/κ-carrageenan membranes reinforced with activated bentonite for adsorption of methylene blue. *Mater. Chem. Phys.* **2022**, *278*, 125611. [CrossRef]
3. Fayazi, M.; Taher, M.A.; Afzali, D.; Mostafavi, A. Enhanced Fenton-like degradation of methylene blue by magnetically activated carbon/hydrogen peroxide with hydroxylamine as Fenton enhancer. *J. Mol. Liq.* **2016**, *216*, 781–787. [CrossRef]
4. Shi, J.; Bai, X.; Xu, L.; Jin, X.; Shi, X.; Jin, P. Facile preparation of Fe-C_3N_4 heterojunction for enhanced pollutant degradation in Fenton-like process. *J. Water Process. Eng.* **2022**, *46*, 102628. [CrossRef]
5. Sun, J.; Yu, M.; Kang, R.; Sun, H.; Zhang, Y.; Wang, N. Self-assembled graphene aerogels for removal of methylene blue and copper from aqueous solutions. *J. Hazard. Mater.* **2021**, *4*, 100026. [CrossRef]
6. Kumar, K.V.; Ramamurthi, V.; Sivanesan, S. Modeling the mechanism involved during the sorption of methylene blue onto fly ash. *J. Colloid Interface Sci.* **2005**, *284*, 14–21. [CrossRef]
7. Makhado, E.; Pandey, S.; Nomngongo, P.N.; Ramontja, J. Fast microwave-assisted green synthesis of xanthan gum grafted acrylic acid for enhanced methylene blue dye removal from aqueous solution. *Carbohydr. Polym.* **2017**, *176*, 315–326. [CrossRef]
8. Zhang, Y.; Tan, H.; Wang, C.; Li, B.; Yang, H.; Hou, H.; Xiao, C. TiO_2-coated glass hollow fiber membranes: Preparation and application for photocatalytic methylene blue removal. *J. Eur. Ceram. Soc.* **2022**, *42*, 2496–2504. [CrossRef]
9. Wang, K.S.; Wei, M.C.; Peng, T.H.; Li, H.C.; Chao, S.J.; Hsu, T.F.; Lee, H.S.; Chang, S.H. Treatment and toxicity evaluation of methylene blue using electrochemical oxidation, fly ash adsorption and combined electrochemical oxidation-fly ash adsorption. *J. Environ. Manag.* **2010**, *91*, 1778–1784. [CrossRef]
10. Tang, X.; Ran, G.; Li, J.; Zhang, Z.; Xiang, C. Extremely efficient and rapidly adsorb methylene blue using porous adsorbent prepared from waste paper: Kinetics and equilibrium studies. *J. Hazard. Mater.* **2021**, *402*, 123579. [CrossRef]
11. Wang, Z.; Li, Y.; Xie, X.; Wang, Z. Bifunctional $MnFe_2O_4$/chitosan modified biochar composite for enhanced methyl orange removal based on adsorption and photo-Fenton process. *Colloids Surf. A Physicochem. Eng. Asp.* **2021**, *613*, 126104. [CrossRef]
12. Zhang, L.; Zhang, F.; Yang, X.; Long, G.; Wu, Y.; Zhang, T.; Leng, K.; Huang, Y.; Ma, Y.; Yu, A.; et al. Porous 3D graphene-based bulk materials with exceptional high surface area and excellent conductivity for supercapacitors. *Sci. Rep.* **2013**, *3*, 1408. [CrossRef]
13. Wang, F.; Zhang, L.; Wang, Y.; Liu, X.; Rohani, S.; Lu, J. Fe_3O_4@SiO_2@CS-TETA functionalized graphene oxide for the adsorption of methylene blue(MB) and Cu(II). *Appl. Surf. Sci.* **2017**, *420*, 970–981. [CrossRef]
14. Momina; Mohammad, S.; Suzylawati, I. Study of the adsorption/desorption of MB dye solution using bentonite adsorbent coating. *J. Water Process. Eng.* **2020**, *34*, 101155. [CrossRef]
15. Ullah, N.; Ali, Z.; Ullah, S.; Khan, A.S.; Adalat, B.; Nasrullah, A.; Alsaadi, M.; Ahmad, Z. Synthesis of activated carbon-surfactant modified montmorillonite clay-alginate composite membrane for methylene blue adsorption. *Chemosphere* **2022**, *309*, 136623. [CrossRef]
16. Rahmi; Ishmaturrahmi; Mustafa, I. Methylene blue removal from water using H_2SO_4 crosslinked magnetic chitosan nanocomposite beads. *Microchem. J.* **2019**, *144*, 397–402. [CrossRef]
17. Chen, J.; Ouyang, J.; Cai, X.; Xing, X.; Zhou, L.; Liu, Z.; Cai, D. Removal of ciprofloxacin from water by millimeter-sized sodium alginate/H_3PO_4 activated corncob-based biochar composite beads. *Sep. Purif. Technol.* **2021**, *276*, 119371. [CrossRef]
18. Dilamian, M.; Noroozi, B. Rice straw agri-waste for water pollutant adsorption: Relevant mesoporous super hydrophobic cellulose aerogel. *Carbohydr. Polym.* **2021**, *251*, 117016. [CrossRef]
19. Liu, Z.; Xu, Z.; Xu, L.; Buyong, F.; Chay, T.C.; Li, Z.; Cai, Y.; Hu, B.; Zhu, Y.; Wang, X. Modified biochar: Synthesis and mechanism for removal of environmental heavy metals. *Carbon Res.* **2022**, *1*, 8. [CrossRef]
20. Lucaci, A.R.; Bulgariu, D.; Ahmad, I.; Lisă, G.; Mocanu, A.M.; Bulgariu, L. Potential use of biochar from various waste biomass as biosorbent in Co(II) removal processes. *Water* **2019**, *11*, 1565. [CrossRef]
21. Ouyang, J.; Chen, J.; Chen, W.; Zhou, L.; Cai, D.; Ren, C. H_3PO_4 activated biochars derived from different agricultural biomasses for the removal of ciprofloxacin from aqueous solution. *Particuology* **2023**, *75*, 217–227. [CrossRef]

22. Chen, H.; Gao, Y.; Li, J.; Fang, Z.; Bolan, N.; Bhatnagar, A.; Gao, B.; Hou, D.; Wang, S.; Song, H.; et al. Engineered biochar for environmental decontamination in aquatic and soil systems: A review. *Carbon Res.* **2022**, *1*, 8. [CrossRef]
23. Luo, M.; Lin, H.; Li, B.; Dong, Y.; He, Y.; Wang, L. A novel modification of lignin on corncob-based biochar to enhance removal of cadmium from water. *Bioresour. Technol.* **2018**, *259*, 312–328. [CrossRef]
24. Liu, C.; Zhang, H.X. Modified-biochar adsorbents(MBAs) for heavy-metal ions adsorption: A critical review. *J. Environ. Chem. Eng.* **2022**, *10*, 107393. [CrossRef]
25. Chen, H.; Yang, X.; Liu, Y.; Lin, X.; Wang, J.; Zhang, Z.; Li, N.; Li, Y.; Zhang, Y. KOH modification effectively enhances the Cd and Pb adsorption performance of N-enriched biochar derived from waste chicken feathers. *Waste Manag.* **2021**, *130*, 82–92. [CrossRef]
26. Ma, Y.; Chen, S.; Qi, Y.; Yang, L.; Wu, L.; He, L.; Li, P.; Qi, X.; Gao, F.; Ding, Y.; et al. An efficient, green and sustainable potassium hydroxide activated magnetic corn cob biochar for imidacloprid removal. *Chemosphere* **2022**, *291*, 132707. [CrossRef]
27. Bedia, J.; Belver, C.; Ponce, S.; Rodriguez, J.; Rodriguez, J.J. Adsorption of antipyrine by activated carbons from FeCl$_3$-activation of Tara gum. *Chem. Eng. J.* **2018**, *333*, 58–65. [CrossRef]
28. Wu, S.; Zhao, X.; Li, Y.; Zhao, C.; Du, Q.; Sun, J.; Wang, Y.; Peng, X.; Xia, Y.; Wang, Z.; et al. Adsorption of ciprofloxacin onto biocomposite fibers of graphene oxide/calcium alginate. *Chem. Eng. J.* **2013**, *230*, 389–395. [CrossRef]
29. Mao, W.; Yue, W.; Xu, Z.; Chang, S.; Hu, Q.; Pei, F.; Huang, X.; Zhang, J.; Li, D.; Liu, G.; et al. Development of a synergistic activation strategy for the pilot-scale construction of hierarchical porous graphitic carbon for energy storage applications. *ACS Nano* **2020**, *14*, 4741–4754. [CrossRef]
30. Thommes, M.; Kaneko, K.; Neimark, A.V.; Olivier, J.P.; Rodriguez-Reinoso, F.; Rouquerol, J.; Sing, K.S.W. Physisorption of gases, with special reference to the evaluation of surface area and pore size distribution(IUPAC Technical Report). *Pure Appl. Chem.* **2015**, *87*, 1051–1069. [CrossRef]
31. Zhong, Z.Y.; Yang, Q.; Li, X.M.; Luo, K.; Liu, Y.; Zeng, G.M. Preparation of peanut hull-based activated carbon by microwave-induced phosphoric acid activation and its application in Remazol Brilliant Blue R adsorption. *Ind. Crops Prod.* **2012**, *37*, 178–185. [CrossRef]
32. Liu, C.; Wang, W.; Wu, R.; Liu, Y.; Lin, X.; Kan, H.; Zheng, Y. Preparation of acid-and alkali-modified biochar for removal of methylene blue pigment. *ACS Omega* **2020**, *5*, 30906–30922. [CrossRef]
33. Zhou, X.Y.; Xie, F.; Jiang, M.; Ke-Ao, L.; Tian, S.G. Physicochemical properties and lead ion adsorption of biochar prepared from Turkish gall residue at different pyrolysis temperatures. *Microsc. Res. Tech.* **2021**, *84*, 1003–1011. [CrossRef]
34. Yin, Z.; Xu, S.; Liu, S.; Xu, S.; Li, J.; Zhang, Y. A novel magnetic biochar prepared by K$_2$FeO$_4$-promoted oxidative pyrolysis of pomelo peel for adsorption of hexavalent chromium. *Bioresour. Technol.* **2020**, *300*, 122680. [CrossRef]
35. Cheng, L.; Ji, Y.; Liu, X.; Mu, L.; Zhu, J. Sorption mechanism of organic dyes on a novel self-nitrogen-doped porous graphite biochar: Coupling DFT calculations with experiments. *Chem. Eng. Sci.* **2021**, *242*, 116739. [CrossRef]
36. Liu, X.J.; Li, M.F.; Ma, J.-F.; Bian, J.; Peng, F. Chitosan crosslinked composite based on corncob lignin biochar to adsorb methylene blue: Kinetics, isotherm, and thermodynamics. *Colloids Surf. A Physicochem. Eng. Asp.* **2022**, *642*, 128621. [CrossRef]
37. Lin, Q.; Gao, M.; Chang, J.; Ma, H. Adsorption properties of crosslinking carboxymethyl cellulose grafting dimethyldiallylammonium chloride for cationic and anionic dyes. *Carbohydr. Polym.* **2016**, *151*, 283–294. [CrossRef]
38. Yan, L.; Liu, Y.; Zhang, Y.; Liu, S.; Wang, C.; Chen, W.; Liu, C.; Chen, Z.; Zhang, Y. ZnCl$_2$ modified biochar derived from aerobic granular sludge for developed microporosity and enhanced adsorption to tetracycline. *Bioresour. Technol.* **2020**, *297*, 122381. [CrossRef]
39. Qu, J.; Wang, Y.; Tian, X.; Jiang, Z.; Deng, F.; Tao, Y.; Jiang, Q.; Wang, L.; Zhang, Y. KOH-activated porous biochar with high specific surface area for adsorptive removal of chromium(VI) and naphthalene from water: Affecting factors, mechanisms and reusability exploration. *J. Hazard. Mater.* **2021**, *401*, 123292. [CrossRef]
40. Huang, X.Y.; Bu, H.-T.; Jiang, G.B.; Zeng, M.H. Cross-linked succinyl chitosan as an adsorbent for the removal of Methylene Blue from aqueous solution. *Int. J. Biol. Macromol.* **2011**, *49*, 643–651. [CrossRef]
41. Zhang, Z.; Li, H.; Liu, H. Insight into the adsorption of tetracycline onto amino and amino-Fe^{3+} gunctionalized mesoporous silica: Effect of functionalized groups. *J. Environ. Sci.* **2018**, *65*, 171–178. [CrossRef]
42. Wang, H.; Zhong, D.; Xu, Y.; Chang, H.; Shen, H.; Xu, C.; Mou, J.; Zhong, N. Enhanced removal of Cr(VI) from aqueous solution by nano- zero-valent iron supported by KOH activated sludge-based biochar. *Colloids Surf. A Physicochem. Eng. Asp.* **2022**, *651*, 129697. [CrossRef]
43. Jamion, N.A.; Hashim, I. Preparation of activated carbon from tamarind seeds and Methylene blue(MB) removal. *J. Fundam. Appl. Sci.* **2018**, *9*, 102. [CrossRef]
44. Yu, M.; Li, J.; Wang, L. KOH-activated carbon aerogels derived from sodium carboxymethyl cellulose for high-performance supercapacitors and dye adsorption. *Chem. Eng. J.* **2017**, *310*, 300–306. [CrossRef]
45. Liu, Q.S.; Zheng, T.; Li, N.; Wang, P.; Abulikemu, G. Modification of bamboo-based activated carbon using microwave radiation and its effects on the adsorption of methylene blue. *Appl. Surf. Sci.* **2010**, *256*, 3309–3315. [CrossRef]
46. Islam, M.A.; Ahmed, M.J.; Khanday, W.A.; Asif, M.; Hameed, B.H. Mesoporous activated carbon prepared from NaOH activation of rattan(Lacosperma secundiflorum) hydrochar for methylene blue removal. *Ecotoxicol. Environ. Saf.* **2017**, *138*, 279–385. [CrossRef]
47. Reffas, A.; Bernardet, V.; David, B.; Reinert, L.; Lehocine, M.B.; Dubois, M.; Batisse, N.; Duclaux, L. Carbons prepared from coffee grounds by H$_3$PO$_4$ activation: Characterization and adsorption of methylene blue and Nylosan Red N-2RBL. *J. Hazard. Mater.* **2010**, *175*, 779–788. [CrossRef]

48. Gokce, Y.; Aktas, Z. Nitric acid modification of activated carbon produced from waste tea and adsorption of methylene blue and phenol. *Appl. Surf. Sci.* **2014**, *313*, 352–359. [CrossRef]
49. Yu, Y.; Wan, Y.; Shang, H.; Wang, B.; Zhang, P.; Feng, Y. Corncob-to-xylose residue(CCXR) derived porous biochar as an excellent adsorbent to remove organic dyes from wastewater. *Surf. Interface Anal.* **2019**, *51*, 234–245. [CrossRef]
50. Andersson, K.I.; Eriksson, M.; Norgren, M. Removal of lignin from wastewater generated by mechanical pulping using activated charcoal and fly ash: Adsorption isotherms and thermodynamics. *Ind. Eng. Chem. Res.* **2011**, *50*, 7722–7732. [CrossRef]
51. Cheng, D.; Ngo, H.H.; Guo, W.; Chang, S.W.; Nguyen, D.D.; Zhang, X.; Varjani, S.; Liu, Y. Feasibility study on a new pomelo peel derived biochar for tetracycline antibiotics removal in swine wastewater. *Sci. Total Environ.* **2020**, *720*, 137662. [CrossRef]

Disclaimer/Publisher's Note: The statements, opinions and data contained in all publications are solely those of the individual author(s) and contributor(s) and not of MDPI and/or the editor(s). MDPI and/or the editor(s) disclaim responsibility for any injury to people or property resulting from any ideas, methods, instructions or products referred to in the content.

Review

Agricultural Solid Wastes Based Adsorbent Materials in the Remediation of Heavy Metal Ions from Water and Wastewater by Adsorption: A Review

Tushar Kanti Sen

Chemical Engineering Department, College of Engineering, King Faisal University, P.O. Box 380, Al-Ahsa 31982, Saudi Arabia; tsen@kfu.edu.sa

Abstract: Adsorption has become the most popular and effective separation technique that is used across the water and wastewater treatment industries. However, the present research direction is focused on the development of various solid waste-based adsorbents as an alternative to costly commercial activated carbon adsorbents, which make the adsorptive separation process more effective, and on popularising the sustainable options for the remediation of pollutants. Therefore, there are a large number of reported results available on the application of raw or treated agricultural biomass-based alternatives as effective adsorbents for aqueous-phase heavy metal ion removal in batch adsorption studies. The goal of this review article was to provide a comprehensive compilation of scattered literature information and an up-to-date overview of the development of the current state of knowledge, based on various batch adsorption research papers that utilised a wide range of raw, modified, and treated agricultural solid waste biomass-based adsorbents for the adsorptive removal of aqueous-phase heavy metal ions. Metal ion pollution and its source, toxicity effects, and treatment technologies, mainly via adsorption, have been reviewed here in detail. Emphasis has been placed on the removal of heavy metal ions using a wide range of agricultural by-product-based adsorbents under various physicochemical process conditions. Information available in the literature on various important influential physicochemical process parameters, such as the metal concentration, agricultural solid waste adsorbent dose, solution pH, and solution temperature, and importantly, the adsorbent characteristics of metal ion removal, have been reviewed and critically analysed here. Finally, from the literature reviewed, future perspectives and conclusions were presented, and a few future research directions have been proposed.

Keywords: heavy metal pollution; agricultural solid waste-based adsorbents; remediation via adsorption

Citation: Sen, T.K. Agricultural Solid Wastes Based Adsorbent Materials in the Remediation of Heavy Metal Ions from Water and Wastewater by Adsorption: A Review. *Molecules* 2023, 28, 5575. https://doi.org/10.3390/molecules28145575

Academic Editors: Yongchang Sun and Dimitrios Giannakoudakis

Received: 5 June 2023
Revised: 2 July 2023
Accepted: 12 July 2023
Published: 21 July 2023

Copyright: © 2023 by the author. Licensee MDPI, Basel, Switzerland. This article is an open access article distributed under the terms and conditions of the Creative Commons Attribution (CC BY) license (https://creativecommons.org/licenses/by/4.0/).

1. Introduction

The sustainable and cost-effective remediation of water pollutants to produce clean water is a challenging task for scientists, researchers, and engineers worldwide. As per the United Nations World Water Development Report in 2020 [1], around four billion people face severe water scarcity for at least one month per year [2]. Environmental water contamination due to the large release of various potential and toxic pollutants from human activities, such as increased industrialisation, urbanisation, populations, and agricultural activities, into water bodies present a high risk to human life and aquatic environments [3]. The commonly found heavy metals ions include Cu^{2+}, Ni^{2+}, Zn^{2+}, Cd^{2+}, Pb^{2+}, and Hg^{2+} ions [3–5]. Among them, Cd^{2+}, Pb^{2+}, Hg^{2+}, and As^{3+} ions are the most dangerous heavy metal ions that have been identified by the World Health Organisation (WHO) [5]. Heavy metals are not biodegradable and are carcinogenic in nature. About 40% of Earth's surface water, comprising mainly river and lake water, is being polluted by heavy metal ions primarily from the industrial and agricultural activities [6].

The major sources of heavy metal ion pollution comprise discharge from various untreated industrial effluents from refineries, coal-fired power plants, mining industries, alumina refineries, metallurgical industries, heavy chemicals, chloro-alkali industries, battery industries, dyes and pigments, fertilisers, metal smelters, paints and ceramics, tanneries, textiles, etc. [3,7]. The metal ions Cd^{2+}, Pb^{2+}, Hg^{2+} Cr^{3+}, Cu^{2+}, Mn^{2+}, Fe^{3+}, and Zn^{2+} are significantly toxic and pose risks to both humans and the environment [7–9]. There are various adverse health effects, such as diarrhoea, disorderedness, stomach problems, paralysis, various skin deceases, haemoglobinuria, vomiting, etc., that occur due to heavy metal ion contamination [3,10]. Heavy metal ions are highly toxic, hazardous to health, and non-biodegradable, and pose a high threat to the ecosystem if they are left untreated [3,11,12]. Therefore, there is an urgent need to develop ecofriendly, economically feasible technology to remove these potential pollutants from the aqueous phase [11]. Detailed information on heavy metal ion classifications, sources, and their toxicity effects have been detailed in our own previous publications [3].

A number of conventional technologies, such as chemical precipitation, oxidation, advanced oxidation, coagulation/flocculation, electrocoagulation, photo catalysis, membrane processes, reverse osmosis (RO), filtration, adsorption, solvent extraction, electroplating, ion exchange, activated sludge, and aerobic and anaerobic treatment, have been used to remove these potential pollutants from water and wastewater with varying levels of success [2,3,13–20]. All these treatment technologies have their own advantages and disadvantages. Various researchers, including the current author's in their own reported review publication, have critically discussed the advantages and disadvantages of these different metal ion treatment technologies [7,19,21–27]. Among these methods, adsorption-based separation technology is one of the most effective but widely used treatment technologies for heavy metal-contaminated water and wastewater. This is due to its simple operation, design simplicity, high separation efficiency, efficiency at lower pollutant concentrations, high selectivity at the molecular level, low energy consumption, and ability to separate multiple pollutant components with minimal secondary pollution, making it a form of sustainable development [3,20,21,28]. In their previous review article, Afroze and Sen [3] reported statistical data (Figure 1) on the increasing trend of published research papers on inorganic and organic adsorption using various adsorbents since the year of 1995. The adsorptive removal of heavy metals from water and wastewater has become an essential and widely used separation technique in recent times [12].

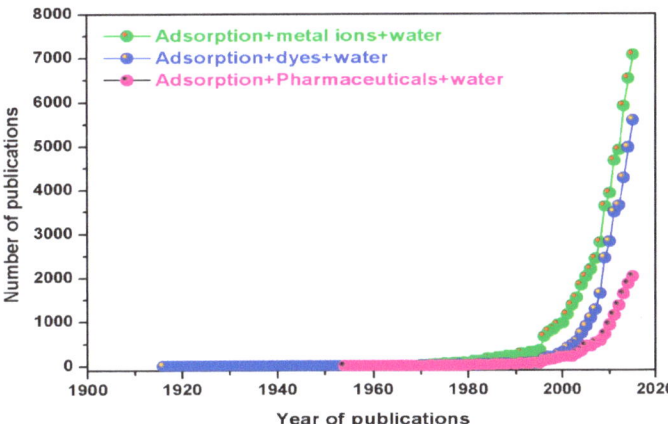

Figure 1. Number of adsorption publications for metal ions and organics removal. Source: taken from [3] with written permission.

Adsorption may be defined as the transfer of one or more solute molecules from the bulk fluid phase to the solid adsorbent surface and getting retained there. The solid that adsorbs a solute component is called the adsorbent, and the solute component that is adsorbed is termed the adsorbate. When the adsorption arises as a result of weak Van der Waals or short-range forces, it is called physical adsorption. In contrast, in the case of chemical adsorption, a chemical covalent or ionic bond formation takes place between the adsorbate and the adsorbent via electronic transfer, which is irreversible in nature [7]. There are three major steps involved at the solid/liquid interface of the adsorption process. These mechanistic steps are as follows: (a) diffusion of the solute adsorbate from the bulk aqueous phase to the surface of the adsorbent by film diffusion; (b) the adsorption at the solid/liquid interface means on the active sites of adsorbent surface; and then (c) the internal diffusion of the solute molecules within the solid adsorbent via pore diffusion or surface diffusion, or both. In simple terms, the adsorption of aqueous phase heavy metal ions involves a solid adsorbent phase and a liquid solvent phase, wherein metal ions are in the dissolved solute adsorbate molecules and are therefore part of the solid/liquid interfacial adsorption separation process. The mechanism underlying this adsorptive separation process involves chemisorption, complexation formation at the solid/liquid interface, adsorption on surface and interior pore structure of the adsorbent, ion exchange, etc., and this is due to the presence of the mass transfer concentration gradient and diffusional processes [3,14,29].

To predict the rate of adsorption and to identify the mechanisms underlying adsorption and the adsorbent's capacity, it is vital to understand the various reported adsorption kinetic models and isotherm model equations [3,29]. In terms of the adsorption process design, the determination of various kinetic parameters is a particularly critical design parameter. Numerous kinetic models, such as the first-order and second-order reversible or irreversible kinetic models, along with the pseudo-first-order or pseudo-second-order adsorption models, have been reported and applied to batch adsorption experimental results by various researchers [2,3,30,31]. The most reported kinetic models are the pseudo-first order (PFO) and pseudo-second order (PSO) kinetic models, in which batch experimental data are fitted to these PFO and PSO models. In their critical review article, Tan and Hameed [30] mentioned that Ho [32] reviewed the applications of second-order models for adsorption systems, while Liu and Liu [33] summarised the useful kinetic models for biosorption. Surface reaction mechanism-based adsorption models have been reviewed by Plazinski et al. [34]. Alberti et al. discussed the batch and dynamic adsorption models [35]. Afroze and Sen [3] presented a compilation of reported batch adsorption results on the applicability of pseudo-second-order kinetic models for heavy metal and dye adsorption using several agricultural solid wastes, and readers are encouraged to go through this review article.

Adsorption isotherm studies are crucial for understanding the mechanisms of adsorption and for finding the maximum adsorption capacity of the adsorbent. Several adsorption isotherm models have been reported in the literature, such as the Langmuir, Freundlich, Redlich–Peterson, Tempkin, and Toth isotherm models. Of these isotherm models, the Freundlich 1906 [36] and Langmuir (1918) [37] models have been widely used in the evaluation of the adsorption process. From this research, readers are encouraged to go through the review article by Afreza and Sen [3], where the applicability of various isotherm models on the batch heavy metal adsorption process using wide ranges of agricultural solid waste-based adsorbents have been reported.

The adsorption process depends on the nature and the types of the adsorbent and adsorbate characteristics. The adsorbate characteristics, such as molecular weight, structure, size, charge, and solution concentration, and the adsorbent characteristics, such as particle size, surface area, surface charge, and surface functional groups are all responsible for effective adsorption [22,38]. Apart from these adsorbent-adsorbate characteristics, many physicochemical process parameters, such as the initial metal ion concentration, adsorbent dosage, contact time, solution pH, temperature, and salt concentration all significantly affect the adsorption process [3,4,27]. In the adsorptive separation process, four commonly used important adsorbents are activated carbon, zeolites or molecular sieves, natural inorganic clay minerals, silica gel, and activated alumina [39,40]. However, commercial activated carbon (CAC) is most used in the water and wastewater treatment industry due to its large porous structure, large surface area, high capacity, and the hydrophobic nature of activated carbon [3,12,38]. But coal-based CAC is costly and possesses significant regeneration issues. Therefore, the current focus of research has been shifted towards the use of various carbonaceous, lignocellulosic, and agricultural by-product solid wastes, such as fruit and vegetable wastes, leaves, seeds, tree waste, fibres, fruit peels, dates, sawdust, bark, etc., for the development of an effective adsorbent alternative to costly coal-based activated carbon. Agricultural biomasses materials, like the shells of wheat, orange peels, sunflower leaves, biochar from plant residues, activated carbon from plant residues, wood waste, bark residues, fruit wastes, and manures have been successfully used in heavy metal ion removal from water by adsorption [31]. Figure 2 shows a few agricultural by-products that are cost-effective, and function as alternative adsorbents that can be used in the adsorption of heavy metal ions.

Figure 2. Several examples of agricultural biomass alternative adsorbents that are used in the adsorption of metal ions. Source: taken from [41] with written permission.

Modified agricultural solid waste has been widely used as an effective adsorbent in the removal of various contaminations from wastewater, and this has ben attributed to their surface properties improvements. Raw biomass can be modified using acids, such as hydrochloric, phosphoric, sulfuric, nitric, citric acids etc., or alkaline solutions, such as sodium hydroxide, potassium hydroxide, zinc chloride, calcium chloride, ammonia etc., or cross-linked with other materials [3]. Chemical treatment removes natural fats, waxes, and low-molecular-weight lignin compounds from agricultural adsorbent surfaces. In recent years, the production of activated carbon, biochar, and charcoal from agricultural solid residuals is emerging as an alternative and cost-effective adsorbent with a high selectivity, porosity, and surface area, and these waste materials have naturally been available in large quantities, requires less processing time, are a renewable source, and have little or no commercial value [3,31]. Biochar is produced via the pyrolysis of biomass residues. The production and properties of these valuable adsorbents depend on the production and treatment methods, which are presented in Figure 3. Figure 4 shows a flowchart for the overall adsorption process for the removal of inorganic/organic compounds using agricultural wastes as adsorbents under various physicochemical process conditions.

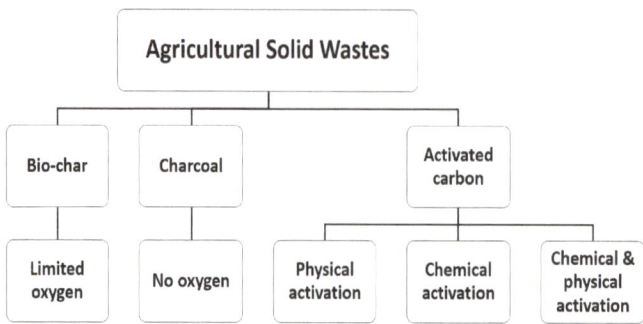

Figure 3. Various treatments of adsorbent materials.

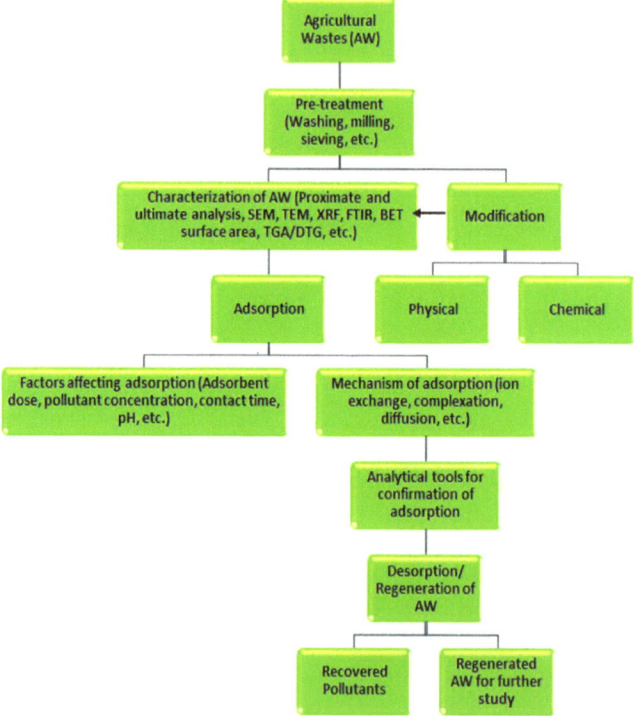

Figure 4. A flowchart presenting the overall adsorption process for pollutant removal from waste water. Source: taken from Ogunlalu et al. [31] with written permission.

In recent times, these agricultural by-products have raised environmental awareness about their safe disposal, and therefore any kind of their utilisation is considered as a win-win situation for effective solid waste management as well. Hence, this review article will provide a comprehensive compilation of all the up-to-date developments of the current state of knowledge on various batch adsorption results using a wide range of raw and modified agricultural solid waste adsorbents in the removal of heavy metal ions from water and wastewater. The significance of this review is not only the compilation and up-to-date developments of the current state of knowledge, but also the critical analysis of the recent research articles that have been published in the directions of agricultural solid waste and modified agricultural solid waste adsorbents. In this review, we have also reported and compiled the various batch heavy metal ion adsorption results under various physicochemical process parameters. Therefore, the structure of this review article began with a general introduction section comprising heavy metal ion water pollution and their sources, toxicity, and treatment methods. Emphasis has been given to agricultural by-product-based adsorbents for the removal of aqueous phase heavy metal ions through adsorption under various process conditions. Finally, the knowledge gap between the future perspectives and the future directions have been presented.

2. Characteristics of the Role of Adsorbents and Agricultural Waste-Based Adsorbents in Heavy Metal Adsorption

The current research has primarily been driven towards using lignocelluloses, and carbonaceous, agricultural, and forest-based adsorbents for water decontamination, including metal decontamination using an adsorption alternative to the costly CAC. These materials are available locally in large quantities and are almost priceless, with a minimum pre-treatment cost for improvements in terms of their effectiveness, efficiency, and environmental friendliness, and are an alternative adsorbent to the costly CAC. Further, agricultural solid waste adsorbent materials require minimal pre-treatment operations, such as washing, drying, grinding, or minor chemical treatments [42]. The adsorption capacity of an adsorbent plays a vital role in the selection of effective adsorbents in the removal of aqueous phase pollutants, which is either determined experimentally or theoretically using various isotherms and kinetic models. The metal adsorption at the solid/liquid interface is highly dependent on many physicochemical process parameters, such as metal ion concentration, solution pH, temperature, adsorbent dose etc., and hence the adsorption capacity was discussed and reviewed in the next section. For example, Gumus et al. [43] reported that the leaf biomass of *Laurus nobilis* is an effective adsorbent in the removal of Cd^{2+}, Cu^{2+}, Pb^{2+}, and Zn^{2+} toxic metal ions from its aqueous solution and strong functions of temperature and solution pH with the adsorption capacity as the pH increases. A theoretical maximum Cr^{6+} adsorption capacity of 70.49 mg/g for data palm empty fruit bunch biomass was obtained at an optimum solution pH of 2 and a temperature of 30 °C [44]. Rice bran and rice straw adsorbents were successfully used to remove aqueous phase Cu^{2+} metal ions and their reported maximum adsorbent capacities were found to be 21 mg/g and 18.4 mg/g, respectively [45]. Similarly, the metal ions Pb^{2+} and Cr^{6+} were also effectively removed from water using the peanut shell residue adsorbent [46,47]. The same peanut shell residue biomass was effectively used to remove the aqueous phase from the Cr^{3+}, Cu^{2+}, and Pb^{2+} ions with an adsorption capacity of 7.7 mg/g, 10.2 mg/g, and 29.1 mg/g, respectively [48]. Afroze et al. [49] successfully developed a eucalyptus bark-based adsorbent for the removal of heavy metal ions from water. Ahmed and Danish [50] reviewed the raw and treated avocado waste-based effective adsorbents used in heavy metal ion removal under various conditions. Anastopoulos et al. [51] reviewed and compiled various coffee adsorbents, such as coffee grounds, coffee residues, spent coffee grains, and coffee husks in the removal of aqueous phase heavy metal ions under various experimental conditions. Hence, while a large number of reported metal adsorption results through various raw or treated/modified agricultural solid waste-based processes have been deemed as effective, cost-effective alternative adsorbents include fruit wastes, such as lemon peel [52], durian peel [53], banana peel, Kuwai peel [54], raw pomegranate peel [55], watermelon shell [56], and coconut coir [57], along with various tree leaves, such as *Artocarpus odoratissimus* leaves [58], and *Colocation esculenta* leaves [59]. All these articles have also reported on the effects of various factors on heavy metal ion adsorption kinetics and equilibrium adsorption by agricultural wastes and their maximum adsorption capacity. Raw and chemically activated various agricultural wastes, such as jackfruit, rice husk, pecan shell, bamboo, pine leaves, pinecone, eucalyptus bark, hazelnut shell, maize cob or husk, castor hull etc., are also reported effective adsorbents in the removal of aqueous phase heavy metal ions [3,31]. There are a couple of reported review articles available in the literature, such as those by Ahmed and Danesh, [50]; Saukat et al. [42]; Ogunlalu et al. [31];

Afroze and Sen; [3]; and Sulyman et al. [60] on aqueous phase heavy metal ion removal through selective agricultural solid waste-derived adsorbents. Their maximum adsorbent capacities have been reported, and readers are encouraged to go through these articles. Table 1 presents the compilation of various reported results on the maximum adsorption capacity of various agricultural by-products in the removal of heavy metals from water during the last 10-year period of 2012–2022.

Table 1. Adsorption capacities q_m (mg/g) of several recently reported raw and modified agricultural waste materials for heavy metal ion adsorption.

Agricultural By-Products Raw and Modified/Treated Adsorbents	Adsorbate Heavy Metal Ions	Maximum Monolayer Adsorption Capacity, q_{max} (mg/g), at Optimum Process Conditions	References
Avocado seed	Cr (VI)	35.5	Ahmet and Danish [50]; Rangel et al. [61]
Jackfruit peels	Cu^{2+} Pb^{2+} Cd^{2+} Mn^{2+}	17.5 10.1 20 76.9	Ibrahim et al. [62]; Ayob et al. [63]
Data palm empty fruit bunch	Cr^{6+}	70.49	Rambabu et al. [44]
Pineapple peel	Cr^{6+}	40	Shakya et al. [64], Yousef et al. [65]
Canola seeds	Pb^{2+} Cd^{2+}	44.25 52.36	Affonso et al. [66]; Ayob et al. [63]
Laurus nobilis leaves	Cu^{2+} Pb^{2+} Cd^{2+} Zn^{2+}	6.04 96.15 8.6 8.74	Gumus et al. [43]; Ogunlalu et al. [31]
Vigna radiata husk biomass	Cu^{2+} Co^{2+} Ni^{2+}	11.05 15.04 19.88	Naseem et al. [67]
Coffee pulp	Cr^{6+}	13.48	Ayob et al. [63]
Cajanus cajan Husk	Cd^{2+}	42.16	Devani et al. [68]; Sazali et al. [69]
Orange peel	Cd^{2+}	170.3	Chen et al. [70]
Litchi peel	Cd^{2+}	230.5	Chen et al. [70]
Date seed biochar	Ni^{2+}	19.54	Mahdi et al. [71]
Avocado peel	Pb (II) Ni (II)	4.93 9.82	Ahmet and Danish [50]; Mallampati, [72]
Modified peanut shell	Hg(II)	30.72	Sulyman et al. [60]
Coconut husk	Cu^{2+} Ni^{2+} Pb^{2+} Zn^{2+}	443.0 404.5 362.2 338.0	Malik and Dahiya [73]
Orange peel	Pb (II)	204	Sulyman et al. [60]
Banana peels	Cu^{2+} Ni^{2+} Pb^{2+}	14.3 27.4 34.5	Thuan et al. [74]; Ayob et al. [63]
Corn straw	Cd^{2+} Pb^{2+}	38.91 28.99	Chi et al. [75], Yousef et al. [65], Yan et al. [76]
Pomegranate peel	Cu^{2+}	30.12	Ben-Ali et al. [55]
Modified activated bamboo	Cd^{2+}	202.55	Zhang et al. [77]; Sazali et al. [69]
Orange peel	Cu^{2+}	63.3	Guiza [78]
Flax fiber tows	Cu^{2+} Pb^{2+} Zn^{2+}	9.92 10.74 8.4	Abbar et al. [79]
Eucalyptus bark	Zn (II)	131.6	Afroze et al. [49]
Banana peel	Cd^{2+} Pb^{2+}	5.71 2.18	Gisi et al. [5]
Sweet potato peel	Pb^{2+}	18	Asuquo et al. [80]

Table 1. Cont.

Agricultural By-Products Raw and Modified/Treated Adsorbents	Adsorbate Heavy Metal Ions	Maximum Monolayer Adsorption Capacity, q_{max} (mg/g), at Optimum Process Conditions	References
Peanut husk	Ni^{2+}	56.82	Abdelfattah et al. [81]
Orange peel	Hg^{2+}	7.46	Chinyelu [82]
Tomato leaf	Ni (II)	58.8	Gutha et al. [83]
Rapeseed waste	Zn (II)	13.9	Paduraru et al. [84]
Jackfruit leaf	Ni (II)	11.5	Boruah et al. [85]
Sorghum hulls	Cu^{2+}	148.93	Imaga, Abia et al. [86]
Coffee residues	Pb^{2+}, Zn^{2+}	9.7 (Pb^{2+}), 4.4 (Zn^{2+})	Wu, Kuo et al. [28], Utomo and Hunter [87]
Modified Okra biomass	Cu^{2+}, Zn^{2+}, Cd^{2+}, Pb^{2+}	72.72 (Cu^{2+}), 57.11 (Zn^{2+}), 121.51 (Cd^{2+}), 273.97 (Pb^{2+})	Singha and Guleria [88]
Sugarcane bagasse	Mn^{2+}	0.423	Anastopoulos et al. [51]
Sugarcane bagasse	Cd^{2+}	0.955	Moubarik and Grimi [89], Anastopoulos et al. [51]
Peanut shell	Pb^{2+}	39	Tasar et al. [47]
Pistachio hull waste	Hg^{2+}	48.78	Rajamohan [90]
Coconut tree sawdust	Cu (II) Pb (II) Zn (II)	3.9 25.0 23.8	Putra et al. [91]
Modified rice husk	Hg^{2+}	89	Song et al. [92], Yousef et al. [65]
Modified Sugarcane bagasse	Cu^{2+}	30.9	Rana et al. [17]
Garcinia cambogia plants	As	704.11	Gautam et al. [93]
Oryza sativa plants	Cd^{2+}	20.70	Gautam et al. [93]
Corn stover	Cr^{2+}	84	Gautam et al. [93]
Palm tree branches	Cr^{+4}	157	Guat et al. [2]
Egyptian mandarin peel (raw)	Hg^{2+}	19.01	Husein et al. [94]; Gisi et al. [5]
Raw sugarcane bagasse	Hg^{2+}	35.71	Khovamzadeh et al. [95]; Anastopoulos et al. [51]
Orange peel	Cu^{2+}, Pb^{2+}, Zn^{2+}	70.73 (Cu^{2+}), 209.8 (Pb^{2+}) and 56.18 (Zn^{2+})	Feng and Guo [16] Gomez-Al [96]
Barley straw (raw)	Cu^{2+}	4.64	Gisi et al. [5]
Garden grass (raw)	Pb^{2+}	58.34	Gisi et al. [5]

The effectiveness and adsorbent capacity depend on the adsorbent's size, shape, and morphological and chemical structure, including surface characteristics such as the surface area, pore volume, point of zero charge (pH_{pzc}), bulk density, and the presence of surface functional groups [49,97]. The presence of surface functional groups in agricultural by-product adsorbent surfaces, such as carbonyl, phenolic, acetamido, alcoholic, amino groups etc., undergo strong interactions with heavy metal ions under physicochemical process conditions to form metal complexes or chelates. Adsorption is a reaction, and the rate of adsorption increases with the adsorbent surface area, shape, and surface charge, respectively. Table 2 represents the effects of various adsorbent characteristic parameters on heavy metal ion adsorption from some of the more recently published research articles [3].

Table 2. The effects of several agricultural solid waste-based adsorbent characteristics on heavy metal ion adsorption.

Adsorbents	Contaminants (Heavy Metals and Dyes)	Characterisation Properties					References
		Specific Surface Area/BET(m^2/g)	Particle Size Distribution	Elemental Analysis (%)	FTIR Analysis	pH_{pzc}	
Pinecone	Cd^{2+}, Cu^{2+}, Pb^{2+}	0.2536	50 μm	-	O-H, C-H, -CH_2, C=O	6.2	Dawood et al. [97] Marawa et al. [98]
Avocado seed	Cr^{6+}	1.75	0.1–1.5 mm		O-H group -CH_2 stretching	6.4	Bazzo et al. [99]; Leite et al. [100] Garcia and Cristiani-Urbina, [101]
HAS avocado shell?	Ni^{2+}		-	43.13 (carbon), 7.17 (hydrogen), 48.35 (oxygen), 0.66 (nitrogen) and 0.89 (sulphur)	C==O, O-H, -CH_2 stretching	6.8	
Raw pomegranate peel	Cu^{2+}	598.78	205 μm, 850 μm and 2375 μm		C=O in carboxylic acid, acetate groups -COO, ketone, C-O groups of carboxylic acid, alcoholic, phenolic, ether and ester groups.		Ben-Ali et al. [55]
Sugarcane bagasse pith (sulphurised activated carbon)	Zn^{2+}	500	-	9.10 (sulphur) and 5.20 (ash)	S=O, and C-S vibrations	4.3	Krishnan et al. [102]
Jack fruit leaf powder	Ni^{2+}	246.9	-	-	-OH groups, -CH_2 group, and CO bonds and C=S bonds.	-	Boruah et al. [85]
Coffee residues	Pb^{2+}, Zn^{2+}	0.19	-	-	-	3.9	Wu et al. [28]
Guava leaves (activated) DateStones $Pd^{2+}Cd^{2+}$ Olive stone Hg^{2+}	Cd^{2+}	100.76 950 950 400–850	Pore volume 0.415 cm^3/g and pore diameter 47.091 Å -		O-H, C-H, C=C and -SO_3 bonds	-	Abdelwahab, Fouad et al. [81] Sulyman et al. [60] Wahby et al. [103]

3. Batch Metal Ion Adsorption by Agricultural Solid Waste Biomass Adsorbents under Various Physicochemical Process Parameters

In this section, the effects of the important process parameters, such as metal ion concentration, contact time, adsorbent load, pH, and temperature on the adsorbent capacity towards metal ion adsorption has been reviewed and discussed below. The identification and optimisation of these process parameters were generally determined through batch adsorption studies prior to pilot-scale continuous adsorption operation.

3.1. The Effects of the Initial Metal Ion Concentration and the Contact Time

To understand the adsorbate load and their optimum load concentration, a wide range of initial adsorbate metal ion concentrations has been examined across various reported batch adsorption studies [3,104]. Generally, with the increase in the initial adsorbate heavy metal ion concentration, the percentage removal efficiency of the carbon-based adsorbents initially increased up to a certain level and then decreased [20,105,106]. A higher solute concentration increases the competition due to the presence of excess solutes in the system to adhere with an adsorbent surface, which subsequently reduces the overall removal efficiency of the system [4,27,49,100,107]. The adsorbate or solute offers the driving force in terms of the concentration gradient to overcome the mass transfer resistance. Increasing the initial adsorbate concentration leads to the decrease in the percentage of adsorbate metal removal and an increase in the amount of heavy metal ions adsorbed per gram of adsorbent (q_t). At lower concentration ranges, the available adsorbent sites are occupied by adsorbate molecules and hence increase the adsorption capacity [49]. Sometimes the adsorption process slows down due to the steric repulsion between the solute molecules [108]. Generally, the higher percentage of heavy metal removal decreases with the metal ion concentration; in this research direction, readers are encouraged to go through these various recently reported review articles [3,11,20,69,104]. The percentage removal of Zn^{2+} metal ions by the sorghum hull adsorbent was found to have decreased from 50.98% to 12.8% for the metal ion concentration range of 10–50 mg/L, respectively [86]. With the increase in the initial metal ion concentration from 25 to 150 mg/L, the percentage of adsorption of the rice husk decreased from 90.8% to 60.85% for Cr^{2+}, 96.12% to 65.42% for Pb^{2+}, and from 94.36% to 66.83% for Zn^{2+}, respectively [20,109]. Similarly, it was reported by Ding et al. [110] that the maximum hickory wood biochar adsorbent capacity for the Cd^{2+}, Zn^{2+}, Ni^{2+}, and Cu^{2+} metal ions was increased with the metal ion concentrations of 2–100 mg/L, respectively [7]. Yargic et al. [111] reported on the batch Cu^{2+} adsorption studies by the chemically-treated tomato waste where the percentage of metal ion removal decreased with the increase in the initial metal ion concentration, and the adsorbed amount of metal (q_e) per gram of adsorbent increased with the initial metal ion concentration. Similarly, Kilic et al. [112] presented the variation between the adsorptive capacities of Ni^{2+} and Co^{2+}, q_e (mg/g), by almond shell biochar with the Ni^{2+} and Co^{2+} metal ion concentration ranges of 50–150 ppm and 100–200 ppm, respectively, under various temperatures, which are presented in Figure 5. As shown in Figure 5, the metal ion adsorption increased with time and followed the three step process with an initial fast reaction rate period followed by a slow rate, ending with the attainment of an equilibrium stage at 240 min [112]. A further amount of metal ion adsorption (q_e (mg/g)) was increased with the increased temperature, which is also shown in Figure 5. The adsorption capacity of the Hass avocado shell (HAS) adsorbent for Ni^{2+} increased from 5.63 to 107.26 mg per gram, respectively, with the increase in the metal ion concentration [50].

Generally, the percentage removal of aqueous phase pollutants by initial adsorption increases with the contact time, and then slowly reaches a steady-state saturation level. It may present in the form of either a two-stage or multistage adsorption process [3,4,14,49,63]. Therefore, adsorption kinetic studies are important for obtaining crucial knowledge on the speed of the reaction and the equilibrium time for maximum adsorption achievement, as well as to know the kinetic parameters required for the adsorber design.

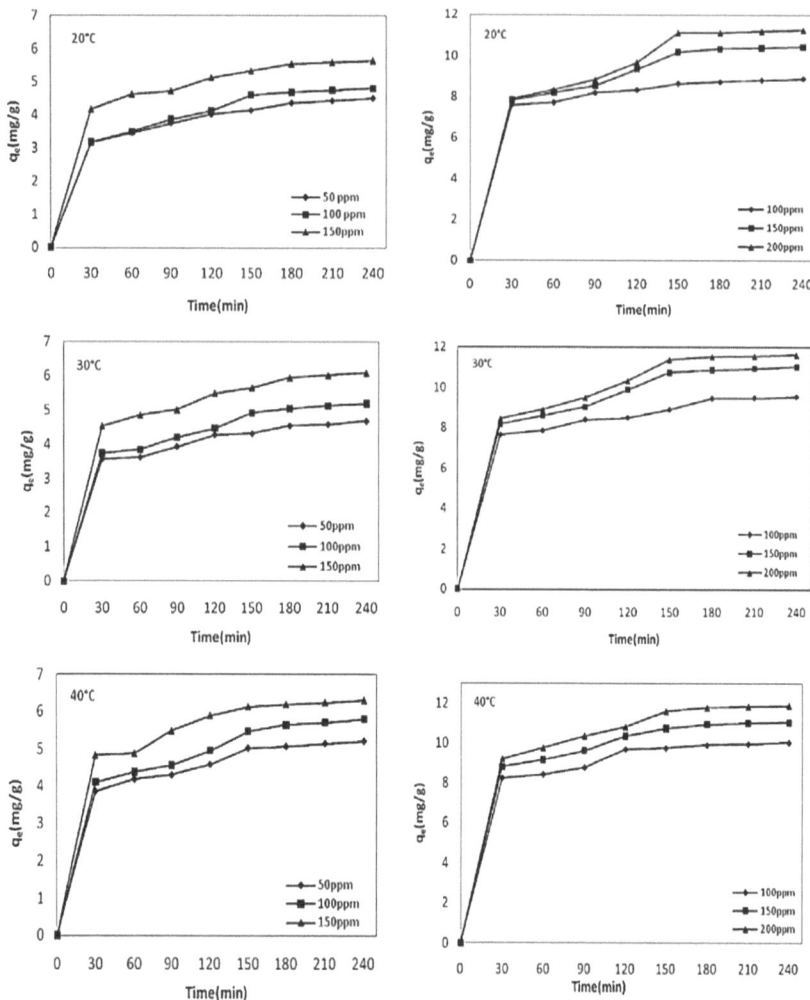

Figure 5. The effects of the contact time and initial metal ion concentrations of Ni^{2+} and Co^{2+} on the amount of adsorption by the almond shell biochar adsorbent at temperatures of 20, 30, and 40 °C, respectively. Source: taken from [112] with written permission.

3.2. Effects of the Adsorbent Dose

For the successful design, development, and scale-up of a continuous adsorption column, the knowledge of the adsorption capacity of the adsorbent is essential. The effect of an adsorbent dose on heavy metal adsorption in a solution indicates its adsorption capacity, which also depends on the available active sites on the adsorbent's surface for adsorption [63,97]. In general, the adsorption capacity q_e (mg/g) decreases with the increase in the adsorbent dose, whereas the percentage removal of metal ions increases along with the increase in the adsorbent dose [97,113]. A high adsorption capacity indicates that the adsorption process is running with a lower adsorbent dose/load. At higher adsorbent doses, there are maximum available active sites for adsorption and hence higher percentage removals of the adsorptive metal ions takes place at higher adsorbent dosages [3]. However, with a lower adsorption capacity, the removal percentage of pollutants increases rapidly and then slows down as the dose is reduced [50,114]. Much of the information presented in the literature supports these findings, such as Kılıç et al. [112], who reported from their batch adsorption study that the percentage adsorptive removal of the Ni^{2+} and Co^{2+} metal ions by the almond shell biochar increased from 10% to 38%, and from 25% to 50%, with the increase in the adsorbent doses from 1 to 10 g/L, respectively. In contrast, Ni^{2+} and Co^{2+} adsorbent's capacities, q_e (mg/g), were decreased from 10 mg/g to 3 mg/g, and from 24 mg/g to 7 mg/g, respectively, for which their results are presented in Figure 6.

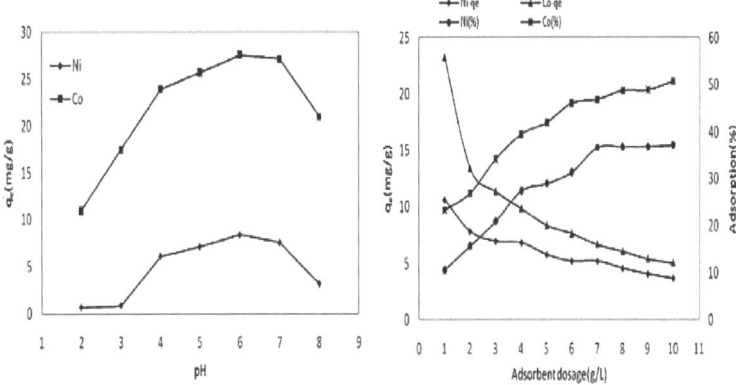

Figure 6. Effects of the solution pH and adsorbent dosages on Ni^{2+} and Co^{2+} adsorption. Source: taken from Kilic et al. [112] with written permission.

Afroze et al. [49] also reported similar results for Zn^{2+} adsorption by modified eucalyptus *sheathiana* bark biomass, and it was found that their adsorbent capacity, q_e (mg/g), decreased from 72.52 mg g^{-1} to 17.57 mg g^{-1} with the increase in the adsorbent doses from 0.01 g to 0.03 g, respectively [4]. There are also a few reported results on the same trends, i.e., with increases in the adsorbent dose accompanied with a decrease in the percentage of metal adsorption [115]. Imran-Shaukat et al. [42] reviewed and presented a compilation list on the variation of the adsorptive capacities of various amounts/loads of different agricultural biomass groups (such as bark, husk, leaves, peels, seeds, and straw) towards heavy metal ion (including Cd^{2+}, Co^{2+}, Cr^{2+}, Cu^{2+}, Mn^{2+}, Ni^{2+}, Pb^{2+}, and Zn^{2+}) adsorption, and critically analysed their comparative results at high, medium, and low adsorbent doses. When the amount of adsorbent mass in a fixed-volume solution is below the optimum value, the removal of metal ions is also low due to the lower number of available active sites for adsorption [69]. Table 3 presents an updated compilation of the selected reported results on the effect of adsorbent dosage in the removal of aqueous phase heavy metals using agricultural waste biomass during the last 10-year period [3].

Table 3. The selected reported list on the effect of changes in the adsorbent dosages on the percentage of adsorptive metal ion removal using several agricultural wastes as adsorbents during the last 10-year period.

Adsorbents	Adsorbates (Heavy Metals)	Adsorbent Dosage	Trend on Percentage (%) Removal Range	References
Brassica campestris agricultural waste	Ni^{2+}, Pb^{2+} Cr^{6+} 0.2–1 g/L		Increase	Shaikh et al. [116]
Mango kernel (bio-composite)	Cr (VI)	0.05–0.3 g/L	Decrease	Akram et al. [117]
Bagasse (activated)	Cr	0.5–1.5 g/L	Increase	Olayebi et al. [118]
Croncob (activate)	Cr	0.5–2.4 g/L	Increase	Olayebi et al. [118]
Bagasse (activated)	Fe^{3+}	0.5–2.5 g/L	Increase	Olayebi et al. [118]
Croncob (activated)	Fe^{3+}		Increase	Olayebi et al. [118]
Banana peel biochar	Pb^{2+}	0.5–3.0 g/L 0.01–0.2 g/L	Increase	Zhou et al. [119]
Eucalyptus sheathiana bark	Zn^{2+}	0.01–0.03 g	Decrease	Afroze et al. [49]
Bagasse pith (sulphurised activated carbon)	Zn^{2+}	0.5–8 g L^{-1}	Increase	Krishnan et al. [102]
Jackfruit leaf powder	Ni^{2+}	1–5 g L^{-1}	Decrease	Boruah et al. [85]
Sugarcane bagasse (sulphuric acid-treated)	Cu^{2+}	0.5–2 gm/100 mL	Increase	Rana et al. [17]
Grapefruit peel	Cd^{2+}, Ni^{2+}	1–4 g L^{-1}	Increase	Torab-mostaedi et al. [120]
Tamarind fruit shell	Ni^{2+}	0.01–0.08 g/10 mL	20–90	Pandharipande and kalnaka [121]
Almond shell biocar	Ni^{2+} Cd^{2+}	0.1–10 g/L 0.1–10 g/L	Increase	Kilic et al. [112
Rice husk	Pb^{2+}, Cd^{2+} Cu^{2+}, Ni^{2+}	0.02–0.06 g/L	Increase Increase	Hegazi [122]

3.3. Influential Effect of the Solution pH

The variation of solution pH plays a major role in changing the adsorbent surface charges, degree of ionisation, and metal speciation in solution, and hence causes changes to the adsorption capacity during the adsorption process [98,123]. Therefore, changes in the solution pH facilitate the adsorbent site dissociation and adsorbate solution chemistry, such as hydrolysis, surface complex formation, redox reactions, and precipitation, which are all strongly influenced by the pH [124]. The protonation and deprotonation of both functional groups in the adsorbent and adsorbate compound will produce different surface charges/zeta potential in the solution depending on the system's pH [125]. Adsorbent capacity depends on its point of zero charge (pH_{pzc}), and hence the surface charge. The point of zero charge (pzc) or the isoelectric point (iep) is defined as a particular pH where the surface charge becomes zero, i.e., where the extent of the adsorption of the positively charged species equals that of the negatively charged species. The point of zero charge (pH_{pz}) of various raw, treated, or modified agricultural biomass-based adsorbents was determined by many investigators to obtain a better understanding of the adsorptive removal mechanism [49,126–128]. Generally, at lower acidic solutions, where pH < pH_{pzc}, the adsorbent surface becomes positively charged and hence less metal cation adsorption takes place due to electrostatic repulsion between the positive cations and the positive surface-binding sites. Whereas, at pH > pH_{pzc}, the surface becomes negatively charged and favours metal cation adsorption. However, at a higher basic pH, metal complex formation occurs resulting in precipitative separation instead of adsorptive metal ion separation [60]. For example, at a solution pH < 6.0, Pb $(NO_3)_2$ in solution predominately exists as Pb^{2+} ions. Meanwhile, with an increasing solution pH, for example at pH = 8, Pb $(OH)^+$ formation occurs, and at pH = 11, it will precipitate as Pb $(OH)_2$ [49,129]. Therefore, cationic species adsorption is favoured at pH > pH_{pzc} due to the presence of the functional groups, such as the OH^-, and COO^- groups, while anionic adsorbate adsorption is favoured at pH < pH_{pzc} due to the presence of H^+ ions [113,130]. An electrical double layer at the solid/liquid interface is formed by the adsorbing counter ions from the aqueous solution to its adsorbent surface. Overall, the adsorbent surface functional groups/surface charges and the chemical nature of adsorbates at a solution pH strongly influence the adsorption behaviour and capacity. In their review article, Ahmad and Danish [50] mentioned that Mallampati et al. [72,131] reported the results of the solution pH effect on the adsorptive removal of the aqueous phase Pb^{2+}, Ni^{2+}, and $Cr_2O_7^{2-}$ ions with the avocado peel adsorbent. They found that the percentage removal of the cationic Pb^{2+} and Ni^{2+} adsorption was increased with the increase in the solution pH, whereas the adsorption of anionic $Cr_2O_7^{2-}$ was decreased with the same increasing solution pH. Abbar et al. [79] presented the batch adsorption experimental results on the effects of the solution pH on Cu^{2+}, Ni^{2+}, and Zn^{2+} adsorption by the flax fibre tows (FFTs) adsorbent in the solution pH range from 1.6 to 8.5, respectively, for all metal ions. To investigate the effects of the solution pH on metal ion precipitation, Abbar et al. [79] presented the experimental results without adsorbent. It was found that the percentage removal of all three metal cations increased with the solution pH and attained a maximum value at an optimum pH, and thereafter decreased with the further increase in the solution pH. The maximum percentage removal of the Pb^{2+} and Cu^{2+} ions occurred in the solution pH range of 4–6, whereas for Zn^{2+} metal ions, the maximum values were observed at the solution pH of 7, respectively. At a higher pH, lead, copper, and zinc metal ions precipitate as hydroxides and reduce the rate of adsorption and hence reduce their removal capacity as well [79]. Similarly, Kilic et al. [112] reported that the amount of Ni^{2+} and Co^{2+} adsorption, q_e (mg/g), by the almond shell biochar was increased from solution pH 2 to 6, and then decreased with the increasing solution pH, as shown in Figure 6. The deprotonation of the agricultural solid waste-based adsorbent typically takes place at a solution pH higher than pH_{zpc}, and the surface becomes more negatively charged due to the presence of the stretching hydroxyl (–OH) and carboxyl (–COOH) functional groups [4]. Therefore, more adsorption of the cationic metal ions takes place mainly through the electrostatic force of attraction mechanism. Dawood and Sen [4] reported a similar trend in Ni^{2+} adsorption using the pinecone biochar adsorbent. At a low solution, the pH tends to decrease the adsorption capacity of the cations onto the adsorbent due to the presence of hydronium (H_3O^+) ions competing with the cationic metal ions for the available adsorption sites [7], accompanied with the fact that similar charges repel each other [50]. However, a lower, acidic solution pH favours anionic ion adsorption more, and this is because of the positively charged adsorbent surface and the opposite counter anions adsorption mechanism.

3.4. Effects of the Temperature and Thermodynamics of Adsorption

Temperature plays an important role in the adsorption of metal ions associated with the thermodynamics of the adsorption process. Temperature was found to be another significant physiochemical process parameter that influences the adsorption/biosorption mechanism and hence the equilibrium adsorbent capacity [3,42,132]. Different metal ions and different adsorbents have different responses to the system's temperature [7,42,133]. Temperature induces various changes in the thermodynamic parameters, such as changes in the Gibb's free energy (ΔG^0), enthalpy (ΔH^0), and entropy (ΔS^0), for the heavy metal ion adsorption by the agricultural solid waste-based adsorbents, which can be determined by the following two equations [134]:

$$\Delta G^0 = \Delta H^0 - T\Delta S^0$$

and

$$log\left(1000\frac{q_e}{C_e}\right) = \frac{\Delta S^0}{2.303R} + \frac{-\Delta H^0}{2.303RT}$$

where q_e is the amount of metal ion adsorbed per unit mass adsorbent (mg/g), C_e is the equilibrium concentration (mg/L), T is the temperature in K, and R is the universal gas constant (8.314 J/molK).

Shaukat et al. [42] recently reviewed and reported the temperature effects on the agricultural waste biomass adsorption efficiency for various heavy metal ions under three temperature levels: high: 45 °C < x ≤ 60 °C, medium: 30 °C < x ≤ 45 °C, and low: 20 °C ≤ x ≤ 30 °C, respectively. At low-temperature levels, the metal ion adsorption increases in the order of $Mn^{2+} > Pb^{2+} > Cu^{2+} > Cr^{6+} > Co^{2+} > Zn^{2+} > Ni^{2+} > Cd^{2+}$ and in the order of $Pb^{2+} > Cd^{2+} > Zn^{2+}$ at the medium level, respectively. Meanwhile, for the high level of temperature, the order was $Cd^{2+} > Pb^{2+} > Cr^{2+} > Cu^{2+}$. Temperature is an important indicator of the exothermic or endothermic nature of the adsorption reaction process [135]. The solution viscosity is reduced with the increase in the solution's temperature and hence increases the diffusive transport of adsorbate species from the bulk phase to the solid/liquid interface and through pore diffusion [136]. An increase in the adsorption capacity at higher solution temperatures indicates the endothermic nature of the adsorption reaction due an increase in the kinetic transport of adsorbate solutes and a higher diffusional rate [38]. However, a decrease in the adsorption capacity with an increase in the temperature indicates that the reaction has become exothermic, and this is due to the heat-induced decrease in the attractive adsorptive forces between the adsorbate and the adsorbent's surface [137].

The temperature effect on the agricultural adsorbent's capacity depends on the surface functional groups [138]. They reviewed and reported the results of many studies, such as mango leaf powder [139], rice husk [139], orange peel [140], and coconut shell [141]), which were all found to increase the percentage adsorption of metal ions with the increase in the temperature range (25–40 °C). In comparison, the adsorption of Cd^{2+} on the cashew nut shell was decreased from 80.13% to 74.32% with the rise in temperature from 30 °C to 60 °C, respectively. Many studies have also reported that metal ion uptake by some adsorbents is reduced with an increasing temperature [133,142].

4. Future Perspectives and Future Challenges

To overcome the high costs of commercial activated carbon (CAC) and to overcome the other operational issues that have been associated with the use of CAC as the adsorbent, raw and modified agricultural biomass residue-based adsorbents have gained a significant level of attention as an alternative, carbon-containing, easily accessible, and cost-effective adsorbent in the removal of aqueous phase heavy metal ions with a high degree of binding capacity. From the extensive literature review on adsorption-based wastewater treatment technology, the following points presented are the challenges and future directions that need to be addressed so that adsorption-based technology may be more effective and popularize this technology for the future remediation of water pollution.

Overall economy: the overall economically feasible operation of an adsorption-based treatment plant depends on many factors. Various costs associated with the operating costs, fixed costs, including the installation cost, adsorbent pre-treatment/preparation costs, and cost of adsorbent regeneration are all especially important for determining the feasibility of the full process. Among them, the adsorbent cost alone, including its procession, is above 60% of the total operating cost. Therefore, the adsorbent material selection is crucial for the adsorptive separation process. Various non-conventional solid waste-based adsorbents may be an alternative, cost-effective solution to this process.

Industrial scale problems and lab-based experiments: due to the introduction of various environmental protection laws and regulations, industries have imposed the discharge of waste into the environment. However, industries sometimes discharge harmful chemical waste at a higher than prescribed limit. Therefore, the industry always looks into some low-cost technology like adsorption, and many industries have already adopted this technology. However, the effectiveness of this adsorption-based technology is mainly judged using the laboratory-based batch adsorption results with limited continuous experimental results. Therefore, more continuous adsorption operation results, if possible, along with the pilot-scale results are required before commercial implementation.

Batch and continuous column analysis: based on the literature review over the last two decades, it has been found that more than 80% of adsorption-based studies are of the batch scale. The challenge of adsorption-based studies lies here. These batch studies are confined to various kinetics, isotherm, and thermodynamic analysis with a very small lab scale. The batch-scale study results cannot be adopted directly for industrial use without continuous operation. Several recent studies have come up with some lab or bench-scale continuous studies of adsorption in a packed bed, fluidised bed, and semi-fluidised system to help in the scale up of this process. More research is required in the field of continuous adsorption systems and scale-up processes.

Adsorption modelling: For large-scale operation and process design adsorption modelling, the procedure for the accurate estimation of various kinetic parameters, isotherm models, and the thermodynamic parameters for the multicomponent system are essential.

Adsorbent regeneration and reuse: it has been mentioned previously in that the 60% cost of an adsorption-based system depends on the cost of the adsorbent. Therefore, in the age of sustainable development, adsorbent regeneration must be given significant priority. To reduce waste production, secondary pollution, and operating costs, and to make the overall technology more cost-effective for further reuse, regeneration of the loaded adsorbents is an essential process. Moreover, the capture adsorbate must be recovered as they may be valuable products or to aid in minimising secondary pollution. Hence, an eco-friendly and low-cost alternative regeneration method must be developed to reduce waste production and cost, as well as maximise the cycle number to use for a greater number of times under industrial operations.

Process optimisation: in adsorption-based studies, process optimisation is required under controlled conditions and for further applications in real-field situations. In most cases, the actual process effluents are multicomponent and compete with the adsorbates. The multicomponent systems always reduce the ideal adsorption capacity, meaning therefore that the modelling and optimisation of these multicomponent systems will be quite complex.

5. Conclusions

Water pollution due to heavy metal ion contamination resulting from various sources, including untreated industrial effluent discharge and agricultural activities, is of global concern and to find out an efficient but sustainable and cost-effective remediation solution to these important global problems imposes a challenging task on scientists, researchers, and practising engineers. Among the various conventional remediation techniques, adsorption-based separation technology is considered to be one of the most effective approaches widely used in treating heavy metal contaminated water and wastewater due to its simple operation, design simplicity, high separation efficiency, efficiency at lower pollutant concentrations, high selectivity at the molecular level, low energy consumption, ability to separate multiple pollutant components, and minimize secondary pollution. This review article presented a compilation of various scattered literature data along with the up-to-date development batch metal cation adsorption results using a wide range of non-conventional and cost-effective agricultural solid waste-based adsorbents under various process conditions. It is clear from the present literature survey in that non-conventional raw or modified agricultural solid waste-based adsorbents are emerging as effective, but low-cost adsorbents for heavy metal ions present decontamination problems. The utilisation of this large amount of agricultural solid waste-based effective adsorbents in the water and wastewater treatment industries is a sustainable and cost-effective pollution control option alternative to the costly CAC adsorbents. The literature has also revealed that in some cases, the modification of the adsorbent increased the removal efficiency of adsorption. The effective metal removal efficiency from the aqueous phase mainly depends on the adsorbent's characteristics and various physicochemical process parameters. Therefore, this review article was compiled to critically analyse the large batch adsorption results on heavy metal ion adsorption by the wide ranges of agricultural solid waste-based adsorbents, specifically the adsorbent's characteristics, and under various influential process parameters, such as the initial adsorbate metal ion concentration, the initial solution pH, the adsorbent doses, and the temperature, respectively.

Funding: This work was supported by the Deanship of Scientific Research, Vice Presidency for Graduate Studies and Scientific Research, King Faisal University, Saudi Arabia (GRANT-3804).

Institutional Review Board Statement: Not applicable.

Informed Consent Statement: Not applicable.

Data Availability Statement: Data sharing not applicable to this article as no datasets were generated or analysed during the current study.

Acknowledgments: The author acknowledges the Deanship of Scientific Research, Vice Presidency for Graduate, Studies and Scientific Research, King Faisal University, Saudi Arabia supported this work [GRANT3804]. The author also acknowledges Sharmeen Afroze and Sara Dawood for their minor assistance in literature information collection.

Conflicts of Interest: The author declares no conflict of interest.

References

1. UNWWDR. United Nations World Water Development Report 2020. Available online: https://www.unwater.org/publications/un-world-water-development-report-2020 (accessed on 22 November 2022).
2. Tee, G.T.; Gok, X.Y.; Yong, W.F. Adsorption of pollutants in wastewater via biosorbents, nanoparticles and magnetic biosorbents: A review. *Environ. Res.* **2022**, *212*, 113248. [CrossRef] [PubMed]
3. Afroze, S.; Sen, T.K. A Review on Heavy Metal Ions and Dye Adsorption from Water by Agricultural Solid Waste Adsorbents. *Water Air Soil Pollut.* **2018**, *229*, 225. [CrossRef]
4. Dawood, S.; Sen, T.K.; Phan, C. Synthesis and characterization of slow pyrolysis pinecone biochar in the removal of organic and inorganic pollutants from aqueous solution by adsorption: Kinetic, equilibrium, mechanism and thermodynamic. *Bioresour. Technol.* **2017**, *246*, 76–81. [CrossRef] [PubMed]
5. Gisi, S.D.; Lofrano, G.; Grassi, M.; Notarnicola, M. Characteristics and adsorption capacities of low-cost sorbents for wastewater treatment: A review. *Sustain. Mater. Technol.* **2016**, *9*, 10–40.
6. Liu, C.; Zhang, H.X. Modified biochar adsorbents (MBAs) for heavy metal ions adsorption: A critical review. *J. Environ. Chem. Eng.* **2022**, *10*, 107393. [CrossRef]
7. Dawood, S. Synthesis and Characterization of Biomass and Clay Minerals-Based Adsorbents for the Removal of Cationic Dye and Metal Ion from Wastewater by Adsorption. Ph.D. Thesis, Curtin University Library, Bentley, WA, Australia, 2018.
8. Chatterjee, S.; Bhattacharjee, I.; Chandra, G. Biosorption of heavy metals from industrial wastewater by *Geobacillus thermodenitrificans*. *J. Hazard. Mater.* **2010**, *175*, 117–125. [CrossRef]
9. Rafatullah, M.; Sulaiman, O.; Hashim, R.; Ahmad, A. Adsorption of copper (II), chromium (III), nickel (II) and lead (II) ions from aqueous solutions by meranti sawdust. *J. Hazard. Mater.* **2009**, *170*, 969–977. [CrossRef]
10. Duruibe, J.; Ogwuegbu, M.; Egwurugwu, J. Heavy metal pollution and human biotoxic effects. *Int. J. Phys. Sci.* **2007**, *2*, 112–118.
11. Garg, R.; Garg, R.; Sillanpää, M.; Alimuddin Khan, M.A.; Mubarak, N.M.; Tan, Y.H. Rapid adsorptive removal of chromium from wastewater using walnut-derived biosorbents. *Sci. Rep.* **2023**, *13*, 6859. [CrossRef]
12. Rajendran, S.; Priya, A.K.; Kumar, P.S.; Hoang, T.K.; Sekar, K.; Chong, K.Y.; Khoo, K.S.; Ng, H.S.; Show, P.L. A critical and recent developments on adsorption technique for removal of heavy metals from wastewater—A review. *Chemosphere* **2022**, *303*, 135146. [CrossRef]
13. Jiang, Q.; Han, Z.; Li, W.; Ji, T.; Yuan, Y.; Zhang, J.; Zhao, C.; Cheng, Z.; Wang, S. Adsorption properties of heavy metals and antibiotics by chitosan from larvae and adulty trypoxylus dichotonus. *Carbohydr. Polym.* **2021**, *276*, 118735. [CrossRef] [PubMed]
14. Sen, T.K. *Air, Gas and Water Pollution Control Using Industrial and Agricultural Solid Wastes Adsorbents*, 1st ed.; CRC-Press/Taylor & Francis: New York, NY, USA, 2017; ISBN 13:978-1-138-19673-5.
15. Qu, J.; Song, T.; Liang, J.; Bai, X.; Li, Y.; Wei, Y.; Huang, S.; Dong, L.; Jin, Y. Adsorption of lead (II) from aqueous solution by modified Auricularia matrix waste: A fixed-bed column study. *Ecotoxicol. Environ. Saf.* **2019**, *169*, 722–729. [CrossRef] [PubMed]
16. Feng, N.; Guo, X.; Liang, S.; Zhu, Y.; Liu, J. Biosorption of heavy metals from aqueous solutions by chemically modified orange peel. *J. Hazard. Mater.* **2011**, *185*, 49–54. [CrossRef] [PubMed]
17. Rana, K.; Shah, M.; Limbachiya, N. Adsorption of Copper Cu (2+) Metal Ion From Wastewater Using Sulphuric Acid Treated Sugarcane Bagasse as Adsorbent. *Int. J. Adv. Eng. Res. Sci.* **2014**, *1*, 55–59.
18. Al-Sahari, M.; Al-Gheethi, A.; Mohamed, R.M.S.R.; Noman, E.; Naushad, M.; Rizuan, M.B.; Vo, D.-V.N.; Ismail, N. Green approach and strategies for wastewater treatment using bioelectrochemical systems: A critical review of fundamental concepts, applications, mechanism, and future trends. *Chemosphere* **2021**, *285*, 131373. [CrossRef] [PubMed]

19. Shaheen, S.M.; Natasha; Mosa, A.; El-Naggar, A.; Hossain, F.; Abdelrahman, H.; Niazi, N.K.; Shahid, M.; Zhang, T.; Tsang, Y.F.; et al. Manganese oxide modified biochar: Production, characterization and application for the removal of pollutants from aqueous environments: A review. *Bioresour. Technol.* **2022**, *346*, 126581. [CrossRef]
20. Chen, X.; Hossain, F.; Duan, C.; Lu, J.; Tsang, Y.F.; Islam, S.; Zhou, Y. Isotherm models for adsorption of heavy metals from water—A review. *Chemosphere* **2022**, *307*, 135545. [CrossRef]
21. Hossain, N.; Bhuiyan, M.A.; Pramanik, B.K.; Nizamuddin, S.; Griffin, G. Waste Materials for Wastewater Treatment and Waste Adsorbents for Biofuel and Cement Supplement Applications: A Critical Review. *J. Clean. Prod.* **2020**, *255*, 120261. [CrossRef]
22. Naef, A.; Qasem, A.; Ramy, H.M.; Dahiru, U.L. Removal of heavy metal ions from wastewater: A comprehensive and critical review. *Clean Water* **2021**, *4*, 36. [CrossRef]
23. Zhao, Q.; Zhao, X.; Cao, J. Advanced nanomaterials for degrading persistent organic pollutants. In *Advanced Nanomaterials for Pollutant Sensing and Environmental Catalysis*; Elsevier: Amsterdam, The Netherlands, 2020; pp. 249–305. [CrossRef]
24. Khan, S.H.; Pathak, B. Zinc oxide based photocatalytic degradation of persistent pesticides: A comprehensive review. *Environ. Nanotechnol. Monit. Manag.* **2020**, *13*, 100290. [CrossRef]
25. Richardson, S.D.; Kimura, S.Y. Emerging environmental contaminants: Challenges facing our next generation and potential engineering solutions. *Environ. Technol. Innov.* **2017**, *8*, 40–56. [CrossRef]
26. Divyapriya, G.; Singh, S.; Martínez-Huitle, C.A.; Scaria, J.; Karim, A.V.; Nidheesh, P.V. Treatment of real wastewater by photoelectrochemical methods: An overview. *Chemosphere* **2021**, *276*, 130188. [CrossRef] [PubMed]
27. Biswas, S.; Meikap, B.C.; Sen, T.K. Adsorptive Removal of Aqueous Phase Copper (Cu^{2+}) and Nickel (Ni^{2+}) Metal Ions by Synthesized Biochar–Biopolymeric Hybrid Adsorbents and Process Optimization by Response Surface Methodology (RSM). *Water Air Soil Pollut.* **2019**, *230*, 197. [CrossRef]
28. Wu, C.-H.; Kuo, C.-Y.; Guan, S.-S. Adsorption Kinetics of Lead and Zinc Ions by Coffee Residues. *Pol. J. Environ. Stud.* **2015**, *24*, 761–767. [CrossRef] [PubMed]
29. Sud, D.; Mahajan, G.; Kaur, M.P. Agricultural waste material as potential adsorbent for sequestering heavy metal ions from aqueous solutions—A review. *Bioresour. Technol.* **2008**, *99*, 6017–6027. [CrossRef]
30. Tan, K.L.; Hameed, B.H. Insight into the adsorption kinetics models for the removal of contaminants from aqueous solutions. *J. Taiwan Inst. Chem. Eng.* **2017**, *74*, 25–48. [CrossRef]
31. Ogunlalu, O.; Oyekunle, I.P.; Iwuozor, K.O.; Aderibigbe, A.D.; Emenike, E.C. Trends in the mitigation of heavy metal ions from aqueous solutions using unmodified and chemically modified agricultural waste adsorbents. *Curr. Res. Green Sustain. Chem.* **2021**, *4*, 100188. [CrossRef]
32. Ho, Y.S. Review of second-order models for adsorption systems. *J. Hazard. Mater.* **2006**, *136*, 681–689. [CrossRef]
33. Liu, Y.; Liu, Y.J. Biosorption isotherms, kinetics and thermodynamics. *Sep. Purif. Technol.* **2008**, *61*, 229–242. [CrossRef]
34. Plazinski, W.; Rudzinski, W. A novel two-resistance model for description of the adsorption kinetics onto porous particles. *Langmuir* **2010**, *26*, 802–808. [CrossRef]
35. Alberti, G.; Amendola, V.; Pesavento, M.; Biesuz, R. Beyond the synthesis of novel solid phases: Review on modelling of sorption phenomena. *Coord. Chem. Rev.* **2012**, *256*, 28–45. [CrossRef]
36. Freundlich, H. Over the adsorption in solution. *J. Phys. Chem.* **1906**, *57*, 385–470.
37. Langmuir, I. The adsorption of gases on plane surfaces of glass, mica, platinum. *J. Am. Chem. Soc.* **1918**, *40*, 1361–1403. [CrossRef]
38. Gopinath, K.P.; Vo, D.-V.; Prakash, D.G.; Joseph, A.; Viswanathan, S.; Arun, J. Environmental applications of carbon-based materials: A review. *Environ. Chem. Lett.* **2021**, *19*, 557–582. [CrossRef]
39. Dutta, B.K. *Principles of Mass Transfer and Separation Processes*, 1st ed.; Prentice Hall of India: New Delhi, India, 2007; Chapter 12; pp. 609–677.
40. Galamboš, M.; Suchánek, P.; Rosskopfová, O. Sorption of anthropogenic radionuclides on natural and synthetic inorganic sorbents. *J. Radioanal. Nucl. Chem.* **2012**, *293*, 613–633. [CrossRef]
41. Chakraborty, R.; Asthana, A.; Singh, A.K.; Jain, B.; Susan, A.B.H. Adsorption of heavy metal ions by various low-cost adsorbents: A review. *Int. J. Environ. Anal. Chem.* **2022**, *102*, 342–379. [CrossRef]
42. Shaukat, M.I.; Wahi, R.; Ngaini, Z. The application of agricultural wastes for heavy metals adsorption: A meta-analysis of recent studies. *Bioresour. Technol. Rep.* **2022**, *17*, 100902. [CrossRef]
43. Gümüş, D.; Gümüş, F. Modeling heavy metal removal by retention on *Laurusnobilis* leaves biomass: Linear and nonlinear isotherms and design. *Int. J. Phytoremediat.* **2020**, *22*, 755–763. [CrossRef]
44. Rambabu, K.; Banat, F.; Pham, Q.M.; Ho, S.H.; Ren, N.Q.; Show, P.L. Biological remediation of acid mine drainage: Review of past trends and current outlook. *Environ. Sci. Ecotechnol.* **2020**, *2*, 100024. [CrossRef]
45. Singha, B.; Das, S.K. Adsorptive removal of Cu (II) from aqueous solution and industrial effluent using natural/agricultural wastes. *Colloids Surf. B Biointerfaces* **2013**, *107*, 97–106. [CrossRef]
46. Ahmad, A.; Ghazi, Z.A.; Saeed, M.; Ilyas, M.; Ahmad, R.; Muqsit Khattak, A.; Iqbal, A. A comparative study of the removal of Cr(vi) from synthetic solution using natural biosorbents. *New J. Chem.* **2017**, *41*, 10799–10807. [CrossRef]
47. Taşar, Ş.; Kaya, F.; Özer, A. Biosorption of lead (II) ions from aqueous solution by peanut shells: Equilibrium, thermodynamic and kinetic studies. *J. Environ. Chem. Eng.* **2014**, *2*, 1018–1026. [CrossRef]
48. Li, Q.; Zhai, J.; Zhang, W.; Wang, M.; Zhou, J. Kinetic studies of adsorption of Pb(II), Cr(III) and Cu(II) from aqueous solution by sawdust and modified peanut husk. *J. Hazard. Mater.* **2007**, *141*, 163–167. [CrossRef] [PubMed]

49. Afroze, S.; Sen, T.K.; Ang, H.M. Adsorption removal of zinc (II) from aqueous phase by raw and base modified Eucalyptus sheathiana bark: Kinetics, mechanism and equilibrium study. *Process Saf. Environ. Prot.* **2016**, *102*, 336–352. [CrossRef]
50. Ahmad, T.; Danish, M. A review of avocado waste-derived adsorbents: Characterizations, adsorption characteristics, and surface mechanism. *Chemosphere* **2022**, *296*, 134036. [CrossRef]
51. Anastopoulos, I.; Bhatnagar, A.; Hameed, B.H.; Ok, Y.S.; Omirou, M. A review on waste-derived adsorbents from sugar industry for pollutant removal in water and wastewater. *J. Mol. Liq.* **2017**, *240*, 179–188. [CrossRef]
52. Nahar, K.; Chowdhury, M.A.K.; Chowdhury, M.A.H.; Rahman, A.; Mohiuddin, K.M. Heavy metals in handloom-dyeing effluents and their biosorption by agricultural by products. *Environ. Sci. Pollut. Res.* **2018**, *25*, 7954–7967. [CrossRef]
53. Ngabura, M.; Hussain, S.A.; Ghani, W.A.; Jami, M.S.; Tan, Y.P. Utilization of renewable durian peels for biosorption of zinc from wastewater. *J. Environ. Chem. Eng.* **2018**, *6*, 2528–2539. [CrossRef]
54. Al-Qahtani, K.M. Water purification using different waste fruit cortexes for the removal of heavy metals. *J. Taibah Univ. Sci.* **2016**, *10*, 700–708. [CrossRef]
55. Ben-Ali, S.; Jaouali, I.; Souissi-Najar, S.; Ouederni, A. Characterization and adsorption capacity of raw pomegranate peel biosorbent for copper removal. *J. Clean. Prod.* **2017**, *142*, 3809–3821. [CrossRef]
56. Gupta, H.; Gogate, P.R. Intensified removal of copper from wastewater using activated watermelon based biosorbent in the presence of ultrasound. *Ultrason. Sonochem.* **2016**, *30*, 113–122. [CrossRef] [PubMed]
57. Asim, N.; Amin, M.H.; Samsudin, N.A.; Badiei, M.; Razali, H.; Akhtaruzzaman, M.; Sopian, K. Development of effective and sustainable adsorbent biomaterial from an agricultural waste material: Cu (II) removal. *Mater. Chem. Phys.* **2020**, *249*, 123128. [CrossRef]
58. Zaidi, N.A.H.M.; Lim, L.B.L.; Usman, A. Enhancing adsorption of Pb (II) from aqueous solution by NaOH and EDTA modified *Artocarpus odoratissimus* leaves. *J. Environ. Chem. Eng.* **2018**, *6*, 7172–7184. [CrossRef]
59. Nakkeeran, E.; Saranya, N.; Giri, M.; Nandagopal, S.; Santhiagu, A.; Selvaraju, N. Hexavalent chromium removal from aqueous solutions by a novel powder prepared from *Colocasia esculenta* leaves. *Int. J. Phytoremediat.* **2016**, *18*, 812–821. [CrossRef] [PubMed]
60. Sulyman, M.; Namiesnik, J.; Gierak, A. Low-cost Adsorbents Derived from Agricultural By-products/Wastes for Enhancing Contaminant Uptakes from Wastewater: A Review. *Pol. J. Environ. Stud.* **2017**, *26*, 479–510. [CrossRef]
61. Rangel, A.V.; Becerra, M.G.; Guerrero-Amaya, H.; Ballesteros, L.M.; Mercado, D.F. Sulfate radical anion activated agro-industrial residues for Cr (VI) adsorption: Is this activation process technically and economically feasible? *J. Clean. Prod.* **2021**, *289*, 125793. [CrossRef]
62. Ibrahim, R.I. Optimization process for removing of copper ions from groundwater of Iraq. Using watermelon shell as natural adsorbent. *IOP Conf. Ser. Mater. Sci. Eng.* **2020**, *737*, 012195. [CrossRef]
63. Ayob, S.; Othman, N.; Ali, W.; Altowayti, H.; Khalid, F.S.; Bakar, N.A. A Review on Adsorption of Heavy Metals from Wood-Industrial Wastewater by Oil Palm. *J. Ecol. Eng.* **2021**, *22*, 249–265. [CrossRef]
64. Shakya, A.; Agarwal, T. Removal of Cr(VI) from water using pineapple peel derived biochars: Adsorption potential and re-usability assessment. *J. Mol. Liq.* **2019**, *293*, 111497. [CrossRef]
65. Yousef, R.; Qiblawey, H.; El-Naas, M.H. Adsorption as a Process for Produced Water Treatment: A Review. *Process* **2020**, *8*, 1657. [CrossRef]
66. Gonçalves, A.C., Jr.; da Paz Schiller, A.; Conradi, E., Jr.; Manfrin, J.; Schwantes, D.; Zimmermann, J.; Klassen, G.J.; Campagnolo, M.A. Removal of Pb^{2+} and Cd^{2+} from Contaminated Water using Activated Carbon from Canola Seed Wastes. In Proceedings of the 5th World Congress on New Technologies (NewTech'19), Lisbon, Portugal, 18–20 August 2019. [CrossRef]
67. Naseem, K.; Huma, R.; Shahbaz, A.; Jamal, J.; Rehman, M.Z.U.; Sharif, A.; Ahmed, E.; Begum, R.; Irfan, A.; Al-Sehemi, A.G.; et al. Extraction of Heavy Metals from Aqueous Medium by Husk Biomass: Adsorption Isotherm, Kinetic and Thermodynamic study. *Z. Phys. Chem.* **2018**, *233*, 201–223. [CrossRef]
68. Devani, M.A.; Oubagaranadin, J.U.K.; Munshi, B.; Lal, B.B.; Mandal, S. BP-ANN Approach for Modeling Cd (II) Bio-Sorption from Aqueous Solutions Using *Cajanus cajan* Husk. *Iran. J. Chem. Eng.* **2019**, *38*, 110–124.
69. Sazali, N.; Harun, Z. A Review on Batch and Column Adsorption of Various Adsorbent Towards the Removal of Heavy Metal. *J. Adv. Res. Fluid Mech. Therm. Sci.* **2020**, *67*, 66–88.
70. Chen, Y.; Wang, H.; Zhao, W.; Huang, S. Four different kinds of peels as adsorbents for the removal of Cd (II) from aqueous solution: Kinetics, isotherm, and mechanism. *J. Taiwan Inst. Chem. Eng.* **2018**, *88*, 146–151. [CrossRef]
71. Mahdi, Z.; Yu, Q.J.; El Hanandeh, A. Competitive adsorption of heavy metal ions (Pb^{2+}, Cu^{2+}, and Ni^{2+}) onto date seed biochar: Batch and fixed bed experiments. *Sep. Sci. Technol.* **2019**, *54*, 888–901. [CrossRef]
72. Mallampati, R.; Xuanjun, L.; Adin, A.; Valiyaveettil, S. Fruit peels as efficient renewable adsorbents for removal of dissolved heavy metals and dyes from water. *ACS Sustain. Chem. Eng.* **2015**, *3*, 1117–1124. [CrossRef]
73. Malik, R.; Dahiya, S. An Experimental and Quantum Chemical Study of Removal of Utmostly Quantified Heavy Metals in Wastewater Using Coconut Husk: A Novel Approach to Mechanism. *Int. J. Biol. Macromol.* **2017**, *98*, 139–149. [CrossRef]
74. Thuan, T.V.; Phuong, B.B.; Nguyen, D. Response surface methodology approach for optimization of Cu^{2+}, Ni^{2+} and Pb^{2+} adsorption using KOH-activated carbon from banana peel. *Surf. Interfaces* **2017**, *6*, 209–217. [CrossRef]
75. Chi, T.; Zuo, J.; Liu, F. Performance and mechanism for cadmium and lead adsorption from water and soil by corn straw biochar. *Front. Environ. Sci. Eng.* **2017**, *11*, 15. [CrossRef]

76. Yan, S.; Yu, W.; Yang, T.; Li, Q.; Guo, J. The Adsorption of Corn Stalk Biochar for Pb and Cd: Preparation, Characterization, and Batch Adsorption Study. *Separations* **2022**, *9*, 22. [CrossRef]
77. Zhang, S.; Zhang, H.; Cai, J.; Zhang, X.; Zhang, J.; Shao, J. Evaluation and Prediction of Cadmium Removal from Aqueous Solution by Phosphate-Modified Activated Bamboo Biochar. *Energy Fuels* **2018**, *32*, 4469–4477. [CrossRef]
78. Guiza, S. Biosorption of heavy metal from aqueous solution using cellulosic waste orange peel. *Ecol. Eng.* **2017**, *99*, 134–140. [CrossRef]
79. Abbar, B.; Alem, A.; Pantet, A.; Marcotte, S.; Ahfir, N.D.; Duriatti, D. Removal of dissolved and particulate contaminants from aqueous solution using natural flex fibres. *Int. J. Civ. Eng.* **2017**, *35*, 656–661.
80. Asuquo, E.D.; Martin, A.D. Sorption of cadmium (II) ion from aqueous solution onto sweet potato (*Ipomoea batatas* L.) peel adsorbent: Characterisation, kinetic and isotherm studies. *J. Environ. Chem. Eng.* **2016**, *4*, 4207–4228. [CrossRef]
81. Abdelfattah, A.; Fathy, I.; Sayed, A.; Almedolab, A.; Aboelghait, K.M. Biosorption of heavy metals ions in real industrial wastewater using peanut husk as efficient and cost-effective adsorbent. *Environ. Nanotechnol. Monit. Manag.* **2016**, *6*, 176–183. [CrossRef]
82. Chinyelu, E. Use of unmodified orange peel for the adsorption of Cd (II), Pb (II) and Hg(II) ions in aqueous solutions. *Am. J. Phys. Chem.* **2015**, *4*, 21–29. [CrossRef]
83. Gutha, Y.; Munagapati, V.S.; Naushad, M.; Abburi, K. Removal of Ni (II) from aqueous solution by *Lycopersicum esculentum* (Tomato) leaf powder as a low-cost biosorbent. *Desalination Water Treat.* **2015**, *54*, 200–208. [CrossRef]
84. Paduraru, C.; Tofan, L.; Teodosiu, C.; Bunia, I.; Tudorachi, N.; Toma, O. Biosorption of zinc (II) on rapeseed waste: Equilibrium studies and thermogravimetric investigations. *Process Saf. Environ. Prot.* **2015**, *94*, 18–28. [CrossRef]
85. Boruah, P.; Sarma, A.; Bhattacharyya, K.G. Removal of Ni (II) ions from aqueous solution by using low cost biosorbent prepared from jackfruit (*Artocarpus heterophyllus*) leaf powder. *Indian J. Chem. Technol.* **2015**, *22*, 322–327.
86. Imaga, C.; Abia, A.; Igwe, J. Adsorption Isotherm Studies of Ni (II), Cu (II) and Zn (II) Ions on Unmodified and Mercapto-Acetic Acid (MAA) Modified Sorghum Hulls. *Int. Res. J. Pure Appl. Chem.* **2015**, *5*, 318–330. [CrossRef]
87. Utomo, H.D.; Hunter, K.A. Adsorption of divalent copper, zinc, cadmium, and lead ions from aqueous solution by waste tea and coffee adsorbents. *Environ. Technol.* **2006**, *27*, 25–32. [CrossRef] [PubMed]
88. Singha, A.S.; Guleria, A. Utility of chemically modified agricultural waste okra biomass for removal of toxic heavy metal ions from aqueous solution. *Eng. Agric. Environ. Food* **2015**, *8*, 52–60. [CrossRef]
89. Moubarik, A.; Grimi, N. Valorization of olive stone and sugar cane bagasse by-products as biosorbents for the removal of cadmium from aqueous solution. *Food Res. Int.* **2015**, *73*, 169–175. [CrossRef]
90. Rajamohan, N. Biosorption of Mercury onto Protonated Pistachio Hull Wastes–Effect of Variables and Kinetic Experiments. *Int. J. Chem. Eng. Appl.* **2014**, *5*, 415–419. [CrossRef]
91. Putra, W.P.; Kamari, A.; Yusoff, S.N.M.; Ishak, C.F.; Mohamed, A.; Hashim, N.; Isa, I.M. Biosorption of Cu (II), Pb (II) and Zn (II) ions from aqueous solutions using selected waste materials. *J. Encapsulation Adsorpt. Sci.* **2014**, *4*, 741. [CrossRef]
92. Song, S.-T.; Saman, N.; Johari, K.; Mat, H. Surface chemistry modifications of rice husk toward enhancement of Hg (II) adsorption from aqueous solution. *Clean Technol. Environ. Policy* **2014**, *16*, 1747–1755. [CrossRef]
93. Gautam, R.K.; Mudhoo, A.; Lofrano, G.; Chattopadhyaya, M.C. Biomass-derived biosorbents for metal ions sequestration: Adsorbent modification and activation methods and adsorbent regeneration. *J. Environ. Chem. Eng.* **2014**, *2*, 239–259. [CrossRef]
94. Husein, D.Z. Adsorption and removal of mercury ions from aqueous solution using raw and chemically modified Egyptian mandarin peel. *Desalination Water Treat.* **2013**, *51*, 6761–6769. [CrossRef]
95. Khoramzadeh, E.; Nasernejad, B.; Halladj, R. Mercury biosorption from aqueous solution by Sugarcane Bagasse. *J. Taiwan Inst. Chem. Eng.* **2013**, *44*, 266–269. [CrossRef]
96. Gomez-Aguilar, D.L.; Miranda, J.P.-R.; Salcedo-Parra, O.J. Fruit Peels as a Sustainable Waste for the Biosorption of Heavy Metals in Wastewater: A Review. *Molecules* **2022**, *27*, 2124. [CrossRef]
97. Dawood, S.; Sen, T.K.; Phan, C. Synthesis and characterisation of novel-activated carbon from waste biomass pine cone and its application in the removal of Congo red dye from aqueous solution by adsorption. *Water Air Soil Pollut.* **2014**, *225*, 1818. [CrossRef]
98. Marwa, B.A.; Khaled, W.; Victoria, S. Valorisation of Pine Cone as an Efficient Biosorbent for the Removal of Pb (II), Cd (II), Cu(II), and Cr(VI). *Adsorpt. Sci. Technol.* **2021**, *2021*, 6678530. [CrossRef]
99. Bazzo, A.; Adebayo, M.A.; Dias, S.L.P.; Lima, E.C.; Vaghetti, J.C.P.; de Oliveira, E.R.; Leite, A.J.B.; Pavan, F.A. Avocado seed powder: Characterization and its application for crystal violet dye removal from aqueous solutions. *Desalination Water Treat.* **2016**, *57*, 15873–15888. [CrossRef]
100. Leite, A.J.B.; Carmalin, S.A.; Thue, P.S.; Reis, G.; Dias, S.; Lima, E.C.; Vaghetti, J.C.; Pavan, F.A.; De Alencar, W.S. Activated carbon from avocado seeds for the removal of phenolic compounds from aqueous solutions. *Desalination Water Treat.* **2017**, *71*, 168–181. [CrossRef]
101. Garcia, E.-A.; Cristiani-Urbina, E. Effect of pH on hexavalent and total chromium removal from aqueous solutions by avocado shell using batch and continuous systems. *Environ. Sci. Pollut. Res.* **2019**, *26*, 3157–3173. [CrossRef]
102. Krishnan, K.A.; Sreejalekshmi, K.; Vimexen, V.; Dev, V.V. Evaluation of adsorption properties of sulphurised activated carbon for the effective and economically viable removal of Zn (II) from aqueous solutions. *Ecotoxicol. Environ. Saf.* **2016**, *124*, 418–425. [CrossRef]

103. Wahby, A.; Abdelouahab-Reddam, Z.; El Mail, R.; Stitou, M.; Silvestre-Albero, J.; Sepúlveda-Escribano, A.; Rodríguez-Reinoso, F. Mercury removal from aqueous solution by adsorption on activated carbons prepared from olive stones. *Adsorption* **2011**, *17*, 603–609. [CrossRef]
104. Yagub, M.T.; Sen, T.K.; Ang, M. Removal of cationic dye methylene blue (MB) from aqueous solution by ground raw and base modified pine cone powder. *Environ. Earth Sci.* **2014**, *71*, 1507–1519. [CrossRef]
105. Hayati, B.; Maleki, A.; Najafi, F.; Daraei, H.; Gharibi, F.; McKay, G. Super High Removal Capacities of Heavy Metals (Pb^{2+} and Cu^{2+}) Using CNT Dendrimer. *J. Hazard. Mater.* **2017**, *336*, 146–157. [CrossRef]
106. Biswas, S.; Mohapatra, S.S.; Kumari, U.; Meikap, B.C.; Sen, T.K. Batch and continuous closed circuit semi-fluidized bed operation: Removal of MB dye using sugarcane bagasse biochar and alginate composite adsorbents. *J. Environ. Chem. Eng.* **2020**, *8*, 103637. [CrossRef]
107. Kamari, S.; Shahbazi, A. Biocompatible Fe_3O_4@SiO_2-NH_2 nanocomposite as a green nanofiller embedded in PES–nanofiltration membrane matrix for salts, heavy metal ion and dye removal: Long–term operation and reusability tests. *Chemosphere* **2020**, *243*, 125282. [CrossRef] [PubMed]
108. Hameed, B.H.; Rahman, A.A. Removal of phenol from aqueous solutions by adsorption onto activated carbon prepared from biomass material. *J. Hazard. Mater.* **2008**, *160*, 576–581. [CrossRef] [PubMed]
109. Priya, A.; Yogeshwaran, V.; Rajendran, S.; Hoang, T.K.; Soto-Moscoso, M.; Ghfar, A.A.; Bathula, C. Investigation of mechanism of heavy metals (Cr^{6+}, Pb^{2+} & Zn^{2+}) adsorption from aqueous medium using rice husk ash: Kinetic and thermodynamic approach. *Chemosphere* **2022**, *286*, 131796. [CrossRef] [PubMed]
110. Ding, Z.; Hu, X.; Wan, Y.; Wang, S.; Gao, B. Removal of lead, copper, cadmium, zinc, and nickel from aqueous solutions by alkali-modified biochar: Batch and column tests. *J. Ind. Eng. Chem.* **2016**, *33*, 239–245. [CrossRef]
111. Yargic, A.S.; Yarbay, R.Z.; Sahin, N.; Onal, O.E. Assessment of toxic copper (II) biosorption from aqueous solution by chemically treated tomato waste. *J. Clean. Prod.* **2015**, *88*, 152–159. [CrossRef]
112. Kılıç, M.; Kırbıyık, Ç.; Çepelioğullar, Ö.; Pütün, A.E. Adsorption of heavy metal ions from aqueous solutions by bio-char, a by-product of pyrolysis. *Appl. Surf. Sci.* **2013**, *283*, 856–862. [CrossRef]
113. Salleh, M.A.M.; Mahmoud, D.K.; Karim, W.A.; Idris, A. Cationic and anionic dye adsorption by agricultural solid wastes: A comprehensive review. *Desalination* **2011**, *280*, 1–13. [CrossRef]
114. Ngulube, T.; Gumbo, J.R.; Masindi, V.; Maity, A. An update on synthetic dyes adsorption onto clay-based minerals. A state-of-art review. *J. Environ. Manag.* **2017**, *191*, 35–57. [CrossRef]
115. Senthil Kumar, P.; Palaniyappan, M.; Priyadharshini, M.; Vignesh, A.M.; Thanjiappan, A.; Sebastina Anne Fernando, P.; Tanvir Ahmed, R.; Srinath, R. Adsorption of basic dye onto raw and surface-modified agricultural waste. *Environ. Prog. Sustain. Energy* **2014**, *33*, 87–98. [CrossRef]
116. Shaikh, R.B.; Saifullah, B.; Rehman, F. Greener Method for the Removal of Toxic Metal Ions from the Wastewater by Application of Agricultural Waste as an Adsorbent. *Water* **2018**, *10*, 1316. [CrossRef]
117. Akram, M.; Bhatti, H.N.; Iqbal, M.; Noreen, S.; Sadaf, S. Biocomposite efficiency for Cr (VI) adsorption: Kinetic, equilibrium and thermodynamics studies. *J. Environ. Chem. Eng.* **2017**, *5*, 400–411. [CrossRef]
118. Olayebi, O.O.; Olagboye, S.A.; Olatoye, R.A.; Olufemi, A.S. Agricultural Waste Adsorbents for Heavy Metals Removal from Wastewater. *J. Phys. Chem. Sci.* **2017**, *5*, 5. [CrossRef]
119. Zhou, N.; Chen, H.; Xi, J.; Yao, D.; Zhou, Z.; Tian, Y. Biochars with excellent Pb (II) adsorption property produced from fresh and dehydrated banana peels via hydrothermal carbonization. *Bioresour. Technol.* **2017**, *232*, 204–210. [CrossRef] [PubMed]
120. Torab-Mostaedi, M.; Asadollahzadeh, M.; Hemmati, A.; Khosravi, A. Equilibrium, kinetic, and thermodynamic studies for biosorption of cadmium and nickel on grapefruit peel. *J. Taiwan Inst. Chem. Eng.* **2013**, *44*, 295–302. [CrossRef]
121. Pandharipande, S.; Kalnake, R.P. Tamarind fruit shell adsorbent synthesis, characterization and adsorption studies for Cr (VI) & Ni(II) ions from aqueous solution. *Int. J. Eng. Sci. Emerg. Technol.* **2013**, *4*, 83–89.
122. Hegazi, H.A. Removal of heavy metals from wastewater using agricultural and industrial wastes as adsorbents. *HBRC J.* **2013**, *9*, 276–282. [CrossRef]
123. Kyzas, G.Z.; Siafaka, P.I.; Pavlidou, E.G.; Chrissafis, K.J.; Bikiaris, D.N. Synthesis and adsorption application of succinyl-grafted chitosan for the simultaneous removal of zinc and cationic dye from binary hazardous mixtures. *Chem. Eng. J.* **2015**, *259*, 438–448. [CrossRef]
124. Ahalya, N.; Chandraprabha, M.N.; Kanamli, R.D.; Ramchandran, T.P. Adsorption of fast green onto coffee husk. *J. Chem. Eng. Res.* **2014**, *2*, 201–207.
125. Karmaker, S.; Uddin, M.N.; Ichikawa, H.; Fukumori, Y.; Saha, T.K. Adsorption of reactive orange 13 onto jackfruit seed flakes in aqueous solution. *J. Environ. Chem. Eng.* **2015**, *3*, 583–592. [CrossRef]
126. Jain, S.; Jayaram, R.V. Removal of basic dyes from aqueous solution by low-cost adsorbent: Wood apple shell (*Feronia acidissima*). *Desalination* **2010**, *250*, 921–927. [CrossRef]
127. Saeed, A.; Sharif, M.; Iqbal, M. Application potential of grapefruit peel as dye sorbent: Kinetics, equilibrium, and mechanism of crystal violet adsorption. *J. Hazard. Mater.* **2010**, *179*, 564–572. [CrossRef] [PubMed]
128. Pavan, F.A.; Camacho, E.S.; Lima, E.C.; Dotto, G.L.; Branco, V.T.; Dias, S.L. Formosa papaya seed powder (FPSP): Preparation, characterization, and application as an alternative adsorbent for the removal of crystal violet from aqueous phase. *J. Environ. Chem. Eng.* **2014**, *2*, 230–238. [CrossRef]

129. Xu, M.; Mckay, G. Removal of Heavy Metals, Lead, Cadmium, and Zinc, Using Adsorption Processes by Cost-Effective Adsorbents. In *Adsorption Processes for Water Treatment and Purification*, 1st ed.; Springer: Berlin/Heidelberg, Germany, 2017; pp. 109–138. [CrossRef]
130. Lin, Z.; Yang, Y.; Liang, Z.; Zeng, L.; Zhang, A. Preparation of Chitosan/Calcium Alginate/Bentonite Composite Hydrogel and Its Heavy Metal Ions Adsorption Properties. *Polymers* **2021**, *13*, 1891. [CrossRef] [PubMed]
131. Cruz-Lopes, L.; Macena, M.; Esteves, B.; Guine, R.P. Ideal pH for the adsorption of metal ions Cr^{6+}, Ni^{2+}, Pb^{2+} in aqueous solution with different adsorbent materials. *Open Agric.* **2021**, *6*, 115–123. [CrossRef]
132. Dawood, S.; Sen, T.K. Removal of anionic dye Congo red from aqueous solution by raw pine and acid-treated pine cone powder as adsorbent: Equilibrium, thermodynamic, kinetics, mechanism and process design. *Water Res.* **2012**, *46*, 1933–1946. [CrossRef]
133. Gupta, V.K.; Rastogi, A.; Nayak, A. Biosorption of nickel onto treated alga (*Oedogonium hatei*): Application of isotherm and kinetic models. *J. Colloid Interface Sci.* **2010**, *342*, 533–539. [CrossRef]
134. Sen, T.K. Adsorptive Removal of Dye (Methylene Blue) Organic Pollutant from Water by Pine Tree Leaf Biomass Adsorbent. *Processes* **2023**, *11*, 1877. [CrossRef]
135. Mondal, M.; Mukherjee, R.; Sinha, A.; Sarkar, S.; De, S. Removal of cyanide from steel plant effluent using coke breeze, a waste product of steel industry. *J. Water Process Eng.* **2019**, *28*, 135–143. [CrossRef]
136. Hameed, B.H.; Krishni, R.R.; Sata, S.A. A novel agricultural waste adsorbent for the removal of cationic dye from aqueous solutions. *J. Hazard. Mater.* **2009**, *162*, 305–311. [CrossRef]
137. Tran, H.N.; You, S.J.; Hosseini-Bandegharaei, A.; Chao, H.P. Mistakes and inconsistencies regarding adsorption of contaminants from aqueous solutions: A critical review. *Water Res.* **2017**, *120*, 88–116. [CrossRef]
138. Khatoon, H.; Rai, J.P.N. Agricultural waste materials as biosorbents for the removal of heavy metals and synthetic dyes—A review. *Octa J. Environ. Res.* **2016**, *4*, 208–229.
139. Kamsonlian, S.; Suresh, S.; Ramanaiah, V.; Majumder, C.; Chand, S.; Kumar, A. Biosorptive behaviour of mango leaf powder and rice husk for arsenic (III) from aqueoussolutions. *Int. J. Environ. Sci. Technol.* **2012**, *9*, 565–578. [CrossRef]
140. Kamsonlian, S.; Balomajumder, C.; Chand, S.; Suresh, S. Biosorption of Cd (II)andAs (III) ions from aqueous solution by teawaste biomass. *Afr. J. Environ. Sci. Technol.* **2011**, *5*, 1–7.
141. Okafor, P.C.; Okon, P.U.; Daniel, E.F.; Ebenso, E.E. Adsorption Capacity of Coconut (*Cocos nucifera* L.) Shell for Lead, Copper, Cadmium and Arsenic from aqueous solutions. *Int. J. Electrochem. Sci.* **2012**, *7*, 12354–12369. [CrossRef]
142. Tuzen, M.; Sari, A.; Mendil, D.; Uluozlu, O.D.; Soylak, M.; Dogan, M. Characterization of biosorption process of As (III) on green algae Ulothrix cylindricum. *J. Hazard. Mater.* **2009**, *165*, 566–572. [CrossRef]

Disclaimer/Publisher's Note: The statements, opinions and data contained in all publications are solely those of the individual author(s) and contributor(s) and not of MDPI and/or the editor(s). MDPI and/or the editor(s) disclaim responsibility for any injury to people or property resulting from any ideas, methods, instructions or products referred to in the content.

Article

Adsorption of Toxic Tetracycline, Thiamphenicol and Sulfamethoxazole by a Granular Activated Carbon (GAC) under Different Conditions

Risheng Li [1,2], Wen Sun [1,2], Longfei Xia [1,2], Zia U [3,*], Xubo Sun [1,2], Zhao Wang [1,2], Yujie Wang [4,*] and Xu Deng [5]

1. Shaanxi Provincial Land Engineering Construction Group Co., Ltd., Xi'an 710075, China
2. Key Laboratory of Degraded and Unused Land Consolidation Engineering, The Ministry of Natural Resources, Xi'an 710075, China
3. Department of Environmental Science and Engineering, School of Energy and Power Engineering, Xi'an Jiaotong University, Xi'an 710049, China
4. Department of Chemistry, College of Resource and Environment, Baoshan University, Baoshan 678000, China
5. School of Basic Medicine, Shaanxi University of Chinese Medicine, XiXian New Area, Xianyang 712046, China
* Correspondence: renjieling@stu.xjtu.edu.cn (Z.U.); wangyujieufo@163.com (Y.W.)

Citation: Li, R.; Sun, W.; Xia, L.; U, Z.; Sun, X.; Wang, Z.; Wang, Y.; Deng, X. Adsorption of Toxic Tetracycline, Thiamphenicol and Sulfamethoxazole by a Granular Activated Carbon (GAC) under Different Conditions. *Molecules* 2022, 27, 7980. https://doi.org/10.3390/molecules27227980

Academic Editors: Yongchang Sun and Dimitrios Giannakoudakis

Received: 11 October 2022
Accepted: 15 November 2022
Published: 17 November 2022

Publisher's Note: MDPI stays neutral with regard to jurisdictional claims in published maps and institutional affiliations.

Copyright: © 2022 by the authors. Licensee MDPI, Basel, Switzerland. This article is an open access article distributed under the terms and conditions of the Creative Commons Attribution (CC BY) license (https://creativecommons.org/licenses/by/4.0/).

Abstract: Activated carbon can be applied to the treatment of wastewater loading with different types of pollutants. In this paper, a kind of activated carbon in granular form (GAC) was utilized to eliminate antibiotics from an aqueous solution, in which Tetracycline (TC), Thiamphenicol (THI), and Sulfamethoxazole (SMZ) were selected as the testing pollutants. The specific surface area, total pore volume, and micropore volume of GAC were 1059.011 m^2/g, 0.625 cm^3/g, and 0.488 cm^3/g, respectively. The sorption capacity of GAC towards TC, THI, and SMZ was evaluated based on the adsorption kinetics and isotherm. It was found that the pseudo-second-order kinetic model described the sorption of TC, THI, and SMZ on GAC better than the pseudo-first-order kinetic model. According to the Langmuir isotherm model, the maximum adsorption capacity of GAC towards TC, THI, and SMZ was calculated to be 17.02, 30.40, and 26.77 mg/g, respectively. Thermodynamic parameters of ΔG^0, ΔS^0, and ΔH^0 were obtained, indicating that all the sorptions were spontaneous and exothermic in nature. These results provided a knowledge base on using activated carbon to remove TC, THI, and SMZ from water.

Keywords: adsorption; antibiotics; activated carbon; water treatment

1. Introduction

Antibiotics can be used to kill or inhibit the growth of bacteria. Although antibiotics are used in humans and animals, roughly 80% of their total usage is on livestock and poultry for human consumption. Antibiotics are routinely added to the food and water of livestock to promote growth and improve feed-use efficiency. In addition, antibiotics are injected into animals when they are sick or at high risk of getting sick. According to recent sales data of the world market, China is the biggest producer and user of antibiotics [1]. When antibiotics are used, the organisms cannot absorb antibiotics fully, so they are released into the environment in an active form [2]. In 2013 in China, more than 50 × 10^3 tons of antibiotics entered into aquifers, as stated in a report. The aquifers include the outflow of sewage processing plants [3], drinking water, groundwater [4], rivers and lakes [5], and seawater. Antibiotics have different half-lives, which some are long-lived [6], and their contagion rates in the environment have increased over the years. In a water-based environment, antibiotics are generally harmful, i.e., they prevent the ability to break down micro-organisms deposits, destructs the development of marine organisms as well encourage maturation in bacterial drug-opposing genetic codes [7]. Various research [8] have shown that any contact with antibiotics (μg/L–mg/L) creates an adverse effect and influence on the lives of water-based creatures, for example, the growth of their body and weight.

The residual of these antibiotics in water-based items comes into the human body with the help of the food chain and later mounts up via biological enhancement [9]. Since the majority of antibiotics are cancerogenic, teratogenic, and mutagenic, along with creating hormone-related issues, so using antibiotics causes serious interference with the anatomy of humans and the immune system [7]. Antibiotics are obtaining the identification of rising environmental contaminants, are classified as fractious bio-accumulative substances [10,11], as well as are considered harmful and toxic chemicals.

The outflow of Municipal treatment plants and pharmaceutical manufacturing plants are the basic sources of discharging antibiotics in water. According to Michael [12] and Rizzo [13], the cities' wastewater treatment plants are considered the main source of releasing antibiotics into the environment. It is important to dispose of residues of antibiotics before discharging wastewater into the environment. There is an urgent need for case studies to provide a cheap solution for eliminating antibiotics. Sera Budi Verinda et al. used ozonation to remove ciprofloxacin in wastewater, and the removal rate reached 83.5% [14]; Shang, K.F. et al. indicate that the combination of DBD plasma and PMS/PDS is an efficient pretreatment method for bio-treatment of refractory SMX [15].

The adsorption process is [16] easy to plan and feasible to function. The adsorbent should be a biomass adsorbent, such as agricultural waste, which is environmentally friendly and economical [17]. The adsorption method is utilized to remove organic pollutants from contaminated waters over the surface of the adsorbent [18]. Its application to eliminating antibiotics for approximately 30 different compounds has been reported so far [19]. The performance of adsorption processes is largely influenced by hydrogen bonding and electrostatic interactions. Different adsorbent materials are used for the elimination of antibiotics from the aqueous solution, for instance, clinoptilolite [20], soil [21], different kinds of activated carbons [22–25], calcium phosphate materials, and core-shell magnetic nanoparticles [26]. Compared with traditional adsorption materials, activated carbon has excellent porosity, large specific surface area, low cost, and environmental friendliness and is reported to be an effective adsorbent for eliminating trace pollutants [27]. Granular activated carbon is divided into stereotyped and unshaped particles. It is mainly made of coconut shell, nut shell, and coal, which is refined through a series of production processes. Its appearance is black amorphous particles; it has developed pore structure, good adsorption performance, high mechanical strength, low cost, etc. Therefore, granular activated carbon is widely used in drinking water, industrial water, wine, waste gas treatment, decolorization, desiccant, gas purification, and other fields [28,29].

At present, there is very little related research on antibiotics in wastewater through activated carbon, In this paper, three different kinds of antibiotics, including tetracycline, thiamphenicol, and sulfamethoxazole, were selected as the target pollutants, and a kind of activated carbon in granular form was used as the adsorbent. The adsorption capacity and mechanism were studied. This paper aimed to evaluate the removal efficiency of adsorption technology in treating antibiotics-loading wastewater and to promote the application of activated carbon in such a field.

2. Results and Discussion

2.1. Characterization of the Granular Activated Carbon

Figure 1A exhibits the outlook of GAC with an average length of 1–2 mm and diameter of 1 mm. Nitrogen adsorption/desorption at 77 k for the granular activated carbon was shown in Figure 1B, in which the GAC sample possessed a type I sorption isotherm. The details of BET are shown in Table 1: These results demonstrated that GAC was of high specific surface area and porous structure. In conclusion, this type of GAC might be an ideal adsorbent for removing antibiotics from wastewater. In Figure 1A, the SEM image of granular activated carbon shows that the external surface of GAC was multi-layer and rigid, which helped to increase the specific surface area of granular activated carbon. The measurement of zeta potential is a technique for calculating the surface charge of activated carbon in a colloidal solution. The graph of activated carbon zeta potential in the solution was shown as a function of pH in Figure 1B. The graph shows that the surface charge of activated carbon was linked with solution pH. The pH_{ZPC} zero-point charge of activated carbon is 4.0, indicating the charge on the activated carbon surface was positive when the solution pH was less than pH_{ZPC} while negative when the solution pH was greater than pH_{ZPC}. It can be seen from Figure 1D that the infrared spectrum of GAC has a C-O characteristic absorption peak near 1000 cm^{-1}, a C=C characteristic absorption peak near 1600 cm^{-1}, and an O-H stretching vibration peak near 3200 cm^{-1}.

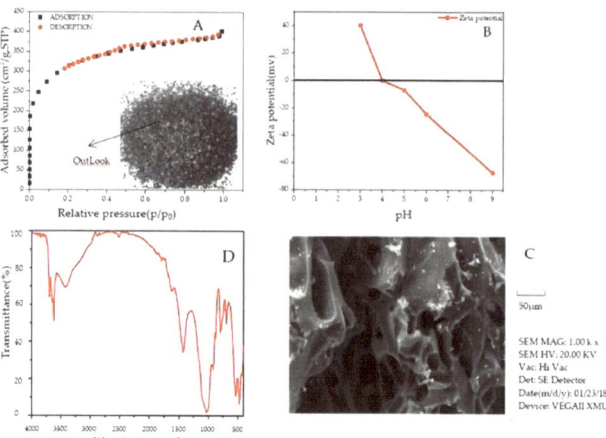

Figure 1. (**A**): Nitrogen (N_2) adsorption isotherm (**B**): zeta potential of the GAC. (**C**): SEM image (**D**): FTIR image.

Table 1. Shows the results of the BET.

BET	Specific Surface Area	Low Pressure ($p/p_0 < 0.1$) Adsorption Capacity	Hysteresis Loop ($p/p_0 = 0.2$)	Total Pore Volume	Micro Pore Volume
	1059.011 m^2/g	increased micropores	closed mesoporous	0.625 cm^3/g	0.488 cm^3/g

2.2. Effect of pH on GAC Adsorption

Under the conditions of environmental conditions of 25 °C, the antibiotic concentration of 25 mg/L, GAC dosage of 8 g/L, TC adsorption time of 100 min, and THI and SMZ adsorption time of 60 min: when the pH of the solution is 7, the effect of GAC on TC, THI, and SMZ the best adsorption efficiencies are 91.76%, 96.34%, and 94.23%, respectively (Figure 2).

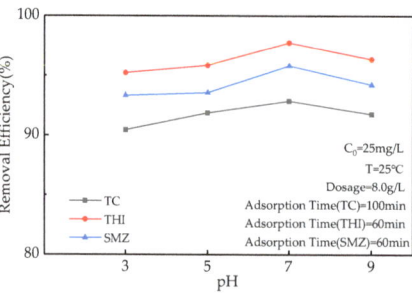

Figure 2. The relationship between pH and adsorption efficiency.

2.3. Effect of GAC Dosage on the Adsorption

Figure 3 shows the removal efficiency versus GAC dosage. When the GAC dosage increased from 2 to 8 mg/g, all the removal efficiencies for the three antibiotics increased remarkably and then slightly from 6 to 10 mg/g. A possible reason was that increased GAC dosage increased the sorbent surface area, the number of sorption sites, and the contact area increased [30–32]. Considering the removal efficiencies and economic benefits, this study selected a GAC dosage of 4 mg/g as an optimum dosage.

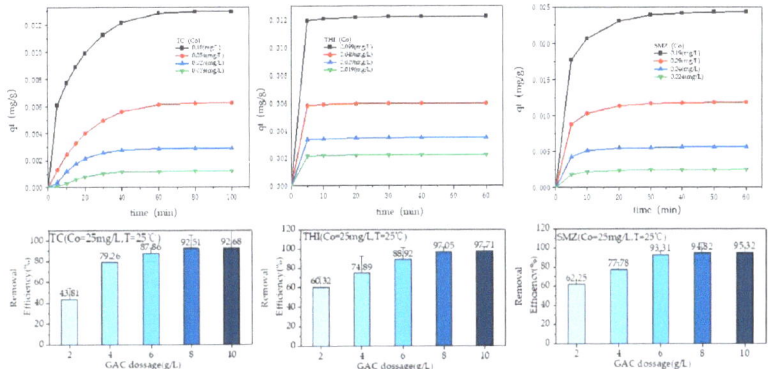

Figure 3. Influence of contact time on the adsorption of the three antibiotics by GAC under different initial concentrations.

2.4. Determination of the Adsorption Equilibrium Time

For the measurement of equilibrium time, q_t versus contact time (t) was represented in Figure 3. During the adsorption of TC, it was seen that the value of q_t increased quickly within the first 30 min and then slightly from 30 to 100 min for all the initial concentrations.

The possible cause was that there were enough adsorption sites on the surface of GAC during the initial stage of adsorption (0–30 min). As the contact time prolonged, the adsorption sites provided by GAC became fewer and fewer. As a result, q_t increased slightly (30–100 min). In the case of SMZ, the value of q_t increased significantly in the first 20 min, and from 20 to 60 min, the q_t value increased slightly. Different from TC and SMT, the adsorption process of THI was faster, in which after 10 min, no increase in the value of q_t was observed. Therefore, the equilibrium times for TC, SMT, and THI were set at 100 min, 60 min, and 10 min. In most cases, when the initial concentration of antibiotic was low while the removal efficiency was high (Figure 4), the possible reason was that at low concentrations, there are more available adsorbing sites for antibiotic molecules to absorb onto.

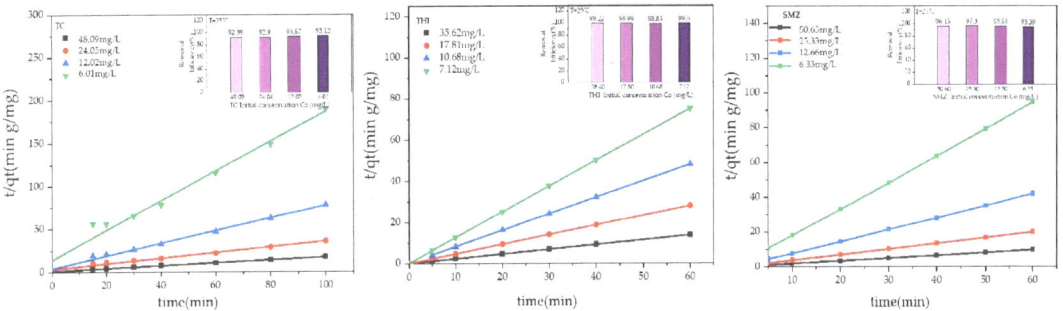

Figure 4. The pseudo-second-order kinetics for the adsorption of the three antibiotics on the GAC at various initial concentrations.

2.5. Sorption Kinetics

The pseudo-first-order and pseudo-second-order models were used to describe all the data shown in Figure 3. The results are given in Table 2, in which Equations (1) and (2) represent the mathematical formula of the two models.

$$\ln(q_{e,\exp} - q_t) = \ln(q_{e,cal}) - k_1 t \quad (1)$$

$$\frac{t}{q_t} = \frac{1}{k_2 q_{e,cal}^2} + \frac{1}{q_{e,cal}} t \quad (2)$$

where

$q_{e,\exp}$ = adsorption amount in (mg/g) at equilibrium
q_t = adsorption amount in (mg/g) at time t
k_1 = rate constant of pseudo-first-order (min^{-1})
k_2 = rate constant of pseudo-second-order [g/(mg min)].

Table 2. Parameters of the pseudo-first-order and pseudo-second-order kinetic models for the sorption of the three antibiotics onto GAC at 25 °C.

Sorbate	C_0 (mg/L)	$q_{e,\exp}$ (mg/g)	Pseudo-First-Order Model				Pseudo-Second-Order Model			
			$q_{e,cal}$ (mg/g)	K_1 (min^{-1})	R^2		$q_{e,cal}$ (mg/g)	K_2 [g/(min mg)]	$K_2 q_{e,cal}^2$ [mg/(min g)]	R^2
TC	6.01	0.52	2.332	0.023	0.77		0.49	0.704	0.169	0.68
	12.02	1.27	0.481	0.043	0.88		1.64	0.028	0.076	0.91
	24.04	2.78	2.773	0.053	0.98		3.46	0.014	0.167	0.98
	48.09	5.75	3.377	0.050	0.94		6.27	0.020	0.790	0.99
THI	7.12	0.79	15.7	0.109	0.73		0.80	5.14	3.306	1.00
	10.68	1.24	14.11	0.100	0.74		1.25	2.39	3.740	1.00
	17.81	2.13	12.23	0.091	0.76		2.14	2.65	12.158	1.00
	35.62	4.35	10.51	0.096	0.75		4.37	1.54	29.409	1.00
SMZ	6.33	0.63	4.57	0.021	0.78		0.65	0.811	0.349	0.99
	12.66	1.42	3.38	0.027	0.77		1.46	0.482	1.033	0.99
	25.32	3	1.11	0.038	0.84		3.10	0.185	1.781	0.99
	50.65	6.16	1.58	0.074	0.95		6.4	0.78	31.948	0.99

Figure 4 shows the linearized graph of t/q_t versus the time of pseudo-second-order kinetic. Table 2 contains all the parameters of these two kinetic models. The value of R^2 obtained from the pseudo-second-order model is given in Table 2, which is close to unity, indicating that the pseudo-second-order kinetic model best fits the adsorption of antibiotics on GAC. Furthermore, the experimental value of $q_{e,\exp}$ in (mg/g) agreed with the calculated value of $q_{e,cal}$ (mg/g). These results indicate that the rate of adsorption on GAC is controlled by chemisorption. At the same time, valency forces with an exchange, or there might be a sharing of the electrons in-between these four antibiotics and the GAC. According to Table 2 primary initial adsorption rate $K^2 q_{e,cal}^2$ gradually increased with the concentration of the primary four antibiotics, indicating that the greater value of concentration enhanced driving forces that can help to overcome the barrier of mass transfer-resistance in between the phases of solid and liquid.

Weber and Morris's intraparticle diffusion model was utilized to analyze the experimental data to determine the rate-limiting step during the adsorption process. All values are shown in Table 2.

$$q_t = k_{id} t^{1/2} + I \quad (3)$$

where

q_t (mg/g) = the removal amount at time t and reaction equilibrium
k_{id} (mg/g min$^{1/2}$) = the particle diffusion rate constant
I (mg/g) = intercept, which gives the information about the boundary layer effect. If the value of I is greater than the boundary layer has a greater effect.

If the plot between q_t versus $t^{1/2}$ is linear, then intraparticle diffusion takes place. When the plot also passes through the origin ($I = 0$), the rate-limiting is controlled by intraparticle diffusion. If the plots showed deviation from linearity, then it indicates the effect of the boundary layer. Figure 5 represents the plot of intraparticle for three antibiotics. This was observed that during whole time plots, the linear portion could not pass through the origin, indicating that both boundary layer and intraparticle diffusion occur during the adsorption of antibiotics on GAC.

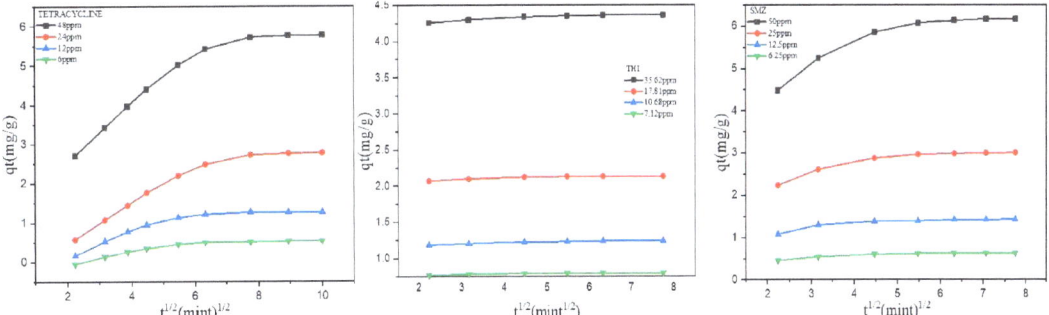

Figure 5. Intraparticle diffusion plots for the antibiotics sorption on the GAC.

In Table 3, the rate constant and intercept values are shown. The intercepts C of the straight lines fitted by TC, THI, and SMZ are not 0, indicating that internal diffusion is not the only step controlling the removal of antibiotics by GAC, and the adsorption rate should be controlled by both external diffusion and intraparticle diffusion. According to Figure 5, the plots of TC and SMZ have two different portions. Firstly, the steeper segment of the plot depicts external surface adsorption, and the second portion, slow adsorption, shows intraparticle diffusion.

Table 3. Intraparticle diffusion kinetic model parameters at 25 °C.

		Intra-Particle Diffusion Model					
Sorbate	C_0 (mg/L)	K_{id1} (mg/g min$^{1/2}$)	I_1 (mg/g)	R^2	K_{id2} (mg/g min$^{1/2}$)	I_2 (mg/g)	R^2
TC	6.01	0.13	−0.3	0.94	0.01	0.44	0.83
	12.02	0.25	−0.29	0.93	0.01	1.10	0.58
	24.04	0.47	−0.4	0.99	0.10	1.84	0.77
	48.09	0.66	1.31	0.98	0.13	4.58	0.77
THI	7.12	0.00	0.75	0.70	—	—	—
	10.68	0.01	1.17	0.85	—	—	—
	17.81	0.01	2.06	0.75	—	—	—
	35.62	0.01	4.23	0.79	—	—	—
SMZ	6.33	0.06	0.33	0.88	0.00	0.56	0.82
	12.66	0.13	0.82	0.76	0.01	1.32	0.86
	25.32	0.28	1.65	0.92	0.03	2.73	0.78
	50.65	0.6	3.21	0.94	0.09	5.49	0.76

Boyd's kinetic model was used to further examine the kinetic data to measure the slowest step involved in the adsorption process.

$$F(t) = 1 - \left(\frac{6}{\pi}\right)\sum_{n=1}^{\infty}\left(\frac{1}{n^2}\right)\exp(-n^2 B_t) \qquad (4)$$

where

$F(t) = q_t/q_e$ = ratio of the antibiotics adsorbed at time t and equilibrium;
B_t = function of $F(t)$

If the $F(t)$ value is higher than 0.85, then

$$B_t = 0.4977 - \ln(1 - F(t)) \tag{5}$$

If the $F(t)$ value is less than 0.85, then

$$B_t = \left(\sqrt{\pi} - \sqrt{\pi - \left(\frac{\pi^2 F(t)}{3}\right)}\right)^2 \tag{6}$$

As for the Boyd kinetic model, if the plot of B_t against t is linear and through the origin, it suggests that intraparticle diffusion controls the process of mass transfer. The sorption rate is controlled through film diffusion when the plot can be seen as nonlinear or linear but does not go through the origin. Figure 6 illustrates the Boyd plots for the three antibiotics on the GAC sample. The fact that Boyd plots were linear even though they were unable to pass through the origin was noticed. These results indicate the information regarding the fact that the major controlling process required for the adsorption procedure was, nonetheless, diffusion with the layer at the border. Table 4 lists the parameters of the Boyd kinetic model.

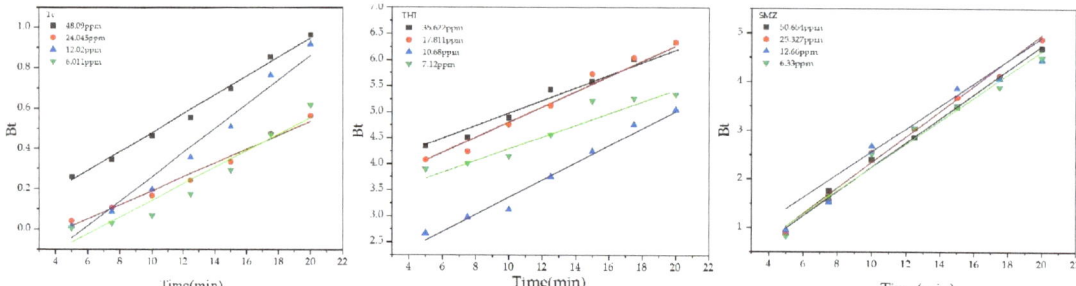

Figure 6. Plots of Bod kinetic model.

Table 4. The Boyd kinetic model parameters at 25 °C.

Sorbate	C_0 (mg/L)	Boyd Plot	
		Intercept	R^2
TC	6.01	−0.026	0.92
	12.02	−0.344	0.97
	24.04	−0.155	0.98
	48.09	0.008	0.99
THI	7.12	2.640	1.00
	10.68	1.700	0.97
	17.81	2.870	0.96
	35.62	3.220	0.94
SMZ	6.33	−0.168	0.97
	12.66	0.206	0.89
	25.32	−0.285	0.99
	50.65	−0.262	0.99

2.6. Sorption Isotherm

The adsorption process can be defined as the process of mass transfer of adsorbate at the boundary layer in-between the solid adsorbent and liquid phase. The adsorption isotherm is defined as the equilibrium relationship between solid adsorbent and adsorbate at a constant temperature. For example, the ratio between the amount of adsorbate absorbed on the solid and the remaining amount left in the aqueous solution at equilibrium. There are several adsorption isotherms models like Langmuir, Freundlich, and Temkin.

Experimental data can be fitted using these models to examine the suitability of the model. The information obtained from these models can be used for designing the adsorption process. The parameters of adsorption isotherm normally estimate the sorption ability of several adsorbents for specific adsorbates with predetermined reaction conditions. The performance of the sorption process depends not only on the rate at which mass transfer occurs but also on the sorbent–sorbate equilibrium concentration [33]. Three different isotherm models, called the Langmuir [34], Freundlich [35], and Temkin [21,22], were used. The mathematical formula of these isotherm models was shown as follows:

$$q_e = \frac{q_m k_L C_e}{1 + k_L C_e} \quad (7)$$

$$q_e = k_F C_e^{1/n} \quad (8)$$

$$q_e = B \ln(k_T C_e) \quad (9)$$

k_L = constant from Langmuir (L/mg)
C_e = adsorbate residual concentration (mg/L)
q_e = The amount of adsorbate per unit mass of sorbent (mg/g)
q_m = The maximum sorption capacity (mg/g)
K_F (mg/L$^{(1-1/n)}$ g) and n are constants of the Freundlich model
B is the Temkin constant of the sorption heat, and K_T (1/mg) stands for the constant of the Temkin isotherm.

Both Table 5 and Figure 7 show that the Langmuir model is best fitted to the experimental data than Freundlich and Temkin. The calculated values of q_{max} from Langmuir indicate that the maximum degree of adsorption capacity of the GAC sample to three antibiotics has been following the trend: THI > SMZ > TC.

Table 5. Parameters of the isotherm models describing the sorption of antibiotics on GAC at 25 °C.

Sorbate	Langmuir			Freundlich			Temkin		
	q_m (mg/g)	K_L (L/mg)	R^2	K_F (mg/L$^{(1-1/n)}$ g)	n	R^2	B	K_T	R^2
TC	17.02	0.154	0.93	2.28	1.2	0.93	1.944	4.093	0.88
THI	30.42	0.530	0.92	12.17	1.1	0.91	—	—	—
SMZ	26.77	0.155	0.95	3.51	1.1	0.94	2.930	3.800	0.97

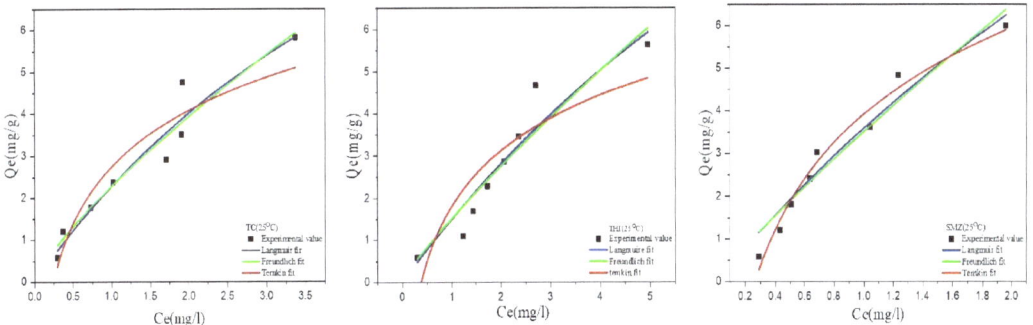

Figure 7. Plots of adsorption isotherms for the sorption of antibiotics on the GAC.

In 1974, Weber and Chakkravorti explained the Langmuir isotherm. The formula is given below.

$$R_L = \frac{1}{1 + k_L C_0} \quad (10)$$

where,

k_L = constant from Langmuir model (L/mg)
C_0 = adsorbate initial concentration (mg/L)

R_L is a separation factor that explains the nature of adsorption, and an explanation of this is given in Tables 5 and 6. According to this formula, the value of R_L in all this experiment is $0 < R_L < 1$, which suggests the favorable nature of the adsorption of antibiotics.

Table 6. Separation factor.

Value of R_L	Adsorption Nature
$0 < R_L < 1$	Favorable
$R_L = 0$	Irreversible
$R_L = 1$	Linear
$R_L > 1$	Unfavorable

2.7. Adsorption Thermodynamics

The Gibbs free energy change (ΔG^0) is an indication of the spontaneity of a chemical reaction and therefore is one of the most important criteria. It is calculated as follows:

$$\Delta G^0 = -RT \ln K \tag{11}$$

where R is the universal gas constant (8.314 J/(molK)), T is the absolute temperature in (K), and K is the thermodynamic equilibrium constant [36].

And

$$\Delta G^0 = \Delta H^0 - T\Delta S^0 \tag{12}$$

After combining Equations (13) and (14), we get

$$\text{Ln} K_L = \frac{-\Delta H^0}{R} \times \frac{1}{T} + \frac{\Delta S^0}{R} \tag{13}$$

By constructing a plot of $\text{Ln} K_L$ versus $1/T$, from the intercept calculated the change in entropy (ΔS^0) and by the slope, it is possible to calculate the change in enthalpy (ΔH^0) [37].

The value of K_L is obtained from the Langmuir isotherm model. The value of K_L is in (L/mg) so first, convert it into (L/g) by multiplying 1000 and then multiply it by antibiotic molecular mass. The value of K_L then becomes dimensionless because, for measuring the correct value of thermodynamics parameters, we need k in the dimensionless unit [38,39].

The thermodynamics parameters ΔG^0, ΔH^0, and ΔS^0 are shown in Table 7, and also plot between $\ln K$ versus $1/T$ is shown in Figure 8. The ΔG^0 negative values verified the process feasibility and the spontaneous nature of adsorption. As a rule of thumb, the decrease in the negative value of ΔG^0 with temperature increase indicates that the adsorption at higher temperatures is more favorable. This may be possible because, with the increase in temperature, the mobility of the adsorbate ion/molecule in the solution increases, and the adsorbate affinity to the adsorbent is high. On the contrary, an increase in the negative value of ΔG^0 with an increase in temperature implies that lower temperature facilitates adsorption.

Table 7. Thermodynamics parameters for the adsorption of three antibiotics on GAC.

Sorbate	T^0C	$\ln K$	ΔG^0 (kJ/mol)	ΔH^0 (kJ/mol)	ΔS^0 (kJ/mol K)
TC	15	11.522	−26.97		
	20	11.368	−27.67	−21.86	19.17
	25	11.215	−27.89		
THI	15	12.521	−29.98		
	20	12.412	−30.23	−33.58	−12.22
	25	12.049	−29.85		
SMZ	15	11.687	−27.98		
	20	11.132	−27.11	−81.39	−185.56
	25	10.545	−26.12		

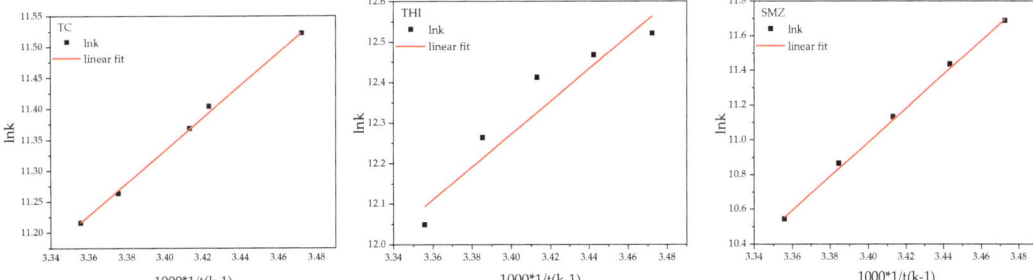

Figure 8. Plots of thermodynamics.

The negative value of enthalpy ΔH suggests that the sorption of three antibiotics is exothermic in nature. The positive value of ΔS^0 indicates high randomness at the solid/liquid phase with some structural changes in the adsorbate and the adsorbent. The negative value of ΔS^0 suggests that the adsorption process is enthalpy driven. The negative value of entropy change (ΔS^0) also means that the disorder of the solid/liquid interface decreases during the adsorption process, resulting in the escape of adsorbed ions/molecules from the solid. Therefore, the amount of adsorbate adsorbed will decrease.

K is the adsorption equilibrium constant of Langmuir isotherms. The value of k is in the unit (L/mg). First, it converts it into (L/g) and then multiplies it by the molecular formula of the antibiotic. Its value then becomes dimensionless, as we required a dimensionless value of k for measuring the value of ΔG (Table 1).

2.8. Regeneration Experiments

Through regeneration experiments, it was found that the adsorption capacity of granular activated carbon for TC, THI, and SMZ decreased with the increase in regeneration times (Figure 9). At an ambient temperature of 25 °C, the initial concentration of antibiotics was 25 mg/L, and the dosage of GAC was 8 g/L. The adsorption of TC on GAC reached saturation after 60 min. The maximum adsorption efficiencies of initial GAC, 1 GAC regeneration, and 5 regeneration GAC were 92.54%, 85.73%, and 62.14%. Under the same conditions, the maximum adsorption efficiencies of initial GAC, 1-time regeneration GAC, and 5-time regeneration GAC for THI and SMZ were: 96.32%, 32.33%; 91.23%, 85.21%; 70.26%, 50.19%, respectively. Compared with other adsorbents, such as attapulgite, granular activated carbon has the characteristics of easy preparation and better regeneration [40].

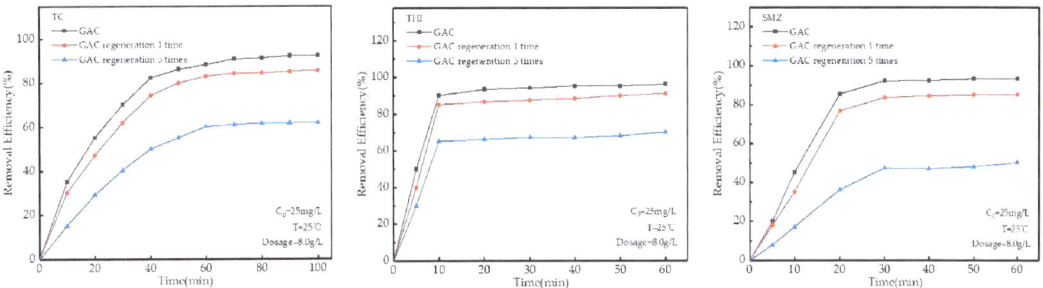

Figure 9. Relationship between GAC regeneration adsorption efficiency and time.

3. Materials and Methods

3.1. GAC Preparation

The Granular activated carbon used in these experiments was prepared using corn Stover collected from Shaanxi agriculture Technology Company (Xi'an, China). First, this corn Stover was put 1 month for air drying. After, it was chopped into small pieces of length nearly 5 cm and positioned inside an electrically operating container resistance furnace (LNB4-13Y; Haozhuang Co., Ltd., Shanghai, China). Under a nitrogen atmosphere, pieces of cornstalk were heated at 500 °C for 2 h and then 700 °C for 2 h.

Physical activation management was carried out in an activation furnace (HHL-1; Huatong Co., Ltd., Zhengzhou, Henan, China) through superheated steam at elevated temperatures (600 °C, 2.0 MPa) for 2 h to obtain the AC. The AC was also cleansed using 1.0 M HCl-HF (1:1, v/v) solution thrice and with ultrapure water a few times till the 7.0 pH value was obtained. After washing, AC drying was carried out at a temperature of 100 °C for a duration of 15 h and then crushed and pushed through a nylon mesh with an entrance of 5.0 mm. Finally, the AC's morphology was granular.

3.2. Chemicals

The physio-chemical characteristics and molecular structures of tetracycline (TC), thiamphenicol (THI), and sulfamethoxazole (SMZ) are given in Table 8. All the antibiotics have chromatographic clarity, purchased from J & K Scientific (Beijing, China). *N,N*-dimethylformamide with more than 99.9% purity was purchased by Sigma-Aldrich (Shanghai, China). Apparatus SPI-11-10T was used to prepare ultra-pure water (ULUPURE, Chengdu, Sichuan, China). The GAC sample used in this paper was supported by Fan et al. [41].

Table 8. Physicochemical properties of the three antibiotics.

Property	Tetracycline (TC)	Thiamphenicol (THI)	Sulfamethoxazole (SMZ)
Molecular formula & Chemical Structure	$C_{22}H_{24}N_2O_8$	$C_{12}H_{15}Cl_2NO_5S$	$C_{10}H_{11}N_3O_3S$
Molar mass	444.43	356.22	253.28
Solubility (25 °C)	1700 mg/L	2270 mg/L	459 mg/L

3.3. Methodology for Selecting Antibiotics

Wastewater influents contain several types of antibiotics; however, due to the limited availability of information, only a few antibiotics will be selected for this work. The criteria for selecting antibiotic classes were defined by considering (1) the relevance of antibiotic class to human medicine, (2) usage amongst the different animal species, and (3) their presence in wastewater treatment plants.

Based on the selection criteria mentioned above, the following antibiotics were selected for this work: (1) Tetracycline, (2) Thiamphenicol, and (3) Sulfamethoxazole.

3.4. Sorption Experiments

Stock solution preparation: 0.1000 g of TC, THI, and SMZ was dissolved in a 50 mL volumetric flask with ultra-pure water and then transferred into a 100 mL volumetric flask to obtain the stock solutions with a final concentration of 1 g/L respectively. The testing solutions with different concentrations were obtained by diluting the stock solutions with ultrapure water.

Effect of GAC dosage: on the sorption of TC by the GAC, tests were conducted in a 100 mL beaker containing 50 mL TC testing solution with 25 mg/L. For GAC dosage, the amount of GAC ranged from 2 to 8 g/L, while the temperature was fixed at 25 °C and the contact time was 100 min. The effect of GAC dosage was also conducted for the sorption of THI and SMZ.

Determination of the adsorption equilibrium time: to investigate the sorption of TC onto the GAC, batch experiments were carried out using a 100 mL beaker containing

0.4000 g GAC and 50 mL TC testing solutions with various preliminary concentrations. Each beaker was put into a thermostatic reciprocating shaker (ZHWY-2102C, Changzhou Guowang Instrument Manufacturing Co., Ltd., Changzhou, Jiangsu, China) at 180 r/min and 25 °C in the dark. The sample was withdrawn by a 5 mL syringe from one beaker after shaking for 5 min and passed through a 0.45 µm filter. The sampling times were 5, 10, 15, 20, 30, 40, 50, 60, 80, and 100 min. The same steps were carried out to investigate the sorption of SMZ and THI on GAC, respectively. Filter liquor was used to determine the residual concentrations of the three antibiotics.

Adsorption isotherm: for measuring the adsorption isotherm, a 250 mL beaker containing 50 mL testing solution with a different initial concentration of TC and 0.4000 g of GAC was agitated in the dark under 180 rpm at 25 °C. Samples were withdrawn after 100 min using a 5 mL pipette tip and filtered using a 0.45 µm filter membrane. This filter-out solution of TC was used for the remaining concentration analysis to find adsorption isotherm representation. For investigating the adsorption isotherm of THI and SMZ, the same procedure was followed.

Adsorption Thermodynamics: a 250 mL beaker containing 50 mL testing solution with 25 mg/L of TC and 0.4000 g of GAC was agitated in the dark under 180 rpm at 15, 20, and 25 °C. Samples were withdrawn after 100 min using a 5 mL pipette tip and filtered using a 0.45 µm filter membrane. This filter-out solution of TC was used for the remaining concentration analysis. For investigating the adsorption isotherm of THI and SMZ, the same procedure was followed.

Analysis: physisorption analyzer (ASAP-2020, Micrometrics, Beijing, China) was used to measure the prepared GAC characteristics, like surface area, pore volume, and pore diameter at 77 K temperature. The Brunauer-Emmett-Teller (BET) method was used to measure surface area, and the Barrett-Joyner-Halenda (BJH) method was used to measure pore size distribution. Scanning electron (TM-1000; Hitachi, Tokyo, Japan) was used to describe the GAC surface morphology and porous structure. The Zeta potential analyzer (Malvern Zetasizer Nano S90, Shanghai, China) was used to monitor GAC Zeta potential value. The GAC was first converted into powder form and then submerged in NaCl solution (1 mmol = litter) to make a blend (0.1 g = 1 L). To adjust the solution's pH value from 3 to 9, HCl or NaOH was used. This solution was placed in an ultrasonic treatment apparatus (25 °C, 40 kHz) for 30 min. After the ultrasonic treatment, this solution was kept for 24 h, and its zeta potential value was measured using a zeta potential analyzer. The Ultraviolet-visible spectrophotometer (model: SP-1915, Spectrum, Shanghai, China) was used to measure the antibiotics' residual concentration. The calibration curves for TC, SMZ, and THI were given as y = 0.0324x + 0.0307, y = 0.0635x + 0.0007, and y = 0.1455x + 0.0068, respectively. For measuring the sorption capacity on GAC following formula can be used.

$$q_t = \frac{C_0 - C}{M} \times V \tag{14}$$

where

q_t = adsorption capacity at time t (mg/g)
C_0 = adsorbate initial concentration in (mg/L)
C = adsorbate residual concentration at time t (mg/L)
V = volume of solution (L)
M = mass of sorbent (g)

The removal efficiency at different initial concentrations of antibiotics was calculated by using the following formula:

$$R.E = \frac{C_0 - C_E}{C_0} \times 100\% \tag{15}$$

where

$R.E$ = removal efficiency (%)
C_0 = initial concentration (mg/L)
C_E = equilibrium concentration (mg/L)

3.5. Regeneration Experiments

Saturated GAC (8.0000 ± 0.0004) g adsorbing TC, THI, and SMZ was placed in a quartz glass reactor, and N_2 was used as a protective gas and placed in a microwave oven for irradiation, microwave power 730 w, microwave time 180 s [42], and carried out microwave regeneration test.

4. Conclusions

This research investigated the experimental results of granular activated carbon adsorbing different antibiotics, including tetracycline, thiamphenicol, and sulfamethoxazole, from an aqueous solution. The BET experiment indicated that GAC had a high specific surface area of approximately 1059 m^2/g and a high pore volume of 0.625 cm^3/g, meaning that it is a very useful adsorbent for antibiotics removal. The equilibrium adsorption data of TC, THI, and SMZ were well expressed by the Langmuir isotherm model, and maximum adsorption capacities were 17.02, 30.40, and 26.77 mg/g, respectively. The kinetic data of sorption were well described by the pseudo-second-order model, indicating the sorption of the three antibiotics onto GAC involving valency forces through sharing or exchange of electrons between sorbent and sorbate. The Weber- Morris intraparticle diffusion model and Boyd kinetic model proved the main controlling step for the adsorption process was diffusion through the boundary layer. Using the adsorption equilibrium constant obtained from Langmuir isotherm, the thermodynamic parameter ΔG^0 was calculated to tell the spontaneity of the adsorption reaction. The values of ΔH^0 and ΔS^0 were also obtained from a slope and intercept of the relationship between lnK and reaction temperature. A negative value of ΔG^0 and a Negative value of ΔH^0 confirmed the spontaneous and exothermic nature of the adsorption process. In conclusion, GAC could be employed as an environmentally friendly adsorbent for the removal of antibiotics from water and wastewater.

Author Contributions: Methodology, R.L.; software, W.S.; validation, R.L. and W.S.; formal analysis, L.X.; investigation, Z.U.; resources, Z.W.; data curation, Y.W.; writing—original draft preparation, W.S.; writing—review and editing, R.L.; visualization, X.D.; supervision, X.S.; project administration, R.L.; funding acquisition, R.L. All authors have read and agreed to the published version of the manuscript.

Funding: This research was funded by Shaanxi Province science and technology resources open sharing platform project (Program No. 2021PT-03) and high-precision determination of lead isotopes in environmental samples by MC-ICP-MS (DJNY2021-31).

Institutional Review Board Statement: Not applicable.

Informed Consent Statement: Not applicable.

Data Availability Statement: Not applicable.

Acknowledgments: This study was funded by the Shaanxi Province science and technology resources open sharing platform project (Program No. 2021PT-03) and high-precision determination of lead isotopes in environmental samples by MC-ICP-MS (DJNY2021-31). The sponsors had no role in the design, execution, interpretation, or writing of the study. We are grateful to the anonymous reviewers whose comments have helped to clarify and improve the text.

Conflicts of Interest: The authors declare no conflict of interest.

Sample Availability: Samples of the GAC are available from the authors.

References

1. Zhou, L.-J.; Ying, G.-G.; Liu, S.; Zhao, J.-L.; Yang, B.; Chen, Z.-F.; Lai, H.-J. Occurrence and fate of eleven classes of antibiotics in two typical wastewater treatment plants in South China. *Sci. Total. Environ.* **2013**, *452–453*, 365–376. [CrossRef] [PubMed]
2. Hirsch, R.; Ternes, T. Occurrence of antibiotics in the aquatic environment. *Sci. Total Environ.* **1999**, *225*, 109–118. [CrossRef]
3. Zhang, Q.-Q.; Ying, G.-G.; Pan, C.-G.; Liu, Y.-S.; Zhao, J.-L. Comprehensive Evaluation of Antibiotics Emission and Fate in the River Basins of China: Source Analysis, Multimedia Modeling, and Linkage to Bacterial Resistance. *Environ. Sci. Technol.* **2015**, *49*, 6772–6782. [CrossRef]
4. Huang, C.H.; Renew, J.E. Assessment of potential antibiotic contaminants in water and preliminary occurrence analysis. *J. Contemp. Water Res. Educ.* **2011**, *1*, 324–332. [CrossRef]

5. Jiang, L.; Hu, X.; Yin, D.; Zhang, H.; Yu, Z. Occurrence, distribution and seasonal variation of antibiotics in the Huangpu River, Shanghai, China. *Chemosphere* **2011**, *82*, 822–828. [CrossRef]
6. Daughton, C.G.; Ternes, T.A. Pharmaceuticals and personal care products in the environment agents of subtle change. *Environ. Health Perspect.* **1999**, *107*, 907–938. [CrossRef] [PubMed]
7. Gao, P.; Mao, D.; Luo, Y.; Wang, L.; Xu, B.; Xu, L. Occurrence of sulfonamide and tetracycline-resistant bacteria and resistance genes in aquaculture environment. *Water Res.* **2012**, *46*, 2355–2364. [CrossRef]
8. Lai, H.-T.; Hou, J.-H.; Su, C.-I.; Chen, C.-L. Effects of chloramphenicol, florfenicol, and thiamphenicol on growth of algae Chlorella pyrenoidosa, Isochrysis galbana, and Tetraselmis chui. *Ecotoxicol. Environ. Saf.* **2009**, *72*, 329–334. [CrossRef]
9. Pruden, A.; Pei, R.; Storteboom, H.; Carlson, K.H. Antibiotic Resistance Genes as Emerging Contaminants: Studies in Northern Colorado. *Environ. Sci. Technol.* **2006**, *40*, 7445–7450. [CrossRef]
10. Chen, K.; Zhou, J. Occurrence and behavior of antibiotics in water and sediments from the Huangpu River, Shanghai, China. *Chemosphere* **2014**, *95*, 604–612. [CrossRef]
11. Michael, I.; Rizzo, L.; McArdell, C.S.; Manaia, C.M.; Merlin, C.; Schwartz, T.; Dagot, C.; Fatta-Kassinos, D. Urban wastewater treatment plants as hotspots for the release of antibiotics in the environment: A review. *Water Res.* **2013**, *47*, 957–995. [CrossRef] [PubMed]
12. Rizzo, L.; Manaia, C.; Merlin, C.; Schwartz, T.; Dagot, C.; Ploy, M.C.; Michael, I.; Fatta-Kassinos, D. Urban wastewater treatment plants as hotspots for antibiotic resistant bacteria and genes spread into the environment: A review. *Sci. Total Environ.* **2013**, *447*, 345–360. [CrossRef] [PubMed]
13. Mehrjouei, M.; Mueller, S. Energy consumption of three different advanced oxidation methods for water treatment: A cost-effectiveness study. *J. Clean. Prod.* **2014**, *65*, 178–183.
14. Sera, B.V.; Muflihatul, M.; Eko, Y. Degradation of ciprofloxacin in aqueous solution using ozone microbubbles: Spectroscopic, kinetics, and antibacterial analysis. *Heliyon* **2022**, *8*, e10537.
15. Shang, K.; Morent, R.; Wang, N.; Wang, Y.; Peng, B.; Jiang, N.; Lu, N.; Li, J. Degradation of sulfamethoxazole (SMX) by water falling film DBD Plasma/Persulfate: Reactive species identification and their role in SMX degradation. *Chem. Eng. J.* **2021**, *431*, 133916. [CrossRef]
16. Ahmaruzzaman, M. Adsorption of phenolic compounds on low-cost adsorbents: A review Advances in Colloid and Interface Science. *Process Biochem.* **2008**, *143*, 148–167.
17. Han, R.; Ding, D.; Xu, Y.; Zou, W.; Wang, Y.; Li, Y.; Zou, L. Use of rice husk for the adsorption of congo red from aqueous solution in column mode. *Bioresour. Technol.* **2008**, *99*, 2938–2946. [CrossRef]
18. Lam, A.; Rivera, A.; Rodrídguez-Fuentes, G. Theoretical study of metronidazole adsorption on clinoptilolite. *Microporous Mesoporous Mater.* **2001**, *49*, 157–162. [CrossRef]
19. Homem, V.; Santos, L. Degradation and removal methods of antibiotics from aqueous matrices—A review. *J. Environ. Manag.* **2011**, *92*, 2304–2347. [CrossRef]
20. Rabølle, M.; Spliid, N.H. Sorption and mobility of metronidazole, olaquindox, oxytetracycline and tylosin in soil. *Chemosphere* **2000**, *40*, 715–722. [CrossRef]
21. Ahmed, M.J. Microwave assisted preparation of microporous activated carbon from Siris seed pods for adsorption of metronidazole antibiotic. *Chem. Eng. J.* **2013**, *214*, 310–318. [CrossRef]
22. Ahmed, M.J.; Theydan, S.K. Microporous activated carbon from Siris seed pods by microwave-induced KOH activation for metronidazole adsorption. *J. Anal. Appl. Pyrolysis* **2012**, *99*, 101–109. [CrossRef]
23. Rivera-Utrilla, J.; Prados-Joya, G.; Sánchez-Polo, M.; Ferro-García, M.; Bautista-Toledo, I. Removal of nitroimidazole antibiotics from aqueous solution by adsorption/bioadsorption on activated carbon. *J. Hazard. Mater.* **2009**, *170*, 298–305. [CrossRef] [PubMed]
24. Ocampo-Pérez, R.; Orellana-Garcia, F.; Sánchez-Polo, M.; Rivera-Utrilla, J.; Velo-Gala, I.; López-Ramón, M.; Alvarez-Merino, M. Nitroimidazoles adsorption on activated carbon cloth from aqueous solution. *J. Colloid Interface Sci.* **2013**, *401*, 116–124. [CrossRef] [PubMed]
25. Dan, C.; Jian, D. Core-shell magnetic nanoparticles with surface-imprinted polymer coating as a new adsorbent for solid phase extraction of metronidazole. *Anal. Methods* **2013**, *5*, 722–728.
26. He, J.; Dai, J.; Zhang, T.; Sun, J.; Xie, A.; Tian, S.; Yan, Y.; Huo, P. Preparation of highly porous carbon from sustainable α-cellulose for superior removal performance of tetracycline and sulfamethazine from water. *RSC Adv.* **2016**, *6*, 28023–28033. [CrossRef]
27. Fan, Y.; Zheng, C.; Hou, H. Preparation of Granular Activated Carbon and Its Mechanism in the Removal of Isoniazid, Sulfamethoxazole, Thiamphenicol, and Doxycycline from Aqueous Solution. *Environ. Eng. Sci.* **2019**, *36*, 1027–1040. [CrossRef]
28. He, Y.; Cheng, F. The Affect Analysis of Microwave Regeneration on the Adsorption Properties of Granular Activated Carbon. *J. Tianjin Inst. Urban Constr.* **2012**, *18*, 6.
29. Ilavsk, J.; Barlokov, D.; Marton, M. Removal of Selected Pesticides from Water Using Granular Activated Carbon. *IOP Conf. Ser. Earth Environ. Sci.* **2021**, *900*, 012011. [CrossRef]
30. Gagliano, E.; Falciglia, P.P.; Zaker, Y. Microwave regeneration of granular activated carbon saturated with PFAS. *Water Res.* **2021**, *15*, 198. [CrossRef]
31. Liu, P.; Wang, Q. Sorption of sulfadiazine, norfloxacin, metronidazole, and tetracycline by grangranular activated carbon: Kinetics, mechanisms, and isotherms. *Water Air Soil Pollut. J.* **2017**, *228*, 1027–1040. [CrossRef]

32. Ho, Y.-S. Review of second-order models for adsorption systems. *J. Hazard. Mater.* **2006**, *136*, 681–689. [CrossRef] [PubMed]
33. El-Khaiary, I.M.; Malash, G.F. On the use of linearized pseudo-second-order kinetic equations for modeling adsorption systems. *Desalination* **2010**, *257*, 93–101. [CrossRef]
34. Mannarswamy, A.; Munson-McGee, S.H.; Steiner, R.; Andersen, P.K. D-optimal experimental designs for Freundlich and Langmuir adsorption isotherms. *Chemom. Intell. Lab. Syst.* **2009**, *97*, 146–151. [CrossRef]
35. Langmuir, I. The adsorption of gases on plane surfaces of glass, mica and platinum. *J. Am. Chem. Soc.* **1918**, *40*, 1361–1403. [CrossRef]
36. Freundlich, H. Ueber die adsorption in Loesungen. *J. Phys. Chem.* **1906**, *57*, 385–470.
37. Atkins, P.; Paula, J. *Physical Chemistry*; W. H. Freeman and Company: New York, NY, USA, 2010.
38. Chang, R.; Thoman, J.W. *Physical Chemistry for the Chemical Sciences*; University Science Books, AIP Publishing LLC.: New York, NY, USA, 2014.
39. Çaliskan, E.; Göktürk, S. Adsorption characteristics of sulfamethoxazole and metronidazole on activated carbon. *Sep. Sci. Technol.* **2010**, *45*, 244–255. [CrossRef]
40. Zhang, S.Y. *Adsorption and Removal of Sulfonamide Antibiotics in Water by Granular Activated Carbon and Modified Attapulgite*; Lanzhou Jiaotong University: LanZhou, China, 2020.
41. Zhu, Y.; Liu, L. Experimental Study of Sewage Plant Tail Water Treatment by Granular Active Carbon Immobilized Catalyst fen-Ton-like. *Shandong Chem. Ind.* **2018**, *8*, 21–24.
42. Qin, Q.; Chen, Y. Optimization of the Modified Components of mn-sn-ce/gac Particle Electrode by Response Surface Method. *Ind. Water Treat.* **2019**, *1*, 54–59.

Article

Study of the Influence of the Wastewater Matrix in the Adsorption of Three Pharmaceuticals by Powdered Activated Carbon

Marina Gutiérrez [1], Paola Verlicchi [1] and Dragana Mutavdžić Pavlović [2,*]

[1] Department of Engineering, University of Ferrara, Via Saragat 1, 44122 Ferrara, Italy
[2] Department of Analytical Chemistry, Faculty of Chemical Engineering and Technology, University of Zagreb, Trg Marka Marulića 19, 10000 Zagreb, Croatia
* Correspondence: dmutavdz@fkit.hr; Tel.: +385-1-4597-204

Abstract: The use of powdered activated carbon (PAC) as an absorbent has become a promising option to upgrade wastewater treatment plants (WWTPs) that were not designed to remove pharmaceuticals. However, PAC adsorption mechanisms are not yet fully understood, especially with regard to the nature of the wastewater. In this study, we tested the adsorption of three pharmaceuticals, namely diclofenac, sulfamethoxazole and trimethoprim, onto PAC under four different water matrices: ultra-pure water, humic acid solution, effluent and mixed liquor from a real WWTP. The adsorption affinity was defined primarily by the pharmaceutical physicochemical properties (charge and hydrophobicity), with better results obtained for trimethoprim, followed by diclofenac and sulfamethoxazole. In ultra-pure water, the results show that all pharmaceuticals followed pseudo-second order kinetics, and they were limited by a boundary layer effect on the surface of the adsorbent. Depending on the water matrix and compound, the PAC capacity and the adsorption process varied accordingly. The higher adsorption capacity was observed for diclofenac and sulfamethoxazole in humic acid solution (Langmuir isotherm, $R^2 > 0.98$), whereas better results were obtained for trimethoprim in the WWTP effluent. Adsorption in mixed liquor (Freundlich isotherm, $R^2 > 0.94$) was limited, presumably due to its complex nature and the presence of suspended solids.

Keywords: adsorption; diclofenac; sulfamethoxazole; trimethoprim; dissolved organic matter; powdered activated carbon; wastewater

Citation: Gutiérrez, M.; Verlicchi, P.; Mutavdžić Pavlović, D. Study of the Influence of the Wastewater Matrix in the Adsorption of Three Pharmaceuticals by Powdered Activated Carbon. *Molecules* 2023, 28, 2098. https://doi.org/10.3390/molecules28052098

Academic Editor: Yongchang Sun

Received: 26 January 2023
Revised: 19 February 2023
Accepted: 20 February 2023
Published: 23 February 2023

Copyright: © 2023 by the authors. Licensee MDPI, Basel, Switzerland. This article is an open access article distributed under the terms and conditions of the Creative Commons Attribution (CC BY) license (https://creativecommons.org/licenses/by/4.0/).

1. Introduction

Pharmaceuticals are one of the most common organic micropollutants found in wastewater. Among pharmaceuticals, nonsteroidal anti-inflammatory drugs (NSAIDs) and antibiotics are in the spotlight due to their high consumption and/or recalcitrant nature [1,2]. In wastewater treatment plants (WWTPs), the core treatment is biological degradation, and even though some pharmaceuticals are highly biodegradable, the concentrations found in WWTP effluent are still an issue, because WWTPs are not designed to remove them [3]. In this way, advanced treatments have gained interest and have been gradually implemented over the last few years [4–6]. These treatments include activated carbon adsorption (in powder or granules), which offers the advantage of being able to remove a wide range of compounds. This is particularly relevant in wastewater treatment, where organic micropollutants often occur as a "cocktail", and tens to hundreds of substances can be found at the same time [7]. Indeed, the removal of many recalcitrant substances relies almost uniquely on sorption processes [8]. Powdered activated carbon (PAC) is known for being a very flexible option that can be added to existing treatment lines (i.e., addition to the biological tank) or as a polishing treatment to treat the secondary effluent (i.e., in a new contact tank) [9,10]. PAC is used to enhance the removal of substances via adsorption and to promote diverse removal mechanisms with the main aim of obtaining synergistic effects (such as enhanced biodegradation).

Adsorption onto activated carbon, which is driven by the properties of the adsorbent and absorbate as well as the water quality, is a complex process that is not fully understood [11]. When considering the application of PAC in WWTPs, the potential enhancement of the removal of pharmaceuticals depends on many factors for which the extent of their influence is challenging to consider altogether [12]. Activated carbon is a porous adsorbent of which the adsorption capacity depends on its surface properties (specific surface area, pore volume, functional chemical groups) [13]. Pharmaceuticals instead depend on their physicochemical characteristics (compound charge, hydrophobicity, molecular weight, etc.) to be adsorbed, which usually leads to competition effects such that some substances tend to adsorb more easily than others. Moreover, the overall adsorption process depends also on the conditions in which it occurs, such as the water matrix. The constituents of the water matrix and, more specifically, the dissolved organic matter (DOM), may influence the adsorption process. DOM is formed by many fractions that differ in size (building blocks, biopolymers, humic acids, low molecular weight organics, etc.), which may limit the adsorption of pharmaceuticals by blocking the pores on the PAC surface or by direct competition for the adsorption sites [14,15]. Pharmaceuticals may also interact with the DOM present in the liquid phase or the DOM that is adsorbed onto the PAC surface. The results of the interaction may enhance or diminish the adsorption onto PAC, depending on the tested compounds and conditions [11,16,17]. In our previous paper [18], the removal efficiencies of a vast selection of organic pollutants at trace levels were compared and discussed in different MBR coupled to PAC treatment configurations. Specifically, the PAC was added either inside the biological tank of the bioreactor (mixed liquor) or in a post-treatment unit to treat the MBR permeate. Results indicated that the effect of the PAC dosage point was dependent on the compound under study. In general, the presence of suspended solids and the complex nature of the mixed liquor requires higher doses of PAC compared to the MBR permeate to achieve equivalent removal efficiencies [19]. Due to the presence of the micro- or ultra-filtration membranes in the bioreactor, the MBR permeate is free of suspended solids [20]. In light of the foregoing information, the use of synthetic water matrices (i.e., humic acid solution) can act as a means to understand the adsorption process under certain DOM constituents [17].

Because the adsorption onto PAC is influenced by the adsorbate's properties, three pharmaceuticals (Figure 1), namely diclofenac (DCF), sulfamethoxazole (SMX) and trimethoprim (TMP), were selected. These compounds have been subjected to several studies due to their low-to-moderate removal in WWTPs and the potentially harmful effects on the environment that they may entail [21,22]. Additionally, they differ in hydrophobicity (octanol–water partition coefficient, K_{ow}) and charge at the pH of the wastewater. These parameters are commonly used to predict the effectiveness of the addition of PAC on the wastewater treatment line [23].

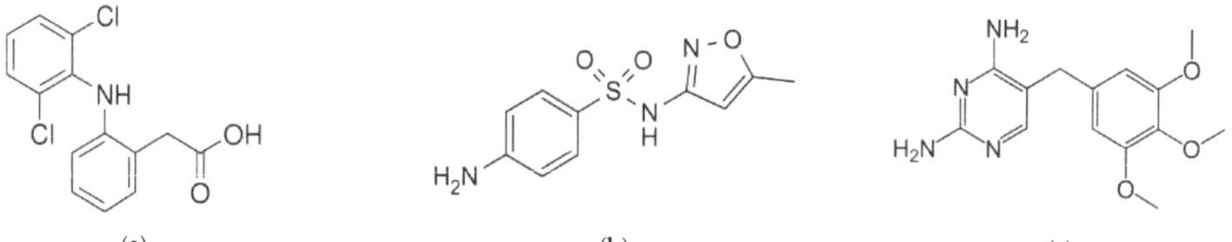

Figure 1. Molecular structure of (**a**) diclofenac, (**b**) sulfamethoxazole and (**c**) trimethoprim.

DCF is a non-steroidal anti-inflammatory drug (NSAID) used to treat pain and inflammatory disorders. Banned in many countries of Southeast Asia [24], DCF was selected for the first Watch List (Decision 2015/495) for Union-wide monitoring in Europe [25]. DCF is a weak electrolyte (Figure 1) with high hydrophobicity (logK_{ow} = 4.3) [26] that predominates in its anionic form in wastewater [27]. Compared to other NSAIDs, DCF shows inefficient and variable removal efficiencies in WWTPs, with great discrepancy among the literature data [28]. In this way, the addition of PAC has been shown to be beneficial, albeit the removal efficiencies found in the literature still show great variability (32–99%) [18].

SMX is a bacteriostatic antibiotic commonly prescribed in combination with TMP. SMX is an anionic compound with very low hydrophobicity ($\log K_{ow}$ = 0.8) [26]. Although these chemical properties are disadvantageous for the direct adsorption of SMX onto PAC, it has been shown that the addition of this adsorbent to the biological tank of a membrane bioreactor (MBR) may increase the removal of this compound [29]. Moreover, batch adsorption isotherms obtained by Li et al. [8] estimated a maximum adsorption of (q_m) 0.017 mg/g.

TMP is an antibiotic that was included in the European Watch List in 2020 (Decision EU 2020/1161) and was maintained in the recent update published in 2022 (Decision 2022/1307) [30,31], for which its monitoring and related research are promoted. It is a relatively hydrophilic compound with a low tendency for sorption onto the sludge of the WWTPs [21]. It has been generally classified as moderately removed in WWTPs, with better removal efficiencies when PAC is added inside the bioreactor compared to when it is added as a post-treatment [18].

Adsorption batch experiments and mathematical models can be useful tools to examine the conditions under which PAC adsorption takes place and to predict adsorbent response to such conditions [32]. In previous research, the application of adsorption models has been of great value to understand the mechanisms of adsorption of certain pollutants on porous adsorbents such as PAC [33]. However, only a few studies have applied these models to study the effect of varying concentrations of DOC [6] and DOM constituents [15–17] in the adsorption of pharmaceuticals in wastewater. Indeed, the potential positive effect of these interactions between DOM and pharmaceuticals has been rarely documented and quantified [11,16]. With regard to the adsorbates, the influence of their physicochemical properties (polarity, charge and hydrophobicity) in adsorption has been the subject of study in the literature [6], but rarely has the literature focused on the subsequent potential competition effect caused by their different affinity towards PAC under realistic conditions of wastewater treatment [34].

For all the above-mentioned reasons, the adsorption of three pharmaceuticals onto PAC is investigated under different conditions using four different approaches. First, the adsorption capacity of PAC for the three target compounds is determined experimentally, and the adsorption process is described by three isotherm models (Linear, Langmuir and Freundlich) and three kinetic models (Lagergren's pseudo-first-order, pseudo-second-order and intraparticle diffusion model (IPD)). Second, the potential competition effect among pharmaceuticals due to their different physicochemical properties (charge, hydrophobicity) is evaluated. Third, the potential influence of the water matrix is assessed by comparing the adsorption process (kinetics, isotherms, experimental adsorption capacity) in ultra-pure water, humic acid solution, permeate of a full-scale membrane bioreactor (MBR) and mixed liquor from the nitrification tank of the same MBR. Finally, the interaction between the pharmaceuticals and the DOM on the adsorption onto PAC is studied.

2. Results and Discussion

2.1. Effect of the Contact Time and Initial Concentration of Pharmaceuticals

In order to determine the time needed to reach the maximum adsorption of the target pharmaceuticals onto PAC, adsorption experiments at various contact times were conducted. For this purpose, individual solutions of each pharmaceutical were tested at three concentrations (5, 15 and 25 mg/L) with two concentrations of PAC (0.1 and 1 g/L) at various contact times (10, 20, 30, 40 and 50 min and 1, 2, 4, 6, 12, 18 and 24 h). Figure 2 shows the removal (in terms of % of adsorption) of the three target compounds over time (10 min–24 h) in Milli-Q water with 1 and 0.1 g/L of PAC. All target compounds reached the equilibrium within 24 h, with very little difference in the adsorption between 18 h and 24 h, indicating that no more molecules could be adsorbed. In this way, 24 h was taken as the equilibrium time for the adsorption isotherms.

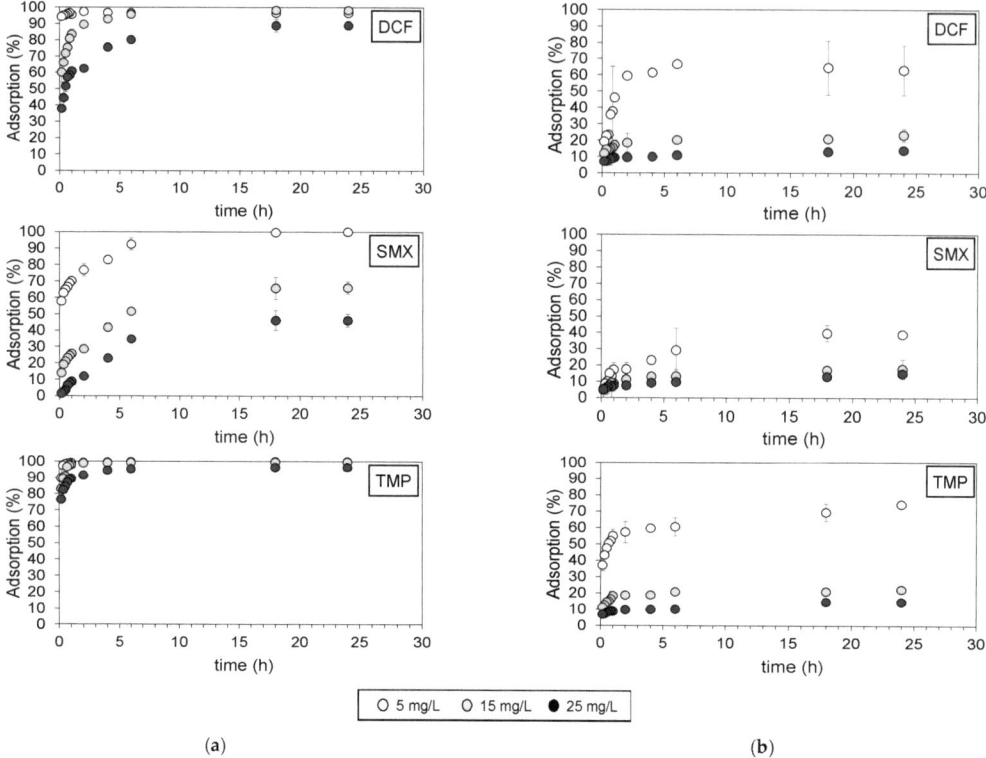

Figure 2. Kinetics of adsorption of DCF, SMX and TMP at three different concentrations in Milli-Q water with (**a**) 1 g/L of PAC and (**b**) 0.1 g/L of PAC at different contact times (10 min–24 h). Error bars indicate the standard deviation.

TMP was almost completely removed by the adsorption onto PAC (1 g/L) at 24 h (96–99.8%), followed by DCF (88–97%) and SMX (46–99.9%). TMP was the compound with the fastest kinetics, with removal from 77% (for the initial concentration of 25 mg/L) to 90% (for the initial concentration of 5 mg/L) in the first 10 min of agitation. SMX instead was the compound with the lowest rates and overall adsorption, depending on the initial concentration. In the first 10 min, 57% of the compound was adsorbed for 5 mg/L (maximum adsorption of 99.9% after 24 h), whereas only 1.5% was adsorbed for 25 mg/L (at 24 h, only 46% of the compound was adsorbed).

Lower adsorption percentages were found when PAC was added at 0.1 g/L for all OMPs in all tested shaking times (Figure 2). At an initial concentration of 5 mg/L, adsorption of 39%, 63% and 74% was obtained at 24 h for SMX, DCF and TMP, respectively. On the other hand, maximum adsorption of approximately 15% was obtained for all OMPs at 25 mg/L. From Figure 2, it can be seen that the adsorption rate was particularly high within the first ten minutes in all tested OMPs with an initial concentration of 15 and 25 mg/L. The adsorption percentage that was reached in 10 min was approximately 50% of the total adsorption that was obtained after 24 h. As an example, the adsorption of DCF at 10 min was 7%, and after 24 h, it was 15% (Figure 2b). After the first ten minutes, the rate of adsorption was considerably low until it reached equilibrium.

Note that adsorption seems to be dependent on the initial concentration of the pharmaceuticals (Figure 2). Higher adsorptions were found at the initial concentration of 5 mg/L compared to 15 and 25 mg/L for DCF, SMX and TMP, indicating that the adsorption of pharmaceuticals onto activated carbon is dependent on their initial concentration.

2.2. Kinetics

Sorption of the tested pharmaceuticals has proved to be a fast process overall. However, the behavior of each compound was different, presumably due to their physicochemical properties and the initial conditions of the experiments (i.e., the concentrations of the adsorbent and the adsorbate).

The kinetics models were applied to all the tested concentrations of pharmaceuticals and PAC, even though the behavior should be the same regardless of the initial concentration ratios. In this way, a vast data set was covered, and the reliability of the results obtained was assured. The kinetics followed a pseudo-second-order model for the three target compounds at the two tested PAC concentrations (1 and 0.1 g/L). The sorption rate constants (k_1 and k_2), $q_{e,\,calc.}$, $q_{e,\,exp.}$ and correlation coefficients (R^2) are shown in Table 1. The correlation coefficients of the adjustments were very close to the unity ($R^2 > 0.98$), with no significant differences between the experimental q_e ($q_{e,\,exp.}$) and calculated values ($q_{e,\,calc.}$), suggesting that the sorption is governed by the number of available active sites [34,35]. The lowest $q_{e,\,exp.}$ values were obtained via SMX in all tested concentrations. The maximum amounts of adsorbed pharmaceuticals onto PAC ($q_{e,\,exp.}$) were the highest at the lowest PAC concentration and vice versa. The values obtained were in the range of 4826–24,083 µg/g for 1 g/L of PAC and 19,398–37,184 µg/g for 0.1 g/L of PAC based on the three tested OMPs. Furthermore, higher initial concentrations (C_0) of tested pharmaceuticals led to higher values of $q_{e,\,exp.}$. The results indicate that PAC adsorption capacity in the equilibrium increases when it is found at low concentrations with high concentrations of the absorbate (i.e., pharmaceutical) in the solution.

Table 1. Sorption kinetic parameters of DCF, SMX and TMP in ultra-pure water with 1 g/L and 0.1 g/L of added PAC. C_0 indicates the initial concentration of the pharmaceutical, and $q_{e,\,exp.}$ indicates the values of q_e obtained experimentally.

Compound	PAC (g/L)	C_0 (mg/L)	$q_{e,\,exp.}$ (µg/g)	Pseudo-First Order			Pseudo-Second Order		
				$q_{e,\,calc.}$ (µg/g)	k_1 (1/min)	R^2	$q_{e,\,calc.}$ (µg/g)	k_2 (g/µg·min)	R^2
DCF	1	5	4826	206	1.61×10^{-4}	0.135	5000	4.00×10^{-3}	1.000
	1	15	14,729	3185	2.07×10^{-3}	0.806	14,286	6.13×10^{-6}	1.000
	1	25	22,240	11,163	1.15×10^{-3}	0.851	25,000	1.14×10^{-6}	0.993
	0.1	5	31,442	127,321	6.91×10^{-5}	0.743	33,333	1.13×10^{-6}	0.999
	0.1	15	34,852	29,971	4.61×10^{-4}	0.430	33,333	1.29×10^{-6}	0.996
	0.1	25	34,869	229,192	4.61×10^{-4}	0.877	33,333	6.92×10^{-7}	0.995
SMX	1	5	4999	2085	5.07×10^{-3}	0.987	5000	8.16×10^{-6}	0.999
	1	15	9910	11,527	6.91×10^{-4}	0.902	11,111	8.71×10^{-7}	0.992
	1	25	11,549	23,206	4.61×10^{-4}	0.877	14,286	1.88×10^{-7}	0.979
	0.1	5	19,398	43,813	2.30×10^{-4}	0.868	20,000	4.55×10^{-7}	0.992
	0.1	15	26,490	138,038	9.21×10^{-5}	0.784	25,000	7.41×10^{-8}	0.996
	0.1	25	37,016	233,830	6.91×10^{-5}	0.940	33,333	3.83×10^{-8}	0.984
TMP	1	5	4992	82	2.07×10^{-3}	0.598	5000	4.00×10^{-7}	1.000
	1	15	14,933	606	1.84×10^{-3}	0.543	14,286	4.90×10^{-5}	1.000
	1	25	24,083	3151	1.15×10^{-3}	0.657	25,000	8.00×10^{-6}	1.000
	0.1	5	37,184	25,439	4.61×10^{-4}	0.844	33,333	1.13×10^{-6}	0.997
	0.1	15	33,416	126,765	6.91×10^{-5}	0.561	33,333	1.5×10^{-6}	0.999
	0.1	25	36,425	229,826	6.91×10^{-5}	0.917	33,333	6.43×10^{-7}	0.989

As anticipated in Figure 2, the fastest kinetics (k_2) were obtained with the lowest pharmaceutical concentration (5 mg/L) for all the tested compounds except for TMP at 1 g/L PAC. Depending on the initial concentration, k_2 changes by at least one order of magnitude, indicating that the initial OMP concentration seems to have a significant role in the sorption kinetics.

In parallel with pseudo-first and second-order models, the data were fit into the IPD. Previous studies have reported that the removal of pharmaceuticals via adsorption onto PAC does not fit IPD because the rate of adsorption is controlled by one or more stages [34,36,37]. Nevertheless, although the model does not fit, it is known that in porous adsorbents such as PAC, intraparticle diffusion plays a major role in the adsorption process [36]. The IPD model may be useful for predicting the reaction pathways and the rate-controlling step in the transport from the water matrix to the active sites [38]. For porous adsorbents such as PAC, the adsorption process is differentiated into four stages, as stated originally by Walter and Weber [39]. The first stage is the transfer of the target pollutant to the solution (bulk transport); the second is the film diffusion, in which the adsorbate is transported from the bulk phase to the external surface of the PAC; the third stage is the diffusion of the adsorbate molecules along the adsorbent surface or through the pores (i.e., intraparticle diffusion), which is defined as the rate-controlling step in the IPD model; and the fourth stage is when the adsorption bond is formed between the OMP and the active site. When the adsorption onto PAC is controlled via intraparticle diffusion, stages 1, 2 and 4 occur very quickly, and the intraparticle diffusion is the only rate-controlling step. As a result, the IPD model adjustment should show a linear relationship between $t^{1/2}$ and q_t with a null intercept ($C = 0$). In the original linear form of the IPD [40], only the second, third and fourth stages are considered because bulk transport does not directly relate to the solid–liquid sorption process.

In this study, the q_t versus $t^{1/2}$ plot showed multi-linearity with three different slopes, indicating that the adsorption process is governed by a multistep mechanism, which is differentiated via the three abovementioned stages [38]. The fitting data for the model are shown in Table 2. First of all, it can be seen that the values of the rate constant (k_{id}) follow the following order: $k_{id1} > k_{id2} > k_{id3}$, for all the samples tested. k_{id} values are also at a higher C_0. The fact that the third stage is the lowest is due to it corresponding to the equilibrium state in which intraparticle diffusion gradually slows down; the OMPs come into contact with the active sites, and the final equilibrium is reached, resulting in the corresponding plots being nearly horizontal lines [41,42]. Regarding constant C, the results show that $C \neq 0$ in all samples tested, and increasing values from C_1 to C_3 were found for DCF and TMP. Constant C is associated with the thickness of the boundary layer, which implies that there is a higher boundary layer effect within the pores (and active sites) of the activated carbon compared to the outer surface. According to Rudzinski and Plazinski [43], negative values of intercept C observed for SMX can be explained by the presence of a "subsurface" region close to the surface of PAC on which the concentration of the adsorbate is different from that in the bulk phase, which affects the rate of the surface reactions (pseudo-second-order kinetics) at the initial times.

Table 2. Intraparticle diffusion model constants and correlation coefficients for DCF, SMX and TMP sorption at different initial concentrations (C_0), together with the respective regression coefficients (R^2). The PAC concentration used for the model is 1 g/L.

Compound	C_0 (mg/L)	Intraparticle Diffusion								
		First Phase			Second Phase			Third Phase		
		k_{p1} (µg/g min$^{1/2}$)	C_1	R^2	k_{p2} (µg/g min$^{1/2}$)	C_2	R^2	k_{p3} (µg/g min$^{1/2}$)	C_3	R^2
DCF	5	0.402	93.03	0.921	0.078	95.45	1.000	−0.006	96.751	0.979
	15	15.657	129.49	0.996	2.322	242.66	0.999	−0.013	295.08	1.000
	25	26.047	109.34	0.976	11.310	192.21	0.962	0.029	443.70	1.000
SMX	5	2.596	50.37	0.985	1.926	54.95	0.962	0.074	97.16	1.000
	15	7.479	20.77	0.977	8.644	−8.99	1.000	0.187	191.09	1.000
	25	8.524	−23.18	0.947	14.061	−97.14	0.991	0.033	229.71	1.000
TMP	5	3.813	78.36	0.889	0.348	96.39	0.995	0.020	99.15	0.781
	15	12.055	211.25	0.958	0.958	285.80	0.998	0.072	296.07	0.938
	25	15.330	337.67	0.982	3.291	420.88	0.999	0.321	470.19	0.933

Although the adsorption onto PAC is governed via a multi-step mechanism, and intraparticle diffusion is not the only rate-limiting stage in the adsorption process, the IPD model was useful for understanding the sorption mechanisms of the three target pharmaceuticals. In general, it can be deduced that once the compound passes through the boundary layer from the bulk phase to the external surface of the PAC, it slowly moves from the macropores to the active sites, decreasing the adsorption rate. The adsorption also seems to be determined by a boundary layer effect that increases its relevance in the latter stages of the adsorption process.

2.3. Sorption Isotherms in Ultra-Pure Water and Competition Effect

Pharmaceutical concentrations tested for isotherm determination were in the range of 5–25 mg/L, whereas PAC concentration was between 0.1 and 1 g/L. The equilibrium time was set at 24 h. PAC concentrations were selected in accordance with the literature [8,29,44]. The pharmaceutical concentrations were the lowest allowed by the analytical method. Due to the high adsorption capacity of the PAC, lower concentrations would be almost completely adsorbed and would not be detectable. The sorption coefficient of the linear sorption, together with the sorption parameters derived from the Langmuir and Freundlich models, and regression coefficients (R^2) are listed in (Table 3, individual solutions). From the analysis of the results obtained, it emerges that regression coefficients for linear sorption (0.783–0.96) were significantly lower than the Langmuir and Freundlich models ($p < 0.05$) for all three tested compounds, which means that the model does not fit the adsorption data very well. On the other hand, no significant differences were found between Langmuir and Freundlich for DCF and TMP, whereas the Freundlich model provided better R^2 coefficients for SMX. This finding is in agreement with previous studies in the literature [36,37,45], where very similar R^2 values were obtained, and no statistical analyses were performed to determine the best-fitting equation. Langmuir and Freundlich isotherms are the most used for describing the adsorption of porous adsorbents in wastewater, but further investigations on isotherm modelling may be needed to best describe the adsorption process.

Considering K_d, q_m and K_F parameters, the results observed in the kinetic studies were confirmed once again, and the pharmaceuticals that were better adsorbed in PAC are as follows: TMP, DCF and SMX. On the other hand, the term $1/n$ of Freundlich isotherm represents the intensity of adsorption. Because the values found for all compounds are less than 1, it can be assumed that there is a good affinity between the adsorbates and the adsorbent and that chemical adsorption occurs.

Complex mixtures of pharmaceuticals are usually found in urban wastewater [7]. The diversity of the nature and target use of these substances is usually reflected in their physicochemical properties (e.g., hydrophobicity, solubility, charge, molecular weight). When PAC is applied for the removal of pollutants in wastewater, adsorption depends on the interactions between the compound and the adsorbent surface, and the aforesaid pharmaceutical properties may be the key to understanding and predicting the adsorption tendency of the compound. For these reasons, it is of great importance to understand the competitive effect among pharmaceuticals when considering adsorption onto activated carbon. The target compounds are expected to be adsorbed to varying degrees, and the competition for the adsorption sites may vary depending on the initial concentration and physicochemical properties of the compound.

To evaluate the competitive effect of DCF, SMX and TMP, the results of adsorption isotherms of the mixture (Table 3) and kinetic studies (Table 4) are presented. As for individual solutions, no statistical differences among isotherm models were found, except for the significantly lower R^2 of linear isotherm in the case of DCF ($p < 0.05$). Despite the lack of significance, the regression coefficients for the Langmuir isotherm are slightly higher, indicating that monolayer adsorption on the PAC surface is assumed and that the differences in adsorption among pharmaceuticals depend on the affinity of the compound to the PAC surface. Although there were no differences between the maximum adsorption capacity (q_m) among the pharmaceuticals, the Langmuir adsorption constants (K_L) were significantly lower for SMX ($p = 0.018$). Similarly, K_d and K_F showed significant differences among tested compounds ($p < 0.05$), with higher coefficient values in the following order: TMP > DCF > SMX.

Table 3. Distribution coefficient (K_d), Langmuir and Freundlich isotherm constants obtained in individual solutions of each pharmaceutical (DCF, SMX and TMP) and the mixture of the three pharmaceuticals in ultra-pure water. N.A. (not applicable) indicates that the parameters could not be obtained, as the residual concentration found in the liquid phase was too low to conduct the modelling.

Compound	PAC Conc. (g/L)	Linear Sorption		Langmuir Isotherm			Freundlich Sorption		
		K_d (mL/g)	R^2	q_m (μg/g)	K_L (L/mg)	R^2	$1/n$	K_F (mg/g) (mL/mg)$^{1/n}$	R^2
				Individual solutions					
DCF	0.1	1777.9	0.895	33,333	0.300	0.963	0.281	12,673.9	0.925
	0.25	1949.2	0.836	33,333	0.429	0.979	0.215	14,368.6	0.953
	0.5	2980.6	0.783	25,000	2.000	0.978	0.271	14,099.6	0.991
	1	7167.1	0.855	20,000	5.000	0.946	0.574	10,802.1	0.999
SMX	0.1	1896.0	0.960	50,000	0.100	0.915	0.439	8918.7	0.959
	0.25	1634.0	0.947	33,333	0.150	0.936	0.392	7972.1	0.967
	0.5	1756.3	0.902	25,000	0.444	0.956	0.380	7667.1	0.985
	1	1417.6	0.937	16,667	0.300	0.912	0.520	3947.0	0.990
TMP	0.1	2618.9	0.833	50,000	0.400	0.951	0.178	23,576.4	0.801
	0.25	3712.3	0.820	50,000	0.667	0.972	0.249	21,407.6	0.961
	0.5	5939.4	0.852	33,333	1.500	0.967	0.393	16,565.9	0.998
	1	19,820.0	0.910	25,000	4.444	0.939	N.A.	N.A.	N.A.
				Mixture					
DCF	0.1	1806.2	0.852	33,333	0.375	0.987	0.203	16,008.9	0.955
	0.25	1063.6	0.763	16,667	1.000	1.000	0.124	11,356.0	0.900
	0.5	1348.8	0.785	16,667	1.000	0.995	0.212	9531.0	0.962
	1	1390.1	0.707	12,500	1.000	0.996	0.125	9464.5	0.823
SMX	0.1	423.03	0.935	50,000	0.010	0.031	0.587	1222.7	0.423
	0.25	385.32	0.965	14,286	0.054	0.924	0.670	1012.2	0.950
	0.5	280.47	0.976	10,000	0.053	0.869	0.709	629.0	0.998
	1	162.16	0.868	3333	0.375	0.968	0.137	1783.6	0.652
TMP	0.1	2442.7	0.901	50,000	0.200	0.832	0.257	17,243.7	0.597
	0.25	2036.5	0.733	25,000	2.000	0.999	0.128	19,171.9	0.964
	0.5	2716.2	0.730	25,000	2.000	0.999	0.151	17,870.4	0.955
	1	5636.6	0.843	25,000	2.000	0.995	0.239	14,485.5	1.000

When comparing isotherm coefficients between individual solutions and the mixture, only K_F and K_d were found to be significantly lower in the mixture compared to the individual solution in SMX. In this sense, although no significant differences were found for the other parameters (q_m, K_L) and compounds (DCF, TMP), higher values were found in the individual solutions, indicating that there is some competition effect, especially for SMX.

Kinetics studies were used to evaluate whether the rate and mechanism of adsorption of each compound in the mixture (Table 4) varied in comparison with individual solutions (Table 1). In this regard, the same experimental conditions were applied to compare the results with accuracy. In the mixture, the results show that the compounds followed a pseudo-second order equation (Table 4), with no significant differences between $q_{e,exp}$ and $q_{e,calc}$ ($p > 0.05$). Despite there being no differences between the kinetic coefficients (k_2) for the individual solutions and the mixture, the $q_{e,exp}$ values were overall greater in the individual solutions compared to the mixture ($p = 0.01$). Indeed, considering the removal of the compounds in the liquid phase, removal efficiencies were found to be between 23% and

27% higher in the individual solutions at 5 mg/L of the three tested compounds compared to the mixture (e.g., 62.9% versus 36.9% for DCF).

Table 4. Sorption kinetic parameters for the mixture of DCF, SMX and TMP in ultra-pure water with 0.1 g/L of added PAC.

Compound	C_0 (mg/L)	$q_{e, exp.}$ (µg/g)	Pseudo-First Order			Pseudo-Second Order		
			$q_{e, calc.}$ (µg/g)	k_1 (1/min)	R^2	$q_{e, calc.}$ (µg/g)	k_2 (g/µg·min)	R^2
DCF	5	18,467	136,395	9.21×10^{-5}	0.878	16,667	1.33×10^{-6}	0.991
	15	28,362	40,272	1.84×10^{-4}	0.851	33,333	4.09×10^{-7}	0.993
	25	15,957	242,493	2.30×10^{-5}	0.387	16,667	1.2×10^{-6}	0.990
SMX	5	5716	48,865	6.909×10^{-5}	0.801	10,000	1.81×10^{-7}	0.890
	15	4742	147,809	1.382×10^{-5}	0.633	5000	2.72×10^{-6}	0.991
	25	35,771	237,684	6.909×10^{-5}	0.740	33,333	2.81×10^{-7}	0.997
TMP	5	25,531	32,464	2.30×10^{-4}	0.820	25,000	1.45×10^{-6}	0.999
	15	25,310	134,122	4.61×10^{-5}	0.435	25,000	1.23×10^{-6}	0.990
	25	25,948	239,111	4.61×10^{-5}	0.874	25,000	5.71×10^{-7}	0.941

In general, TMP was the compound that adsorbed best at PAC. TMP is the only tested pharmaceutical that is found mainly in its cationic form at the pH of water and wastewater (pH 6–8) (Figure 3). Regardless of their other physicochemical properties, cationic compounds are proven to be well removed on PAC hybrid systems, due to the electrostatic interactions with the negatively charged surface of most manufactured PACs [5,6]. The charge of ionizable compounds is the conducting parameter that determines their adsorption onto PAC [12]. In water and wastewater, DCF and SMX are present mainly in their anionic form, and the expected removal via PAC is lower. In the absence of positive electrostatic interactions, hydrophobicity (measured logK_{ow}) becomes the critical factor for predicting adsorption. SMX is an anionic compound with very low hydrophobicity (logK_{ow} = 0.79) compared to that of DCF (logK_{ow} = 4.26). Both properties are responsible for the lower adsorption of SMX onto PAC in the tested conditions.

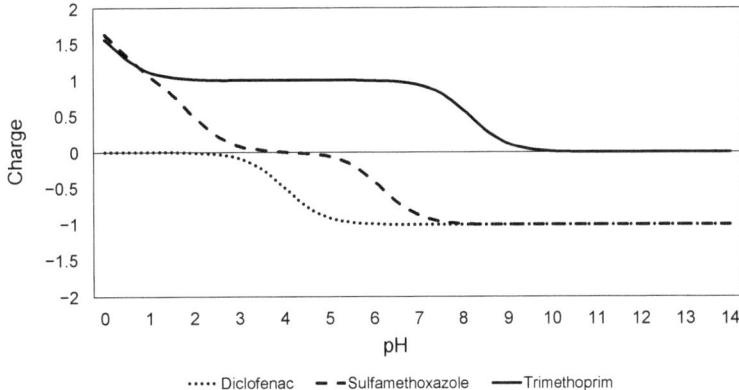

Figure 3. Changes in the ionization state of DCF, SMX and TMP as a function of the pH. J Chem for Office (20.11.0, ChemAxon, https://www.chemaxon.com, accessed on 11 June 2021) was used for calculating the ionization state.

2.4. Influence of the Water Matrix

In wastewater treatment, the water matrix influences the adsorption process as well as the physicochemical properties of the of the adsorbates. In hybrid systems combining biological treatment with adsorption, PAC can be added in the biological tank (in contact with the mixed liquor) or as a polishing treatment for the secondary effluent [9,32]. Because the constituents and quality of the wastewater change along with the treatment step, it is essential to study the influence of the water matrix on the adsorption of contaminants. One of the most important parameters to consider is the presence of dissolved organic matter (DOM) [46]. DOM is constituted of fractions of different sizes (i.e., building blocks, humic and fulvic acids, biopolymers and low molecular weight organics) which may interfere with the adsorption to varying degrees [15] by blocking the PAC pores or competing with the pollutants of interest for adsorption sites. Indeed, the addition of fresh PAC is required to maintain high removal efficiencies, because the PAC surface becomes saturated over time mainly due to the adsorption of the DOM present in the wastewater [9,46]. In addition, the effect of PAC saturation is more pronounced for anionic compounds, because DOM is negatively charged at the overall pH of wastewater and interferes with the adsorption of anionic compounds through electrostatic repulsion [6]. However, the effect of the presence of DOM is still unclear. Many studies report that DOM has no significant effect or may even have a positive effect on the adsorption of some pharmaceuticals, depending on the experimental conditions [11,14,47].

The influence on the water matrix was studied by performing adsorption batch experiments in ultra-pure water, humic acid (HA) solution, MBR permeate and mixed liquor and comparing the obtained experimental results and isotherm modelling. Although the composition of DOM in the MBR permeate and the mixed liquor was not determined, the total DOC concentration was measured for the HA solution (29.35 mg/L), MBR permeate (4.1 mg/L) and mixed liquor (4.7 mg/L). It should be noted that the DOC concentration in the MBR permeate and that in the mixed liquor are quite similar, despite their different nature. Mixed liquor possesses a high concentration of total suspended solids (6 g/L) compared to MBR permeate (5.4 mg/L). In this case, the solid phase mixed liquor was included in the adsorption experiments, because it can act as an adsorbent and influence the interactions between pharmaceuticals and PAC.

Experimental equilibrium adsorption capacities of DCF, SMX and TMP for each water matrix are depicted in Figure 4. Sorption parameters from isotherm models and regression coefficients for each water matrix are listed in Table 5.

The adsorption mechanisms and, therefore, the isotherm models that describe them may vary from compound to compound, as described in the literature [48]. Similarly, they appear to depend on the water matrix in which adsorption occurs. As mentioned earlier, both the Langmuir and the Freundlich models fitted the results of DCF and TMP in ultra-pure water very well, whereas for SMX, the Freundlich model provided a better fit. Nonetheless, the regression coefficients of the Langmuir model for SMX are very high ($R^2 > 0.956$). As for ultra-pure water, both Langmuir and Freundlich isotherms had very similar regression coefficients in MBR permeate, and there was not a model that fitted the results better for any of the compounds tested. None of the Langmuir parameters (K_L and q_m) differed significantly between the pharmaceuticals. Instead, the Langmuir isotherm clearly fitted the q_e versus C_e plot in the humic acid solution, whereas the Freundlich isotherm had significantly higher R^2 values in the mixed liquor. In the Langmuir isotherm, monolayer adsorption onto the PAC surface is assumed with a fixed number of energetically equivalent sites, whereas the Freundlich isotherm is considered to be an empirical expression for multilayer adsorption with different energy in the active sites [35]. Mixed liquor is expected to represent a much more complex matrix because it was extracted from the biological reactor, where most of the biological and chemical transformations take place for the removal of contaminants. In previous studies, it has been observed that given similar DOC-pharmaceutical concentrations, DOM composition may induce a stronger adsorption competition effect depending on the type of water (i.e., drinking water compared to WWTP effluent) [14]. In this way, the results are not surprising and confirm that adsorption mechanisms change depending on experimental conditions.

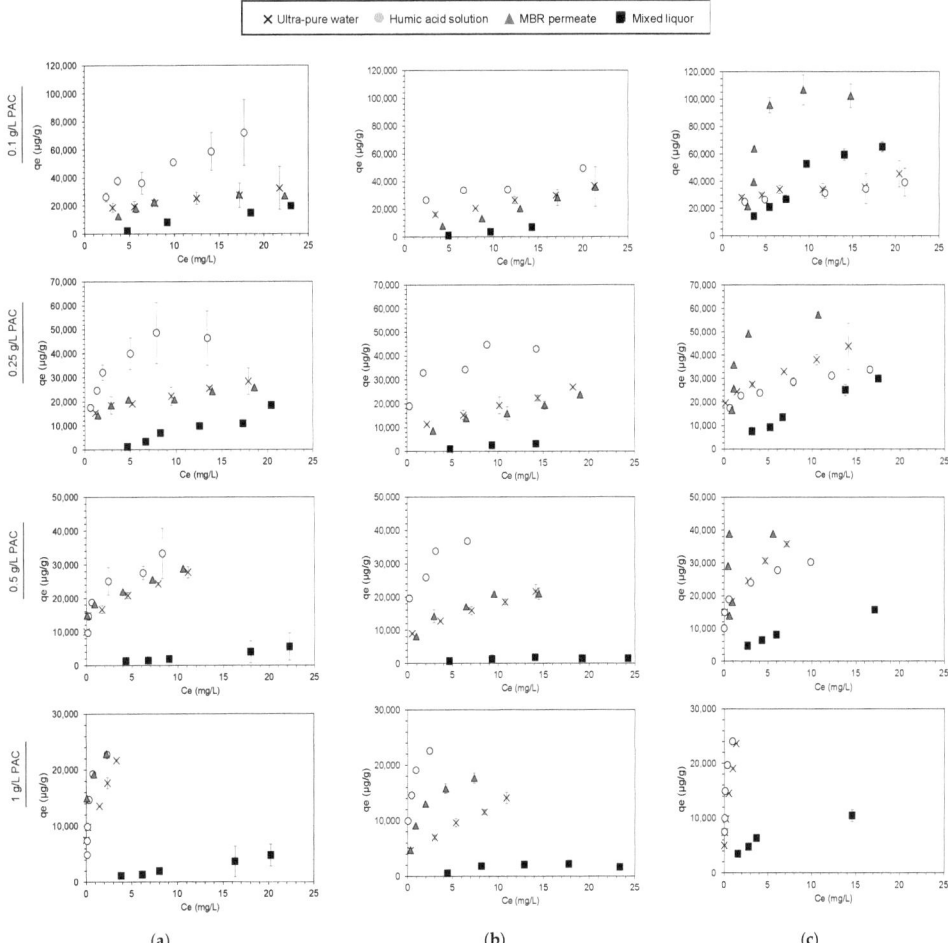

Figure 4. Experimental equilibrium adsorption capacity of (**a**) DCF, (**b**) SMX and (**c**) TMP at four different PAC concentrations (0.1, 0.25, 0.5 and 1 g/L) in ultra-pure water (×), humic acid solution (○), MBR permeate (▲) and mixed liquor (■). Error bars indicate the standard deviation.

Assuming that the Freundlich isotherm had the best fit for all the water matrices, the higher average K_F values were found as follows: HA solution, ultra-pure water, MBR permeate and mixed liquor. Higher K_F values correspond to a higher adsorption capacity of the PAC (q_e) for the same equilibrium concentration (C_e) for all three compounds. As shown in Figure 4, higher PAC loads were obtained in the humic acid solution for DCF and SMX, followed by ultra-pure water and MBR permeate, with very similar results ($p > 0.05$). On the other hand, PAC loads were found to be the lowest in the mixed liquor for all pharmaceuticals. For TMP instead, the best results were obtained in the MBR permeate, followed by ultra-pure water, humic acid solution and mixed liquor. Indeed, for 1 g/L of PAC, the remaining concentrations of TMP in the MBR permeate were too low to perform the isotherm modelling. For 0.1 g/L of PAC, an unexpected increase in the adsorption capacity was achieved at higher TMP concentrations in the mixed liquor, not following the trend in the other PAC concentrations. Although the overall results are not consistent with other studies [11,49], in which the adsorption capacity in wastewater was systematically lower compared to that in ultra-pure water, it is possible that positive interactions between the humic acids and MBR effluent DOM lead to an increased adsorption capacity of PAC. Moreover, in real wastewater systems, DOM is present at a concentration of three to six orders of magnitude higher than organic micropollutants (mg/L compared to μg/L—ng/L). In our experimentation, the extent of the effect of DOM may be limited or altered because the C_0 of the tested pharmaceuticals ranged from 5 to 25 mg/L. In all water matrices, the highest PAC loadings (q_e) were observed at the lowest PAC concentration (0.1 g/L) and maximum pharmaceutical concentration (25 mg/L) for all the water matrixes and compounds (Figure 4).

Table 5. Distribution coefficient (K_d), Langmuir and Freundlich isotherm constants in different water matrices (humic acid solution, MBR permeate and mixed liquor). Results for humic acid solutions were considered without pre-contact time between the HAs and the pharmaceuticals. N.A. (not applicable) indicates that the parameters could not be obtained, as the residual concentration found in the liquid phase was very low to conduct the modelling.

Compound	PAC Conc. (g/L)	Linear Sorption		Langmuir Isotherm			Freundlich Sorption		
		K_d (mL/g)	R^2	q_m (μg/g)	K_L (L/mg)	R^2	$1/n$	K_F (mg/g) (mL/mg)$^{1/n}$	R^2
Humic acid solution									
DCF	0.1	4521.6	0.941	100,000	0.125	0.908	0.4568	18,012.1	0.929
	0.25	4802.4	0.783	50,000	1.000	0.994	0.2799	24,760.4	0.896
	0.5	4600.6	0.768	33,333	1.500	0.984	0.2000	20,607.3	0.781
	1	12,308.0	0.718	100,000	1.429	0.994	N.A.	N.A.	N.A.
SMX	0.1	2856.7	0.878	50,000	0.250	0.919	0.2630	20,426.7	0.863
	0.25	3957.7	0.792	50,000	1.000	0.983	0.1408	29,673.2	0.651
	0.5	6994.4	0.801	33,333	3.000	0.983	0.2731	22,606.7	0.807
	1	11,372.0	0.763	25,000	5.000	0.991	N.A.	N.A.	N.A.
TMP	0.1	2287.9	0.860	50,000	0.286	0.976	0.2116	19,150.8	0.900
	0.25	2600.1	0.791	33,333	0.750	0.992	0.1891	19,424.7	0.958
	0.5	3824.5	0.720	33,333	3.000	0.994	0.1960	19,358.8	0.998
	1	31,430.0	0.740	25,000	10.000	0.998	N.A.	N.A.	N.A.
MBR permeate									
DCF	0.1	1553.7	0.880	33,333	0.150	0.978	0.4160	8206.6	0.865
	0.25	1785.4	0.802	25,000	0.667	0.989	0.2066	14,004.0	0.925
	0.5	3273.8	0.776	50,000	1.000	0.985	0.2785	14,831.4	0.997
	1	12,011.0	0.734	25,000	1.000	0.995	N.A.	N.A.	N.A.
SMX	0.1	1642.8	0.999	1,000,000	0.002	0.028	0.9527	1843.1	0.988
	0.25	1349.2	0.962	33,333	0.100	0.924	0.4976	5154.4	0.978
	0.5	1874.2	0.870	25,000	0.444	0.993	0.2650	10,690.4	0.932
	1	3009.7	0.837	20,000	1.000	0.999	0.2310	11,178.0	0.996
TMP	0.1	9370.2	0.875	250,000	0.057	0.225	0.8102	15,532.7	0.647
	0.25	6616.8	0.690	50,000	1.000	0.974	0.2822	31,351.0	0.754
	0.5	8417.5	0.535	50,000	1.000	0.937	N.A.	N.A.	N.A.
	1	N.A.	N.A.	N.A.	N.A.	N.A.	N.A.	N.A.	N.A.
Mixed liquor									
DCF	0.1	827.9	0.993	−25,000	−0.019	0.466	1.3563	299.7	0.957
	0.25	766.7	0.963	−10,000	−0.033	0.407	1.6076	148.1	0.903
	0.5	235.9	0.995	50,000	0.005	0.038	0.9064	295.6	0.952
	1	234.1	0.998	33,333	0.008	0.268	0.8891	312.6	0.979
SMX	0.1	431.5	0.965	−3333	−0.048	0.907	1.8270	55.0	1.000
	0.25	233.2	0.990	−33,333	−0.006	0.085	1.0983	186.3	0.954
	0.5	84.0	0.892	2000	0.172	0.873	0.4847	384.3	0.707
	1	109.4	0.858	2500	0.118	0.552	0.6615	300.6	0.594
TMP	0.1	3988.7	0.976	1,250,000	0.004	0.015	1.0055	4011.8	0.939
	0.25	1785.7	0.995	125,000	0.020	0.538	0.8440	2659.3	0.980
	0.5	1002.7	0.960	33,333	0.060	0.986	0.6493	2484.4	1.000
	1	822.7	0.868	14,286	0.233	0.996	0.4847	2980.3	0.967

It has been observed that the adsorption of some pharmaceuticals is promoted by the presence of humic acid in soils and sediments, suggesting that the presence of these substances may positively influence the sorption affinity for the adsorbent. Humic substances, which are also commonly found in wastewater, are known to act as carriers of organic micropollutants such as pharmaceuticals [50]. Due to their mobility and ability to form complexes with organic and inorganic species, commercial HAs may contain trace elements (e.g., ions, heavy metals) that contribute to the adsorption of further organic compounds (i.e., diclofenac) in adsorption experiments [50]. In another study, the formation of ciprofloxacin–HA complexes has been reported as a "false positive adsorption" when testing the sorption capacity of various adsorbents [17]. According to Behera et al. [33], the pharmaceutical–HA complex would be able to adsorb onto the surface of the adsorbent. These authors also suggest that the free pharmaceuticals in the solution could adsorb onto the already adsorbed HA, leading to an increase in adsorption [17]. On the other hand, the high concentrations of HAs in our study (29.35 mg/L) may enhance the sorption of some pharmaceuticals via hydrophobicity. Even if the interaction between DOM and pharmaceuticals is not expected, the presence of HAs may promote the adsorption through the PAC in the solution. The adsorption of dissolved humic substances has been proved to reduce the aggregation of carbon nanotubes, thus increasing the surface area available for adsorption by two orders of magnitude, increasing the change in the hydrophobic interactions between the adsorbent and SMX [47]. This could explain the increased adsorption of DCF and SMX, two anionic compounds for which the electrostatic interactions with the DOM would not be primarily considered. For the aforementioned reasons, the increased adsorption capacity of PAC in the HA solution is not surprising. Although there is no single phenomenon that explains the observed results, the literature data confirm that the presence of humic substances can affect the adsorption of organic compounds such as pharmaceuticals in several ways.

In the case of the MBR permeate, the results show that the presence of DOM had no negative effect on drug adsorption, with no statistical differences from ultra-pure water for DCF and SMX ($p > 0.05$) and with an increase in the adsorption capacity of PAC for TMP ($p < 0.05$). Because the concentration of the pharmaceutical influences the experimental adsorption values (with the highest q_e values at C_0 of 25 mg/L in all water matrices), it may be that DOC is not high enough in the solution to cause a decrease in adsorption compared with ultra-pure water. In any case, the results show that the adsorption of TMP in the MBR permeate was enhanced, probably due to the above-mentioned reasons related to HAs and, in particular, to the fact that TMP is positively charged, which could favor the interactions with negatively charged DOM. PAC added to the secondary effluent of full-scale WWTPs has been proved to provide a better quality effluent (i.e., lower TMP concentration) compared to PAC added in the biological reactor, in contact with the mixed liquor, indicating that the DOM constituents of the MBR permeate have a different effect on the adsorption of TMP onto PAC [18]. Indeed, TMP was not the only compound with lower adsorption in the mixed liquor (Figure 4). Even with the very similar DOC concentration, the differences in the adsorption capacity of PAC between the MBR permeate and mixed liquor indicate that the DOM constituents play a significant role in the adsorption process. Although HAs appeared to favor adsorption, low molecular weight organics have been demonstrated to limit the process due to direct competition for the adsorption sites [35]. However, it should be noted that the experiments conducted aimed to reproduce the adsorption process under real WWTP conditions, and, therefore, the solid fraction of the mixed liquor was included in the adsorption batch experiments. Because some pharmaceuticals are also able to adsorb onto the sludge [27], additional adsorption experiments were performed without the addition of PAC to quantify the adsorption onto the solid phase of the mixed liquor (dried sludge). The results of the experimental q_e and C_e values were highly variable, and no modelling could be performed (data not shown). However, the resulting q_e values were very low compared to PAC adsorption (e.g., the maximum q_e found was 530 µg/g for SMX), and thus, the adsorption onto the mixed liquor can be neglected for the pharmaceuticals under study [27]. However, the presence of additional suspended material (with a concentration of 6 g/L) could limit the ability of the pharmaceuticals to reach the PAC adsorption sites and, thus, physically reduce the adsorption of pharmaceuticals.

2.5. Influence of the Pre-Equilibrium Time between Pharmaceuticals and DOM on Adsorption

The influence on the interaction between DOM and the pharmaceuticals before the adsorption onto PAC was studied by using the HA solution. Humic acids are one of the most common DOM fractions found in wastewater [50], and they were chosen because of their commercial availability and ease of use in the laboratory. Because the objective was to study the interaction between DOM and the pharmaceuticals, DOC concentration does not have to be identical to the one found in the biological tank of the WWTP (4.7 mg/L). In fact, the experiments were performed with the highest possible DOC concentration, in order to produce the largest difference between DOM and the pharmaceutical concentration. The pre-contact time between HAs and OMPs was set at 24 h because it has already been tested as sufficient to evaluate the influence of the interaction between them [11].

The results of the adsorption isotherm parameters and correlation coefficients of the three pharmaceuticals with 24 h of pre-contact time are shown in Table 6, whereas the results of adsorption without a pre-equilibrium time are depicted in Table 5. Langmuir isotherm is the model that better fits the results in the HA solution in both conditions, and no statistical differences were found between them for the maximum adsorption capacity (q_m) and Langmuir coefficient (K_L). Regarding removal efficiencies (data not shown), no statistical differences were found between no pre-contact time and 24 h of pre-contact time with the HA solution, although a slight increment was observed for the condition without pre-contact time (3% for SMX, 7% for TMP and 8% for DCF). As explained earlier, the presence of HA in the solution had a neutral to positive effect for the three pharmaceuticals tested, which may be attributed to the high adsorption of the HAs and the interaction between the HA and the pollutants. However, the pre-contact time had no significant effect on the adsorption. The long shaking times of the adsorption experiments (24 h) were already sufficient to observe the potential beneficial effects of the presence of HA in the solution (e.g., formation of pharmaceutical–HA complexes, increased dispersion of the PAC), without the need for additional pre-contact time. In this way, in a previous study, it was found that the 24 h pre-contact time between DOM and various pharmaceuticals favored adsorption only at short contact times (i.e., 30 min) and had no effect once the equilibrium between the adsorbent and adsorbates was reached (i.e., 72 h) [11].

Table 6. Distribution coefficient (K_d) and Langmuir and Freundlich isotherm parameters, together with the corresponding regression coefficients (R^2) for the adsorption DCF, SMX and TMP onto PAC in a humic acid solution with 24 h pre-contact time between the pharmaceuticals and the humic acids. Not applicable (N.A.) indicates that the parameters could not be obtained, as the residual concentration found in the liquid phase was too low to conduct the modelling.

Compound	PAC Conc. (g/L)	Linear Isotherm		Langmuir Isotherm			Freundlich Isotherm		
		K_d (mL/g)	R^2	q_m (µg/g)	K_L (L/mg)	R^2	1/n	K_F (mg/g) (mL/mg)$^{1/n}$	R^2
DCF	0.1	2931.5	0.8482	50,000	0.400	0.983	0.2194	23,435.3	0.951
	0.25	3196.9	0.8713	50,000	0.333	0.980	0.3342	15,739.0	0.988
	0.5	2932.1	0.7000	25,000	2.000	0.999	0.0962	20,854.0	0.959
	1	2403.6	0.6190	16,667	6.000	0.990	0.0825	13,418.6	0.614
SMX	0.1	2378.1	0.8972	50,000	0.250	0.985	0.2619	18,030.1	0.953
	0.25	3434.0	0.8685	50,000	0.500	0.981	0.3040	18,300.8	0.970
	0.5	5491.8	0.8108	33,333	3.000	0.988	0.1885	23,086.4	0.948
	1	7530.7	0.7846	20,000	6.250	0.988	N.A.	N.A.	N.A.
TMP	0.1	2340.0	0.8716	50,000	0.250	0.989	0.2932	16,000.9	0.9487
	0.25	2567.3	0.7911	33,333	0.750	0.991	0.2272	17,667.8	0.9735
	0.5	3896.4	0.7282	33,333	3.000	0.996	0.1366	22,092.6	0.9872
	1	N.A.	N.A.	N.A.	N.A.	N.A.	N.A.	N.A.	N.A.

3. Materials and Methods

3.1. Adsorbent and Adsorbates

PAC (ACTISORBE 700, Brenntag S.p.A, Italy) was used for all the adsorption experiments. The PAC characteristics were supplied by the manufacturer as follows: iodine number 750 mg/g, methylene blue 12 mL, BET specific surface area 850 m^2/g, bulk density 430 kg/m^3, ash content 10%, humidity 5% and alkaline pH. The surface properties of the selected PAC are in agreement with the literature on adsorption of organic pollutants [18,51–53]. After its purchase, the PAC was not treated in order to emulate real conditions for which the adsorbent is directly added to the wastewater treatment line.

The DCF, SMX and TMP properties are listed in Table 7. J Chem for Office (20.11.0, ChemAxon, https://www.chemaxon.com, accessed on 11 June 2021) was used for calculating the physicochemical properties (logK_{ow}, molecular weight) and the ionization state (Figure 3). The calculation method for logK_{ow} is based on a modified version of the algorithm published by Viswanadhan et al. [54]. In this publication, the K_{ow} is the sum of the assigned values of the individual atomic contributions of a molecule. Molecular weight was based on the data published by IUPAC on the atomic weights of elements [55]. To calculate the ionization state, the software conducts a weighted sum of the net charges of the microspecies comprising the molecule as a function of the pH in aqueous solution. More information about the software functioning is available online.

Diclofenac and sulfamethoxazole (≥98% TLC) were purchased from Sigma-Aldrich (St. Louis, MO, USA), and trimethoprim (≥98% TLC) was purchased from Acros Organics (Thermo Fisher Scientific Inc., Trenton, NJ, USA). To prepare the pharmaceutical solutions, exact amounts of the target compounds were weighed and added to the corresponding water matrix (Section 3.2). To ensure that the compounds were completely dissolved, a maximum of 1% of methanol was added, and the solutions were sonicated in an ultrasonic bath (Sonorex Digital 10P, Bandelin electronic, Berlin, Germany) for 5 min.

Table 7. Physicochemical properties of the selected pharmaceuticals. J Chem for Office (20.11.0, ChemAxon, https://www.chemaxon.com, accessed on 11 June 2021) was used for calculating the physicochemical properties (molecular weight and logK_{ow}). Values for pK_{a1} and pK_{a2} were obtained from the literature [56,57].

Compound	Molecular Formula	Molecular Weight (g/mol)	logK_{ow} [1]	pK_{a1}	pK_{a2}
Diclofenac	$C_{14}H_{10}Cl_2NNaO_2$	318.13	4.26	4.21 [2]	
Sulfamethoxazole	$C_{10}H_{11}N_3O_3S$	253.28	0.79	1.83 [2]	5.57 [2]
Trimethoprim	$C_{14}H_{18}N_4O_3$	290.32	1.28	7.10 ± 0.02 [3]	

[1] Octanol–water partition coefficient. [2] Obtained from [56]. [3] Obtained from [57].

3.2. Water Matrices

Four different water matrices were used to prepare pharmaceutical solutions: ultrapure water (Milli-Q), humic acid (HA) solution and effluent and mixed liquor from a WWTP. The preparation method of each water matrix is described below.

Milli-Q water was obtained from the Millipore Simplicity UV system (Millipore Corporation, Billerica, MA, USA).

Commercially available humic acids (CAS 1415-93-6, Sigma-Aldrich, St. Louis, MO, USA) were used to prepare the HA solution (50 mg/L), with a dissolved organic carbon (DOC) concentration of 29.35 mg/L. The solution was prepared following the method described by [48]. Briefly, to prepare a volume of 100 mL, 5 mL of 1M NH_4OH were added to a 100 mL flask. Then, 0.005 g of HAs were weighed, and the Milli-Q water was added to a maximum of 85 mL. The pH of the solution was then adjusted to 5.34 with 1 M formic acid and prepared to the desired volume (100 mL).

The effluent and mixed liquor were collected from the permeate and the nitrification tank, respectively, of a full-scale MBR located in northern Italy, and frozen at −20 °C until their use. Both the MBR permeate and mixed liquor were autoclaved at 121 °C to reduce any potential biological activity and subsequently filtered through paper filters (Lab Expert, KEFO d.o.o, Croatia) to remove any particulate matter. Filters from the mixed liquor were air dried for 24 h and scrapped to obtain dry sludge. To ensure that all the glass beakers on which the adsorption experiments were conducted contained the same amount of mixed liquor suspended solids (MLSS), a certain amount (120 mg) of dry sludge was added to each glass baker. The resulting MLSS concentration in the mixed liquor was 6 g/L, a concentration commonly found in real WWTPs.

3.3. Batch Adsorption Experiments

Experiments were conducted in triplicate using 20 mL of pharmaceutical solutions in each glass beaker. The glass beakers were sealed with parafilm to avoid evaporation. All experiments were performed in triplicate using an incubator shaker at 150 rpm and a constant temperature of 25 °C (Innova 4080, New Brunswick Scientific, Edison, NJ, USA), which enabled continuous contact between the compounds and the activated carbon. To avoid photodegradation, all experiments were performed in darkness.

Preliminary experiments were conducted to determine the contact time necessary to reach the equilibrium between the PAC and the target pharmaceutical in ultra-pure water. Three different concentrations of target pollutants were tested (5, 15 and 25 mg/L). The PAC was agitated in the solutions for 10, 20, 30, 40 and 50 min and 1, 2, 4, 6, 12, 18 and 24 h at a constant temperature (25 °C). Two PAC concentrations (0.1 g/L and 1 g/L) were tested in each target compound individually, and 0.1 g/L of PAC was also tested in the mixture of the three pharmaceuticals. The results of the preliminary experiments determined 24 h to be sufficient time to reach the equilibrium for all three compounds and the mixture. Based on the results obtained, the sorption kinetics were determined. Kinetics studies were conducted by applying three different kinetics models: Lagergren pseudo-first-order [58] (1), pseudo-second-order (2) and intraparticle diffusion model (IPD) (3) [40].

$$\frac{dq_e}{dt} = k_1(q_e - q_t) \tag{1}$$

$$\frac{t}{q_t} = \frac{1}{k_2 q_e^2} + \frac{1}{q_e}t \tag{2}$$

$$q_t = k_{id}t^{1/2} + C \tag{3}$$

where q_e and q_t are the quantity of solute adsorbed onto the PAC surface (µg/g) at the equilibrium (q_e) and at time t (q_t); k_1 (1/min), k_2 (µg/g min) and k_{id} (µg/g·min$^{1/2}$) are considered the Lagergren pseudo-first order, pseudo-second order and IPD rate constants, respectively; and intercept C provides information about the thickness of the boundary layer.

The batch sorption experiments were conducted in 20 mL of pharmaceutical solutions. For each water matrix, concentrations of pharmaceuticals ranging from 5 to 25 mg/L were tested to determine the sorption isotherms. PAC was added to the solutions at 0.1, 0.25, 0.5 and 1 g/L in each experiment and placed into agitation for 24 h. Equilibrium adsorption was studied by applying linear (4), Langmuir (5) and Freundlich (6) isotherm models to the experimental data,

$$q_e = K_d C_e \tag{4}$$

$$q_e = K_F C_e^{1/n}$$
$$q_e = K_F C_e^{1/n} \tag{5}$$

$$\frac{1}{q_e} = \frac{1}{q_m} + \frac{1}{K_L q_m C_e} \tag{6}$$

where q_e is the amount of adsorbed compound per mass unit of adsorbent at the equilibrium (µg/g); C_e is the equilibrium concentration of the pharmaceutical (mg/mL); K_d is the distribution coefficient; K_F is the Freundlich adsorption constant ((µg/g) (mL/mg)$^{1/n}$);

1/n is the heterogeneity constant; q_m is the equilibrium sorption capacity, that is, the maximum amount of OMP to be adsorbed by the activated carbon (μg/g); and K_L is the adsorption constant for Langmuir isotherms and is related to the sorption bonding energy (L/mg). Based on the four water matrices previously described, different experiments were conducted. Firstly, the pharmaceuticals were tested individually in each water matrix to compare the effect of the DOM (measured as DOC) in the adsorption process (ultra-pure water, humic acid solution, MBR permeate and mixed liquor). Secondly, sorption experiments were conducted in ultra-pure water with a mixture of the three target compounds (DCF, SMX and TMP) at the previously selected concentrations to evaluate the interaction and competition among the pharmaceuticals. Then, the HA solution was used to study the influence of a pre-equilibrium contact time between the DOM and the pharmaceuticals prior to the adsorption onto PAC. Pharmaceuticals were added to the HA solution 24 h before the addition of PAC to simulate their interactions in the sewer and inside the WWTP. Finally, mixed liquor experiments were performed with the addition of PAC and without PAC to assess the adsorption of the pharmaceuticals to the MLSS (i.e., added dried sludge).

3.4. HPLC Analysis

Prior to the quantitative analysis of the OMP concentration, glass beakers were decanted, and samples were centrifuged at 3500 rpm for 5 min (Hettich EBA 20, Westphalia, Germany) to subsequently be filtered with a 0.45 μm Nylon syringe filter (Filter-Bio, Nantong, China). Blank samples containing the corresponding water matrices were also included in the analysis to act as controls.

The residual pharmaceutical concentration was determined via high-performance liquid chromatography coupled to a photodiode array detection (HPLC-PDA) (Waters 2795 Separation Module and Waters 2996, Waters Corporation, Milford, MA, USA). A Kinetex C18 column was used (Phenomenex, 150 × 4.6 mm, 5 μm particle size, 100 Å pore size). The mobile phase contained eluent A, which was composed of 0.1% of formic acid in Milli-Q water, and solvent B, with 0.1% of formic acid in acetonitrile. The flow rate was 0.5 mL/min for all the experiments. The column temperature was 20 °C. The injection volume for each sample was 20 μL. Peak wavelengths are 276.9 nm for DCF, 269.8 for SMX and 270.8 nm for TMP. Isocratic methods were used to determine the concentrations of individual target pollutants. For DCF, the volume proportion of eluent A was 35%, and that of eluent B was 65%. For SMX, the proportions were 65% A and 35% B, whereas for TMP, the proportions were 85% A and 15% B. The total elution time was 10 min. The retention time was 6.5 min, 6 min and 5.6 min for DCF, SMX and TMP, respectively.

For the solution containing the mixture of pharmaceuticals, a method with gradient elution was developed. The total run time was 25 min, and the flow was kept constant at 0.5 mL/min. It started with a 1 min step gradient with 85% A and 15% B, which was then maintained as linear for another 5 min. Then, the flow was continued with a 1 min linear gradient with 65% A and 35% B, which was maintained for another 3 min; a 5 min gradient with 35% A and 65% B; and a step gradient of 0.1 min back to 85% A and 15% B, which was maintained for another 4.9 min. The retention time of each compound in the mixture was 6.2 min for TMP, 12.9 min for SMX and 20.2 min for DCF in the gradient elution method.

4. Conclusions

The adsorption of three pharmaceuticals (namely DCF, SMX and TMP) onto PAC was studied through the use of kinetic and isotherm models in different water and wastewater matrices. Sorption of the tested pharmaceuticals was proven to be an overall fast process in ultra-pure water. Kinetics followed a pseudo-second order, suggesting that the sorption rate is governed by the number of available active sites. Additionally, the boundary layer effect seems to decrease the adsorption rate as compounds gradually reach the active sites at the equilibrium. Compared to individual solutions, the rate of the adsorption of the compounds in a mixture did not differ; however, a greater adsorption capacity of the PAC was observed in the individual solutions.

Adsorption of pharmaceuticals onto the PAC surface is a complex process that greatly depends on physicochemical properties of the investigated compounds and on the matrix where it takes place. Charge, followed by hydrophobicity, determined the rate and the extent of the adsorption in all the tested matrices, with better results obtained via TMP (cationic compound), followed by DCF (anionic, hydrophobic) and SMX (anionic, hydrophilic). The effect of the water matrix varied from compound to compound. Humic acids appeared to positively affect the affinity for the adsorbent in DCF and SMX, presumably by forming pharmaceutical–HA complexes and by reducing the aggregation of PAC. Mixed liquor gave the lowest adsorption capacities of PAC, probably due to its complex nature and the presence of additional suspended solids. The adsorption isotherms also varied among water matrices. Only Langmuir isotherm explained adsorption in humic acid solution and Freundlich isotherm in the mixed liquor, whereas both isotherms fitted the results in ultra-pure water and MBR permeate very well. In this way, DOM and specifically HAs proved to be beneficial for the adsorption of the selected pharmaceuticals. However, the effects of the interaction of these elements prior to the addition of the adsorbent did not have an effect after long contact times (24h). In this way, future work should be focused on the understanding of the potential interactions between the organic components of the wastewater that may favor the adsorption of pharmaceuticals onto PAC.

Author Contributions: Conceptualization, M.G., D.M.P. and P.V.; validation, M.G. and D.M.P.; formal analysis, M.G.; investigation, M.G.; writing—original draft preparation, M.G.; writing—review and editing, M.G., P.V. and D.M.P.; visualization, M.G.; supervision D.M.P. and P.V.; project administration, P.V.; funding acquisition, P.V. and D.M.P. All authors have read and agreed to the published version of the manuscript.

Funding: This work is supported by the European Union's Horizon 2020 Research and Innovation Program under the Marie Sklodowska-Curie Grant Agreement No 812880—Nowelties ITN-EJD project.

Institutional Review Board Statement: Not applicable.

Informed Consent Statement: Not applicable.

Data Availability Statement: Not applicable.

Acknowledgments: The authors would like to thank the HERA company for providing the wastewater and the mixed liquor samples from their facilities to conduct the experiments.

Conflicts of Interest: The authors declare no conflict of interest. The funders had no role in the design of the study; in the collection, analyses, or interpretation of data; in the writing of the manuscript; or in the decision to publish the results.

Sample Availability: Samples of the compounds are not available from the authors.

References

1. Luo, Y.; Guo, W.; Ngo, H.H.; Nghiem, L.D.; Hai, F.I.; Zhang, J.; Liang, S.; Wang, X.C. A Review on the Occurrence of Micropollutants in the Aquatic Environment and Their Fate and Removal during Wastewater Treatment. *Sci. Total Environ.* **2014**, *473–474*, 619–641. [CrossRef] [PubMed]
2. Verlicchi, P.; Galletti, A.; Petrovic, M.; BarcelÓ, D. Hospital Effluents as a Source of Emerging Pollutants: An Overview of Micropollutants and Sustainable Treatment Options. *J. Hydrol.* **2010**, *389*, 416–428. [CrossRef]
3. Rizzo, L.; Malato, S.; Antakyali, D.; Beretsou, V.G.; Đolić, M.B.; Gernjak, W.; Heath, E.; Ivancev-Tumbas, I.; Karaolia, P.; Lado Ribeiro, A.R.; et al. Consolidated vs New Advanced Treatment Methods for the Removal of Contaminants of Emerging Concern from Urban Wastewater. *Sci. Total Environ.* **2019**, *655*, 986–1008. [CrossRef] [PubMed]
4. Khan, N.A.; Khan, S.U.; Ahmed, S.; Farooqi, I.H.; Yousefi, M.; Mohammadi, A.A.; Changani, F. Recent Trends in Disposal and Treatment Technologies of Emerging-Pollutants- A Critical Review. *TrAC Trends Anal. Chem.* **2020**, *122*, 115744. [CrossRef]
5. Mailler, R.; Gasperi, J.; Coquet, Y.; Derome, C.; Buleté, A.; Vulliet, E.; Bressy, A.; Varrault, G.; Chebbo, G.; Rocher, V. Removal of Emerging Micropollutants from Wastewater by Activated Carbon Adsorption: Experimental Study of Different Activated Carbons and Factors Influencing the Adsorption of Micropollutants in Wastewater. *J. Environ. Chem. Eng.* **2016**, *4*, 1102–1109. [CrossRef]

6. Margot, J.; Kienle, C.; Magnet, A.; Weil, M.; Rossi, L.; de Alencastro, L.F.; Abegglen, C.; Thonney, D.; Chèvre, N.; Schärer, M.; et al. Treatment of Micropollutants in Municipal Wastewater: Ozone or Powdered Activated Carbon? *Sci. Total Environ.* **2013**, *461–462*, 480–498. [CrossRef]
7. Verlicchi, P.; Al Aukidy, M.; Galletti, A.; Petrovic, M.; Barceló, D. Hospital Effluent: Investigation of the Concentrations and Distribution of Pharmaceuticals and Environmental Risk Assessment. *Sci. Total Environ.* **2012**, *430*, 109–118. [CrossRef]
8. Li, X.; Hai, F.I.; Nghiem, L.D. Simultaneous Activated Carbon Adsorption within a Membrane Bioreactor for an Enhanced Micropollutant Removal. *Bioresour. Technol.* **2011**, *102*, 5319–5324. [CrossRef]
9. Alvarino, T.; Torregrosa, N.; Omil, F.; Lema, J.M.; Suarez, S. Assessing the Feasibility of Two Hybrid MBR Systems Using PAC for Removing Macro and Micropollutants. *J. Environ. Manag.* **2017**, *203*, 831–837. [CrossRef]
10. Löwenberg, J.; Zenker, A.; Baggenstos, M.; Koch, G.; Kazner, C.; Wintgens, T. Comparison of Two PAC/UF Processes for the Removal of Micropollutants from Wastewater Treatment Plant Effluent: Process Performance and Removal Efficiency. *Water Res.* **2014**, *56*, 26–36. [CrossRef]
11. Guillossou, R.; Le Roux, J.; Mailler, R.; Pereira-Derome, C.; Varrault, G.; Bressy, A.; Vulliet, E.; Morlay, C.; Nauleau, F.; Rocher, V.; et al. Influence of Dissolved Organic Matter on the Removal of 12 Organic Micropollutants from Wastewater Effluent by Powdered Activated Carbon Adsorption. *Water Res.* **2020**, *172*, 115487. [CrossRef]
12. Gutiérrez, M.; Ghirardini, A.; Borghesi, M.; Bonnini, S.; Mutavdžić Pavlović, D.; Verlicchi, P. Removal of Micropollutants Using a Membrane Bioreactor Coupled with Powdered Activated Carbon — A Statistical Analysis Approach. *Sci. Total Environ.* **2022**, *840*, 156557. [CrossRef]
13. Alves, T.C.; Cabrera-Codony, A.; Barceló, D.; Rodriguez-mozaz, S.; Pinheiro, A.; Gonzalez-olmos, R. Influencing Factors on the Removal of Pharmaceuticals from Water with Micro-Grain Activated Carbon. *Water Res.* **2018**, *144*, 402–412. [CrossRef]
14. Zietzschmann, F.; Stützer, C.; Jekel, M. Granular Activated Carbon Adsorption of Organic Micro-Pollutants in Drinking Water and Treated Wastewater - Aligning Breakthrough Curves and Capacities. *Water Res.* **2016**, *92*, 180–187. [CrossRef]
15. Zietzschmann, F.; Worch, E.; Altmann, J.; Ruhl, A.S.; Sperlich, A.; Meinel, F.; Jekel, M. Impact of EfOM Size on Competition in Activated Carbon Adsorption of Organic Micro-Pollutants from Treated Wastewater. *Water Res.* **2014**, *65*, 297–306. [CrossRef]
16. Hernandez-Ruiz, S.; Abrell, L.; Wickramasekara, S.; Chefetz, B.; Chorover, J. Quantifying PPCP Interaction with Dissolved Organic Matter in Aqueous Solution: Combined Use of Fluorescence Quenching and Tandem Mass Spectrometry. *Water Res.* **2012**, *46*, 943–954. [CrossRef]
17. Jin, J.; Feng, T.; Gao, R.; Ma, Y.; Wang, W.; Zhou, Q.; Li, A. Ultrahigh Selective Adsorption of Zwitterionic PPCPs Both in the Absence and Presence of Humic Acid: Performance and Mechanism. *J. Hazard. Mater.* **2018**, *348*, 117–124. [CrossRef]
18. Gutiérrez, M.; Grillini, V.; Mutavdžić Pavlović, D.; Verlicchi, P. Activated Carbon Coupled with Advanced Biological Wastewater Treatment: A Review of the Enhancement in Micropollutant Removal. *Sci. Total Environ.* **2021**, *790*, 148050. [CrossRef]
19. Boehler, M.; Zwickenpflug, B.; Hollender, J.; Ternes, T.; Joss, A.; Siegrist, H. Removal of Micropollutants in Municipal Wastewater Treatment Plants by Powder-Activated Carbon. *Water Sci. Technol.* **2012**, *66*, 2115–2121. [CrossRef]
20. Streicher, J.; Ruhl, A.S.; Gnirß, R.; Jekel, M. Where to Dose Powdered Activated Carbon in a Wastewater Treatment Plant for Organic Micro-Pollutant Removal. *Chemosphere* **2016**, *156*, 88–94. [CrossRef]
21. Verlicchi, P.; Al Aukidy, M.; Zambello, E. Occurrence of Pharmaceutical Compounds in Urban Wastewater: Removal, Mass Load and Environmental Risk after a Secondary Treatment—A Review. *Sci. Total Environ.* **2012**, *429*, 123–155. [CrossRef] [PubMed]
22. Fatta-Kassinos, D.; Meric, S.; Nikolaou, A. Pharmaceutical Residues in Environmental Waters and Wastewater: Current State of Knowledge and Future Research. *Anal. Bioanal. Chem.* **2011**, *399*, 251–275. [CrossRef] [PubMed]
23. McArdell, C.S.; Kovalova, L.; Siegrist, H. *Input and Elimination of Pharmaceuticals and Disinfectants from Hospital Wastewater. Final Project Report*; Eawag: Das Wasserforschungs-Institut des ETH-Bereichs: Dübendorf, Switzerland, 2011.
24. Oaks, J.L.; Gilbert, M.; Virani, M.Z.; Watson, R.T.; Meteyer, C.U.; Rideout, B.A.; Shivaprasad, H.L.; Ahmed, S.; Chaudhry, M.J.I.; Arshad, M.; et al. Diclofenac Residues as the Cause of Vulture Population Decline in Pakistan. *Nature* **2004**, *427*, 630–633. [CrossRef] [PubMed]
25. European Commission European Commission. Commission Implementing Decision (EU) 2015/495 of 20 March 2015 Establishing a Watch List of Substances for Union-Wide Monitoring in the Field of Water Policy Pursuant to Directive 2008/105/EC of the European Parliament and of T. *Off. J. Eur. Union L 78* **2015**, *2015*, 40–42.
26. Ternes, T.; Joss, A. *Human Pharmaceuticals, Hormones and Fragrances—The Challenge of Micropollutants in Urban Water Management*; IWA Publishing: Longon, UK, 2015; Volume 5, ISBN 9781780402468. [CrossRef]
27. Salvestrini, S.; Fenti, A.; Chianese, S.; Iovino, P.; Musmarra, D. Diclofenac Sorption from Synthetic Water: Kinetic and Thermodynamic Analysis. *J. Environ. Chem. Eng.* **2020**, *8*, 104105. [CrossRef]
28. Radjenović, J.; Petrović, M.; Barceló, D. Fate and Distribution of Pharmaceuticals in Wastewater and Sewage Sludge of the Conventional Activated Sludge (CAS) and Advanced Membrane Bioreactor (MBR) Treatment. *Water Res.* **2009**, *43*, 831–841. [CrossRef]
29. Alvarino, T.; Komesli, O.; Suarez, S.; Lema, J.M.M.; Omil, F. The Potential of the Innovative SeMPAC Process for Enhancing the Removal of Recalcitrant Organic Micropollutants. *J. Hazard. Mater.* **2016**, *308*, 29–36. [CrossRef]
30. European Commission. Commision Implementing Decision (EU) 2020/1161-4 August 2020-Establishing a Watch List of Substances for Union-Wide Monitoring in the Field of Water Policy Pursuant to Directive 2008/105/EC of the European Parliament and of the Council. *Off. J. Eur. Union* **2020**, *257*, 32–35.

31. European Commission. Commission Implementing Decision (EU) 2022/1307 of 22 July 2022 Establishing a Watch List of Substances for Union-Wide Monitoring in the Field of Water Policy Pursuant to Directive 2008/105/EC of the European Parliament and of the Counci. *Off. J. Eur. Union* **2022**, *197*, 117–121.
32. Paredes, L.; Alfonsin, C.; Allegue, T.; Omil, F.; Carballa, M. Integrating Granular Activated Carbon in the Post-Treatment of Membrane and Settler Effluents to Improve Organic Micropollutants Removal. *Chem. Eng. J.* **2018**, *345*, 79–86. [CrossRef]
33. Behera, S.K.; Oh, S.Y.; Park, H.S. Sorption of Triclosan onto Activated Carbon, Kaolinite and Montmorillonite: Effects of PH, Ionic Strength, and Humic Acid. *J. Hazard. Mater.* **2010**, *179*, 684–691. [CrossRef]
34. Delgado, N.; Capparelli, A.; Navarro, A.; Marino, D. Pharmaceutical Emerging Pollutants Removal from Water Using Powdered Activated Carbon: Study of Kinetics and Adsorption Equilibrium. *J. Environ. Manag.* **2019**, *236*, 301–308. [CrossRef]
35. Mutavdžić Pavlović, D.; Glavač, A.; Gluhak, M.; Runje, M. Sorption of Albendazole in Sediments and Soils: Isotherms and Kinetics. *Chemosphere* **2018**, *193*, 635–644. [CrossRef]
36. Çalişkan, E.; Göktürk, S. Adsorption Characteristics of Sulfamethoxazole and Metronidazole on Activated Carbon. *Sep. Sci. Technol.* **2010**, *45*, 244–255. [CrossRef]
37. Torrellas, S.A.; Rodriguez, A.R.; Escudero, G.O.; Martin, J.M.G.; Rodriguez, J.G. Comparative Evaluation of Adsorption Kinetics of Diclofenac and Isoproturon by Activated Carbon. *J. Environ. Sci. Heal.—Part A Toxic/Hazardous Subst. Environ. Eng.* **2015**, *50*, 1241–1248. [CrossRef]
38. Tran, H.N.; You, S.J.; Hosseini-Bandegharaei, A.; Chao, H.P. Mistakes and Inconsistencies Regarding Adsorption of Contaminants from Aqueous Solutions: A Critical Review. *Water Res.* **2017**, *120*, 88–116. [CrossRef]
39. Walter, J.; Weber, J. Evolution of a Technology. *J. Environ. Eng.* **1984**, *110*, 899–917.
40. Weber, W.J.; Morris, J.C. Kinetics of Adsorption on Carbon from Solution. *J. Sanit. Eng. Div.* **1963**, *89*, 31–60. [CrossRef]
41. Suriyanon, N.; Punyapalakul, P.; Ngamcharussrivichai, C. Mechanistic Study of Diclofenac and Carbamazepine Adsorption on Functionalized Silica-Based Porous Materials. *Chem. Eng. J.* **2013**, *214*, 208–218. [CrossRef]
42. Xiang, L.; Xiao, T.; Mo, C.H.; Zhao, H.M.; Li, Y.W.; Li, H.; Cai, Q.Y.; Zhou, D.M.; Wong, M.H. Sorption Kinetics, Isotherms, and Mechanism of Aniline Aerofloat to Agricultural Soils with Various Physicochemical Properties. *Ecotoxicol. Environ. Saf.* **2018**, *154*, 84–91. [CrossRef]
43. Rudzinski, W.; Plazinski, W. Kinetics of Solute Adsorption at Solid/Solution Interfaces: On the Special Features of the Initial Adsorption Kinetics. *Langmuir* **2008**, *24*, 6738–6744. [CrossRef] [PubMed]
44. Nguyen, L.N.; Hai, F.I.; Kang, J.; Nghiem, L.D.; Price, W.E.; Guo, W.; Ngo, H.H.; Tung, K.-L. Comparison between Sequential and Simultaneous Application of Activated Carbon with Membrane Bioreactor for Trace Organic Contaminant Removal. *Bioresour. Technol.* **2013**, *130*, 412–417. [CrossRef] [PubMed]
45. Kim, S.H.; Shon, H.K.; Ngo, H.H. Adsorption Characteristics of Antibiotics Trimethoprim on Powdered and Granular Activated Carbon. *J. Ind. Eng. Chem.* **2010**, *16*, 344–349. [CrossRef]
46. Aschermann, G.; Schröder, C.; Zietzschmann, F.; Jekel, M. Organic Micropollutant Desorption in Various Water Matrices - Activated Carbon Pore Characteristics Determine the Reversibility of Adsorption. *Chemosphere* **2019**, *237*, 124415. [CrossRef] [PubMed]
47. Pan, B.; Zhang, D.; Li, H.; Wu, M.; Wang, Z.; Xing, B. Increased Adsorption of Sulfamethoxazole on Suspended Carbon Nanotubes by Dissolved Humic Acid. *Environ. Sci. Technol.* **2013**, *47*, 7722–7728. [CrossRef]
48. Mutavdžić Pavlović, D.; Tolić Čop, K.; Barbir, V.; Gotovuša, M.; Lukač, I.; Lozančić, A.; Runje, M. Sorption of Cefdinir, Memantine, Praziquantel and Trimethoprim in Sediment and Soil Samples. *Environ. Sci. Pollut. Res.* **2022**, *29*, 66841–66857. [CrossRef]
49. Kovalova, L.; Siegrist, H.; von Gunten, U.; Eugster, J.; Hagenbuch, M.; Wittmer, A.; Moser, R.; McArdell, C.S. Elimination of Micropollutants during Post-Treatment of Hospital Wastewater with Powdered Activated Carbon, Ozone, and UV. *Environ. Sci. Technol.* **2013**, *47*, 7899–7908. [CrossRef]
50. Anielak, A.M.; Styszko, K.; Kłeczek, A.; Łomińska-Płatek, D. Humic Substances—Common Carriers of Micropollutants in Municipal Engineering. *Energies* **2022**, *15*, 8496. [CrossRef]
51. Mailler, R.; Gasperi, J.; Coquet, Y.; Deshayes, S.; Zedek, S.; Cren-Olivé, C.; Cartiser, N.; Eudes, V.; Bressy, A.; Caupos, E.; et al. Study of a Large Scale Powdered Activated Carbon Pilot: Removals of a Wide Range of Emerging and Priority Micropollutants from Wastewater Treatment Plant Effluents. *Water Res.* **2015**, *72*, 315–330. [CrossRef]
52. Burchacka, E.; Pstrowska, K.; Beran, E.; Fałtynowicz, H.; Chojnacka, K.; Kułażyński, M. Antibacterial Agents Adsorbed on Active Carbon: A New Approach for *S. Aureus* and *E. Coli* Pathogen Elimination. *Pathogens* **2021**, *10*, 1066. [CrossRef]
53. Giannakoudakis, D.A.; Kyzas, G.Z.; Avranas, A.; Lazaridis, N.K. Multi-Parametric Adsorption Effects of the Reactive Dye Removal with Commercial Activated Carbons. *J. Mol. Liq.* **2016**, *213*, 381–389. [CrossRef]
54. Viswanadhan, V.N.; Ghose, A.K.; Revankar, G.R.; Robins, R.K. Atomic Physicochemical Parameters for Three Dimensional Structure Directed Quantitative Structure-Activity Relationships. 4. Additional Parameters for Hydrophobic and Dispersive Interactions and Their Application for an Automated Superposition of Certain. *J. Chem. Inf. Model.* **1989**, *29*, 163–172. [CrossRef]
55. Meija, J.; Coplen, T.B.; Berglund, M.; Brand, W.A.; De Bièvre, P.; Gröning, M.; Holden, N.E.; Irrgeher, J.; Loss, R.D.; Walczyk, T.; et al. Atomic Weights of the Elements 2013 (IUPAC Technical Report). *Pure Appl. Chem.* **2016**, *88*, 265–291. [CrossRef]
56. Babić, S.; Horvat, A.J.M.; Mutavdžić Pavlović, D.; Kaštelan-Macan, M. Determination of PKa Values of Active Pharmaceutical Ingredients. *TrAC—Trends Anal. Chem.* **2007**, *26*, 1043–1061. [CrossRef]

57. Zrnčić, M.; Babić, S.; Mutavdžić Pavlović, D. Determination of Thermodynamic p K a Values of Pharmaceuticals from Five Different Groups Using Capillary Electrophoresis. *J. Sep. Sci.* **2015**, *38*, 1232–1239. [CrossRef]
58. Lagergren, S. About the Theory of So-Called Adsorption of Soluble Substances. *K. Sven. Vetensk. Handl.* **1898**, *24*, 1–39.

Disclaimer/Publisher's Note: The statements, opinions and data contained in all publications are solely those of the individual author(s) and contributor(s) and not of MDPI and/or the editor(s). MDPI and/or the editor(s) disclaim responsibility for any injury to people or property resulting from any ideas, methods, instructions or products referred to in the content.

Article

Modification of Multiwalled Carbon Nanotubes and Their Mechanism of Demanganization

Yuan Zhou [1,2], Yingying He [1,2], Ruixue Wang [1,2], Yongwei Mao [1,2], Jun Bai [1,2] and Yan Dou [1,2,*]

1. School of Water and Environment, Chang'an University, No. 126 Yanta Road, Xi'an 710054, China
2. Key Laboratory of Subsurface Hydrology and Ecological Effects in Arid Region of the Ministry of Education, Chang'an University, No. 126 Yanta Road, Xi'an 710054, China
* Correspondence: douyan@chd.edu.cn

Abstract: Multiwalled carbon nanotubes (MWCNTs) were modified by oxidation and acidification with concentrated HNO_3 and H_2SO_4, and the modified multiwalled carbon nanotubes (M-MWCNTs) and raw MWCNTs were characterized by several analytical techniques. Then the demanganization effects of MWCNTs and M-MWCNTs were well investigated and elucidated. The experimental data demonstrated that the adsorption efficiency of Mn(II) could be greatly promoted by M-MWCNTs from about 20% to 75%, and the optimal adsorption time was 6 h and the optimal pH was 6. The results of the kinetic model studies showed that Mn(II) removal by M-MWCNTs followed the pseudo-second-order model. Isothermal studies were conducted and the results demonstrated that the experimental data fitted well with the three models. The reliability of the experimental results was well verified by PSO–BP simulation, and the present conclusion could be used as a condition for further simulation. The research results provide a potential technology for promoting the removal of manganese from wastewater; at the same time, the application of various mathematical models also provides more scientific ideas for the research of the mechanism of adsorption of heavy metals by nanomaterials.

Keywords: M-MWCNTs; Mn(II) removal; kinetic model; isotherm model; PSO-BP model

Citation: Zhou, Y.; He, Y.; Wang, R.; Mao, Y.; Bai, J.; Dou, Y. Modification of Multiwalled Carbon Nanotubes and Their Mechanism of Demanganization. *Molecules* **2023**, *28*, 1870. https://doi.org/10.3390/molecules28041870

Academic Editors: Yongchang Sun and Dimitrios Giannakoudakis

Received: 13 January 2023
Revised: 5 February 2023
Accepted: 8 February 2023
Published: 16 February 2023

Copyright: © 2023 by the authors. Licensee MDPI, Basel, Switzerland. This article is an open access article distributed under the terms and conditions of the Creative Commons Attribution (CC BY) license (https://creativecommons.org/licenses/by/4.0/).

1. Introduction

Manganese, a heavy metal, is abundant in nature and also plays a significant role in many important industries [1,2]. However, due to factors such as artificial mining of mineral resources, illegal discharge of pollutants from factories, and dissolution of manganese minerals in aquifers caused by changes in ecological environment, the problem of pollution has become increasingly prominent [3–5]. In recent years, the concentration of manganese in underground wells in many countries has far exceeded the standards of the World Health Organization [6–9]. Overexposure to manganese can cause a variety of negative health effects for humans [10,11], and high manganese content in plants will lead to crop necrosis and cotton wrinkling, thus affecting the food and textile industry [12]. Studies have shown that in manganese-polluted areas, the soil and groundwater are usually containing acidic organic matter, so manganese mostly exists in the form of divalent ions [13–15].

There have been some studies on methods for the removal of manganese from a solution, such as the oxidation precipitation method [16], the ion exchange method [17], the reverse osmosis method [18] and the adsorption method, which is the more commonly used removal method [19–21]. Recently, with the development of nanomaterials, carbon nanotubes have been reported as the new adsorbents for the removal of heavy metal and organic pollution, such as chlorobenzenes, herbicides, heavy metal ions (Pb^{2+} and Ca^{2+}), and inorganic nonmetallic ions, including F^- [22,23].

Due to lower synthesis and purification costs, and easy application to water treatment, multi-walled carbon nanotubes (MWCNTs) are more widely used than single-walled carbon nanotubes (SWCNTs) [24]. To improve the removal of heavy metals by carbon nanotubes, the original MWCNTs usually need to be modified. The ordinary modification method is to use a strong oxidant to oxidize the MWCNTs under reflux or ultrasonic conditions [25,26]. This oxidation endows carbon nanotubes with rich, oxygen-containing groups and exposes the adsorption site [17,27]. Meanwhile, in order to solve complex non-linear problems in practice, some neural network models, such as artificial neural networks (ANN) and backpropagation-based training optimization neural networks (BPNN), have been gradually applied in the field of pollutant removal [28–30].

It has been reported that MWCNTs can adsorb manganese [31], but the removal efficiency, adsorption mechanism, and related experimental verification work have not been clearly carried out. Therefore, the objectives of this work are as follows: (1) to modify raw MWCNTs and to characterize the MWCNTs and M-MWCNTs by SEM, XPS, FT-IR, etc.; (2) to study the effects of demanganization on MWCNTs and M-MWCNTs with varying pH, contact time, and temperature; (3) to describe the characteristics of the Mn(II) removal by the MWCNTs and M-MWCNTs with an adsorption kinetic model and an isotherm model; and (4) to simulate the adsorption process by using PSO-BP modeling.

2. Results and Discussion
2.1. Characterization of MWCNTs and M-MWCNTs
2.1.1. FT-IR

The most important use of infrared spectroscopy (FTIR) is the structural analysis of organic compounds [32]. In this research, FT-IR was used to verify the structural analysis of MWCNTs and M-MWCNTs, and M-MWCNTs were modified by oxidation and acidification. The FTIR spectra of Figure 1 illustrated that there were functional groups, −OH groups (3200~3600 cm^{-1}), −C=O− groups (1600 cm^{-1}), and −C–C− groups (1150 cm^{-1}), on the external and internal surface of MWCNTs and M-MWCNTs [33]. The transmittance (%) of the −OH groups and −C=O− groups in M-MWCNTs were stronger than in raw MWCNTs; the modification increased the active sites on the surface and further altered the surface polarity and charges. It is reported that the stretching vibration absorption peak of −C=O− usually appears in 1755–1670 cm^{-1}. In this study, an absorption peak of −C=O− is observed at 1623 cm^{-1}. This phenomenon has also shown up in the work of other researchers [34], possibly because the conjugate effect of carbon nanotubes makes the absorption of −C=O move to the shortwave direction [35].

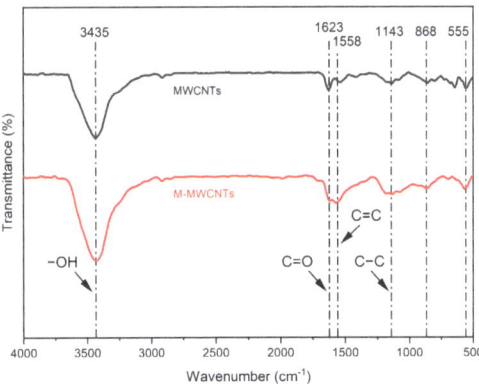

Figure 1. FTIR spectra of MWCNTs and M-MWCNTs.

2.1.2. XPS

XPS can be used to analyze elements present on the surface of the sample and bonding species [36]. The XPS wide-scan spectrum (Figure 2a) shows that the elements present on the surface of MWCNTs and M-MWCNTs were mainly C and O. New active sites are provided, as evidenced by the increased oxygen content of M-MWCNTs. This indicates that the adsorption capacity of M-MWCNTs for heavy metals will be enhanced [37]. The increase in O content (from 3.5% to 13.66%) showed that the modification was successful. Figure 2b shows the fitted XPS spectra of the O1s of M-MWCNTs compared to the reference XPS of MWCNTs and M-MWCNTs; the experimental data showed that $-C=O-$ groups (530.7 eV) and $-C-O-$ groups (533.6 eV) were on the surface of the M-MWCNTs.

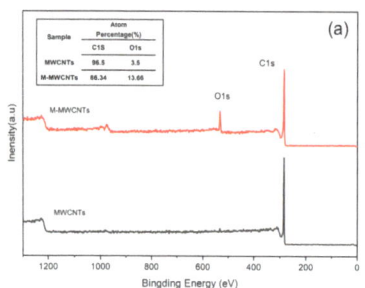 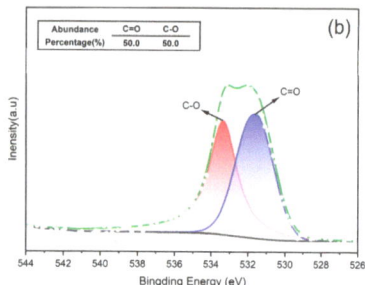

Figure 2. XPS spectra. (**a**) XPS of MWCNTs and M-MWCNTs (**b**) O1s XPS of M-MWCNTs.

2.1.3. SEM

The morphology investigation of MWCNTs and M-MWCNTs was performed using SEM (Figure 3). Figure 3b shows that the MWCNTs and M-MWCNTs were about 20 nm in diameter, M-MWCNTs were shorter than MWCNTs, and reunions were more likely to occur in M-MWCNTs particles, a phenomenon that was consistent with the Zeta potential results.

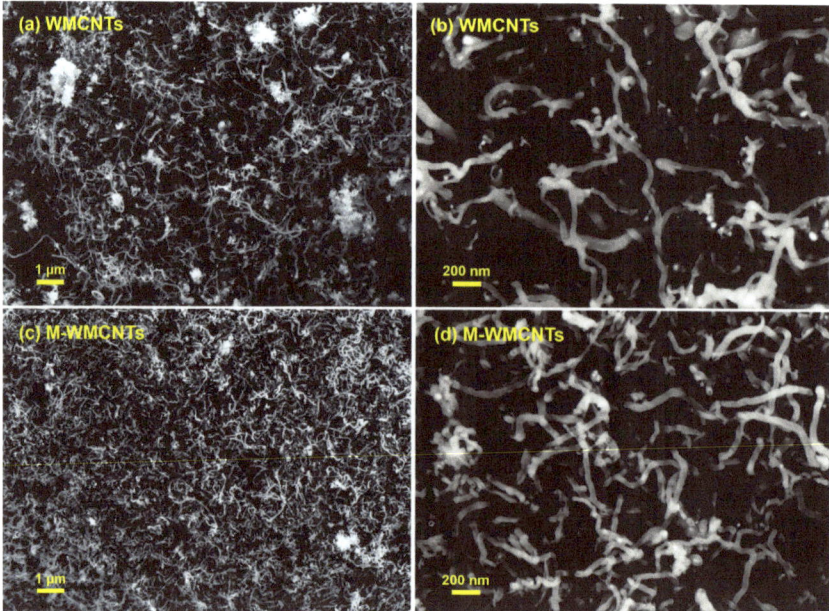

Figure 3. SEM of MWCNTs (**a**,**b**) and M-MWCNTs (**c**,**d**).

2.1.4. Zeta Potentials

The zeta potentials were measured as a function of pH to determine the pH_{PZC} of MWCNTs and M-MWCNTs [38]. The results (Figure 4) showed that the zeta potentials of

MWCNTs and M-MWCNTs decreased with the increasing pH, whereas the zeta potentials of M-MWCNTs became more negative after the treatment. The zeta potentials of M-MWCNTs were all less than 0. It can be speculated that the reason for this phenomenon was the influence of certain groups (−COOH, −OH) [31]. Surface negativity is favorable for the adsorption of heavy metal ions from the solution by the adsorbent [39].

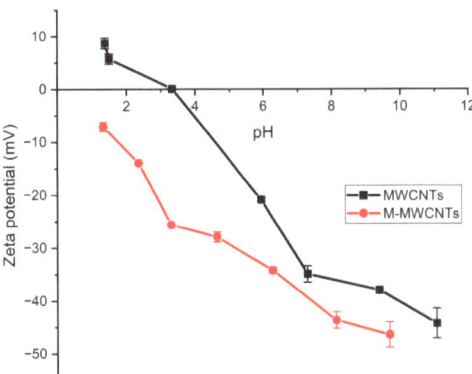

Figure 4. Zeta potentials of MWCNTs and M-MWCNTs.

2.1.5. Size

Figure 5 shows the size distribution of MWCNTs and M-MWCNTs at pH 6.2. In general, the dynamic light-scattering results from MWCNTs represent agglomerations rather than individual nanomaterials [40]. In the measurement process, the agglomeration of the original MWCNTs leads to a wider particle size distribution and an increase in the mean value (Figure 5a). This phenomenon becomes more obvious over time (Figure 5a blue line). The average particle size of M-MWCNTs treated with mixed acid decreased, and the result of the three measurements was close to 193.5, indicating that the stability of carbon nanotubes in an aqueous solution was significantly improved. It is remarkable that, compared with the raw M-MWCNTs, the particle size of the M-MWCNTs with adsorbed Mn(II) was increased significantly; the average hydrated particle size increased from 193.5 nm to 320 nm.

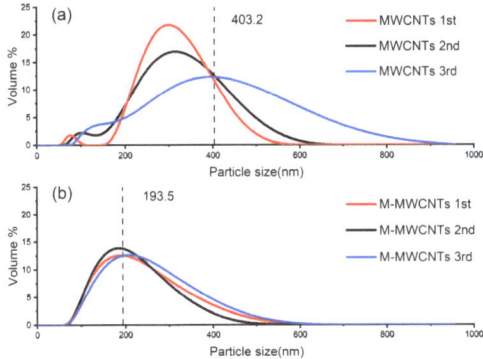

Figure 5. Results of nano-particle sizing (**a**) MWCNTs (**b**) M-MWCNTs.

2.2. Effect of Contact Time

The effect of contact time was studied under the following conditions: agitation speed, 180 rpm; adsorbent, 20 mg; initial concentration, 5 mg/L; pH, 5.6; temperature, 25 °C; and the mass ratio of adsorbent in solution was 1 g/L. The results are presented in Figure 6.

It can be observed that the adsorption of Mn(II) onto the MWCNTs reached equilibrium rapidly within 1 h, but the removal efficiency of Mn(II) was low (27.9%). The adsorption of Mn(II) by M-MWCNTs increased rapidly in the first 30 min, then increased at a slower rate and reached equilibrium at 6 h. This result is corroborated in the literature [31].

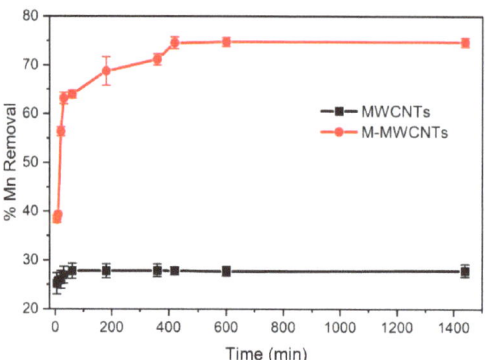

Figure 6. Contact time for Mn(II) adsorption by MWCNTs and M-MWCNTs.

2.3. Effect of pH

The effect of pH on the demanganization by MWCNTs and M-MWCNTs is shown in Figure 7. The pH value varied from 2 to 9, contact time was 10 h, mass of adsorbent was 20 mg, concentration of Mn(II) was 5 mg/L, agitation speed was 180 rpm, the mass ratio of adsorbent in solution was 1 g/L, and the experimental temperature was 25 °C. The results indicated that the removal efficiency of Mn(II) increased. When the pH value is higher than 9, Mn(II) should be formed and Mn(OH)$_2$ can even be precipitated [41]. In addition, under acidic conditions, H$^+$ ions compete for the active sites with Mn(II) ions. As the pH increases, the concentration of H$^+$ ions decreases, leaving more adsorption sites for Mn(II) ions. As the pH continues to increase, the Mn(II) is hydrolyzed, forming Mn(OH)$^+$, Mn(OH)$_2$, Mn$_2$(OH)$^+_3$, and Mn(OH)$^-_4$. When pH > 8, the Mn(II) begins to form a precipitate [42].

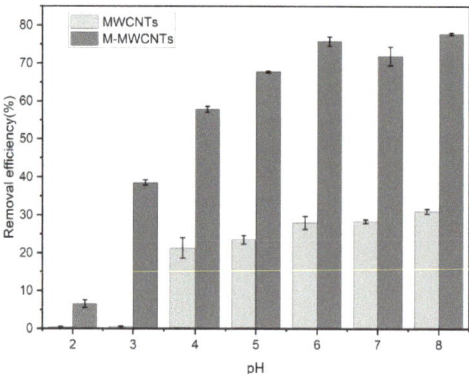

Figure 7. pH and Mn(II) adsorption by MWCNTs and M-MWCNTs.

When pH < 3, the adsorption rate of Mn(II) onto raw MWCNTs is close to 0. In the strong acid solution, the adsorption effect of raw MWCNTs on Mn(II) was inhibited, mainly due to a competitive effect between H$^+$ and Mn^{2+} at the active site. The zeta potential results also support this conclusion; experimental zeta potential results showed that the zeta potential of the raw MWCNTs was positive in a solution of pH < 3 [43].

2.4. Kinetic Modeling

The results of the three kinetic models are shown in Figure 8a,b, and the parameters are shown in Table 1. The results revealed that the adsorption of Mn(II) onto MWCNTs and M-MWCNTs followed second-order kinetics, which suggests that the adsorption process before both is chemical adsorption. The fitting line of the Weber–Morris model (Figure 8c) showed that the adsorption of Mn(II) onto MWCNTs and M-MWCNTs could be described in two or three stages, indicating that the Mn(II) adsorption process is controlled by some diffusion mechanisms, probably including an intra-particle diffusion mechanism, bulk diffusion, and film diffusion mechanisms.

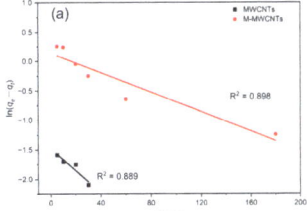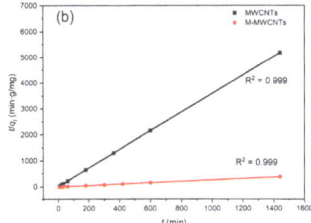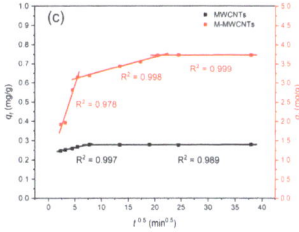

Figure 8. Adsorption kinetics. (**a**) Pseudo-first order, (**b**) pseudo-second order, and (**c**) intraparticle diffusion for Mn(II) adsorption by MWCNTs and M-MWCNTs.

Table 1. Pseudo-first-order, pseudo-second-order and intra-particle diffusion model constants for Mn(II) adsorption by MWCNTs and M-MWCNTs.

Model	Parameter	MWCNTs	M-MWCNTs
Pseudo-first-order model	$q_{e.experiment}$	0.278	3.74
	$q_{e.calculate}$ (mg/g)	0.229	1.16
	k_f (1/min)	0.0433	0.019
	R^2	0.889	0.898
Pseudo-second-order model	$q_{e.calculate}$ (mg/g)	0.279	3.76
	k_s (1/min)	4.12	0.030
	R^2	0.999	0.999
Weber–Morris	k_w	/	/
	C	/	/
	R^2	/	/

2.5. Isotherm Modeling

The results of experimental data fitting to three adsorption isotherm models for Mn(II) are presented in Table 2 and Figure 9.

Table 2. Isotherm parameters and determination coefficients for Mn(II) adsorption by MWCNTs and M-MWCNTs.

Model	Parameter	MWCNTs	M-MWCNTs
Langmuir isotherm	q_{max1} (mg/g)	0.585	5.78
	b (L/mg)	0.249	1.19
	R_L (L/mg)	0.801	0.456
	R^2	0.989	0.999
Freundlich isotherm	K_f (mg/g (L/mg)$^{1/n}$)	0.159	3.96
	$1/n$	0.344	0.009
	R^2	0.978	0.872
Dubinin–Radushkevich	q_{max2} (mg/g)	1.4083	7.43
	β	0.0030	0.0009
	E_S	12.95	23.64
	R^2	0.989	0.913

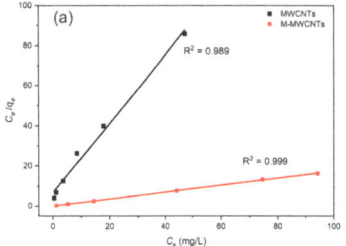 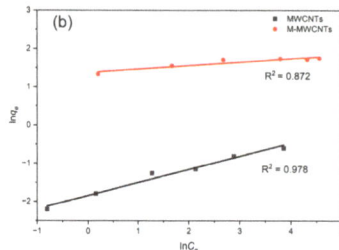 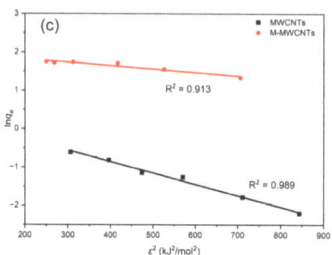

Figure 9. Isotherm models for Mn(II) adsorption by MWCNTs and M-MWCNTs. (**a**) Langmuir, (**b**) Freundlich, and (**c**) D–R.

The adsorption of Mn(II) by MWCNTs simultaneously conformed to three models; all the values of R^2 were >0.97 in the three isotherms. This indicated that the adsorption of Mn(II) by MWCNTs was complex and contained a variety of mechanisms under experimental conditions; the adsorption process had both monolayer and inhomogeneous surface chemisorption. As for M-MWCNTs, the R^2 of the Langmuir model was close to 1, which is better than others. This indicated that the adsorption process of Mn(II) by M-MWCNTs was monolayer adsorption. As the Mn(II) ions attached to the site of M-MWCNTs (−COOH, −OH), no further adsorption occurred at that site [44]. Consequently, the adsorption capacity reached the maximum when the monolayer adsorption of Mn(II) ions was completely formed on the surface. The value of theoretical adsorption capacity calculated for M-MWCNTs is approximately 10 times higher than that of MWCNTs. The values of R_L were from 0 to 1 for the studied concentration range, which means the adsorption of Mn(II) by M-MWCNTs is considered favorable.

Notably, the MWCNT and M-MWCNT results followed the D–R model well (R^2 = 0.989 and 0.913, respectively). The value of E_S indicated that the MWCNT adsorption process was mainly ion-exchange adsorption, while that of M-MWCNTs was mainly chemical adsorption [45].

2.6. PSO–BP Modeling

In this study, 35 and 39 sets of experiments were designed for MWCNTs and M-MWCNTs, respectively, and each set of experiments was conducted thrice. All data (35 × 3, 39 × 3, Table A1) were used for model fitting in order to reveal the inherent mechanisms in the process [46]. Imported data were normalized, randomly shuffled, and divided into three groups for crossover verification (70% for training, 15% testing, and 15% validation).

In Figure 10, two models were developed for adsorption of MWCNTs (Model A) and M-MWCNTs (Model B), to choose the optimal network structure, 1–14 neurons were applied in the hidden layer, RMSE and R2 were used to evaluate the effectiveness of PSO-BP, and in the selected range, Figure 10 shows that the calculated results of the model were consistent with the experimental data, and for the value of R2, 11 of the 14 neurons had values around 0.95 in Figure 10a, and 12 of the 14 neurons of values ranged from 0.95 to 1.00 in Figure 10b. For the value of RMSE, in Figure 10a, 11 values were between 2.5–3, and in Figure 10b, the 10 values ranged from 3–5. The parameters of the PSO–BP model are listed in Table 3. With this fixed parameter, the optimal weights and biases obtained from the two PSO–BP models are shown in Table A2. Figure A1 shows the comparison between the input and output data of two models. The values of R for the training, testing, validation, and all data of both models were better than 0.97, which demonstrated that the predicted data agreed well with the experimental data using the PSO–BP model. Most of the data were distributed on the line of Y = T, indicating good compatibility of the experimental data with the PSO–BP-forecasted data. To further validate the performance of the model predictions, the four non-linear statistics of the model were evaluated. The results of the evaluation (Table 4) demonstrated that the predictions of both models were statistically significant.

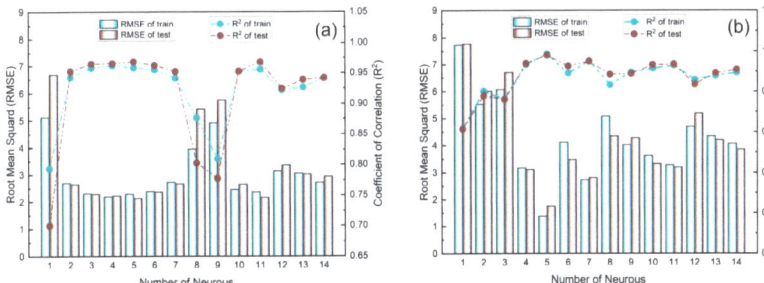

Figure 10. Variation in RMSE and R2 with number of neurons in hidden layer for Model A (**a**) and Model B (**b**).

Table 3. The parameters of the optimum PSO–BP model.

Parameter	Model A: MWCNTs	Model B: M-MWCNTs
Number of samples	117	105
Number of hidden layers	1	1
Hidden nodes	4	5
PSO swarm size	50	50
Cognitive component (c_1)	1.5	1.49554
Social component (c_2)	1.5	1.49554
Number of iterations	1000	1000

Table 4. Nonlinear statistical metrics for validating efficacy of model predictions.

Statistical Metric	Model A: MWCNTs	Model B: M-MWCNTs
R^2 (train/test)	0.962/0.966	0.997/0.995
RMSE (train/test)	2.20/2.26	1.38/1.76
MAE (train/test)	1.404/1.58	0.769/0.909
MAPE	0.074/0.067	0.0202/0.0223

3. Materials and Methods

3.1. Materials

The raw MWCNTs were purchased from COCC (Chengdu Institute of Organic Chemistry, Chinese Academy of Science). Using carbon gas as the carbon source, this material was produced via the CVD (catalytic vapor decomposition) method; its outer diameter was between 5 and 15 nm, the surface area was in the range of 220–300 m^2/g, and the purity of the material was above 95%, with low metal impurity content and high electrical conductivity.

Mn(II) stock solution (1000 mg/L) was prepared by dissolving 3.6010 g MnCl$_2$•4H$_2$O (GR) in 1000 mL of deionized water. All the manganese-containing solutions in later experiments were diluted from the stock solution. The concentration of Mn(II) was determined by atomic absorption spectrometry (WFX120).

3.2. Preparation of M-MWCNTs

A total of 5 g of raw MWCNTs was added to a 120 mL mixture of HNO$_3$ (68%) and H$_2$SO$_4$ (98%) (v/v 1:3). The mixture was sonicated at 40 °C for 4 h and stirred at room temperature for 24 h, so that the MWCNTs were fully in contact with the oxidant and were oxidized. The resulting MWCNTs were separated from the solution using a 0.22 μm membrane filter, rinsed thoroughly with deionized water to remove excess acid, and then rinsed with anhydrous ethanol to accelerate drying. The obtained sample was dried at 65 °C for 24 h. Figure 11 shows the preparation of M-MWCNTs.

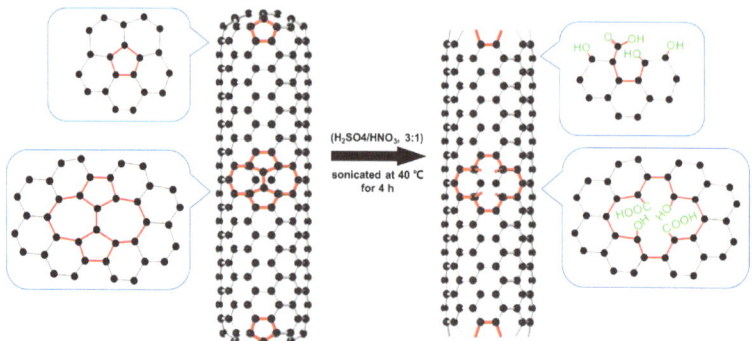

Figure 11. Schematic diagram of the preparation of M-MWCNTs.

3.3. Characterization Analysis Methods

The FT-IR spectra of the dried samples were determined by an IRTracer 100 FTIR spectrometer (Shimadzu, Japan), using the potassium bromide powder method. The range of scanning was 4000–350 cm^{-1} and the resolution was set to 4 cm^{-1}.

XPS (X-ray photoelectron spectroscopy) was performed to investigate the creation of ionic networks between MWCNTs and M-MWCNTs on ESCALAB Xi+. The main parameters were spatial resolution (1 μm) and energy resolution (0.48 eV). The sample needed to be pressed before testing.

SEM (scanning electron microscopy, Zeiss Ultra Plus) was used to detect the surface morphology of the MWCNTs and M-MWCNTs [47] and the micro area constitution of materials was analyzed using Oxford X-Max 50 mm Energy Disperse Spectroscopy (EDS).

A Malvern Zetasizer Ultra was used to measure the particle sizes and zeta potentials of MWCNTs and M-MWCNTs. The samples were dispersed by ultrasonication for 20 min before measurement.

3.4. Batch Experiments

Adsorption experiments. A total of 20 mg MWCNTs and 20 mg M-MWCNTs were separately mixed with 20 mL of Mn(II) solution in a polypropylene centrifuge tube of 50 mL. The initial Mn(II) concentration was 5, 10, 20, 50, 80, and 100 mg/L, while the initial pH value was from 5.6 to 5.2. The mixture was shaken for 10 h and 25 °C at 180 rpm, then filtered using a 0.22 μm membrane filter.

Adsorption thermodynamics experiments. The temperature was set at 25, 35, 45, and 55 °C, and the contact time for adsorption kinetics was controlled from 5 min to 24 h. To study the effect of pH, the pH value was adjusted using sodium hydroxide solution (0.01~0.1 mol/L) and hydrochloric acid solution (0.01~0.1 mol/L).

3.5. Theory

3.5.1. Kinetic Modeling

The pseudo-first-order model, pseudo-second-order model, and Weber–Morris model were the kinetic models used to describe the mechanism of Mn(II) adsorption onto MWC-NTs and M-MWCNTs. The pseudo-first-order kinetic model can be used to fit the materials with fewer active sites, higher initial adsorption concentration and the adsorption process in the early stage, while the pseudo-second-order kinetic model is just the opposite [48]. Pseudo-first-order and pseudo-second-order dynamics models belong to empirical models. They are simple to calculate, but they fail to explain the adsorption process in terms of mechanism. A Weber–Morris dynamics model has been put to use in the adsorption of metal ions by many scholars [49]. It can be used to analyze the rate-controlling steps of the adsorption reaction, which can guide the adsorption process.

The pseudo-first-order model, which is widely used in kinetic adsorption simulation, is based on the assumption that adsorption is controlled by diffusion steps [50]. Equation (1) describes this model, where q_e is the equilibrium absorption capacity, q_t is the absorption capacity at time t, and k_f is the rate constant. Equation (2) is the integrated form.

$$(dq_t)/dt = k_f(q_e - q_t) \tag{1}$$

$$\log(q_e - q_t) = \log(q_e) - k_f t / 2.303 \tag{2}$$

The pseudo-second-order kinetic model assumes that the adsorption rate is determined by the square value of the number of vacant adsorption spots on the adsorbent surface, and the adsorption process is controlled by the chemisorption mechanism, which involves electron sharing or electron transfer between the adsorbent and the adsorbate [51]. Equation (3) describes this model, in which q_e is the equilibrium absorption capacity, q_t is the absorption capacity at time t, and k_s is the constant. Equations (4) and (5) are the integrated forms.

$$(dq_t)/dt = k_s(q_e - q_t)^2 \tag{3}$$

$$1/(q_e - q_t) = 1/q_e + k_s t \tag{4}$$

$$t/q_t = 1/k_s q_e^2 + t/q_e \tag{5}$$

The Weber–Morris dynamics model is a particle diffusion model (Equation (6)) [52,53]. The model is the most suitable for describing the dynamics of the material diffusion process inside the particle, but it is often not suitable for describing the diffusion process outside the particle and inside the liquid film.

$$q_t = k_w t^{1/2} \tag{6}$$

In Equation (6), q_t is the absorption capacity and k_w is the rate constant.

3.5.2. Isotherm Modeling

The adsorption isotherm model is a mathematical model that expresses the relationship between the adsorption capacity and the solution concentration under the condition of fixed temperature. Different types of mathematical expression have been proposed and each has its own scope of application; commonly used expressions are the Langmuir, Freundlich, and Dubinin–Radushkevich isotherm models.

The Langmuir model assumes that the appearance of the adsorbent is uniform and there is no interaction between adsorbents. The adsorption is monolayer adsorption, and the adsorption occurs only on the outer surface of the adsorbent. The Langmuir isothermal adsorption model was the first model to vividly describe the adsorption mechanism [54]. Equations (7) and (8) describe the linearized form [55], where q_e is the equilibrium absorption capacity, C_e is the equilibrium concentration, q_m is the maximum equilibrium absorption capacity, and b is the Langmuir rate constant.

$$q_e = \frac{q_m b C_e}{1 + b C_e} \tag{7}$$

$$\frac{C_e}{q_e} = \frac{1}{q_m b} + \frac{1}{q_m} C_e \tag{8}$$

Equation (9) is the essential expression for the Langmuir model, and it is easy to estimate the characteristics of adsorption.

$$R_L = \frac{1}{1 + b C_0} \tag{9}$$

R_L is a dimensionless constant and C_0 is the concentration of the Mn(II) solution. R_L values between 0 and 1 indicate favorable adsorption [56].

The Freundlich model can be applied to monolayer adsorption and heterogeneous surface adsorption. Not only can the Freundlich adsorption equation describe the adsorption mechanism of an uneven surface, it is also more suitable for adsorption at low concentrations. It can explain the experimental results over a wider range of concentration. The Freundlich isotherm (Equations (10) and (11)) [57] can be expressed as:

$$q_e = K_f C_e^{1/n} \tag{10}$$

$$ln q_e = ln K_f + \frac{1}{n} ln C_e \tag{11}$$

where q_e is the equilibrium absorption capacity, C_e is the equilibrium concentration, K_f is the Freundlich constant, and the n value ranges from 1 to 10.

In order to describe the relative pressure–adsorption capacity characteristics of microporous filling, Dubinin and Radushkevich proposed a method, based on the Polanyi adsorption potential theory, to calculate the adsorption characteristics based on the adsorption isotherm in the low-pressure region. The Dubinin–Radushkevich (D-R) isotherm is shown in Equation (12) [58]:

$$ln q_e = ln q_m - K\varepsilon^2 \tag{12}$$

$$\varepsilon = RT ln(1 + 1/C_e) \tag{13}$$

where q_e is the absorption capacity, q_m is the theoretical adsorption capacity, C_e is the equilibrium concentration, and K is the D-R constant.

In order to describe the calculation results of this model more easily, the relationship between the free adsorption energy, E_s, and the constant, K, is given as Equation (14). The value of E_s determines the mechanism of isothermal adsorption.

$$E_s = (2K)^{-0.5} \tag{14}$$

If the value of E_s is between approximately 8 and 16 kJ/mol, it shows that the main process of adsorption is ion exchange, whereas if the value E_s is more than 18 kJ/mol, it means the adsorption is chemisorptive in nature.

3.5.3. PSO-BP Modeling

Since Warren McCulloch and Walter Pitts introduced the concept of artificial neural networks (ANN) in 1943, ANN have evolved rapidly and have been successfully applied in many fields [59]. Backpropagation-based training-optimization neural networks (BPNN) are the most extensively utilized neural networks in practice and are capable of solving complex nonlinear problems. BPNN also have a wide range of applications in the field of pollutant removal [28–30]. BPNN require constant exploration of different combinations of weights and biases to obtain ideal results. Particle swarm optimization (PSO) algorithms are widely used for this work. PSO is a stochastic search algorithm that simulates the predatory behavior of birds. A set of optimal solutions was obtained from the PSO algorithm by tracking the constantly moving particles.

In this study, a BPNN model optimized by PSO (PSO–BP) was used to predict the adsorption of Mn onto two carbon materials. This process was implemented with Matlab2016a software. A standard BPNN network with one input layer, one hidden layer, and one output layer was created by calling on the software's own "newff" function. The backpropagation algorithm, which was applied to determine the most favorable network structure, selected Levenberg–Marquardt (trainlm) with 1000 iterations. The input layer contains four neurons (pH, manganese concentration, time, and temperature). The output is the percentage of Mn(II) removed. Owing to the small amount of data, the network uses just one hidden layer to avoid overfitting [60] and uses a tangent sigmoid function (tansig) as the activation function, whereas the output layer uses a linear function (purlin) as the transfer function. PSO uses the weights and MSE of the network as the particle and fitness functions, respectively, to learn and iterate. After each iteration, the new particles are transformed into the weights of the neural network; the network computes a new MSE and continues to iterate until the result is ideal or the number of iterations reaches a maximum.

The results of the model calculations are evaluated by the following equations (Equations (15)–(18)):

$$R^2 = 1 - \frac{\sum_{i=1}^{n}\left(y_{prd,i} - y_{exp,i}\right)^2}{\sum_{i=1}^{n}\left(y_{prd,i} - y_m\right)^2} \quad (15)$$

$$RMSE = \sqrt{\frac{1}{n}\sum_{i=1}^{n}\left(y_{prd,i} - y_{exp,i}\right)^2} \quad (16)$$

$$MAE = \frac{1}{n}\sum_{i=1}^{n}\left|y_{prd,i} - y_{exp,i}\right| \quad (17)$$

$$MAPE = \frac{100\%}{n}\sum_{i=1}^{n}\left|\frac{y_{prd,i} - y_{exp,i}}{y_{exp,i}}\right| \quad (18)$$

where $y_{prd,i}$ is the value calculated using the PSO–BP model, $y_{exp,i}$ is the experimental value, n is the number of data, and y_m is the average of the experimental value.

4. Conclusions

Compared with previous studies, the Mn(II)-removal performance by M-MWCNTs was improved [31]. Through the comparative analysis of several analytical techniques results before and after modification, as well as the fitting of thermodynamic and kinetic adsorption models, it was concluded that the increased number of adsorption sites (carboxyl groups and hydroxyl groups) after modification was the key to improving the removal rate. Additionally, the simulation verification of the experimental results by PSO–BP model also provided a scientific guarantee for the reliability of the entire research work.

Author Contributions: Conceptualization, Y.D.; methodology, Y.D.; software, Y.Z. and Y.H.; validation, Y.D. and Y.Z.; investigation, Y.M. and J.B.; data curation, Y.Z., Y.H. and R.W.; writing—original draft preparation, Y.Z. and Y.H.; writing—review and editing, R.W. and Y.D.; visualization, Y.Z.; supervision, Y.D.; project administration, Y.D.; funding acquisition, Y.D. All authors have read and agreed to the published version of the manuscript.

Funding: This research was funded by the Natural Science Foundation of Shaanxi Province, grant number 2021SF-443, and the College Students' Innovative Entrepreneurial Training Plan Program, grant number S202210710227.

Institutional Review Board Statement: Not applicable.

Informed Consent Statement: Not applicable.

Data Availability Statement: We can provide the data of this journal through the corresponding author's email.

Conflicts of Interest: The authors declare no conflict of interest.

Sample Availability: Samples of the compounds are available from the authors.

Appendix A

Table A1. (a) Data used for model A. (b) Data used for model B.

(a)					
No.	pH	C/mg/L	Time/h	T/K	Adsorption Rate %
1	2.18	5	180	298	0.4
2	3.23	5	180	298	0.6
3	4.17	5	180	298	20.4
4	5.15	5	180	298	23.5
5	6.17	5	180	298	28.5
6	7.18	5	180	298	28.31
7	8.25	5	180	298	30.4
8	5.58	1	180	298	58
9	5.58	2	180	298	42
10	5.58	5	180	298	28.6
11	5.58	10	180	298	9.6
12	5.58	20	180	298	11
13	5.58	50	180	298	6.5
14	5.58	5	5	298	22.2
15	5.58	5	10	298	25.8
16	5.58	5	20	298	26.8
17	5.58	5	30	298	28
18	5.58	5	60	298	27
19	5.58	5	180	298	28.6
20	5.58	5	180	298	28.2
21	5.58	10	180	298	15.99
22	5.58	20	180	298	11.01
23	5.58	50	180	298	5.45
24	5.58	5	180	308	29.34
25	5.58	10	180	308	16.61
26	5.58	20	180	308	12.38
27	5.58	50	180	308	5.89
28	5.58	5	180	318	29.6
29	5.58	10	180	318	18.01
30	5.58	20	180	318	14.95
31	5.58	50	180	318	6.7
32	5.58	5	180	328	32.45
33	5.58	10	180	328	18.59
34	5.58	20	180	328	15.22
35	5.58	50	180	328	7.03
36	2.18	5	180	298	0.4
37	3.23	5	180	298	0.58

Table A1. Cont.

(a)					
No.	pH	C/mg/L	Time/h	T/K	Adsorption Rate %
38	4.17	5	180	298	19.4
39	5.15	5	180	298	23.18
40	6.17	5	180	298	27.5
41	7.18	5	180	298	28.5
42	8.25	5	180	298	31
43	5.58	1	180	298	59
44	5.58	2	180	298	43
45	5.58	5	180	298	26.2
46	5.58	10	180	298	29.6
47	5.58	20	180	298	10.75
48	5.58	50	180	298	6
49	5.58	5	5	298	26.4
50	5.58	5	10	298	25.2
51	5.58	5	20	298	23.8
52	5.58	5	30	298	27.6
53	5.58	5	60	298	27
54	5.58	5	180	298	26.2
55	5.58	5	180	298	27.8
56	5.58	10	180	298	15.45
57	5.58	20	180	298	11.05
58	5.58	50	180	298	5.59
59	5.58	5	180	308	28.5
60	5.58	10	180	308	17.12
61	5.58	20	180	308	12.51
62	5.58	50	180	308	6.05
63	5.58	5	180	318	29.8
64	5.58	10	180	318	17.8
65	5.58	20	180	318	14.5
66	5.58	50	180	318	6.58
67	5.58	5	180	328	32.5
68	5.58	10	180	328	18.51
69	5.58	20	180	328	15.1
70	5.58	50	180	328	7.22
71	2.18	5	180	298	0.4
72	3.23	5	180	298	0.51
73	4.17	5	180	298	24.2

Table A1. Cont.

(a)					
No.	pH	C/mg/L	Time/h	T/K	Adsorption Rate %
74	5.15	5	180	298	23.8
75	6.17	5	180	298	28.1
76	7.18	5	180	298	28.2
77	8.25	5	180	298	31.6
78	5.58	1	180	298	49
79	5.58	2	180	298	40.5
80	5.58	5	180	298	28.6
81	5.58	10	180	298	8.6
82	5.58	20	180	298	11.5
83	5.58	50	180	298	4
84	5.58	5	5	298	25.6
85	5.58	5	10	298	24.8
86	5.58	5	20	298	27
87	5.58	5	30	298	24.8
88	5.58	5	60	298	29.6
89	5.58	5	180	298	28.6
90	5.58	5	180	298	27.6
91	5.58	10	180	298	16.5
92	5.58	20	180	298	10.9
93	5.58	50	180	298	5.5
94	5.58	5	180	308	28.2
95	5.58	10	180	308	16.98
96	5.58	20	180	308	12.63
97	5.58	50	180	308	6.09
98	5.58	5	180	318	28.6
99	5.58	10	180	318	17.1
100	5.58	20	180	318	14
101	5.58	50	180	318	6.04
102	5.58	5	180	328	32.49
103	5.58	10	180	328	18.58
104	5.58	20	180	328	14.8
105	5.58	50	180	328	7.31
(b)					
No.	pH	C/mg/L	Time/h	T/K	Adsorption Rate %
1	2.36	5	1440	298	5.6
2	3.38	5	1440	298	37.8
3	4.32	5	1440	298	57
4	5.31	5	1440	298	74.3

Table A1. Cont.

(b)					
No.	pH	C/mg/L	Time/h	T/K	Adsorption Rate %
5	6.3	5	1440	298	76.1
6	7.37	5	1440	298	75.6
7	8.32	5	1440	298	77.8
8	5.58	5	1440	298	75.8
9	5.58	10	1440	298	48.8
10	5.58	20	1440	298	27.6
11	5.58	50	1440	298	12.4
12	5.58	80	1440	298	11.8
13	5.58	100	1440	298	5.69
14	5.58	5	5	298	38.2
15	5.58	5	10	298	39.2
16	5.58	5	20	298	53.6
17	5.58	5	30	298	64
18	5.58	5	60	298	64.4
19	5.58	5	180	298	68.8
20	5.58	5	300	298	69.8
21	5.58	5	420	298	75.8
22	5.58	5	600	298	75.8
23	5.58	5	1440	298	75.8
24	5.58	5	1440	298	75.1
25	5.58	10	1440	298	47.1
26	5.58	20	1440	298	27.5
27	5.58	50	1440	298	11.4
28	5.58	5	1440	308	77.2
29	5.58	10	1440	308	49.8
30	5.58	20	1440	308	29
31	5.58	50	1440	308	12.4
32	5.58	5	1440	318	80.4
33	5.58	10	1440	318	50.9
34	5.58	20	1440	318	30
35	5.58	50	1440	318	12.4
36	5.58	5	1440	328	81
37	5.58	10	1440	328	52
38	5.58	20	1440	328	30.4
39	5.58	50	1440	328	13
40	2.36	5	1440	298	7.6
41	3.38	5	1440	298	38.6
42	4.32	5	1440	298	58.4
43	5.31	5	1440	298	73.8

Table A1. Cont.

(b)					
No.	pH	C/mg/L	Time/h	T/K	Adsorption Rate %
44	6.3	5	1440	298	76.3
45	7.37	5	1440	298	75.4
46	8.32	5	1440	298	77.9
47	5.58	5	1440	298	77
48	5.58	10	1440	298	47
49	5.58	20	1440	298	27.4
50	5.58	50	1440	298	10.4
51	5.58	80	1440	298	10.2
52	5.58	100	1440	298	5.8
53	5.58	5	5	298	37.8
54	5.58	5	10	298	39.2
55	5.58	5	20	298	62.4
56	5.58	5	30	298	61.8
57	5.58	5	60	298	63.5
58	5.58	5	180	298	68.5
59	5.58	5	300	298	72
60	5.58	5	420	298	72.5
61	5.58	5	600	298	74.2
62	5.58	5	1440	298	74.2
63	5.58	5	1440	298	74.8
64	5.58	10	1440	298	47.8
65	5.58	20	1440	298	27.29
66	5.58	50	1440	298	11.41
67	5.58	5	1440	308	77.8
68	5.58	10	1440	308	49.8
69	5.58	20	1440	308	28.8
70	5.58	50	1440	308	11.9
71	5.58	5	1440	318	79.8
72	5.58	10	1440	318	51.2
73	5.58	20	1440	318	30.1
74	5.58	50	1440	318	12.5
75	5.58	5	1440	328	81.5
76	5.58	10	1440	328	51.1
77	5.58	20	1440	328	30.5
78	5.58	50	1440	328	13.1
79	2.36	5	1440	298	6.4
80	3.38	5	1440	298	39.2
81	4.32	5	1440	298	58.2
82	5.31	5	1440	298	74.5
83	6.3	5	1440	298	75.9

Table A1. Cont.

(b)					
No.	pH	C/mg/L	Time/h	T/K	Adsorption Rate %
84	7.37	5	1440	298	75.2
85	8.32	5	1440	298	77.8
86	5.58	5	1440	298	74.6
87	5.58	10	1440	298	46.2
88	5.58	20	1440	298	28
89	5.58	50	1440	298	11.3
90	5.58	80	1440	298	10.6
91	5.58	100	1440	298	5.71
92	5.58	5	5	298	39.2
93	5.58	5	10	298	39.6
94	5.58	5	20	298	53
95	5.58	5	30	298	63.6
96	5.58	5	60	298	64.4
97	5.58	5	180	298	69
98	5.58	5	300	298	71.6
99	5.58	5	420	298	74.6
100	5.58	5	600	298	74.4
101	5.58	5	1440	298	74.5
102	5.58	5	1440	298	74.9
103	5.58	10	1440	298	47.1
104	5.58	20	1440	298	27.5
105	5.58	50	1440	298	11.39
106	5.58	5	1440	308	77.3
107	5.58	10	1440	308	49.8
108	5.58	20	1440	308	29.1
109	5.58	50	1440	308	11.7
110	5.58	5	1440	318	78.9
111	5.58	10	1440	318	51
112	5.58	20	1440	318	30
113	5.58	50	1440	318	12.3
114	5.58	5	1440	328	81.2
115	5.58	10	1440	328	51.6
116	5.58	20	1440	328	30.6
117	5.58	50	1440	328	13.1

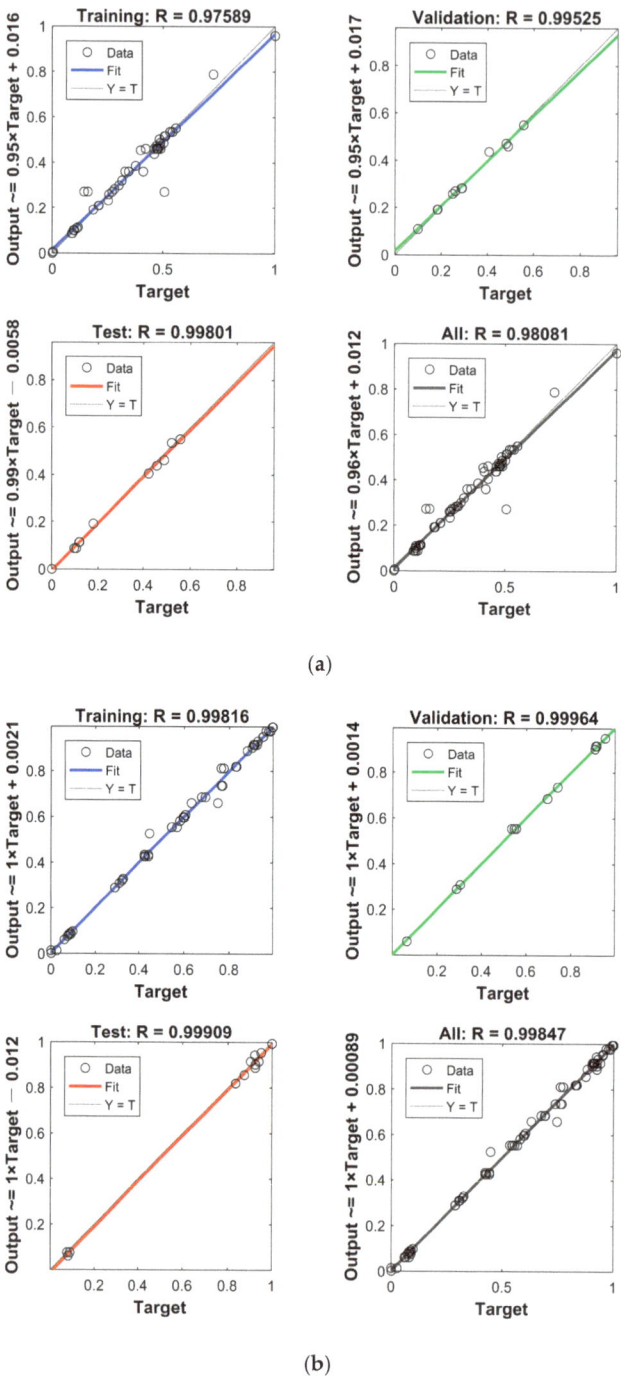

Figure A1. (**a**) Performances of model A on training, testing, validation, and combined data sets. (**b**) Performances of model B on training, testing, validation, and combined data sets.

Table A2. (a) Model A 4-4-1 topology—optimal weights and biases. (b) Model B 4-5-1 topology—optimal weights and biases.

(a)							
Input (4)-Hidden Layer (4)					Hidden Layer (4)-Output (1)		
Weights				Bias	Weights		Bias
−0.0021	−0.0998	0.0628	0.1912	−0.4088	−0.5668		
−0.3308	−0.3013	0.0358	−0.0908	−0.3962	0.2999		−0.1690
−0.1177	0.7891	0.1748	0.2355	−0.3528	−0.3022		
−0.2683	0.1825	0.0773	−0.4901	0.1691	−0.0722		
(b)							
Input (4)-Hidden Layer (5)					Hidden Layer (5)-Output (1)		
Weights				Bias	Weights		Bias
0.8852	−0.6705	−0.3712	0.0612	0.2464	0.1556		
0.2801	0.1780	0.1084	−0.4348	0.3055	1		
−0.1308	0.4948	−0.0209	0.0709	−0.0502	0.3046		−0.3376
−0.9012	−0.2605	−0.4838	−0.8644	0.2783	−0.4782		
0.6520	−1	−0.7253	0.1455	0.6593	−0.2843		

References

1. Clarke, C.; Upson, S. A global portrait of the manganese industry—A socioeconomic perspective. *NeuroToxicology* **2017**, *58*, 173–179. [CrossRef] [PubMed]
2. Singh, T. A Review of Low Grade Manganese Ore Upgradation Processes. *Miner. Process. Extr. Metall. Rev.* **2020**, *41*, 417–438. [CrossRef]
3. Jandieri, G. A generalized model for assessing and intensifying the recycling of metal-bearing industrial waste: A new approach to the resource policy of manganese industry in Georgia. *Resour. Policy* **2022**, *75*, S0301420721004700. [CrossRef]
4. Ghosh, S.; Mohanty, S.; Akcil, A.; Sukla, L.B.; Das, A.P. A greener approach for resource recycling: Manganese bioleaching. *Chemosphere* **2016**, *154*, 628–639. [CrossRef] [PubMed]
5. Marks, A. Aquatic environmental risk assessment of manganese processing industries. *Neurotoxicology* **2017**, *58*, 187–193. [CrossRef]
6. Vetrimurugan, E.; Brindha, K.; Elango, L.; Ndwandwe, O.M. Human exposure risk to heavy metals through groundwater used for drinking in an intensively irrigated river delta. *Appl. Water Sci.* **2017**, *7*, 3267–3280. [CrossRef]
7. Therdkiattikul, N.; Ratpukdi, T.; Kidkhunthod, P.; Chanlek, N.; Siripattanakul-Ratpukdi, S. Manganese-contaminated groundwater treatment by novel bacterial isolates: Kinetic study and mechanism analysis using synchrotron-based techniques. *Sci. Rep.* **2020**, *10*, 13391. [CrossRef]
8. Masoud, A.A.; Koike, K.; Mashaly, H.A.; Gergis, F. Spatio-temporal trends and change factors of groundwater quality in an arid area with peat rich aquifers: Emergence of water environmental problems in Tanta District, Egypt. *J. Arid. Environ.* **2016**, *124*, 360–376. [CrossRef]
9. Yu, D.; Zhou, J.; Zhang, J.; Sun, Y.; Zeng, Y. Hydrogeochemistry and evolution of iron and manganese in groundwater in Kashgar, Xinjiang. *Acta Sci. Circumst.* **2021**, *41*, 2169–2181.
10. Abubakar, B.; Jiun, O.D.; Ismail, M.; Imam, M.U. Nutrigenomics and Antioxidants. In *Nutraceuticals and Natural Product Pharmaceuticals*; Academic Press: Cambridge, MA, USA, 2019; pp. 33–70.
11. Blanc, P.D. The Early History of Manganese and the Recognition of its Neurotoxicity, 1837–1936. *Neurotoxicology* **2017**, *64*, 5–11. [CrossRef]
12. Dey, S.; Tripathy, B.; Kumar, M.S.; Das, A.P. Ecotoxicological consequences of manganese mining pollutants and their biological remediation. *Environ. Chem. Ecotoxicol.* **2023**, *5*, 55–61. [CrossRef]
13. Hou, Q.; Zhang, Q.; Huang, G.; Liu, C.; Zhang, Y. Elevated manganese concentrations in shallow groundwater of various aquifers in a rapidly urbanized delta, south China. *Sci. Total Environ.* **2020**, *701*, 134777. [CrossRef] [PubMed]
14. Champ, D.R.; Gulens, J.; Jackson, R.E. Oxidation–reduction sequences in ground water flow systems. *Can. J. Earth Sci.* **1979**, *16*, 12–23. [CrossRef]
15. Dou, Y.; Li, M.; Liu, Z.; Fang, R.; Yan, K. Simulation of magnesium chloride vertical transport in column experiments. *Hum. Ecol. Risk Assess. Int. J.* **2020**, *26*, 2407–2419. [CrossRef]
16. Szymoniak, L.; Claveau-Mallet, D.; Haddad, M.; Barbeau, B. Improving remineralization and manganese-removal of soft waters using a mixed $CaCO_3$/MgO contactor. *J. Water Process Eng.* **2022**, *49*, 102995. [CrossRef]

17. Bikiaris, D.; Vassiliou, A.; Chrissafis, K.; Paraskevopoulos, K.M.; Jannakoudakis, A.; Docoslis, A. Effect of acid treated multi-walled carbon nanotubes on the mechanical, permeability, thermal properties and thermo-oxidative stability of isotactic polypropylene. *Polym. Degrad. Stab.* **2008**, *93*, 952–967. [CrossRef]
18. Stein, S.; Sivan, O.; Yechieli, Y.; Kasher, R. Redox condition of saline groundwater from coastal aquifers influences reverse osmosis desalination process. *Water Res.* **2021**, *188*, 116508. [CrossRef]
19. Li, M.; Kuang, S.; Kang, Y.; Ma, H.; Dong, J.; Guo, Z. Recent advances in application of iron-manganese oxide nanomaterials for removal of heavy metals in the aquatic environment. *Sci. Total Environ.* **2022**, *819*, 153157. [CrossRef]
20. Nassar, M.Y.; El-Shahat, M.F.; Osman, A.; Sobeih, M.M.; Zaid, M.A. Adsorptive Removal of Manganese Ions from Polluted Aqueous Media by Glauconite Clay-Functionalized Chitosan Nanocomposites. *J. Inorg. Organomet. Polym. Mater.* **2021**, *31*, 4050–4064. [CrossRef]
21. Vasile, E.; Covaliu, C.I.; Stoian, O.; Paraschiv, G.; Ciric, A. Manganese Ions Removal from Industrial Wastewater. *Rev. De Chim.-Buchar. Orig. Ed.* **2020**, *71*, 391–396. [CrossRef]
22. Dehghani, M.H.; Haghighat, G.A.; Yetilmezsoy, K.; McKay, G.; Heibati, B.; Tyagi, I.; Agarwal, S.; Gupta, V.K. Adsorptive removal of fluoride from aqueous solution using single- and multi-walled carbon nanotubes. *J. Mol. Liq.* **2016**, *216*, 401–410. [CrossRef]
23. Rodríguez, C.; Leiva, E. Enhanced Heavy Metal Removal from Acid Mine Drainage Wastewater Using Double-Oxidized Multiwalled Carbon Nanotubes. *Molecules* **2020**, *25*, 111. [CrossRef] [PubMed]
24. Mishra, S.; Sundaram, B. Efficacy and Challenges of Carbon Nanotube in Wastewater and Water Treatment. *Environ. Nanotechnol. Monit. Manag.* **2022**, *28*, 100764. [CrossRef]
25. Aslam, M.; Kuo, H.W.; Den, W.; Usman, M.; Ashraf, H. Functionalized Carbon Nanotubes (CNTs) for Water and Wastewater Treatment: Preparation to Application. *Sustainability* **2021**, *13*, 5717. [CrossRef]
26. Ihsanullah; Abbas, A.; Al-Amer, A.M.; Laoui, T.; Al-Marri, M.J.; Nasser, M.S.; Khraisheh, M.; Atieh, M.A. Heavy metal removal from aqueous solution by advanced carbon nanotubes: Critical review of adsorption applications. *Sep. Purif. Technol.* **2016**, *157*, 141–161. [CrossRef]
27. Hong, C.-E.; Lee, J.-H.; Kalappa, P.; Advani, S.G. Effects of oxidative conditions on properties of multi-walled carbon nanotubes in polymer nanocomposites. *Compos. Sci. Technol.* **2007**, *67*, 1027–1034. [CrossRef]
28. Aghav, R.M.; Kumar, S.; Mukherjee, S.N. Artificial neural network modeling in competitive adsorption of phenol and resorcinol from water environment using some carbonaceous adsorbents. *J. Hazard. Mater.* **2011**, *188*, 67–77. [CrossRef]
29. Tariq, R.; Abatal, M.; Bassam, A. Computational intelligence for empirical modeling and optimization of methylene blue adsorption phenomena using available local zeolites and clay of Morocco. *J. Clean. Prod.* **2022**, *370*. [CrossRef]
30. Du, J.; Shang, X.; Shi, J.; Guan, Y. Removal of chromium from industrial wastewater by magnetic flocculation treatment: Experimental studies and PSO-BP modelling. *J. Water Process Eng.* **2022**, *47*, 102822. [CrossRef]
31. Ganesan, P.; Kamaraj, R.; Sozhan, G.; Vasudevan, S. Oxidized multiwalled carbon nanotubes as adsorbent for the removal of manganese from aqueous solution. *Environ. Sci. Pollut. Res.* **2013**, *20*, 987–996. [CrossRef]
32. Mohamed, I.M.A.; Dao, V.-D.; Yasin, A.S.; Barakat, N.A.M.; Choi, H.-S. Design of an efficient photoanode for dye-sensitized solar cells using electrospun one-dimensional GO/N-doped nanocomposite SnO_2/TiO_2. *Appl. Surf. Sci.* **2017**, *400*, 355–364. [CrossRef]
33. Basahel, S.N.; Al Thabaiti, S.A.; Obaid, A.Y.; Mokhtar, M.; Salam, M.A. Chemical modification of multi-walled carbon nanotubes using different oxidising agents: Optimisation and characterisation. *Int. J. Nanopart.* **2009**, *2*, 200–208. [CrossRef]
34. Pillay, K.; Cukrowska, E.M.; Coville, N.J. Multi-walled carbon nanotubes as adsorbents for the removal of parts per billion levels of hexavalent chromium from aqueous solution - ScienceDirect. *J. Hazard. Mater.* **2009**, *166*, 1067–1075. [CrossRef] [PubMed]
35. Ali, I.; Kuznetsova, T.S.; Burakov, A.E.; Burakova, I.V.; Pasko, T.V.; Dyachkova, T.P.; Mkrtchyan, E.S.; Babkin, A.V.; Tkachev, A.G.; Albishri, H.M.; et al. Polyaniline Modified CNTs and Graphene Nanocomposite for Removal of Lead and Zinc Metal Ions: Kinetics, Thermodynamics and Desorption Studies. *Molecules* **2022**, *27*, 5623. [CrossRef]
36. Mohamed, I.M.A.; Yasin, A.S.; Liu, C. Synthesis, surface characterization and electrochemical performance of ZnO @ activated carbon as a supercapacitor electrode material in acidic and alkaline electrolytes. *Ceram. Int.* **2020**, *46*, 3912–3920. [CrossRef]
37. Sadri, R.; Hosseini, M.; Kazi, S.N.; Bagheri, S.; Zubir, N.; Solangi, K.H.; Zaharinie, T.; Badarudin, A. A bio-based, facile approach for the preparation of covalently functionalized carbon nanotubes aqueous suspensions and their potential as heat transfer fluids. *J. Colloid Interface Sci.* **2017**, *504*, 115–123. [CrossRef]
38. Zhao, W.; Wang, M.; Yang, B.; Feng, Q.; Liu, D. Enhanced sulfidization flotation mechanism of smithsonite in the synergistic activation system of copper–ammonium species. *Miner. Eng.* **2022**, *187*, 107796. [CrossRef]
39. Lemya, B.; Zohra, S.F.; Houari, S.; Esma, C.-B.; Boumediene, B.; Ould, K.S.; Farouk, Z. Removal of Zn(II) and Ni(II) heavy metal ions by new alginic acid-ester derivatives materials. *Carbohydr. Polym.* **2021**, *272*, 118439. [CrossRef]
40. Silvestro, L.; Ruviaro, A.S.; de Matos, P.R.; Pelisser, F.; Mezalira, D.Z.; Gleize, P.J.P. Functionalization of multi-walled carbon nanotubes with 3-aminopropyltriethoxysilane for application in cementitious matrix. *Constr. Build. Mater.* **2021**, *311*, 125358. [CrossRef]
41. Al-Wakeel, K.Z.; Abd El Monem, H.; Khalil, M.M.H. Removal of divalent manganese from aqueous solution using glycine modified chitosan resin. *J. Environ. Chem. Eng.* **2015**, *3*, 179–186. [CrossRef]
42. Li, Y.; Shang, H.; Cao, Y.; Yang, C.; Feng, Y.; Yu, Y. Quantification of adsorption mechanisms distribution of sulfamethoxazole onto biochar by competition relationship in a wide pH range. *J. Environ. Chem. Eng.* **2022**, *10*, 108755. [CrossRef]
43. Stafiej, A.; Pyrzynska, K. Adsorption of heavy metal ions with carbon nanotubes. *Sep. Purif. Technol.* **2007**, *58*, 49–52. [CrossRef]

44. Tsai, Y.P.; Doong, R.A.; Yang, J.C.; Chuang, P.C.; Chou, C.C.; Lin, J.W. Removal of Humic Acids in Water by Carbon Nanotubes. *Adv. Mater. Res.* **2013**, *747*, 221–224. [CrossRef]
45. Özcan, A.S.; Erdem, B.; Özcan, A. Adsorption of Acid Blue 193 from aqueous solutions onto BTMA-bentonite. *Colloids Surf. A* **2005**, *266*, 73–81. [CrossRef]
46. Jun, L.Y.; Karri, R.R.; Yon, L.S.; Mubarak, N.M.; Bing, C.H.; Mohammad, K.; Jagadish, P.; Abdullah, E.C. Modeling and optimization by particle swarm embedded neural network for adsorption of methylene blue by jicama peroxidase immobilized on buckypaper/polyvinyl alcohol membrane. *Environ. Res.* **2020**, *183*, 109158. [CrossRef] [PubMed]
47. Piao, Y.; Tondare, V.N.; Davis, C.S.; Gorham, J.M.; Walker, A. Comparative Study of Multiwall Carbon Nanotube Nanocomposites by Raman, SEM, and XPS Measurement Techniques. *Compos. Sci. Technol.* **2021**, *208*, 108753. [CrossRef]
48. Jwa, B.; Xuan, G.A. Adsorption kinetic models: Physical meanings, applications, and solving methods—ScienceDirect. *J. Hazard. Mater.* **2020**, *390*, 122156.
49. Wang, J.; Guo, X. Rethinking of the intraparticle diffusion adsorption kinetics model: Interpretation, solving methods and applications. *Chemosphere* **2022**, *309*, 136732. [CrossRef]
50. Lagergren, S.K. About the theory of so-called adsorption of soluble substances. *Sven. Vetenskapsakad. Handingarl* **1898**, *24*, 1–39.
51. Ho, Y.S.; McKay, G. Sorption of dye from aqueous solution by peat. *Chem. Eng. J.* **1998**, *70*, 115–124. [CrossRef]
52. Allen, S.; Mckay, G.; Khader, K. Intraparticle diffusion of a basic dye during adsorption onto sphagnum peat. *Environ. Pollut.* **1989**, *56*, 39–50. [CrossRef] [PubMed]
53. Weber, W.J.; Morris, J.C. Proceedings of International Conference Water pollution symposium Pergamon. *Oxford* **1962**, *2*, 231–262.
54. Langmuir, I. The adsorption of gases on plane surfaces of glass, mica and platinum. *J. Am. Chem. Soc.* **1918**, *40*, 1361–1403. [CrossRef]
55. McKay, G.; Blair, H.S.; Gardner, J.R. Adsorption of dyes on chitin. I. Equilibrium studies. *J. Appl. Polym. Sci.* **1982**, *27*, 3043–3057. [CrossRef]
56. Hameed, B.H.; El-Khaiary, M.I. Malachite green adsorption by rattan sawdust: Isotherm, kinetic and mechanism modeling. *J. Hazard. Mater.* **2008**, *159*, 574–579. [CrossRef]
57. Freundlich, H. Uber die adsorption in losungen. *Z. Für Phys. Chem.* **1907**, *57*, 385–470. [CrossRef]
58. Foo, K.Y.; Hameed, B.H. *Insights into the Modeling of Adsorption Isotherm Systems*; Elsevier: Amsterdam, The Netherlands, 2010.
59. Yulia, F.; Chairina, I.; Zulys, A. Multi-objective Genetic Algorithm Optimization with an Artificial Neural Network for CO_2/CH_4 Adsorption Prediction in Metal-Organic Framework. *Therm. Sci. Eng. Prog.* **2021**, *25*, 100967. [CrossRef]
60. Khan, H.; Hussain, S.; Hussain, S.F.; Gul, S.; Ahmad, A.; Ullah, S. Multivariate modeling and optimization of Cr(VI) adsorption onto carbonaceous material via response surface models assisted with multiple regression analysis and particle swarm embedded neural network. *Environ. Technol. Innov.* **2021**, *24*. [CrossRef]

Disclaimer/Publisher's Note: The statements, opinions and data contained in all publications are solely those of the individual author(s) and contributor(s) and not of MDPI and/or the editor(s). MDPI and/or the editor(s) disclaim responsibility for any injury to people or property resulting from any ideas, methods, instructions or products referred to in the content.

Article

Synthesis and Characterization of MIPs for Selective Removal of Textile Dye Acid Black-234 from Wastewater Sample

Maria Sadia [1], Izaz Ahmad [1], Zain Ul-Saleheen [1], Muhammad Zubair [1], Muhammad Zahoor [2,*], Riaz Ullah [3], Ahmed Bari [4] and Ivar Zekker [5]

1. Department of Chemistry, University of Malakand, Chakdara 18800, Lower Dir, Khyber Pakhtunkhwa, Pakistan
2. Department of Biochemistry, University of Malakand, Chakdara 18800, Lower Dir, Khyber Pakhtunkhwa, Pakistan
3. Department of Pharmacognosy, College of Pharmacy, King Saud University, Riyadh 11451, Saudi Arabia
4. Department of Pharmaceutical Chemistry, College of Pharmacy, King Saud University, Riyadh 11451, Saudi Arabia
5. Institute of Chemistry, University of Tartu, 14a Ravila St., 50411 Tartu, Estonia
* Correspondence: mohammadzahoorus@yahoo.com

Abstract: Herein, a molecularly imprinted polymer (MIP) was prepared using bulk polymerization and applied to wastewater to aid the adsorption of targeted template molecules using ethylene glycol dimethacrylate (EGDMA), methacrylic acid (MAA), acid black-234 (AB-234), 2,2′-azobisisobutyronitrile (AIBN), and methanol as a cross linker, functional monomer, template, initiator, and porogenic solvent, respectively. For a non-molecularly imprinted polymer (NIP), the same procedure was followed but without adding a template. Fourier-transform infrared spectroscopy (FT-IR), scanning electron microscopy (SEM), and a surface area analyzer were used to determine the surface functional groups, morphology and specific surface area of the MIP and NIP. At pH 5, the AB-234 adsorption capability of the MIP was higher (94%) than the NIP (31%). The adsorption isotherm data of the MIP correlated very well with the Langmuir adsorption model with Qm 82, 83 and 100 mg/g at 283 K, 298 K, and 313 K, respectively. The adsorption process followed pseudo–second-order kinetics. The imprinted factor (IF) and Kd value of the MIP were 5.13 and 0.53, respectively. Thermodynamic studies show that AB-234 dye adsorption on the MIP and NIP was spontaneous and endothermic. The MIP proved to be the best selective adsorbent for AB-234, even in the presence of dyes with similar and different structures than the NIP.

Keywords: adsorption; acid black-234 dye; environment; selectivity

1. Introduction

The annual productivity of dyes is estimated to be one million tons. Textile industry produces large amount of dirty effluent through dyeing, washing, and other procedures. Synthetic dyes are the most dangerous compounds in wastewater because they are frequently made synthetically and have intricate aromatic structures that demonstrate light, oxidation, heat, and water stability. Dyes induce a variety of conditions, including cancer, allergies, mutation, dermatitis, and skin irritation. Therefore, removing dyes and other pollutants from the environment is critical for preventing contamination [1]. The quantity of dyes released into the water, on the other hand, prevents deoxygenating capacity and sunshine; therefore, aquatic life and biological activities are affected [2]. The dyes used nowadays are generally cancer-causing and have negative environmental consequences [3]. These dyes are made up of an aromatic chemical and a metal, and their photosynthetic activities are harmful. The majority of mutagenic activities are linked to colors (dye) used in the textile industry [4]. For the treatment of textile wastewater, both physical and chemical methods are used. These procedures include oxidation, membrane technology, flocculation, coagulation, and adsorption, all of which are expensive and may result in secondary contamination as a result of excessive chemical usage. Other less expensive procedures for decolorization include ozonation, electrochemical destruction, and photo catalysis [5]. The most practical way to remove dye is biological therapy, which uses a large number of microorganisms in the declaration and mineralization of a variety of colors. It is quite inexpensive, and the end result of biological treatment is non-toxic. However, because of the limited biodegradability of dyes, there is less flexibility in design and operation [6]. As a result, adsorbents such as activated carbon are employed for dye, although they are not commonly used because of their expensive cost. Peat plum kernels, wood coal resin, and chitosan fiber are some of the adsorbents employed in different industrial pollutants. Few of these adsorbents are widely accessible and inexpensive, but they cannot completely remove dyes such as activated carbon; therefore, it is necessary to develop low-cost adsorbents that may be utilized in place of activated carbon [7–9].

To conclude, a molecularly imprinted polymer (MIP) is the best choice for removing dyes from a variety of sources because of its specificity and selectivity, as well as its low cost and simplicity of preparation. Making a molecularly imprinted polymer (MIP) involves combining a template molecule with a functional monomer in the presence of an initiator and a cross linker to generate a polymer that is extremely specific and selective to the template molecules. After washing, cavities comparable in shape and size to the template molecules are generated from the polymerization of monomers and the cavities left in the polymer matrix. Because of their great selectivity even in complicated samples, MIPs used in dyes are utilized as a sorbent for solid phase extraction [10].

The most commercial application of an MIP is in the sample preparation for environmental, food analysis and environmental analysis. Clenbuterol solid-phase extraction (SPE) material is currently available from a Swedish manufacturer [11]. MIPs are popular recognition elements in sensors, and many transducers are employed in conjunction with it [12]. The quartz crystal microbalance, an acoustic transducer sensor, has gained a lot of popularity due to its inexpensive cost and ease of use [13]. The most often utilized universal functional monomer for the preparation of an MIP is methacrylic acid (MAA), and its binding capacity is determined by the bond sites and second pore size of polymeric substances [14]. Ethylene glycol dimethacrylate (EGDMA) is widely utilized as a cross linker, and the cross linker influences the hardness, strength, and selectivity of an MIP. The type and quantity of cross linkers have significant impacts on the polymerization process. If the amount is modest, an unstable polymer will be developed, while a larger amount will lower the number of recognition sites [15]. Acetonitrile, chloroform, dichloroethane, and methanol are among the most often-used solvents for MIP synthesis [16]. The imprinting efficiency, structural adsorption, and interaction between the functional monomer and template will all be affected by the porogen solvent. The use of a less polar porogen solvent promotes the formation of functional-monomer–template complexes, whereas using a more polar porogen solvent disrupts complex interactions [17]. While azo and analogue compounds are applied for the synthesis of an MIP, azobisisobutyronitrile (AIBN) is the best initiator since its decomposition temperature ranges from 50 to 70 °C [18].

There are various methods for MIP preparation, such as suspension polymerization, precipitation, etc., but the most commonly used method for MIP synthesis is bulk polymerization, in which the template is printed in the polymer matrix and the template monomer must be completely removed after polymerization. To convert an MIP to a tiny powder, mechanical breakup and crushing using a mortar and pestle are required [19]. The purpose of this study was to develop an extremely selective MIP adsorbent for acid black-234 (AB-234) dye, as well as to explore the selectivity, rebinding, and use of MIPs in various effluents. The MIP and NIP were produced using bulk polymerization for the rapid determination of AB-234 dye in water samples, but the AB-234 dye showed more selectively toward the MIP than NIP due to the recognition property of the MIP network. The adsorptive mechanism of AB-234 removal by MIP is summarized in Figure 1.

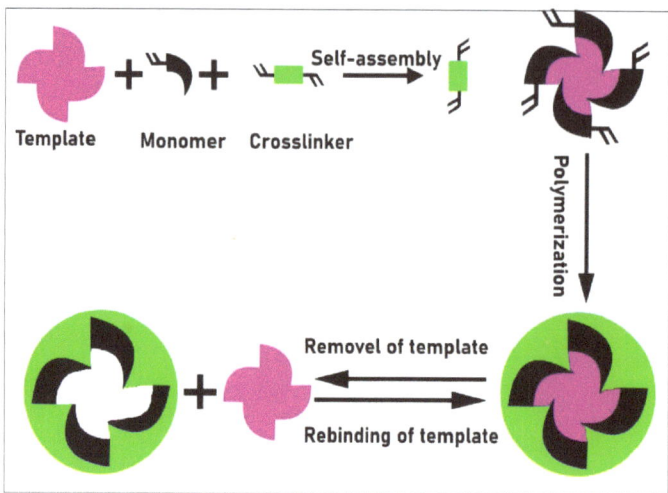

Figure 1. Adsorptive mechanism scheme of AB-234 dye on an MIP.

2. Results and Discussion

2.1. Choice of the Reagents

For the synthesis of a high-affinity molecularly imprinted polymer (MIP) the following conditions are necessary:

Polar porogen solvent and template, high nominal level of cross linker for stable polymer, and one or more functional monomers [20].

2.2. Characterization

2.2.1. Characterization by SEM

Scanning electron microscopy (SEM) may be used to determine the size and shape of the MIP and NIP. The use of SEM to analyze the MIP and NIP particles has been reported in several studies [21,22]. For careful consideration of sample morphology and smoothness, SEM is an essential and valued analytical approach. The morphologies of unwashed and washed MIP and NIP samples are shown in Figure 2. The non-covalent precipitation polymerization process of the MIP was responsible for permeability smoothness, with an extremely small, uniform, spherical, and equal size and shape, while no smoothness was observed for the NIP, as shown in the figures. Because the binding kinetic was exposed to the surface, the consistent size of the MIP revealed that the sample enabled the removal of an efficient template. The unwashed sample dye bonded to the polymer without a clear and uniform size and behaved like crystal reagents.

2.2.2. Characterization by FTIR

FTIR study was carried out within range of 4000–500 cm^{-1} and describes the surface groups of a polymer (MIP) as shown in Figure 3. The starting materials of the MIP and NIP, such as the functional monomer, cross linker, initiator, etc., were the same. Therefore, the overall data of both graphs has an approximate similarity. The peaks at 3400 cm^{-1} and 2900 cm^{-1} were caused by the presence of OH and CH, respectively, whereas stretching at 1720 cm^{-1} and 1200 cm^{-1} was caused by the presence of C=O and C-O, respectively [23,24]. Additionally, the peaks stretching at ~1400 cm^{-1}, ~1300 cm^{-1}, ~1200 cm^{-1} are caused by the presence of -CH_2, -CH_3, and C-O, respectively. The MAA and EGDMA C=C double bond stretching 1637 cm^{-1} peak was absent in the MIP and NIP, indicating that polymerization was successfully carried out.

Figure 2. SEM micrographs after adsorption (**a**,**b**) and washing (**c**,**d**) of an MIP and (**e**) NIP.

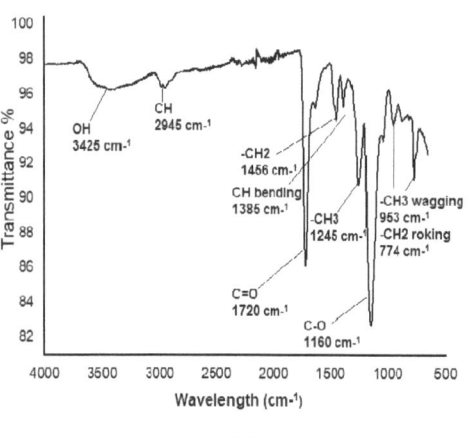 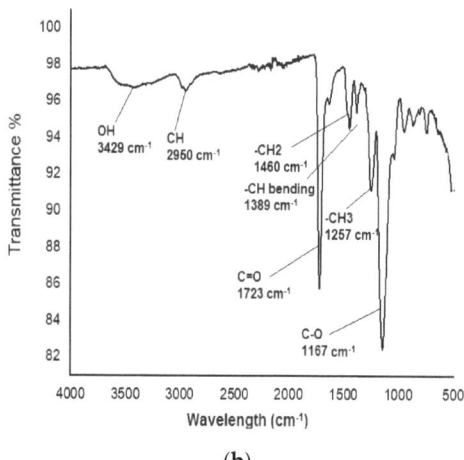

Figure 3. FTIR spectra of the MIP (**a**) and NIP (**b**).

2.2.3. Brunauer–Emmett–Teller Analysis

The porosity and specific area of a chosen AB-234 molecularly imprinted polymer was determined via BET analysis. The specific surface area of the MIP was 232.321 m^2/g, with a pore volume of 0.056 cc/g and a pore radius of 12.245 (Å), while the NIP has specific surface area, volume and pore radius 32.034 m^2/g, 0.0074 cc/g, and 3.441 (Å), respectively as given in Table 1. For the MIP, a greater surface area suggests that unique cavities were generated for AB-234 identification. The MIP has an enhanced adsorption capability due to its larger surface area [25]. The volume ratio of the open pore to total volume is referred to as a particle's porosity [26]. The template in polymerization has the greatest impact on the surface area and porosity of the MIP. The wider pore of the MIP indicates that the structure of the MIP is not more compact compared to the NIP. The typical pore diameter of the MIP is between 2 and 5 nm, indicating that the polymer is mesoporous [27].

Table 1. BET analysis of MIP.

Polymer	Specific Surface Area (m^2/g)	Pore Volume (cc/g)	Pore Radius (Å)
MIP	232.321	0.056	12.245
NIP	32.034	0.0074	3.441

2.3. Effect of Adsorbent Mass and pH

The effect of the adsorbent on the adsorption of acid black-234 dye was investigated, and the polymer dosage changed from 2 to 12 mg. The adsorption rapidly increased from 2 to 6 mg at beginning for both the MIP and NIP, but no influence on dye adsorption was observed after 8 mg. Therefore, maximum adsorption (94%) occurred at 8 mg. The pH of the solution was found to range from 2 to 7. It was observed that adsorption was small at a low pH and increased with increasing pH; hence, the maximum adsorption (94%) was recorded at pH 5 as shown in Figure 4. Therefore, this pH was chosen for further study.

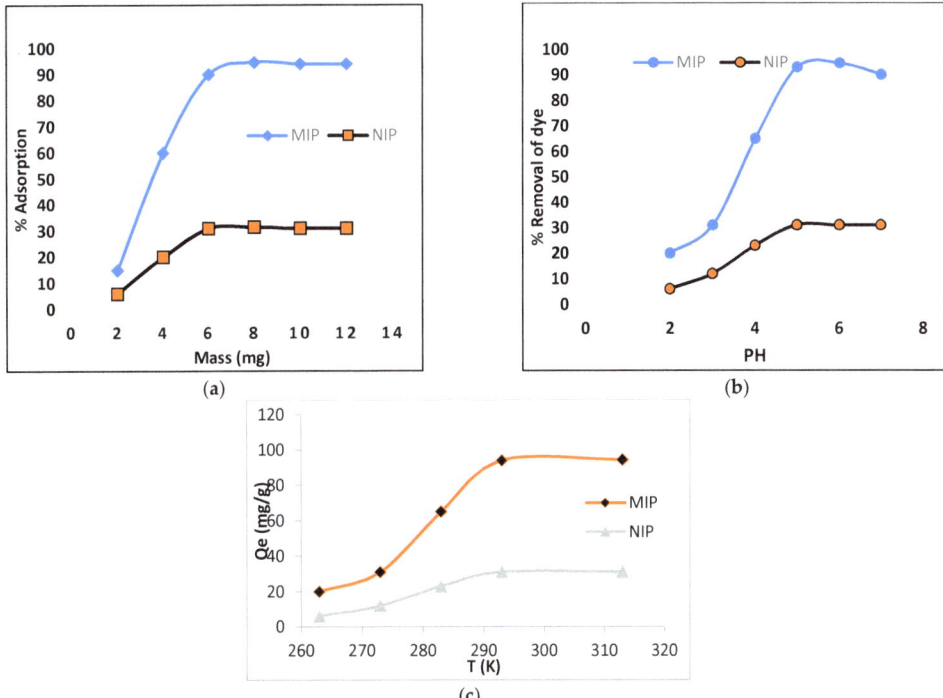

Figure 4. Effect of mass (**a**), pH (**b**) and temperature on adsorption of AB-234 dye on the MIP and NIP (**c**).

2.4. Contact Time Effect of AB-234 Dye Adsorption on the MIP and NIP

Kinetics is important to for obtaining information regarding the rate controlling and binding mechanism. Therefore, the contact time was studied as a function of temperature. The effect of contact time (Figure 5) on adsorption was determined at different times (min), while keeping other parameters constant (pH, dose, volume). The adoption was studied at 283K, 298K and 313K, which indicates that the removal of dye is time-dependent for both the MIP and NIP. The results show a gradual increase in the adsorption of dye in the MIP when the contact time increased from 5 to 20 min, followed by a considerable increase in adsorption (94%) at 40 min. After that, no change was observed; therefore, this time was used throughout the study. In general, the dye adsorption process is divided into two phases: a rapid initial sorption phase, followed by a protracted period of relatively slower adsorption [28]. Therefore, initially, uptake of dye was fast, especially during the first 35 min, most likely due to the exposure of most of the binding sites on the MIP.

2.4.1. Pseudo First Order Kinetic Model

A pseudo–first-order kinetic model provides information about the rate of occupied and unoccupied sites, in which different parameters were calculated using the following equation [29]:

$$\log (q_e - q_t) = \log q_e - k_1 \frac{t}{2.303} \tag{1}$$

The amounts of AB-234 dye (mg g^{-1}) adsorbed at time t (min) and at equilibrium are represented by q_t and q_e, respectively. The pseudo–first-order constant k_1 (min^{-1}) was calculated using a graphing log ($q_e - q_t$) against "t". Figure 6a describes a pseudo-first order kinetic model, with various parameters listed in Table 2. The qe (cal) and qe (exp) do not match; therefore, the values show that the adsorption of acid black-234 dye in the

MIP does not follow first-order kinetics because the regression coefficient $R^2 = 0.7037$ is far from unity.

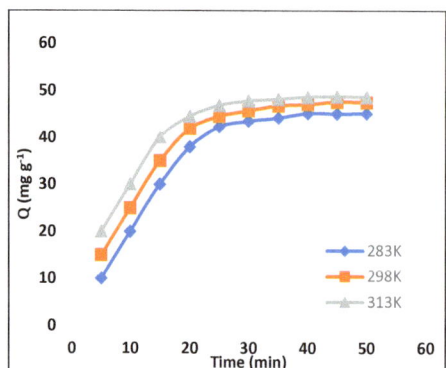

Figure 5. Adsorption kinetics of AB-234 dye in the MIP, polymer mass 8 mg, V = 10 mL, pH = 5, AB-234 = 100 mg/g.

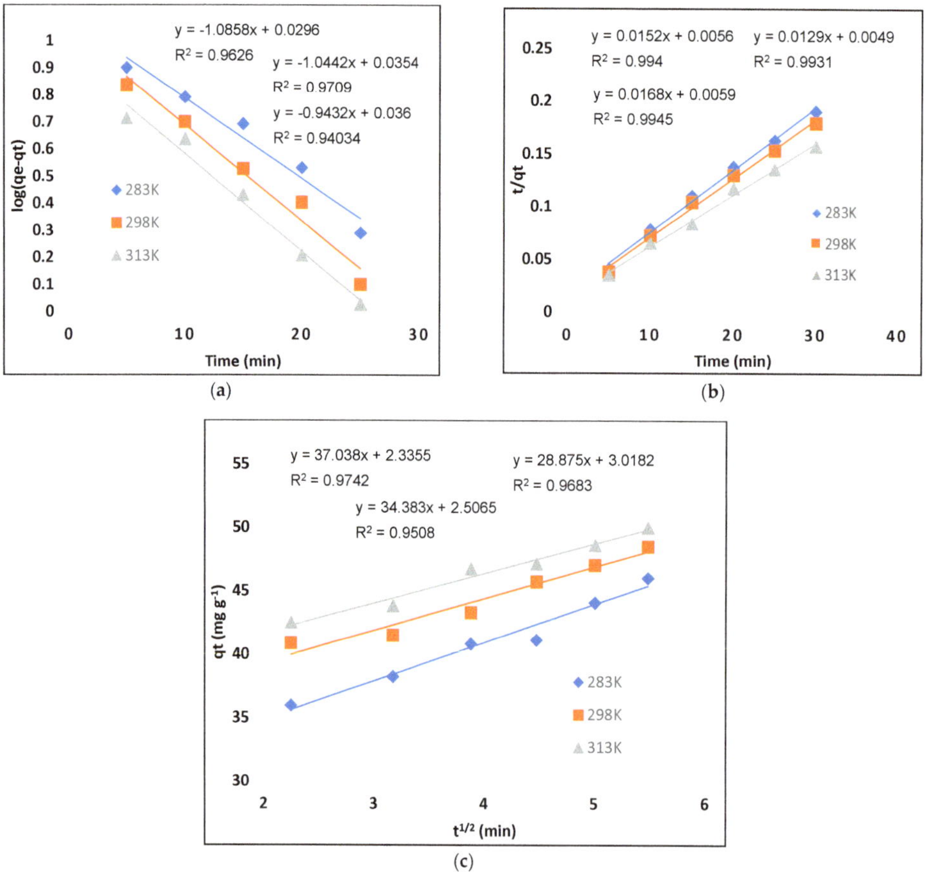

Figure 6. Kinetic model for AB-234 dye adsorption: (**a**) pseudo–first- and (**b**) pseudo–second-order kinetics (**c**) intraparticle models.

Table 2. Different kinetic models' parameters for the adsorption of AB-234 dye on MIP.

Parameters	283 K	c	313 K
	Pseudo first-order		
K1	0.0643	0.0729	0.0841
Qe (calculated)	9.342	11.543	14.768
Qe (experimental)	41.654	42.545	42.653
R^2	0.9626	0.9709	0.9799
	Pseudo second-order		
K2	0.0069	0.0127	0.0141
Qe (calculated)	43.785	44.654	44.876
Qe (experimental)	41.654	42.545	42.653
R^2	0.9945	0.994	0.9783
	Intraparticle diffusion		
K_{id} (mg/g min$^{-1/2}$)	3.04	2.32	1.96
C	28.87	34.38	37.083
R^2	0.9764	0.9508	0.9683

2.4.2. Pseudo–Second-Order Kinetic Model

Pseudo–second-order kinetic model was used to further analyze the kinetic data. The given equation shows the linear version of pseudo–second-order kinetic model [30].

$$\frac{t}{qt} = \frac{1}{k_2 q_e^2} + \frac{t}{q_e} \quad (2)$$

The pseudo–second-order constant k_2 (mg g^{-1} min^{-1}) is computed from the plot of t/qt vs. "t" in the above equation. The pseudo–second-order model was used to determine the adsorption of AB-234 dye in the MIP. The R^2 value for second-order kinetics is larger than the value of pseudo–first-order kinetics, and qe (cal) and qe (exp) have a similar relationship. As a result, we may infer that the pseudo–second-order model's adsorption data have the best fit. These kinetic data reveal that AB-234 dye adsorption is affected by both the adsorbent and the adsorbate because the regression coefficient R^2 value 0.9667 is close to unity. The various parameters obtained from this plot are shown in Table 2 and Figure 6b, illustrate that the acid black-234 dye follows pseudo–second-order kinetics.

2.4.3. Intraparticle Model

Weber and Morris claim that, instead of the contact period "t", the sorption capacity varies according to $t^{1/2}$ Equation (4) contains the linear expression [31].

$$qt = k_{id} \, t^{1/2} + C \quad (3)$$

where C is intercept and Kid is the rate constant. Multiple stages are involved in the adsorption of dye (AB-234) from an aqueous fluid onto the polymer surface. This process includes the molecular diffusion of sorbate molecules from the bulk phase to the adsorbent outer surface, also known as film or external diffusion. Internal diffusion occurs in the second stage, in which sorbate molecules travel from the MIP surface to the interior locations. The adsorption of sorbate molecules from interior locations to inner pores is the third phase [32].

Intra particle diffusion plot as shown in Figure 6c; however, it fails to pass from its origin due to a difference in the rate of mass transfer between the beginning and final temperatures. Furthermore, such a large divergence from the origin indicates that pore diffusion is not the primary rate control step [33]. When the value of "C" is compared to the rate constant, it is clear that intraparticle diffusion is not just a rate-limiting process, while AB-234 dye adsorption on the MIP is a complicated process governed by surface sorption and intraparticle diffusion.

2.5. Binding Isotherm Models of Acid Black-234 Dye Adsorption

The adsorption isotherm provides information about the mechanism of the adsorption process'. The isotherms are used to classify adsorption systems because they demonstrate the adsorption process. The monolayer formation is generally defined by the Langmuir isotherm, and the non-covalent behavior of the MIP is characterized by the Freundlich model [34]. As a result, the Langmuir and Freundlich isotherms were used to assess the adsorption data.

2.5.1. Langmuir Model

The Langmuir model describes a monolayer as an adsorbate surface that is uniform and homogeneous. Adsorption happens at a particular homogenous spot inside the body of the adsorbent. Equation (4) describes the Langmuir model [35]:

$$\frac{C_e}{Q_e} = \frac{1}{K_L Q_m} + \frac{C_e}{Q_m} \quad (4)$$

where C_e (mg·L^{-1}) represents the dye's liquid-phase equilibrium concentration; Q_m (mg·g^{-1}) represents the adsorbent's maximum adsorption capacity; K_L (L·mg^{-1}) represents the amount of dye adsorbed, the energy or net enthalpy of adsorption; and Q_e (mg·g^{-1}) represents the quantity of dye adsorbed. C_e/Q_e and C_e must have a linear connection with a slope of $1/Q_m$ and an intercept of $1/(Q_m K_L)$. Table 3 shows the K_L, Q_m, and R^2 values derived from the curve (Figure 7a). The MIP has a maximal adsorption capacity (Qmax) 100 mg g^{-1} at 313K. The Langmuir model provides the R^2 value that best fits the experimental result as given in Table 3.

Table 3. Parameters of Langmuir and Freundlich models for the adsorption of AB-234 on the MIP.

Parameters	283 K	298 K	313 K
	Langmuir model		
Qm (mg·g^{-1})	82.23	83.04	100
KL	0.207	0.125	0.789
R^2	0.9862	0.978	0.9844
	Freundlich model		
Kf (mg·g^{-1}) (L mg·g^{-1})	3.008	10.543	12.68
1/n	0.6497	0.653	0.7893
R^2	0.9358	0.9406	0.9355

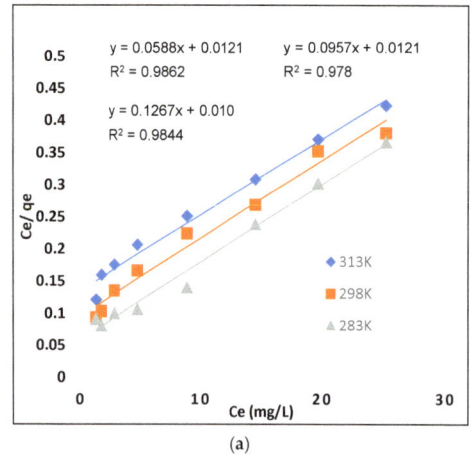
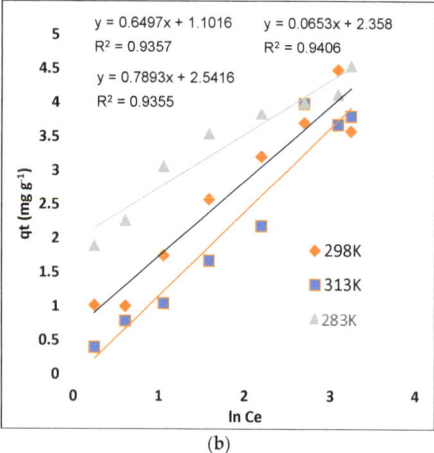

Figure 7. Isotherm models: (**a**) Langmuir and (**b**) Freundlich.

2.5.2. Freundlich Model

The adsorption characteristics of multilayer and heterogeneous surfaces with uneven adsorption sites and unusually accessible adsorption energy were determined using the Freundlich isotherm. The Freundlich isotherm model equation is shown below [36].

$$\ln q_e = \ln Kf + \frac{1}{n} \ln C_e \quad (5)$$

where C_e (mg-L^{-1}) is the liquid phase concentration at equilibrium, qe mg-g^{-1} is the dye adsorption quantity, Kf (mg-g^{-1}) is a relative indication of adsorption capacity, and 1/n is the surface heterogeneity factor indicating adsorption type. For favorable adsorption, the 1/n ratio should be less than 1, whereas for unfavorable adsorption, the value should be larger than 1, indicating poor bond adsorption [37]. Figure 7b depicts the Freundlich model, while Table 3 lists the various parameters.

2.6. Thermodynamic Study

Thermodynamic experiments were conducted to examine the dye (AB-234) adsorption process on the MIP. In this study, a significant judgment must be made on whether the mechanism is spontaneous or not. The following equation was used to determine many thermodynamic parameters, including standard free energy (G°), enthalpy (H°), and entropy (S°) [38].

$$\log Kc = \frac{\Delta S°}{2.303R} - \frac{\Delta H°}{2.303RT} \quad (6)$$

$$\Delta G° = \Delta H° - T\Delta S° \quad (7)$$

where T is the specific heat, R is the universal gas constant (8.314 J-mol^{-1}K^{-1}), and K$_c$ (Lg^{-1}) is the thermodynamic equilibrium constant defined by q_e/C_e. The intercept and slope of a plot log K$_c$ vs. 1/T were used to calculate the values of $\Delta S°$ and $\Delta H°$ (Figure 8). Table 4 shows the different thermodynamic parameters that were examined at various temperatures. The value of $\Delta G°$ was found to be negative at all temperatures, indicating that dye (AB-234) adsorption in the MIP was spontaneous [39]. The value of $\Delta G°$ decreased as the temperature increased, indicating that a higher temperature enhances dye (AB-234) adsorption in the MIP. The positive sign of $\Delta H°$ indicates that this adsorption is endothermic, because with the increasing temperature, the rate of adsorbate diffusion on the adsorbent also increased. Additionally, a positive $\Delta S°$ value shows that disorder increased during adsorption.

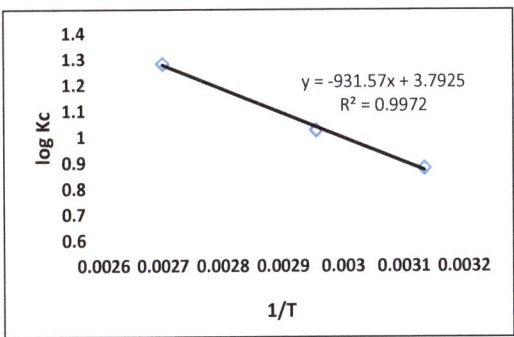

Figure 8. Van't Hoff graph (MIP).

Table 4. Different thermodynamic parameters for AB-234 dye.

Temperature (K)	$\Delta G°$ KJmol^{-1}	$\Delta H°$ KJmol^{-1}	$\Delta S°$ KJmol^{-1}
288	−1439		
298	−1915	7.74	31.52
308	−2516		

2.7. Selectivity Study

A competitive adsorption study was carried out in the presence of dyes, such as AB1, BB3, safranin, and acid yellow 76, that are both comparable and dissimilar in structure to the template (AB-234) to confirm the selective cavities in a polymer. The MIP should be more selective for the specific template (AB-234) than the NIP. At optimum conditions, 100 mg/L of AB2-34 dye and the interfering dyes were added to each Erlenmeyer flask, and the mixture was agitated in a thermostat shaker at 130 rpm. At a certain time, samples were withdrawn, the absorbance was measured using a UV–vis spectrophotometer, and the quantity of dye adsorbed on the polymer was estimated by subtracting the final dye concentration from the starting dye concentration in the mixture. MIP has a 94% selectivity for AB-234, compared to other dyes with adsorption rates ranging from 5% to 23%. Therefore, the selectivity results confirm that MIP cavities are solely selective for the template (AB-234) molecule. The results are graphically shown in Figure 9.

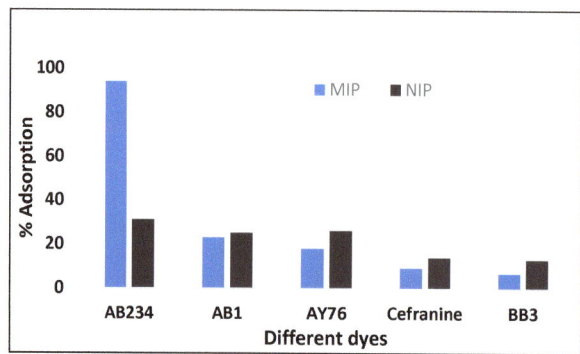

Figure 9. Selectivity of MIP towards different similar structures dyes.

2.8. Distribution Ratio and Imprinting Factor

The contact and strength of the template molecule (AB-234) with the polymer (MIP/NIP) is described by the imprinting factor. It appears that the MIP and NIP have recognition properties for a certain analyte. Equation (9) was used to derive the imprinting factor (IF) for a molecularly imprinted polymer:

$$\text{IF} \propto = \frac{Q_{MIP}}{Q_{NIP}} \qquad (8)$$

where Q_{MIP} denotes MIP adsorption capacity for the dye (AB-234), and Q_{NIP} denotes NIP adsorption capacity for AB-234. The following equation was used to calculate the distribution ratio.

$$K_d = \frac{(C_i - C_f)}{C_f \, m} V \qquad (9)$$

where Kd (L/g) denotes the distribution coefficient, C_i the initial dye concentration, C_f the final dye concentration, V the volume used, and "m" the polymer mass (MIP/NIP) [40]. The obtained results have been summarized in Table 5.

Table 5. Adsorption parameters for dyes (MIP/NIP).

Dyes	% Removal		Adsorption Capacity Q mg/g		Distribution Coefficient Kd (L/g)		Imprinting Factor $IF \propto = \frac{Q_{MIP}}{Q_{NIP}}$	Selectivity $S = \frac{I^{AB234}}{I^{interfering}}$
	MIP	NIP	MIP	NIP	MIP	NIP		
AB-234	94	31	41.1	8.0	0.53	0.06	5.13	-
AB-1	23	25	5.21	6.31	0.42	0.27	0.82	6.25
SEFRANIN	18	26	4.32	6.61	0.02	0.01	0.65	7.89
AY-76	9.2	14	2.35	4.30	0.21	0.15	0.54	9.50
BB-3	7	13	1.38	3.15	0.17	0.11	0.43	11.9

2.9. Application of MIP to Adsorb Dye (AB-234)

The polymer was applied as a solid-phase adsorbent material in the preconcentration of AB-234 from river water and effluent samples in order to test the efficiency of the MIP produced in real environmental samples. In a total volume of 10 mL, aliquots of these three samples were spiked with known volumes of AB-234 dye standard solution at concentrations of 50, 75, and 100 mg/L, with agitation for 40 min. Given the MIP's highest retention capability of 94.67% (±0.1) while the recoveries were between 84 and 94%, as shown in Table 6. Taking into account the complexity of the examined samples, these results reveal that the imprinted material shows an outstanding sorption capability and ability to give unique recognition of the analyte.

Table 6. Recovery test of the MIP using different samples in a province of Pakistan (KPK).

Samples	Added (mg/L)	Found in MIP (mg/L)	Recovery (%)	RSD (%)
Textile industry effluent	50	46.51 ± 0.04	93.02	0.33
	75	71.02 ± 0.07	94.60	0.34
	100	94.8 ± 0.1	94.8	0.39
River 1	50	45.43 ± 0.03	90.86	0.30
	75	71.05 ± 0.1	94.66	0.54
	100	86.62 ± 0.09	86.62	0.77
River 2	50	43.62 ± 0.02	87.24	0.04
	75	66.54 ± 0.1	88.01	0.49
	100	84.89 ± 0.04	84.89	0.61

2.10. Reuse of MIP for AB-234 Dye

The continuous use of the MIP was determined in five cycles of adsorption–desorption under optimal conditions. Table 7 shows that, between the first and fifth cycles, roughly 14% of the AB-234 dye rebinding was lost. According to the results, the MIP could be reused at least five times without significantly lowering its adsorption capability.

Table 7. Repeated use of MIP on adsorption capacity.

Times	1	2	3	4	5
Q (mg g^{-1})	94	92	89	84	80

A comparison present adsorbent capacity with those cited in literature in given in Table 8.

Table 8. Adsorptions capacity of acid black-234 dye on different adsorbents.

Adsorbents	Adsorption Capacity (mg-g^{-1})	Ref
Polyaniline/chitosan (PAn/Ch)	74.1	[32]
Starch (PPy/St)	62.5	[41]
Polyanaline/sugarcane begasse (PAn/SB)	52.6	[41]
MIP	100	Current study

3. Materials and Methods

3.1. Materials

All solvents and dyes were analytical-grade and were supplied by Sigma-Aldrich (Taufkirchen, Germany). The acid black-234 dye (AB-234) as a template, methacrylic acid (MAA) as a functional monomer (Dae-Jung, Korea), azobisisobutyronitrile as a reaction initiator, methanol as a porogenic solvent (Chengdu, China), acetone and methanol mixing solution as a washing solvent (Merck, Darmstadt, Germany), and ethylene glycol dimethacrylate (EGDMA) as a cross linker (J.T. Baker, New York, NY, USA). The MIP's selectivity for AB-234 was tested using AB1, BB3, safranin, and acid yellow-76. The water was deionized using a Milli-Q system (Chennai, India). The chemical structures of the dyes are shown in Figure 10.

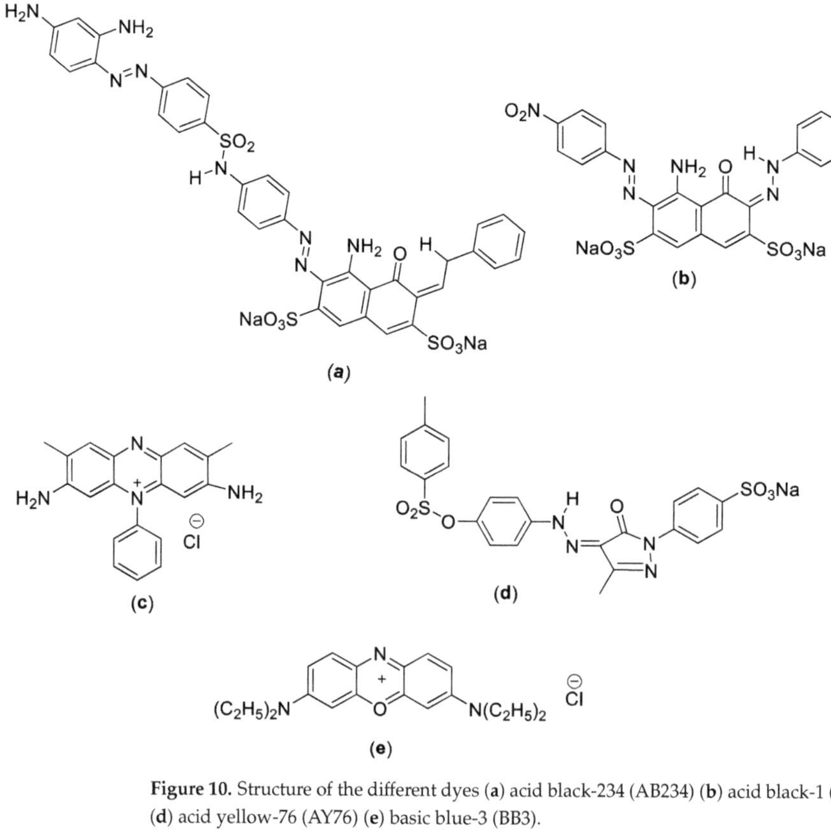

Figure 10. Structure of the different dyes (a) acid black-234 (AB234) (b) acid black-1 (AB1) (c) safranin (d) acid yellow-76 (AY76) (e) basic blue-3 (BB3).

3.2. Characterization of the MIP and NIP

The structure of polymer was studied using FTIR in the range of 4000–500 cm^{-1}; Vertex 70 (Shimadzu, Kyoto, Japan) and SEM (JSM-IT500) were used to assess the size and shape of the MIP and NIP; and a UV–Vis 1800 spectrophotometer was used to quantify absorbance using quartz cuvettes (Shimadzu, Japan). Brunauer–Emmett–Teller theory was used for the surface area calculation from the adsorption of nitrogen (8 mg MIP and NIP, N2, 70 °C, 5 h) (ASAP 2010).

3.3. Preparation of the MIP and NIP
Preparation of the MIP for AB-234 Dye

For the synthesis of the MIP, 0.215 g of acid black-234 dye was dissolved in 10 mL of methanol and stirred for 10 min. Then, 150 mmol of MAA was added and left to rest for two hours, followed by the addition of 225 mmol of EGDMA, and then left for 15 min. Then, 2 mg of initiator (ABIN) was added and maintained in a water bath for 24 h at 60 °C. The NIP was synthesized using same procedure but without dye. The sample was removed from the flask after 24 h of heating, and the polymer was filtered. After filtering, the polymer was washed five to six times in a soxlet system with a methanol/acetone solution (8:2, v/v) to completely remove the template. Finally, a pure polymer was produced, which was then dried at room temperature. The synthesis protocols of the MIP and NIP are shown in Table 9.

Table 9. Polymerization mixture of MIP synthesis.

Reagents	Chemicals	MIP and NIP Composition (Mass and Volume)	
		MIP	NIP
Template	AB-234	0.215 g	-
Monomer	MAA	150 mmol	150 mmol
Cross Linker	EGDMA	225 mmol	225 mmol
Solvent	Methanol and Acetone	10 mL	10 mL
Initiator	AIBN	2.00 g	2.00 g
Template	AB-234		

3.4. Binding Adsorption Analysis

MIP binding was studied using 20 mL vials containing 8 mg of an MIP and 10 mL of 100 mg/L dye by adjusting parameters, such as mass, concentration, pH, and time. After 40 min on a centrifuge at 15,000 rpm, the supernatant was filtered through a 0.45 μm membrane before UV–Vis spectrophotometric measurement. The following equation was used to calculate the binding adsorption capacity:

$$Q = \frac{(C_0 - C_e)}{V} m \qquad (10)$$

The initial dye concentration is Co (mgL^{-1}), the equilibrium dye concentration is Ce (mgL^{-1}), the experimental adsorption quantity is Q (mg·g^{-1}), the volume of solution is V, and the mass of the MIP is m (g).

3.5. Selectivity Study

A competitive adsorption study was carried out in the presence of molecules that are similar and different in structure to AB-234 dye in order to assess the creation of selective cavities in polymer. Different dyes were used in this case, including acid black-1 (AB-1), safranin, acid yellow-76 (AY-76), and BB-3. For each compound's selectivity recognition assays, 8 mg of the MIP was dissolved in a 10 mL solution (pH 5), containing 100 mg/L of each dye and equilibrated for 40 min. After stirring the concentration of these dyes in the supernatant was determined using a UV–Vis spectrophotometer.

4. Conclusions

The current study looks at how to employ a synthetic polymer (MIP) to remove a specific analyte (AB-234) from various water samples under optimal conditions. Compared to other dyes—acid black-1 (AB1), acid yellow-76 (AY-76), safranin, and BB-3, which had 23%, 18%, 9.3%, and 7% selectivity, respectively, toward a specific analyte— AB-234 had about 94% selectivity toward a specific analyte. The adsorption of AB-234 on NIP was only 31%, which indicates that the MIP is more specific for AB2-34 dye due to the formation of complementary cavities. The MIP followed second-order kinetics and the Langmuir model, according to kinetic and isotherm analyses. Therefore, a high adsorption (94%) was found on 313 K. The negative value of $\Delta G°$ shows the process to be spontaneous, while the $\Delta H°$ and $\Delta S°$ demonstrates the endothermic and feasible nature of the process. The MIP can be easily and repeatedly be recovered from the solution using the centrifugation process, without the immense loss of its selectivity. Therefore, the MIP synthesized here could be presented as a possible material for the separation of AB-234 dye from wastewater with immense selectivity and great recovery.

Author Contributions: M.S. and M.Z. (Muhammad Zahoor): methodology, revision of the paper, conceptualization, visualization, investigation, supervision, validation, I.A.: experiments and writing—original draft. Z.U.-S. and M.Z. (Muhammad Zubair): methodology visualization, investigation, R.U., I.Z. and A.B.: review, supervision, and conceptualization. All authors have read and agreed to the published version of the manuscript.

Funding: The authors extend their appreciation to the researchers supporting project number RSP2023R346, King Saud University, Riyadh, Saudi Arabia, regarding financial support.

Institutional Review Board Statement: Not Applicable.

Informed Consent Statement: Not Applicable.

Data Availability Statement: The raw/processed data required to reproduce these findings cannot be shared at this time due to technical or time limitations.

Conflicts of Interest: The authors declare no conflict of interest.

Sample Availability: These are commercially available compounds.

References

1. Nikesh, B.S.; Vinayak, K.; Prakash, A.M.; Ajay, V.R.; Krishnan, K.A. A Historical Perspective and the Development of Molecular Imprinting Polymer—A Review. *Chem. Int.* **2015**, *1*, 202–210.
2. Zollinger, H. Azo dyes and pigments. In *Colour Chemistry-Synthesis, Properties and Applications of Organic Dyes and Pigments*; Wiley: New York, NY, USA, 1987; pp. 10–92.
3. Hameed, H.B.; Mahmoud, K.D.; Ahmed, L.A. Equilibrium modeling and kinetic studies on the adsorption of basic dye by a low-cost adsorbent: Coconut bunch waste. *J. Hazard. Mater.* **2008**, *158*, 65–72. [CrossRef] [PubMed]
4. Ponnusami, V.; Vikram, S.; Shrivastva, N.S. Guava leaf powder, novel adsorbent for methylene removal of blue from aqueous solution. *J. Hazard. Mater.* **2008**, *152*, 276–286. [CrossRef] [PubMed]
5. Okutucu, B.; Akkaya, A.; Kasikara, N.; Pazarlioglu. Molecularly imprinted polymers for some reactive dyes. *Prep. Biochem. Biotechnol.* **2010**, *40*, 366–376. [CrossRef]
6. Crini, G. Non-conventional low-cost adsorbents for dye removal: A review. *Bioresour. Technol.* **2006**, *97*, 1061–1085. [CrossRef] [PubMed]
7. Robinson, T.; McMullan, G.; Marchant, R.; Nigam, P. Remediation of dyes in textile effluent: A critical review on current treatment technologies with a proposed alternative. *Bioresour. Technol.* **2001**, *77*, 247–255. [CrossRef]
8. Ahmad, A.L.; Azlina, H.; Harris, S.; Seng, O.B. Removal of dye from wastewater of textile industry using membrane technology. *J. Teknolog.* **2002**, *36*, 31–44. [CrossRef]
9. Ying, P.; Congying, R.; Xiaolin, T.; Yun, L.; Ayushi, S.; Abhinav, K.; Baohong, L.; Jianqiang, L. Cobalt-seamed C-methylpyrogallol[4]arene nanocapsules-derived magnetic carbon cubes as advanced adsorbent toward drug contaminant removal. *Chem. Eng. J.* **2022**, *433*, 13385.
10. Qiujin, Z.; Liping, W.; Shengfang, W.; Wasswa, J.; Xiaohong, G.; Jian, T. Selectivity of molecularly imprinted solid phase extraction for sterol compounds. *Food Chem.* **2009**, *113*, 608–615.
11. Ramström, O.; Ye, L.; Mosbach, K. Screening of a combinatorial steroid library using molecularly imprinted polymers. *Anal. Commun.* **1998**, *35*, 9–11. [CrossRef]

12. Lie, L.; Yihua, Y.; Klaus, M. Onwards the development of molecularly imprinted artificial receptors for the screening of estrogenic chemicals. *Analyst* **2001**, *126*, 760–765.
13. Surugiu, I.; Danielsson, B.; Ye, L.; Mosbach, K.; Haupt, K. Chemiluminescence Imaging ELISA Using an Imprinted Polymer as the Recognition Element Instead of an Antibody. *Anal. Chem.* **2001**, *73*, 487–491. [CrossRef]
14. Zhang, Y.; Song, D.; Lanni, M.L.; Shimizu, D.K. Importance of Functional Monomer Dimerization in the Molecular Imprinting Process. *Macromolecules* **2010**, *43*, 6284–6294. [CrossRef]
15. Li, X.; Li, X.; Li, F.; Lei, X.; Su, M.; Liu, P.; Tan, X. Synthesis and characterization of molecularly imprinted polymers with modified rosin as a cross-linker and selective SPE-HPLC detection of basic orange II in foods. *Anal. Methods* **2014**, *6*, 6397–6406. [CrossRef]
16. Prasada, T.; Rao, S.; Daniel, J.G. Tailored materials for preconcentration or separation of metals by ion-imprinted polymers for solid-phase extraction (IIP-SPE). *TrAC Trends Anal. Chem.* **2004**, *23*, 28–35. [CrossRef]
17. Saloni, J.; Walker, K.; Hill, J. Theoretical Investigation on Monomer and Solvent Selection for Molecular Imprinting of Nitrocompounds. *J. Phys. Chem.* **2013**, *117*, 1531–1534. [CrossRef] [PubMed]
18. Tokonami, S.; Shiigi, H.; Nagaoka, H. Review Micro- & nanosized molecularly imprinted polymers for high-throughput analytical applications. *Anal. Chem. Acta* **2009**, *641*, 7–13.
19. Mujahid, A.; Iqbal, N.; Afzal, A. Bioimprinting strategies: From soft lithography to biomimetic sensors and beyond. *Biotechnol. Adv.* **2013**, *31*, 1435–1447. [CrossRef]
20. Włoch, M.; Dattab, J. Chapter Two—Synthesis and polymerisation techniques of molecularly imprinted polymers. *Compr. Anal. Chem.* **2015**, *86*, 166–526.
21. Zhou, T.C.; Shen, X.T.; Chaudhary, S.; Ye, L. Molecularly imprinted polymer beads prepared by pickering emulsion polymerization for steroid recognition. *J. Appl. Polym. Sci.* **2014**, *39606*, 1–7. [CrossRef]
22. Faiz, A.; Zuber, S.; Alamgir, K.; Maria, S.; Zeid, A.A.; Cheong, W.J. Synthesis, column packing and liquid chromatography of molecularly imprinted polymers for the acid black 1, acid black 210, and acid Brown 703 dyes. *RSC Adv.* **2022**, *12*, 19611–19623.
23. Zhang, W.; She, X.; Wang, L.; Fan, H.; Zhou, Q.; Huang, X.; Tang, J.Z. Preparation, Characterization and pplication of Molecularly Imprinted Polymer for Selective Recognition of Sulpiride. *Materials* **2017**, *10*, 475. [CrossRef] [PubMed]
24. Liang, Q.; Fenglan, L.; Yan, L.; Jiana, W.; Shiyuan, G.; Meiyin, W.; Shiling, X.; Manshi, L.; Deyun, M. A 2D Porous Zinc-Organic Framework Platform for Loading of 5-Fluorouracil. *Inorganics* **2022**, *10*, 202.
25. Maria, S.; Izaz, A.; Faiz, A.; Muhammad, Z.; Riaz, U.; Farhat, A.H.; Essam, A.A.; Amir, S. Selective removal of emerging dye basic blue 3 via molecularly imprinting technique. *Molecules* **2022**, *27*, 3276.
26. Asadi, E.; Deilami, S.A.; Abdouss, M.; Kordestani, D.; Rahimi, A.; Asadi, S. Synthesis, recognition and evaluation of molecularly imprinted polymer nanoparticle using miniemulsion polymerization for controlled release and analysis of risperidone in human plasma samples. *Korean J. Chem. Eng.* **2014**, *31*, 1028–1035. [CrossRef]
27. Farrington, K.; Regan, F. Investigation of the nature of MIP recognition development and characterisation of a MIP for Ibuprofen. *Biosens. Bioelectron.* **2007**, *22*, 1138–1146. [CrossRef]
28. Cormack, P.A.G.; Elorza, A.Z. Molecularly imprinted polymers: Synthesis and characterization. *J. Chromatogr.* **2004**, *804*, 173–182. [CrossRef]
29. Sun, X.F.; Wang, S.G.; Liu, X.W.; Gong, W.X.; Bao, N.; Gao, B.Y. Competitive Biosorption of Zinc (II) and Cobalt (II) in Single- and Binary-Metal Systems by Aerobic Granules. *J. Colloid Interface Sci.* **2008**, *324*, 1–8. [CrossRef]
30. Maryam, F.; Mohammad, A.T.; Daryoush, A.; Mostafavi, A. Preparation of molecularly imprinted polymer coated magnetic multi-walled carbon nanotubes for selective removal of dibenzothiophene. *Mater. Sci. Semicond. Process.* **2015**, *40*, 501–507.
31. Ho, Y.S.; McKay, C. Pseudo second order model for sorption processes. *Proc. Biochem.* **1999**, *34*, 451–465. [CrossRef]
32. Noreen, S.; Bhatti, H.N.; Iqbal, M.; Hussain, F.; Sarim, F.M. Chitosan, starch, polyaniline and polypyrrole biocomposite with sugarcane bagasse for the efficient removal of Acid Black dye. *Int. J. Biol. Macromol.* **2019**, *147*, 439–452. [CrossRef]
33. Poots, V.J.P.; McKay, G.; Healy, J.J. Removal of basic dye from effluent using wood as an adsorbent. *J. Water Pollut. Control Fed.* **1978**, *50*, 926–935.
34. Zhang, Y.L.; Zhang, J.; Dai, C.M.; Zhou, X.F.; Liu, S.G. Sorption of carbamazepine from water by magnetic molecularly imprinted polymers based on chitosan-Fe_3O_4. *Carbohydr. Polym.* **2013**, *97*, 809–816. [CrossRef] [PubMed]
35. Gautam, R.K.; Rawat, V.; Banerjee, S.; Sanroman, M.A.; Soni, S.; Singh, S.K.; Chattopadhyaya, M.C. Synthesis of bimetallic Fe–Zn nanoparticles and its application towards adsorptive removal of carcinogenic dye malachite green and Congo red in water. *J. Mol. Liq.* **2015**, *212*, 227–236. [CrossRef]
36. Pandian, C.J.; Palanivel, R.; Dhananasekaran, S. Green synthesis of nickel nanoparticles using Ocimum sanctum and their application in dye and pollutant adsorption. *Sep. Sci. Technol.* **2015**, *23*, 1307–1315. [CrossRef]
37. Hameed, H.; Ahmad, A.A. Batch adsorption of methylene blue from aqueous solution by garlic peel, an agricultural waste biomass. *J. Hazard. Mater.* **2009**, *164*, 870–875. [CrossRef]
38. Gautam, R.K.; Gautam, P.K.; Banerjee, S.; Soni, S.; Singh, S.K.; Chattopadhyaya, M.C. Removal of Ni (II) by magnetic nanoparticles. *J. Mol. Liq.* **2015**, *204*, 60–69. [CrossRef]
39. Jain, R.; Gupta, V.K.; Sikarwar, S. Adsorption and desorption studies on hazardous dye Naphthol Yellow S. *J. Hazard. Mater.* **2010**, *182*, 749–756. [CrossRef]

40. Harsini, N.N.; Ansari, M.; Kazemipour, M. Synthesis of molecularly imprinted polymer on magnetic core-shell silica nanoparticles for recognition of congo red. *Eurasian J. Anal. Chem.* **2018**, *13*, 1–13. [CrossRef]
41. Haq, N.B.; Yusra, S.; Sobhy, M.; Yakout, O.H.; Shair Munawar, I.; Arif, N. Efficient removal of dyes using carboxymethyl cellulose/alginate/ polyvinyl alcohol/rice husk composite: Adsorption/desorption, kinetics and recycling studies. *Int. J. Bio. Macromol.* **2020**, *150*, 861–870.

Disclaimer/Publisher's Note: The statements, opinions and data contained in all publications are solely those of the individual author(s) and contributor(s) and not of MDPI and/or the editor(s). MDPI and/or the editor(s) disclaim responsibility for any injury to people or property resulting from any ideas, methods, instructions or products referred to in the content.

Article

New Process for the Sulfonation of Algal/PEI Biosorbent for Enhancing Sr(II) Removal from Aqueous Solutions—Application to Seawater

Mohammed F. Hamza [1,2], Eric Guibal [3,*], Khalid Althumayri [4], Thierry Vincent [3], Xiangbiao Yin [1], Yuezhou Wei [1] and Wenlong Li [1,*]

1. School of Nuclear Science and Technology, University of South China, HengYang 421001, China
2. Nuclear Materials Authority, P.O. Box 530, El-Maadi, Cairo 4710030, Egypt
3. Polymers Composites and Hybrids, IMT—Mines Ales, F-30360 Ales, France
4. Department of Chemistry, College of Science, Taibah University, Al-Madinah Al-Munawarah 30002, Saudi Arabia
* Correspondence: eric.guibal@mines-ales.fr (E.G.); liwenlong@usc.edu.cn (W.L.); Tel.: +33-0-466782734 (E.G.); +86-18845568076 (W.L.)

Citation: Hamza, M.F.; Guibal, E.; Althumayri, K.; Vincent, T.; Yin, X.; Wei, Y.; Li, W. New Process for the Sulfonation of Algal/PEI Biosorbent for Enhancing Sr(II) Removal from Aqueous Solutions—Application to Seawater. *Molecules* 2022, 27, 7128. https://doi.org/10.3390/molecules27207128

Academic Editors: Yongchang Sun and Dimitrios Giannakoudakis

Received: 22 September 2022
Accepted: 17 October 2022
Published: 21 October 2022

Publisher's Note: MDPI stays neutral with regard to jurisdictional claims in published maps and institutional affiliations.

Copyright: © 2022 by the authors. Licensee MDPI, Basel, Switzerland. This article is an open access article distributed under the terms and conditions of the Creative Commons Attribution (CC BY) license (https://creativecommons.org/licenses/by/4.0/).

Abstract: Sulfonic resins are highly efficient cation exchangers widely used for metal removal from aqueous solutions. Herein, a new sulfonation process is designed for the sulfonation of algal/PEI composite (A*PEI, by reaction with 2-propylene-1-sulfonic acid and hydroxylamine-O-sulfonic acid). The new sulfonated functionalized sorbent (SA*PEI) is successfully tested in batch systems for strontium recovery first in synthetic solutions before investigating with multi-component solutions and final validation with seawater samples. The chemical modification of A*PEI triples the sorption capacity for Sr(II) at pH 4 with a removal rate of up to 7% and 58% for A*PEI and SA*PEI, respectively (with SD: 0.67 g L^{-1}). FTIR shows the strong contribution of sulfonate groups for the functionalized sorbent (in addition to amine and carboxylic groups from the support). The sorption is endothermic (increase in sorption with temperature). The sulfonation improves thermal stability and slightly enhances textural properties. This may explain the fast kinetics (which are controlled by the pseudo-first-order rate equation). The sulfonated sorbent shows a remarkable preference for Sr(II) over competitor mono-, di-, and tri-valent metal cations. Sorption properties are weakly influenced by the excess of NaCl; this can explain the outstanding sorption properties in the treatment of seawater samples. In addition, the sulfonated sorbent shows excellent stability at recycling (for at least 5 cycles), with a loss in capacity of around 2.2%. These preliminary results show the remarkable efficiency of the sorbent for Sr(II) removal from complex solutions (this could open perspectives for the treatment of contaminated seawater samples).

Keywords: sulfonation of composite algal/PEI sorbent; strontium recovery; uptake kinetics and sorption isotherms; metal desorption and sorbent recycling; sorption selectivity; application to seawater; composite characterization; functionalization for enhanced performance

1. Introduction

Strontium is part of alkaline-earth metals; it is mainly extracted from celestite and strontianite minerals. Most of its uses are associated with optical and color-specific characteristics for fireworks (red-colored), glow-in-the-dark paints, and plastics (as Sr-aluminate). Strontium (as chloride salt) is also used in specialized toothpaste [1]. However, strontium mainly retains attention as a radioelement (^{90}Sr) for its high-energy beta-emitter specificity. As a by-product of nuclear reaction and due to its similarities with calcium (meaning readily absorption and accumulation in bones and tissues), it remains a very hazardous compound for human and animal beings (as shown by the strong contamination of local seawater after Fukushima Daiichi disaster) [2–6].

These hazardous impacts have motivated for the last decades a strong research for elaborating new treatment processes. Metal removal may involve different techniques; however, sorption and biosorption processes have retained great attention for low-concentration effluents [7–10]. Though precipitation and flotation techniques can be used for pre-treating strontium-bearing effluents [11,12], it is generally useful to couple different techniques [13] for reaching high levels of metal recovery. For the specific treatment of complex solutions (such as brines and seawater), it is thus necessary to design new sorbents having high selectivity against alkaline and alkaline-earth competitor ions.

Inorganic sorbents have been widely investigated for strontium recovery using for example zeolites [14–18], or directly as nanostructured inorganic salts [19–21]. The association of ion-exchangers (IEs) with inorganic supports offers high selectivity for Sr(II) or Cs(I) associated with the cage effect of some IEs (such as Prussian blue, PB, and analogues) [22]. Similar concepts have been extended to PB immobilization onto carbon nanotubes [23], activated charcoal [24], polymer [25], or biopolymers [22,23,26]. Strontium removal was investigated using a wide range of bio-sorbents such as agriculture by-products [27], living microorganisms [28], yeast [29], and functionalized biopolymers [30,31]. Algal-based materials and sub-products have retained great attention for their potential to bind organic [32] and inorganic [9] contaminants.

Ion-exchange and chelating resins have been successfully tested for strontium binding from synthetic and for seawater solutions [33–36]. Functionalized polymers were designed as super-adsorbents for cesium and strontium, through the amidoximation of polyacrylonitrile/silica composites [37], or algal/PEI (polyethyleneimine) composite (A*PEI) [38]. Strong cation exchangers, such as sulfonic acid bearing resins (Dowex 50W series), have particularly retained attention for the sorption of strontium [16,39,40] or other solutes [41–48]. The efficiency of these sulfonated reactive groups for bearing strontium motivated the current research based on the functionalization of algal/PEI beads. Indeed, a new concept of composite support (obtained by the reaction of algal biomass with PEI with crosslinking and calcium ionotropic gelation, interpenetrated polymer network) was recently developed [49–54]. The high density of amine and hydroxyl groups in the composite offers many possibilities for functionalization by grafting specific reactive groups such as quaternary ammonium salt [51], amidoxime [38], and phosphoryl [53]. The sulfonation of algal/PEI beads revealed very efficient for improving the sorption properties for rare earth elements (i.e., Sc(III), Ce(III), and Ho(III) [55]). Actually, the grafting of new sulfonic groups brings not only additional functional groups with different affinities for target metal ions but also some complementary facilities associated with the bi-functionality of the sorbent (co-existing amine and sulfonic groups). Herein, a new process is designed for sulfonating algal/PEI beads (SA*PEI), based on the one-pot reaction of 2-propene-1-sulfonic acid and hydroxylamine-o-sulfonic acid with pristine A*PEI beads. This original method leads to the co-existence of different types of sulfonic groups.

In order to evaluate the potential of this sulfonic-functionalized sorbent, the materials are characterized using SEM coupled with EDX facilities, BET characterization, FTIR spectrometry, elemental analysis, and titration. The sorption properties are investigated through standard criteria such as the effect of pH, uptake kinetics, sorption isotherms, and competitive sorption (multi-component equimolar solutions). Metal desorption and sorbent recycling are studied and compared for A*PEI and SA*PEI. Strontium recovery is also characterized in very complex environmental such as seawater samples. The insights of this study count on: (a) the ecofriendly and cost-effective synthesis of a support based on alginate and algal biomass, (b) the production of a sorbent of highly physicochemical stability, (c) with outstanding efficiency for strontium removal (compared with alternative sorbents), and (d) fine selectivity from seawater treatment (based on the complementarity of reactive groups on the functionalized sorbent). Another merit of this work holds in the extensive characterization of the material and its interactions with strontium ions.

2. Results

2.1. Characterization of Materials

The main characteristics are summarized below. A detailed discussion appears in Appendix A. The sorbents are roughly spherical (Figure A1) and their average sizes are 2.46 ± 0.21 mm and 2.14 ± 0.15 mm for A*PEI and SA*PEI, respectively. The sulfonation contributes to the weak shrinking of the beads. This is confirmed by the enlarged SEM observation of the surface of the sorbents: SA*PEI is characterized by a more wrinkled surface than A*PEI. The semi-quantitative EDX analysis of the surface of A*PEI and SA*PEI is reported in Figure A2. Substantial differences are observed considering the increases in N content (from 3.1% to 6.46%, At.%) and S content (from 0.95% to 4.28%, At.%). Notably, the functionalization of A*PEI leads to a significant decrease in Ca content (from 9.2% to 3.22%, At.%) while K and Na elements appear (up to 2.05% and 2.17%, At.%, respectively); concomitantly the atomic percentage of Cl element decreases from 7.5% to 1.98%.

The isotherms of N_2 sorption and desorption show that the sulfonation of A*PEI increases the specific surface area of the pristine sorbent (from 2.1 to 7.4 m^2 g^{-1}) (Figure A3). The profiles show a substantial increase in the relative importance of the hysteresis loop. This functionalization also increases the porous volume (from ≈0.005 to 0.05 cm^3 g^{-1}), which is associated with a substantial increase in the pore width from 94–99 Å to 199–332 Å. Apparently, the chemical modification improves the ability of the sorbent for mass transfer. Despite low specific surface area (probably due to the large size of pores), the functionalized sorbent shows outstanding sorption properties for Sr(II) (see below).

The thermal degradation also shows some differences in the shape of the TGA profiles (different weight-loss transitions) and some shifts in the specific temperatures (for DrTG profiles) (Figure 1). Apparently, the sulfonation increases the stability of the composite: Shifts in temperature and increased residue (about 17% vs. 7%). Apart from the same initial step (corresponding to water release, which represents different weight fractions), the profile for SA*PEI shows a supplementary transition compared with A*PEI, which may correspond to the specific degradation of sulfonate groups (Figure A4).

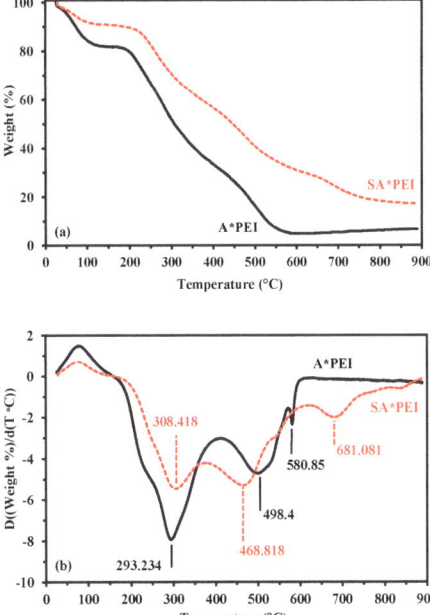

Figure 1. Thermal degradation of A*PEI and SA*PEI sorbents: (**a**) TGA, and (**b**) DrTG curves.

The FTIR spectra for A*PEI and SA*PEI show: The presence of carbohydrate ring and more specifically uronic-based groups (mannuronic and guluronic acid moieties, 1300–1000 cm^{-1}) and carbonyl from carboxyl groups (at 1767 cm^{-1}) (Figure 2, Table A1). The interaction of alginate-based carboxylic acid with amine groups onto PEI is characterized by the formation of amide bonds ($\approx$1620, $\approx$1400, and 1256 cm^{-1}), which are partially superposed with amine bending vibration (Figure A5). The grafting of sulfonic acid is confirmed by the appearance (or reinforcement) of a series of bands at 1159, 850, and 602 cm^{-1}.

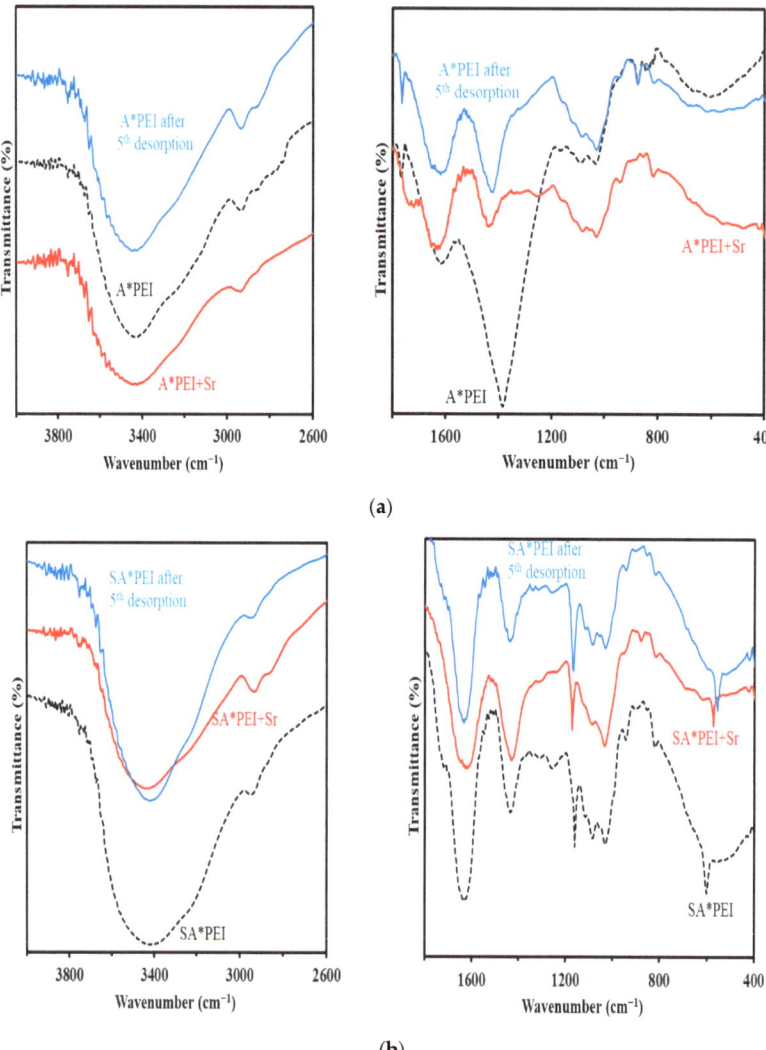

Figure 2. FTIR spectra of APEI (**a**) and SA*PEI (**b**) before and after Sr(II) sorption and after the fifth desorption (in recycling tests).

The sorption of Sr(II) involves some changes in the FTIR spectra that may derive from the interactions of specific reactive groups with metal cations. Figure A6 isolates the main variations in FTIR spectra that confirm the interaction of Sr(II) with carboxylic and amine/amide groups in the case of A*PEI, while for SA*PEI, in addition, the chemical environment of sulfonate groups is also affected by metal binding. This binding is reversible (see Section 2.2.7), while using 0.3 M HCl solution as the eluent. As a consequence of metal desorption, the FTIR spectra are partially restored; however, the characteristic bands of reactive groups are not completely reestablished to their neat form (this may be the effect of protonation of reactive groups due to acidic elution) (Figure A7).

In Figure A7, the spectra of the materials at different stages of use are compared with the spectra of the sorbent exposed to solution (without strontium) at the pH of the sorption step and with the eluent solution (for metal-free sorbents). These spectra allow evaluation of the contributions of the environmental conditions and separate their effects on the spectra to specific changes associated with metal sorption and metal sorption/desorption cycles. The comparison of the spectra of pristine sorbents and those exposed to pH 5 solution (selected pH for strontium sorption) does not show significant differences (Figure A8): The changes in FTIR spectra after Sr(II) binding reported above can be specifically attributed to the interactions of functional groups with Sr^{2+}. After conditioning the sorbent with 0.3 M HCl solutions (i.e., identical to the elution process for sorption/desorption cycles), the FTIR spectrum of A*PEI is more significantly changed at the level of carbonyl and amine groups (with stronger variations compared to the spectrum obtained after the 5 cycles). These functional groups are influenced by protonation and/or by metal binding to different extents). In the case of SA*PEI, the strong protonation affects the FTIR spectrum more significantly than successive cycles of sorption/desorption (the final rinsing step reduces the changes induced by protonation). In addition, the fine restoration of the FTIR profile confirms the good stability of the sorbent. Scheme 1 proposes the interaction modes of Sr(II) with functional groups held on A*PEI and SA*PEI. The grafting of 2-propene-1-sulfonic acid and hydroxylamine-o-sulfonic acid enriches the sulfonic and amine contents, which are involved in the binding of Sr^{2+} ions through chelation (i.e., hydroxyl and amine) and ion-exchange (through replacement of Ca^{2+} with Sr^{2+} and the protons from sulfonic groups). The functionalization opens to a wider variety of sorption mechanisms and a larger number of reactive groups than the pristine sorbent.

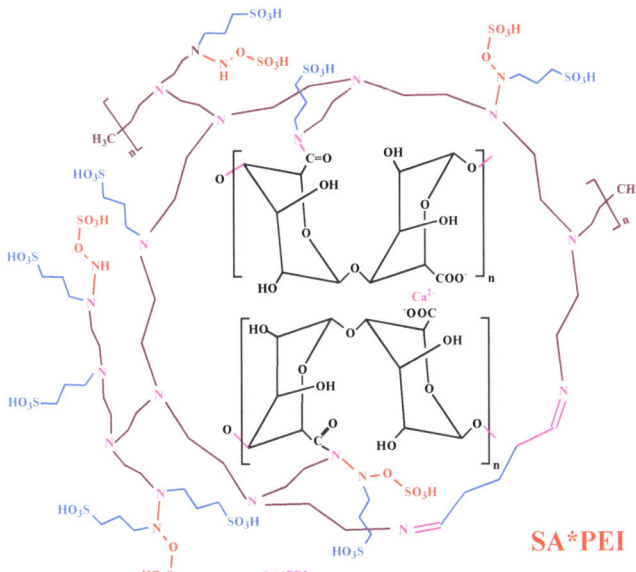

Scheme 1. Structure of SA*PEI beads.

The elemental analysis of the two sorbents confirms the effective grafting of sulfonic groups (reaching ≈ 1.04 mmol S g^{-1}) (Table A2). Based on the structure of grafted moieties and the elemental analysis, the density of sulfonate groups is close to 1 mmol g^{-1}. Nitrogen content remains unchanged (≈2.14 mmol N g^{-1}), while O content increases by 4.68 mmol O g^{-1} after the functionalization of A*PEI. In 2-propene-1-sulfonic acid (2P1SA) and hydroxylamine-O-sulfonic acid (HOSA), the increase in O content is 3 and 4 times the S content. The relative decrease of A*PEI fraction in the sulfonated derivative is compensated by hydroxylamine-O-sulfonic acid grafting. It is noteworthy that the increase in O content exceeds the expected stoichiometric ratio O/S in the insertion of sulfonic derivatives: The ratio reaches 4.52:1, while the ratio would be expected between 3:1 (for 2P1SA) and 4 (for HOSA). This may be due to traces of reagents; semi-quantitative EDX analysis shows traces of K$^+$ and Na$^+$ from potassium persulfate and sodium bisulfite; different levels of hydration (water absorption) may also explain theses discrepancies.

Figure A9 reports the evaluation of pH$_{PZC}$ values for the two sorbents: Data were collected from two sets of experiments using different concentrations of background salt. The profiles hardly changed. The grafting of sulfonic acid onto amine groups (of PEI) logically gives a stronger acid character to the sorbent: The pH$_{PZC}$ is shifted from ≈7.8 to ≈4.6. In the case of sulfonation of similar algal/PEI beads, Hamza et al. [55] reported a stronger decrease in pH$_{PZC}$ values (from 7.35 to 2.86); in the previous work, sulfosuccinic acid was used for the functionalization of the neat sorbent. Obviously, these acid–base characteristics strongly affect the surface charge of the sorbent: The global charge is positive in acidic solutions for SA*PEI (below pH 4.6) while it is necessary to increase the pH to 7.8 for losing the cationic charge in the case of A*PEI. This may have a direct impact on the attraction/repulsion of divalent cations such as Sr^{2+}.

2.2. Sorption Properties—Synthetic Solutions

2.2.1. pH Effect on Sr(II) Sorption

Figure 3 shows the comparison of the average values (triplicated series) of the pH-profiles for the sorption of Sr(II) using A*PEI and SA*PEI. Under selected experimental conditions (C$_0$: 1.17 mmol Sr L^{-1}, and sorbent dosage, SD: 0.67 g L^{-1}), the sorption capacity linearly increases with pH$_{eq}$ (up to pH 6) but remains quite low; never exceeding 0.21 mmol Sr g^{-1}. The beneficial effect of sulfonation is clearly demonstrated: The sorption capacity strongly increases for SA*PEI from 0.122 to 0.98 mmol Sr g^{-1}, when pH$_{eq}$ increases from 1.2 to 4.1. Above pH$_{eq}$ 4, the sorption capacity sharply decreases (down to 0.6 mmol Sr g^{-1} at pH$_{eq}$ 5.2). The effect of pH may be expressed by two main reasons associated with (a) the change in the speciation of the metal (charge of the solute, formation of complexes including polynuclear species or colloids, depending on the metal), and/or (b) the change in the charge of the sorbent (protonation or deprotonation of reactive groups). Under selected conditions, the speciation diagram shows in Figure A10 that strontium is only present as free species (i.e., Sr^{2+}) above pH 3, while below pH 3 a small fraction is present as SrNO$_3$$^+$. Therefore, strontium remains cationic and poorly affected in the pH range investigated herein; the strong impact of pH on sorption capacity is not driven by the speciation characteristics. In acidic solutions, the strong protonation of reactive groups (especially for A*PEI) limits metal cation sorption and it is necessary to increase the pH for enabling Sr^{2+} binding. The main functional groups present on A*PEI are carboxylic acid (alginate fraction of algal biomass) and amino groups (from PEI). PEI bears primary, secondary, and tertiary amino groups with pK$_a$ values close to 4.5, 6.7, and 11.6, respectively [56], while mannuronic and guluronic constituents of alginate have pK$_a$ values close to 3.38 and 3.65, respectively [57]. In acidic solutions, most of the amine groups remain protonated repulsing metal cations. On the opposite side, carboxylic groups progressively deprotonate (mannuronic before guluronic acid) and carboxylate groups may bind strontium by electrostatic attraction or ion-exchange with calcium ions bound to carboxylate groups. Hong et al. [58] also reported that the sorption of Sr(II) onto alginate beads increases with pH and tends to stabilize at a pH higher than 4.

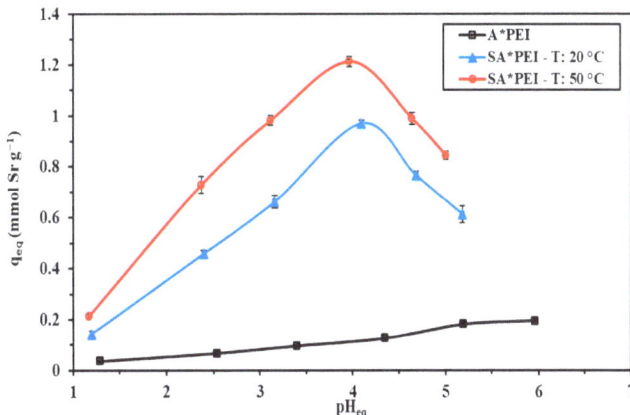

Figure 3. pH effect on Sr(II) sorption using A*PEI and SA*PEI beads—sorption capacity vs. pH$_{eq}$ (C$_0$: 1.17 mmol Sr L^{-1}; Sorbent dose, SD: 0.666 g L^{-1}; T: 20 ± 1 °C (except for SA*PEI test at T: 50 ± 1 °C); time: 48 h; averaged values and standard deviations).

The sulfonation of A*PEI strongly changes the pH-profile with maximum sorption close to pH 4. Sulfonic acids are strong acids; the negative apparent pK$_a$ values vary with the type of substituent (as shown by Dong et al. [59] for benzenesulfonic derivatives). This means that anionic sulfonate groups may coexist on the sorbent with weakly acidic carboxylic groups and alkaline amine groups; the pH$_{PZC}$ is close to 4.6: The global charge of the sorbent remains positive in acidic conditions and progressively decrease. Strontium cations can bind first to sulfonate groups before carboxylate groups begin to contribute with increasing the pH. The competition of protons with Sr^{2+} cations can also explain the weak sorption in strongly acidic pH regions. Surprisingly, the sorption capacity steeply decreases above pH$_{eq}$ 4, despite the favorable conditions for sorbent charge (and simultaneous deprotonation of sulfonic, carboxylic, and primary amine groups). The semi-quantitative analysis of the sorbent surface (using EDX) showed the presence of Na$^+$ and K$^+$, completed by the increase of sodium content due to pH control; this may explain a shielding effect at the surface of the material. This effect limits the ability of negatively-charged functional groups to bind strontium cations under slightly alkaline conditions. It is noteworthy that the same experiment was performed at increased temperature (i.e., T: 50 °C), the pH-edge curve shows the same trends: A sharp optimum is found again close to pH$_{eq}$ 4. However, the curve is shifted toward higher sorption capacities: The sorption of Sr(II) onto SA*PEI is endothermic.

Figure A11 compares initial and equilibrium pH values after strontium sorption. For A*PEI, the pH tends to weakly increase (especially at pH close to 2, by 0.37–0.55 pH unit): Protons are bound; at pH 6, the equilibrium pH remains stable. The sulfonation affects the pH change: The final pH increases by less than 0.4 pH unit between pH 1 and pH 3, the pH remains unchanged up to pH 4 and tends to decrease at pH above 4 (associated with proton release dye ion-exchange with Sr(II) binding. This breakpoint is close to the pH$_{PZC}$ value of functionalized beads. In Figure A12, the log$_{10}$ plots of the distribution ratio (D = q$_{eq}$/C$_{eq}$, L g^{-1}) vs. pH$_{eq}$ show a good correlation but the slopes cannot be clearly associated with conventional ion-exchange ratio (see Appendix B.2).

At acidic pH (i.e., pH 1–2), the protonation of hydroxyl groups (from algal backbone in A*PEI) and both amine and sulfonic groups (in SA*PEI). Under these conditions, Sr(II) is bound through an ion-exchange mechanism. With the increase of the pH, the reactive groups progressively deprotonate and some reactive groups become negatively-charged making possible the binding of Sr(II) ion through a binary mechanism of ion-exchange and chelation. Associated with the decrease of ionic repulsion, the sorption of Sr(II) is enhanced. At pH above the pH_{PZC} value, the sorption capacity for SA*PEI strongly decreases probably due to the shielding and competition effects of Na^+, K^+, and Ca^{2+} (as appearing in semi-quantitative EDX analysis) and in relation to the progressive increase in the negatively-charged surface of the sorbent (sulfonate groups). Similar phenomena were reported for the sorption of U(VI) using two chelating resins [60], and for the development of superabsorbent polymers [61,62]. The screening of anionic charges with Na^+ affects the configuration and packing of the chains but also contributes to the decrease in the availability of reactive groups. This decrease is not observed for A*PEI sorbent, where binding is limited to free amine and carboxylic groups. Further experiments were performed at optimum pH conditions: pH_0 4 for SA*PEI and pH_0 5 for A*PEI.

2.2.2. Uptake Kinetics

The kinetic profiles for Sr(II) sorption using A*PEI and SA*PEI are compared in Figure 4 (under the same experimental conditions, except pH values adjusted to their optimal values). For A*PEI, 90–120 min are necessary for reaching the equilibrium; apparently, there is a break in the slope of the concentration decay at around 30–35 min, possibly associated with the hydration of the sorbent. The kinetics is faster and more homogeneous in the case of SA*PEI: The equilibrium time is reduced to 30–40 min. this faster mass transfer may be correlated with the textural characteristics of the sorbents: SA*PEI has a slightly greater specific surface area and larger pores than the precursor (i.e., A*PEI). The sorption kinetics are controlled by mechanisms of resistance to diffusion (essentially film and intraparticle diffusions) and by the reaction rate (which can be described by pseudo-first, pseudo-second-order rate equations; PFORE and PSORE, Table A3 (a)). The resistance to film diffusion was negligible (due to the appropriate pre-selection of stirring speed); in addition, the resistance to film diffusion is also minimized by the appropriate choice of agitation conditions (herein: 210 rpm). This mechanism is mainly active in the overall control of uptake kinetics within the first minutes of contact. The initial section corresponds, under these conditions, to the fitting of the experimental profile with a first-order rate equation. With a proper agitation rate, the resistances to bulk diffusion and film diffusion can be neglected.

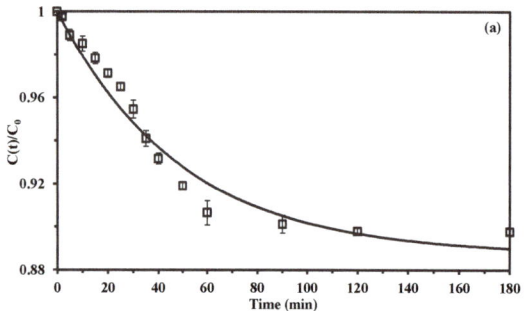
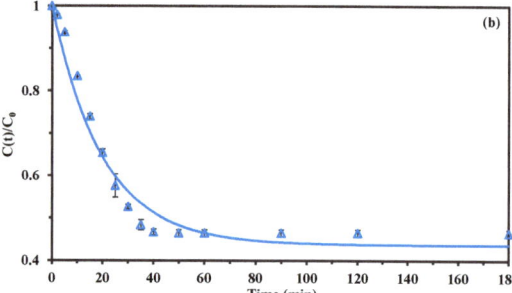

Figure 4. Cont.

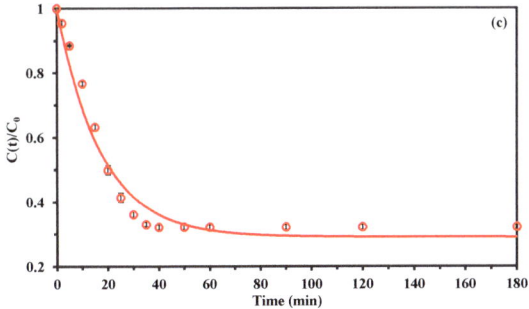

Figure 4. Sr(II) uptake kinetics for A*PEI (**a**) and SA*PEI (**b**,**c**) beads—Modeling with PFORE (C_0: 1.17 mmol Sr L^{-1}; SD: 0.666 g L^{-1}; T: 20 ± 1 °C (**a**,**b**) and at T: 50 ± 1 °C (for SA*PEI, c); pH$_0$: 5 and 4 for A*PEI and SA*PEI, resp.; averaged values/std. dev.).

Table 1 summarizes the apparent rate coefficients for PFORE and PSORE (k_1 and k_2, respectively); the statistical parameters are also compiled for the resistance to intraparticle diffusion (i.e., RIDE, Table A3 (a)). Table A4 reports the parameters and statistical data for individual replicates.

Table 1. Sr(II) uptake kinetics using A*PEI and SA*APEI—parameters of PFORE, PSORE, and RIDE models.

	Sorbent		A*PEI	SA*PEI	
Model	Parameter	Temperature	T: 20 ± 1 °C	T: 20 ± 1 °C	T: 50 ± 1 °C
		Unit			
Experimental	q_{eq}	mmol Sr g^{-1}	0.179	0.926	1.17
PFORE	$q_{eq,1}$	mmol Sr g^{-1}	0.198	0.974	1.23
	$k_1 \times 10^2$	min^{-1}	2.04	5.01	5.80
	R^2	-	0.969	0.976	0.977
	AIC	-	−144	−96	−91
PSORE	$q_{eq,2}$	mmol Sr g^{-1}	0.268	1.15	1.43
	$k_2 \times 10^2$	g mmol^{-1} min^{-1}	26.8	4.93	4.85
	R^2	-	0.952	0.938	0.937
	AIC	-	−139	−83	−77
RIDE	$D_e \times 10^9$	m^2 min^{-1}	6.86	7.13	6.36
	R^2	-	0.944	0.940	0.936
	AIC	-	−133	−82	−75

The comparison of the determination coefficient (R^2) and Akaike Information Coefficient (AIC) clearly demonstrates that the PFORE best fits experimental profiles than the PSORE (and even better the RIDE). In the literature, the mathematical preference for a given model is frequently used for discriminating between physical and chemical sorption; however, this interpretation requires some conditions to be fulfilled (such as a negligible variation of the sorbate concentration in the solution), which are rarely verified; making the interpretation questionable [63]. Herein, except for A*PEI (for which the weak sorption limits concentration change), the conditions are not respected for appropriate interpretation. However, the parameters are useful for comparing the sorbents and evaluating the effect of temperature (in the case of SA*PEI). The solid lines in Figure 4 represent the simulated profiles with PFORE: triplicated curves confirm the good reproducibility of sorption tests. It is noteworthy that increasing the temperature slightly increases the sorption kinetics and the sorption capacity (consistently with the observation reported in the study of the pH effect). The faster sorption of Sr(II) by SA*PEI is confirmed by the ranging of apparent rate coefficients (k_1, min^{-1}) according to:

A*PEI ($2.05 \pm 0.13 \times 10^{-2}$ min^{-1}) < SA*PEI ($5.02 \pm 0.23 \times 10^{-2}$ min^{-1}).

These values are of the same order of magnitude as the apparent rate coefficients reported for Dowex 50W-X8 (i.e., 0.0214 min^{-1}, [36]) but higher than for surfactant-conditioned polyacrylonitrile (i.e., 0.0022 min^{-1}, [64]) or for biochar and magnetic biochar (i.e., 0.0047 and 0.0077 min^{-1}, respectively, [65]). Bezhin et al. [66] reported much lower values for Sr(II) sorption kinetics using a series of resins and ion-exchangers (i.e., in the range 0.0015–0.0024 min^{-1}). It is noteworthy that in most of these studies the PSORE gave better fitting of kinetic curves than the PFORE. The calculated values of sorption capacities at equilibrium general overestimate the experimental results (see Appendix B.3).

The intraparticle diffusivity coefficient (deduced from the application of the RIDE to experimental profiles) hardly varies with the sorbent: $6.84 \pm 0.22 \times 10^{-9}$ m^2 min^{-1} for A*PEI and 7.2–8.1 $\times 10^{-9}$ m^2 min^{-1} for SA*PEI (depending on the temperature). This means that the intraparticle transfer is little enhanced into the slightly more porous sorbent (i.e., SA*PEI). These values are comparable to the diffusivity coefficient (i.e., 7.2×10^{-9} m^2 min^{-1}) reported by Morig and Gopala Rao [67] in the case of Dowex 50W-X8 sulfonic resin. These values are only one order of magnitude lower than the self-diffusivity of Sr(II) in water (i.e., 4.76×10^{-8} m^2 min^{-1}, [68]). Strontium diffusion in A*PEI and SA*PEI sorbents is facilitated compared with porous polymer-coated hydroxyapatite (i.e., 1.8×10^{-10} m^2 min^{-1}) [69].

2.2.3. Sorption Isotherms

Sorption isotherms are summarized in Figure 5 (q_{eq} vs. C_{eq}); the triplicate series are superposed: The sorption tests are reproducible. Consistently with previous results, the sulfonation of A*PEI strongly enhances the sorption of strontium: Both (a) the saturation plateau (i.e., the maximum sorption capacity for $C_{eq} \approx 5$ mmol Sr g^{-1}) from 0.587 ± 0.015 to 1.870 ± 0.029 mmol Sr g^{-1}, and (b) the affinity coefficient (which is correlated with the initial slope of the curve). In A*PEI, carboxylate and amine groups potentially contribute to Sr(II) binding with low affinity: The sorption capacity weakly progresses with increasing metal concentration. Based on the nominal values of their pK_a, the deprotonation of the reactive groups makes carboxylate groups more available than primary amine groups present on PEI for Sr(II) binding. The sulfonation brings additional reactive groups that are good cation exchangers and remain fully deprotonated at pH 4. These strong reactive groups may explain the sharp increase in sorption capacity: The initial slope is much steeper (stronger affinity) and the density of reactive groups increases.

Figure 5 also presents the sorption isotherm at T: 50 °C for SA*PEI. Both the maximum sorption capacity (from 1.870 ± 0.029 to 2.340 ± 0.049 mmol Sr g^{-1}) and the affinity increase with sorbent functionalization: This is consistent with the results obtained in the study of pH effect and uptake kinetics. Increasing the temperature enhances metal binding; the sorption of Sr(II) onto SA*PEI is endothermic. This probably means that the mechanism of Sr(II) binding to sulfonate groups is endothermic. Wang et al. [70] modified the macroporous styrene chelating resin (with diglycolamidic acid functional group) by grafting sulfonic groups for Pb(II) binding: They found that metal sorption is endothermic. Shin et al. [65] also reported endothermic sorption of Sr(II) onto biochar (and magnetic biochar). A similar conclusion was obtained in the case of Sr(II) binding onto sulfactant-functionalized polyacrylonitrile [64].

The sorption isotherms were fitted with different models (appearing in Table A3 (b)). In Figure 5, the solid lines represent the fitting of isotherms for SA*PEI with the Langmuir equation at pH_0 4 (and the Freundlich equation for A*PEI at pH_0 5). The parameters of the different models are reported in Table 2 (combining the triplicated series; Table A5 (a–c) report the individualized treatment of the experimental series). Maximum sorption at saturation of the monolayer (Langmuir model) substantially increases after functionalization (from 0.997 ± 0.113 to 2.057 ± 0.052 mmol Sr g^{-1}), as does also the affinity coefficient (b_L: From 0.282 ± 0.070 to 1.943 ± 0.164 L mmol^{-1}). With the endothermic nature of Sr(II) sorption onto SA*PEI, the Langmuir parameters increase at T: 50 °C up to 2.327 ± 0.045 mmol Sr g^{-1} and 4.760 ± 0.099 L mmol^{-1}.

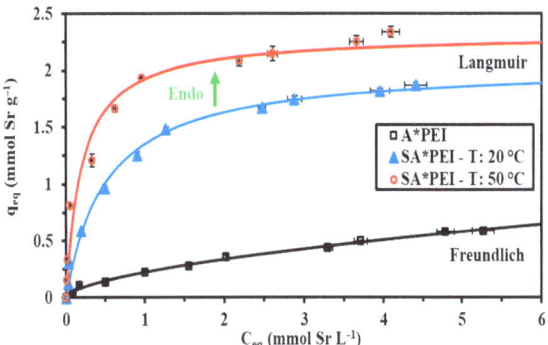

Figure 5. Sr(II) sorption isotherms on A*PEI and SA*PEI beads—Modeling with Langmuir equation or Freundlich equation (C_0: 0.12–5.82 mmol Sr L^{-1}; SD: 0.666 g L^{-1}; T: 20 ± 1 °C (for A*PEI and SA*APEI) and at T: 50 ± 1 °C (for SA*PEI, c); pH$_0$: 5 and 4 for A*PEI and SA*PEI, resp.; average values/std.dev.; fitting on cumulated series).

Table 2. Sr(II) sorption isotherms using A*PEI and SA*PEI—parameters of Langmuir, Freundlich, Sips, and Temkin models.

Sorbent		A*PEI	SA*PEI	
Temperature		20 ± 1 °C	20 ± 1 °C	50 ± 1 °C
Model	Parameter			
Experim.	$q_{m,exp.}$	0.607	1.91	2.40
Langmuir	$q_{eq,L}$	0.977	2.05	2.33
	b_L	0.278	1.94	4.88
	R^2	0.981	0.992	0.969
	AIC	−214	−164	−110
Langmuir dual site	$q_{eq,L1}$	-	0.192	0.903
	b_{L1}	-	63.4	34.0
	$q_{eq,L2}$	-	1.94	1.65
	b_{L2}	-	1.41	1.36
	R^2	-	0.994	0.986
	AIC	-	−180	−136
Freundlich	k_F	0.221	1.18	1.66
	n_F	1.67	2.87	3.61
	R^2	0.986	0.968	0.956
	AIC	−231	−128	−104
Sips	$q_{eq,S}$	-	2.33	2.83
	b_S	-	1.29	1.83
	n_S	-	1.29	1.62
	R^2	-	0.994	0.984
	AIC	-	−177	−134
Temkin	A_T	9.37	46.4	162
	b_T	18,440	7028	7552
	R^2	0.899	0.981	0.984
	AIC	−168	−145	−118
Temkin-II	q_T	-	0.434	0.395
	K_T	-	18.2	92.2
	R^2	-	0.991	0.984
	AIC	-	−168	−137

Units: q, mmol g^{-1}; b and K_T, L mmol^{-1}; n, dimensionless; k_F, mmol$^{1-1/nF}$ L$^{-1/nF}$ g^{-1}; A_T, L mmol^{-1}; b_T, J mmol^{-1}; fitting processed on cumulative triplicates.

Alternate fittings are presented in Figure A13. The comparison of the different models shows that for A*PEI sorbent the Freundlich equation fits better experimental data than the Langmuir and the Sips equations. This model supposes that sorption occurs with possible interactions between sorbed molecules and with heterogeneous energies of sorption. On the opposite hand, for SA*PEI sorbent the Sips equation (with a combination of Langmuir and Freundlich equations) is more appropriate. The Langmuir equation is a mechanistic equation based on the homogeneous sorption of sorbate at the surface of the sorbent, as a monolayer without interactions between sorbate molecules. The Sips equation introduces a third-adjustable parameter that may contribute to better mathematical fit (at the expense of a loss in physical significance) (Figure A13a). The Temkin equation assumes that the heat of adsorption of the molecules bound onto the monolayer linearly decreases with its relative coverage and that there is a uniform distribution of heterogeneous binding sites. It is usually inappropriate to fit the two extreme regions of the isotherm (i.e., low concentration and saturation zone) [71]. This may explain the lower quality of curve fitting (with the remarkable exception of the isotherm at 50 °C for SA*PEI. Chu [71] reported and corrected the dimensional inconsistency commonly found in publications applying the Temkin equation to solid/liquid sorption. In addition, several derived equations one of these forms allows accommodating the Temkin equation to low concentration ranges (called Temkin-II equation (Equation (1)), by opposition to the conventional equation reported in Table 2):

$$\text{Temkin-II } q_{eq} = q_T \ln(1 + K_T C_{eq}) \tag{1}$$

Herein, the Temkin-II equation improves the quality of the fit for SA*PEI at T: 20 °C; while the improvement for T: 50 °C is less significant (Figure A13b).

The functionalized sorbent bears sulfonate groups in additional to the reactive groups present on A*PEI. These reactive groups may have different affinities for the target metal; this introduces heterogeneities at the surface of the sorbent, which can be accounted for using the Langmuir dual site equation (LDS) [72]:

$$q_{eq} = \frac{q_{L1} b_{L1} C_{eq}}{1 + b_{L1} C_{eq}} + \frac{q_{L2} b_{L2} C_{eq}}{1 + b_{L2} C_{eq}} \tag{2}$$

with q_{Li} (mmol g^{-1}), b_{Li} (L mmol^{-1}) as the maximum sorption capacities, and Langmuir affinity coefficients for sites i = 1 and 2.

The LDS equation fits well experimental profiles (Figure A13c and Table 2, lowest values of the AIC). One of the reactive groups is characterized by high sorption capacity with low affinity (Site 1), while the other has a strong affinity for Sr(II) but with lower density (sorption capacity) (Site 2). It is noteworthy that increasing the temperature differently affects these two reactive groups: For Site 2, the parameters are weakly decreased at t: 50 °C, while for Site 1, the maximum sorption capacity significantly increases but the affinity is halved. Figure A14 compares the contributions of Sites 1 and 2 at 20 and 50 °C. With increasing temperature, the relative contribution of Site 1 substantially increases (from ≈10% at T: 20 °C to 39% at T: 50 °C) [72].

Table A6 reports Sr(II) sorption performances of a series of alternative sorbents. *Bacillus pumilus* bacteria show outstanding sorption capacity (about 3.4 mmol Sr g^{-1}) under environmental conditions (i.e., at pH 7) with weak affinity and slow kinetics (equilibrium requires about five days of contact) [28]. Similarly, granular manganese oxide shows relatively good sorption capacity, very weak affinity, and long equilibrium time [73]. The sorption properties of SA*PEI, based on the compromise between sorption capacity and affinity, kinetics, and pH, are comparable to those reported for amidoximated algal/PEI beads [38], functionalized silica beads [74], functionalized graphene oxide [75], and graphene oxide [76]. Compared with a commercial ion-exchanger (i.e., sulfonated Dowex 50W-X8 resin) that also bears sulfonic groups, the sorption capacity, determined at lower pH (i.e., 3.7), is of the same order (as well as the equilibrium time) while the affinity coefficient is significantly higher (up to 206 L mmol^{-1}). Based on the selected criteria, SA*PEI is part of the most efficient sorbents for Sr(II), as an alternative to Dowex 50W-X8 resin.

2.2.4. Binding Mechanisms

Elemental analysis, titration, and FTIR characterization have shown the effective sulfonation of the pristine sorbent. The comparison of sorption properties of A*PEI and SA*PEI showed a strong improvement in sorption properties. This is due to the supplementary binding of Sr(II) to the oxygen-based moieties of sulfonic groups; this is characterized by the decrease in the relative intensities of –OH and sulfone groups bands in FTIR characterization after metal binding. The chelation of strontium cations (related to metal speciation) onto –OH and amine groups results from a favorable balance between the charge of the sorbent (resulting from pH$_{PZC}$ characterization and the study of pH effect on metal binding) and the metal charge. On the other side, Ca^{2+} ions may be replaced with Sr^{2+} through an ion-exchange mechanism. Scheme 2 shows the expected mechanism for the sorption of Sr(II) ions onto pristine beads (A*PEI) and the sulfonated sorbent (SA*PEI).

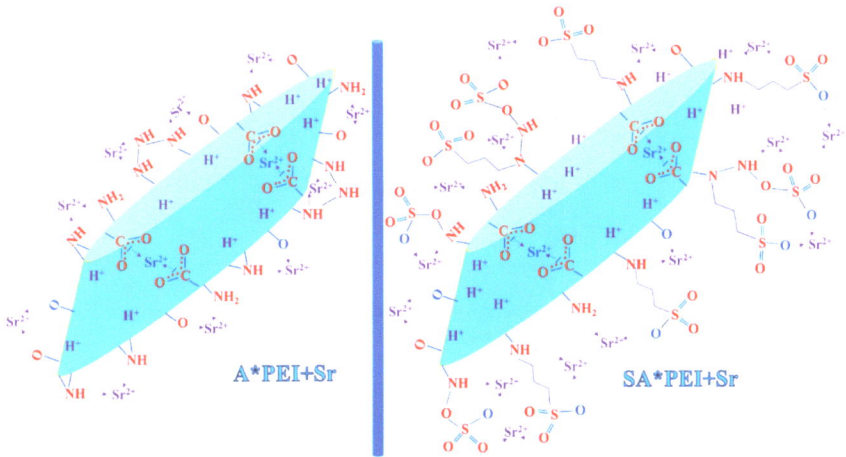

Scheme 2. Suggested mechanism of Sr(II) interactions with A*PEI and SA*PEI.

2.2.5. Sorption Selectivity

The sorption of Sr(II) onto SA*PEI may be influenced by the presence of other metal ions which may have a certain affinity for the reactive groups present on the multi-functional sorbent (carboxylic, amine and sulfonic groups). Obviously, the competition effect exerted by their presence depends on the pH (tough the cross effects of protonation/deprotonation of reactive groups and metal speciation). In order to evaluate this potential effect, the sorption capacity of Sr(II) is determined at different pH values in multi-component equimolar solutions containing Na(I) alkaline element), Ca(II) and Mg(II) (alkaline-earth elements), Fe(III), and Al(III) (heavy metal elements). Under these conditions, it is meaningful and easy to calculate the selectivity coefficient for Sr(II) against other metal ions (i.e., SC$_{Sr/metal}$; ratio of relevant distribution ratios) as:

$$SC_{Sr/metal} = \frac{D_{Sr}}{D_{metal}} = \frac{q_{eq,Sr} \times C_{eq,metal}}{C_{eq,Sr} \times q_{eq,metal}} \quad (3)$$

Figure 6 shows that the selectivity of SA*PEI for Sr(II) against competitor metal ions increases with the pH of the solution (for Fe(III) almost equivalent between pH$_{eq}$ 3.21 and 4.78). At pH$_{eq}$ 4.78, the selectivity ranking for SC$_{Sr/metal}$ follows the order:

Cs(I) (2.9) > Fe(III) (5.6) > Al(III) (11.8) > Ca(II) (12.2) > Mg(II) (13.5) > Na(I) (25.0).

In the case of Dowex 50W-X8 sulfonic resin, Bonner [77] reported the selectivity of the resin according: Cs(I) > Na(I) and Sr(II) > Ca(II) > Mg(II) (also confirmed for divalent cations by Iyer et al. [78]). This is roughly consistent with the specific observation collected herein. Pelhivan and Altun [79] established that the selectivity of sulfonic functions on Dowex 50W increases with atomic number, valence, and degree of ionization of exchanged metal ions. On the other side, Gupta [80] reported the effect of hydration enthalpy of monovalent ions. These different correlations are reported in Figure A15. The data are grouped by ionic charge and show that in each family the selectivity coefficient for Sr(II) against competitor ions: (a) decreases with increasing the atomic number, and the ionic radius and (b) increases with the hydration enthalpy (absolute value). Comparing the competitor effects of Ca(II) and Mg(II) with that of K(I) on Sr(II) sorption by resorcinol-formaldehyde resin, Nur et al. [34] highlighted the effect of charge difference and the inner-sphere complexation behavior of divalent cations.

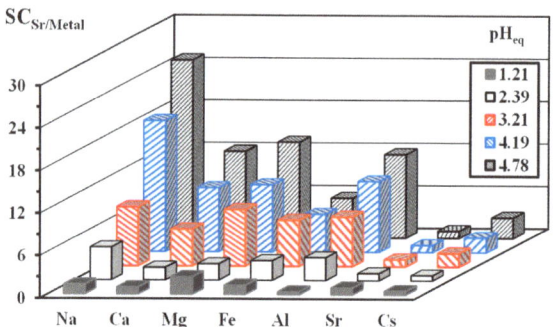

Figure 6. Effect of pH on the selectivity coefficient $SC_{Sr/Metal}$ for multi-component equimolar solutions (C_0: 1 mmol L^{-1}; SD: 1.6 g L^{-1}; T: 20 ± 1 °C; time: 24 h).

The sorbent has clearly a marked preference for Sr(II) and Cs(I). Figure A16 shows the $SC_{Cs/metal}$ for the different metals with varying pH. The response to pH is less "monotonous" than for Sr(II) selectivity. Hence, the optimum separation of Cs(I) from Sr(II), Fe(III) occurs at pH_{eq} 1.21 and 2.39, respectively; while for other metal ions, the highest values for SC are reported at pH_{eq} 4.19–4.78.

Figure A17 tests the correlation of the affinity of the metals for SA*PEI with their physicochemical properties: The individual distribution coefficients are plotted against the ionic index (i.e., I.I. = Z^2/r) and the covalent index (i.e., C.I. = $X_m^2 r$), where Z and X_m are the charge and the Pauling electronegativity of metal ions. Figure A17 clearly confirms the out-of-trend behavior of SA*PEI for Sr(II) and Cs(I) compared with the other competitor ions: Their distribution ratio is much larger for Sr(II) > Cs(I) and much more for the other mono-, di-, tri-valent cations. For the competitor ions, the distribution ratio follows a linear trend against C.I (higher correlation than for I.I.).

2.2.6. Effect of Salinity (NaCl) on Sr(II) Sorption

In an attempt to evaluate the potential of SA*PEI to be applied for seawater treatment, the effect of NaCl concentration is investigated: Figure A18 compares the effect of pH on Sr(II) capacity with increasing Na,Cl concentrations (from 1 to 4 M; the curve without NaCl addition is reported as a reminder, though processed under different experimental conditions). The salt concentration does not change the global behavior of the sorbent: The sorption capacity increases up to pH_{eq} 4 before decreasing. The slopes of the two sections are parallel and hardly changed by salt concentration. The decrease in sorption capacity becomes significant above 2 M. However, even with a Na,Cl concentration as high as 4 M, the sorption capacity, at pH_{eq} ≈4, only decreases by 26% (from 0.996 mmol Sr g^{-1}, average value for 0–2 M data, vs. 0.735 mmol Sr g^{-1}). The little decrease in sorption capacities may be explained by the co-sorption of Na(I) (not shown, up to 20 mmol Na g^{-1} at the highest Na,Cl concentration and higher pH value); however, despite this strong sorption, the decrease is relatively limited, meaning that different functional groups are involved in Na(I) and Sr(II) binding. The mineral charge in seawater does not exceed 3.5% (w/w); with Na,Cl content close to 0.65 M. Despite an excess of NaCl as high as 2 M (even higher than in seawater), the sorption of Sr(II) is hardly affected: Sorption capacity remains close to 0.97 mmol Sr g^{-1} (under selected experimental conditions). This stability is remarkable because of the risk for alginate to be partially dissolved in a such high concentration of sodium (ion-exchange of Ca^{2+} with Na$^+$); this means that the interpenetrating network (alginate/PEI with ionotropic gelation and glutaraldehyde crosslinking) contributes to stabilizing the material. This outstanding stability in sorption performance in complex solutions contrasts with the significant decrease in strontium sorption observed by Yin et al. [81], while using biogene-derived aerogels (imprinted dopamine-alginate materials): The sorption efficiency was divided by three even with Na,Cl concentration as low as 0.1 M. In the case of alginate microsphere, the sorption capacity of Sr(II) was decreased by 60% in presence of NaCl (0.65 M) [58]. In the presence of high concentrations of NaCl, Ca-alginate may be converted into Na-alginate, which can be dissolved and involve the de-structuration of the hydrogel. However, apparently under concentrations as high as 3–4 M NaCl, the multiple mechanisms of crosslinking involved in the preparation of the sorbent (alginate interaction with amine groups (at appropriate pH), amine crosslinking with glutaraldehyde, and ionotropic gelation of carboxylate groups with Ca(II) lead to stable interpenetrating networks. Therefore, the reduction in sorption performance at the highest concentrations may be due to a partial degradation of the hydrogel; however, the material shows a good overall stability.

The distribution ratios (i.e., D, L g^{-1}, not shown) increase with pH for Na(I), contrary to Sr(II) (which showed a maximum for pH 4), and decrease with increasing Na,Cl concentration (from 0.0093 to 0.0029 L g^{-1}). At pH 4, the D ratio for Sr(II) is 215 to 465 times higher than the value for Na(I). This is consistent with competitor studies (Sections 2.2.5 and 2.2.6) and with the remarkable preference of SA*PEI for Sr(II) against Na(I); this is a good indication of the interest of the sorbent for application in seawater (see Section 2.3).

2.2.7. Sr(II) Desorption from Metal-Loaded SA*PEI and Sorbent Recycling

Strontium desorption was operated using an acidic solution; the weak sorption in acidic solutions is the first incentive for using acidic solutions. Hong et al. [58] reported the successful desorption of strontium from alginate beads; however, they pointed out substantial degradation at repeated use. Therefore, they used 0.1 M HCl/CaCl$_2$ solutions; the presence of calcium allows for re-structuring of the hydrogel beads by ionotropic gelation. A combined 0.2 M HCl/0.5 M CaCl$_2$ solution was used for recovering strontium from amidoximated algal/PEI beads, with complete desorption of the metal and stable re-use. In the case of granular manganese oxide, strontium was recovered with 1 M HNO$_3$ solution as the eluent [73]. They obtained very low desorption rates with this sorbent where the metal is tightly bound.

Figure A19 compares the kinetic profiles for Sr(II) desorption using the metal-loaded materials recovered at the end of uptake kinetics and 0.3 M HCl solution as the eluent. Some discrepancies are observed in the repetition tests; however, in all cases, the desorption is fully achieved within 20–40 min. Apparently, the desorption is slightly faster for SA*PEI compared with A*PEI at 20 °C. It is noteworthy that the slowest desorption kinetics are obtained for the sorbent loaded at 50 °C: The thermal activation leads to a stronger interaction between the sorbent and Sr(II) (consistently with the affinity coefficients).

The re-use of the sorbents (over 5 cycles) is reported in Table 3. The two sorbents show remarkable stability in terms of desorption efficiency along the five cycles. However, the comparison of sorption efficiencies shows marked differences between A*PEI and SA*PEI. In the case of A*PEI, the sorption progressively decreases during the recycling tests. Since the desorption is complete, the loss in sorption capacity, which reaches up to 12% in the fifth cycle, is probably associated with a degradation of the sorbent. Some changes have been observed in the FTIR analysis (see Section 2.1), as a confirmation of this hypothesis. On the opposite hand, the sulfonated material shows outstanding stability in sorption efficiencies (the loss at the fifth cycle is less than 2%).

Table 3. Sorbent recycling—Sorption efficiency (SE, %) and desorption efficiency (DE, %) for five successive cycles.

Sorbent	A*PEI				SA*PEI			
Cycle	SE (%)		DE (%)		SE (%)		DE (%)	
	Aver.	St. dev.	Aver.	St. dev.	Aver.	St. dev.	Aver.	St. dev.
#1	10.73	0.85	100.5	0.4	58.7	1.8	99.9	0.2
#2	10.22	0.85	104.9	7.6	58.6	1.8	100.0	0.1
#3	9.86	0.61	100.1	0.6	58.3	1.7	100.0	0.1
#4	9.75	0.65	99.7	0.1	57.8	1.8	100.0	0.0
#5	9.46	1.01	100.2	0.3	57.7	1.8	100.0	0.0
Loss (5th/1st)	11.8%		S and C		1.7%		S and C	

Different types of eluent have been reported for strontium elution, for example, sodium carbonate in the case of imprinted bio-hydrogel (with a loss in sorption that reaches up to 17% at the fifth cycle) [81]. In the case of analogous sulfonic resin (Dowex 50W-X8), Hafizi et al. [36] reported that EDTA is more efficient than acid solutions for Sr(II) elution; however, the recycling of the sorbent was not qualified in their contribution.

Note: Complementary interpretations and analyses of experimental data on sorption properties are reported in Appendix B.

2.3. Application to Complex Solution—Seawater

Section 2.2.5 confirmed that SA*PEI has a marked preference for Sr(II) against a series of alkaline, alkaline-earth, and heavy metals (and Cs(I) to a lesser extent). The next step consists in testing the sorption performances in even more complex systems such as seawater. Table A7 reports the composition of two samples collected from the Mediterranean and the Red Sea coasts. Strontium is present in concentrations as low as 4–6 mg Sr L^{-1} with a very large excess of Na(I) (up to 14 g Na L^{-1}), Mg(II) (up to 1.5 g Mg L^{-1}). As a corollary, chloride ions are in huge excess (about 19 g Cl L^{-1}). These tests are necessary for evaluating the potential of this material for the treatment of contaminated seawater (such as resulting from the Fukushima Daiichi disaster, [82]).

Figure A20 reports the uptake kinetics for Sr(II) using SA*PEI sorbent for both the Mediterranean and Red Sea water samples. The complexity of the solutions strongly decreases the mass transfer properties: (a) for Sr(II) and B(III) a short lag phase lasting for 2–3 h occurs before the concentrations decrease (not observed with uranyl), and (b) reaching the equilibrium requires about 24 h of contact. Bezhin et al. [66] compared different sorbents for the recovery of Cs(I) and Sr(II) from seawater; they also reported quite slow kinetics (in most cases equilibrium was reached after more than 24 h of agitation, especially for strontium removal); the faster Sr(II) uptake was reported for copper potassium ferrocyanide supported on phosphorylated wood. In the case of Sr(II) sorption from seawater using alginate microspheres, Hong et al. observed also slow kinetics: 6 h are necessary (with high SD: 2 g L^{-1}; herein the SD: 0.2 g L^{-1}).

The sorption capacities reached under selected experimental conditions (which do not cover the saturation of the sorbent) are reported in Figure A21. The direct comparison of sorption capacities in seawater with the values reached with synthetic solutions is not possible; the pH is not adjusted (the "natural" pH is close to 7.5; $pH_{eq} \approx 7.6$); and the sorbent dosages are not the same. However, in synthetic solution, for Sr(II) residual concentrations in the range 2.41–2.76 mg Sr L^{-1} the sorption capacities onto SA*PEI at pH 5.19 range between 2.90 and 4.37 mg Sr g^{-1} (i.e., 33–50 μmol Sr g^{-1}; SD: 0.67 g L^{-1}). In the case of seawater samples, for residual concentrations ranging between 2.99 and 4.57 mg Sr L^{-1}, the sorption capacities reach 6.14–6.99 mg Sr g^{-1} (i.e., 70–80 μmol Sr g^{-1}, SD: 0.2 g L^{-1}). This means that despite the complexity of the background solution, which affects strontium speciation, the sorption is weakly affected. Kirishima et al. [82] discussed the speciation of strontium in seawater; they concluded that the metal may coexist under two forms sulfate "aquo" complex (i.e., $SrSO_{4aq}$, ≈60%) and free Sr^{2+} (≈40%). Bezhin et al. [66] summarized the maximum sorption capacities of three sorbents for Sr(II) removal from seawater. In the case of nanostructured crystalline hydroxyapatite, they found a capacity close to 0.21 mg Sr g^{-1}; for biogenic amorphous hydroxyapatite they cite a capacity close to 2.3 mg Sr g^{-1}. The maximum sorption capacity reaches about 7 mg Sr g^{-1} for birnessite-type sorbent (i.e., a layered manganese oxide conditioned under sodium form). Therefore, despite the unsaturation of SA*PEI under selected experimental conditions, the sorption capacities sound very attractive and competitive.

Figure A21 also shows the data for boron and uranyl ions (at trace levels, apart from major metals). The sorption levels for boron range between 4.8 and 5.8 mg B g^{-1} (i.e., 446–533 μmol B g^{-1}) are higher than those reached for uranyl (i.e., 0.078–0.098 μmol g^{-1}). The weak sorption of uranyl may be explained by the low concentration of uranium in seawater but also perhaps by the effect of metal speciation under complex composition. Indeed, Amphlett et al. [83] showed that in seawater uranyl ions are mainly present as anionic species (more specifically: $UO_2(SO_4)_3^{4-}$), which may be poorly available for binding onto sulfonic groups. The distribution ratios for Sr(II), B(III), and U(VI) are close to D: 1.53–2.05, 1.66–1.94, and 3.04–3.79 L g^{-1}, respectively (i.e., much higher than the D values obtained for major elements, below 0.06 L g^{-1}). The drastic differences in the relative concentrations of major and trace elements strongly affect the distribution ratios and then the significance of selectivity coefficients ($SC_{Sr/metal}$). However, a qualitative ranking can be evaluated:

Na(I) (72–86) > Mg(II) (48–66) > Ca(II) (34–43) > K(I) (26–38) >> B(III) (0.79–1.23) > U(VI) (0.40–0.68).

This ranking is roughly consistent with the concentrating effect (CF = q_{eq}/C_0, L g^{-1}): Na(I) (≈0.023) < Mg(II) (≈0.031) < Ca(II) (≈0.045) < K(I) (≈0.056) << Sr(II) (≈1.31) ≈ B(III) (≈1.32) < U(VI) (≈2.02).

The distribution ratio (i.e., D) increases with the pH (not shown). For Na(I), K(I), Mg(II) and Ca(II), D varies between 0.02 and 0.06 L g^{-1}, this is 2 orders of magnitude lower than the D ratio for Sr(II) (i.e., 1.53), B(III) (i.e., 1.94) and U(VI) (i.e., 3.79). This comparison shows the preference of SA*PEI for Sr(II) (and boron or uranium) against alkaline and alkaline-earth metals.

These preliminary results demonstrate that SA*PEI maintains good sorption performance, concentrating effect for Sr(II) in seawater, despite the complexity of the matrix. Noteworthy, the sorbent shows a remarkable concentrating effect for uranyl ions (despite the effect of metal speciation).

3. Materials and Methods

3.1. Materials

Brown algae (*Laminaria digitata*) was supplied by Setalg (Pleubian-France). Branched polyethylene imine (PEI; 50 %, *w/w* in water), cesium nitrate ($CsNO_3$; 99%), strontium nitrate ($Sr(NO_3)_2$; 99.99 %), sodium hydroxide (NaOH: ≥97.0%), calcium chloride ($CaCl_2$ ≥ 99.9%, in synthesis processes), 2-propene-1-sulfonic acid (≥99.99%), hydroxylamine-O-sulfonic acid (≥99.99%), potassium persulfate (99.99%), sodium metabisulfite (99.99%) and glutaraldehyde were purchased from Sigma-Aldrich (Taufkirchen, Germany). Poly(ethylene glycol diglycidyl ether) (PEGDGE) is used for enhancing the stability through crosslinking process. The salts used in selectivity tests (i.e., NaCl (≥99.98%), $MgCl_2 \cdot 6H_2O$ (99%), $FeCl_3$ (≥99.5%), $AlCl_3 \cdot 6H_2O$ (95%)), were obtained from Guangdong Guanghua, Sci-Tech Co., Ltd. (Guangdong, China). Calcium chloride ($CaCl_2$, 99.1%) was supplied from Shanghai Makclin Biochemical Co., Ltd. (Shanghai, China). All other reagents are Prolabo products, which were used as received.

3.2. Synthesis of Sorbents

The synthesis of A*PEI has already been described. The concept is based on the partial thermal extraction of alginate from algal biomass using Na_2CO_3. In a second step, PEI was added to the mixture, which was further distributed through a thin nozzle into an ionotropic gelation solution (containing both $CaCl_2$ and glutaraldehyde, for creating an interpenetrated network associating carboxylate gelation with calcium and crosslinking of amine groups with glutaraldehyde) (Scheme A1, see Appendix C, which reports the detailed experimental conditions).

The one-pot synthesis of SA*PEI (4 g) consisted of an original reaction involving the reaction of two reagents bearing sulfonic acid groups (i.e., 2-propene-1-sulfonic acid (1.0 g) and hydroxylamine-o-sulfonic acid (0.8 g)) with amine groups from PEI. The reaction took place under reflux (at 90 °C), to produce sulfonic acid derivative (5.64 g; yield: ≈97% on mass balance). Appendix C (and Scheme A2) provides detailed experimental conditions. Scheme 2 shows the expected structure of the sorbent.

3.3. Characterization of Sorbents

The pH of zero charge (pHpzc) was measured using the pH-drift method [84]. The FTIR spectra of samples (incorporated into a KBr disk, after dryness at 60 °C) were collected using an IRTracer-100 FT-IR spectrometer (Shimadzu, Tokyo, Japan). SEM analyses were acquired using a Phenom ProX -SEM; Thermo Fisher Scientific (Eindhoven, The Netherlands), while the chemical composition of the sorbents (raw beads, after functionalization and after metal-loading) was investigated using EDX analysis (energy dispersive X-ray analysis, coupled to the SEM system). The BET surface area and the porosity of the sorbents were determined using adsorption and desorption branches of N_2 isotherms through Micromeritics TriStar- II; Norcross, GA, USA, (system-77 K), while the BET equation was used for surface area determination and the BJH method for porosity characterization. The samples were firstly swept under N2 gas atmosphere for 4 h at 120 °C before testing. The thermal decomposition (thermogravimetric analysis; TGA) of the samples was determined using TG-DTA (Netzsch STA: 449-F3 Jupiter, NETZSCH-Gerätebau HGmbh, Selb, Germany); the analysis was performed under nitrogen atmosphere, with 10 °C min-1 temperature ramp. The pH of the solution was adjusted using a S220 Seven Compact pH/ionometer (Mettler-Toledo, Shanghai, China). Strontium (and other metal ions) concentration were measured using inductively coupled plasma atomic emission spectrometer- ICP-AES (ICPS-7510 Shimadzu, Tokyo, Japan) (after filtration using filter membranes, 1.2 µm pore size). Sodium concentration in the solution was analyzed by flame atomic absorption spectrophotometry (FAAS-AA 7000, Shimadzu, Tokyo, Japan).

3.4. Sorption and Desorption Procedures

Sorption and desorption tests were performed in batches. The suspensions containing a given solid/liquid ratio (sorbent dose, SD = m/L, with m mass of sorbent (g) and V volume of solution (L), at concentration C_0, mmol Sr L^{-1}) were agitated at 210 rpm velocity. Unless specified, the standard temperature was 21 ± 1 °C. For uptake kinetics, homogeneous samples were collected at fixed contact times and the residual concentration (C_{eq} or $C_{(t)}$, mmol Sr L^{-1}) was determined by ICP-AES. For sorption isotherms, the contact time was fixed to 48 h. The initial pH_0 of the solution was fixed using 0.1/1 M NaOH or HNO_3 solutions; the pH was not controlled during sorption, but the equilibrium pH_{eq} was monitored. The sorption capacity q_{eq} (mmol Sr g^{-1}) was deduced from the mass balance equation; $q_{eq} = (C_0 - C_{eq}) \times V/m$. The same method was used for investigating the sorption properties from multi-component equimolar solutions, and for the evaluation of sorption properties from seawater samples (two samples collected from the Mediterranean Sea and the Red Sea).

For the study of desorption properties, the same experimental procedure was used (batch tests); the Sr-loaded sorbents were collected from uptake kinetics. The desorption kinetics using 0.3 M HCl solution were collected. In addition, the recycling of the sorbent was also tested for five successive cycles; a rinsing step (using demineralized water) was systematically processed between each step. By mass balance, it was possible calculating the sorption and desorption efficiencies.

Detailed experimental conditions are directly reported in the caption of the Figures.

4. Conclusions

The successful grafting of two types of sulfonic groups onto A*PEI using simultaneously two sulfonating agents (2-propylene-1-sulfonic acid and hydroxylamine-O-sulfonic acid) triples the maximum sorption capacity ($q_{eq,exp}$) of SA*PEI for Sr(II) from 0.61 to 1.91 mmol Sr g^{-1}, at $pH_{eq} \approx$ 4.5–5. S content reaches about 1 mmol S g^{-1} (maintaining N content close to 2.14 mmol N g^{-1}). The optimum initial pH is strictly found at pH_0 4 for sulfonated sorbent (at pH_0 5 for A*PEI). Sorption isotherms for SA*PEI can be fitted by the Sips and the Langmuir dual site equations (confirmation of the contribution of both sulfonate groups and amine—or carboxyl—groups, consistently with FTIR characterization). The sorption onto SA*PEI is endothermic (shown by the substantial increase in sorption capacity and affinity coefficient with temperature increase from 20 to 50 °C). The sulfonation also improves mass transfer properties: Equilibrium is reached within 40 min (vs. 120 min for pristine beads), consistently with the enhancement of porous properties; the kinetic profiles are fitted by the pseudo-first-order rate equation. The effective diffusivity coefficient (D_e) is about one order of magnitude lower than the self-diffusivity of strontium in water (limited contribution of intraparticle diffusion in the kinetic).

Sorption in multi-component (equimolar) solutions show the marked preference of SA*PEI for Sr(II) (and Cs(I), to a lesser extent) against alkaline, alkaline-earth, and heavy metals, especially at pH_{eq} close to 4. Quantitative structure-activity relationship tools confirm that strontium and cesium keep out the trends followed by other competitor metals. For these competitor metals, the distribution ratio can be relatively well correlated with the covalent index (rather than the ionic index). Selectivity for Sr(II) increases with increasing the absolute value of hydration enthalpy and decreasing the ionic radius and the atomic number (in each group of mono-, di-, and tri-valent cations). In even more complex solutions (i.e., seawater), SA*PEI keeps goods affinity for Sr(II).

Strontium can be readily (and fastly) desorbed from loaded sorbent using 0.3 M HCl solution as the eluent. Total desorption allows re-using the sorbent for successive cycles. The sulfonation notably increased the stability in sorption performance for sulfonated sorbent (loss in sorption efficiency does not exceed 1.7%) at the fifth cycle, compared with A*PEI (loss close to 12%).

Author Contributions: Conceptualization, M.F.H., E.G., Y.W. and X.Y.; methodology, M.F.H., E.G., Y.W. and X.Y.; software, M.F.H., W.L. and K.A.; validation, M.F.H., W.L. and X.Y.; formal analysis, K.A. and T.V.; investigation, K.A., T.V. and M.F.H.; resources, Y.W. and X.Y.; data curation, E.G., X.Y. and K.A.; writing—original draft preparation, M.F.H. and E.G.; writing—review and editing, E.G.; visualization, K.A. and M.F.H.; supervision, Y.W., X.Y. and E.G.; project administration, Y.W.; funding acquisition, Y.W. and M.F.H. All authors have read and agreed to the published version of the manuscript.

Funding: The National Natural Science Foundation of China [U1967218, and 11975082] (Y.W.).

Institutional Review Board Statement: Not applicable.

Informed Consent Statement: Not applicable.

Data Availability Statement: Data can be obtained from the authors on demand.

Conflicts of Interest: The authors declare no conflict of interest.

Sample Availability: Not applicable.

Appendix A. Characterization of Materials

Appendix A.1. SEM and SEM-EDX Characterizations

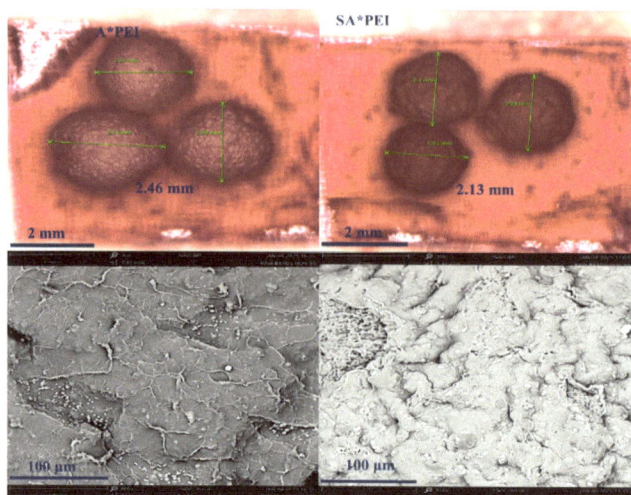

Figure A1. Optical microscope observations (**top**) and SEM observations (**bottom**) of A*PEI and SA*PEI sorbents.

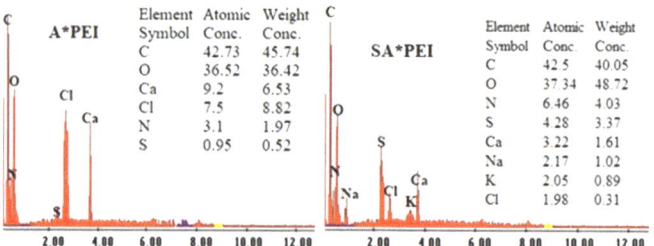

Figure A2. Semi-quantitative EDX analysis of the surface of A*PEI and SA*PEI sorbents.

Appendix A.2. Textural Properties

The isotherms of adsorption and desorption of N_2 are characterized by a Type II profile according to IUPAC classification [85]: The first section corresponds to a concave shape followed by a linear section and terminated with a convex section (Figure A3a). This type of isotherm is usually associated with non-porous or macroporous materials. However, substantial differences are observed between A*PEI and its sulfonated derivative (SA*PEI). First, the specific surface area (SSA) determined by BET analysis is substantially increased from 2.1 to 7.4 m² g⁻¹. In addition, the hysteresis loop (sorption vs. desorption branches) is much less marked for raw sorbent than its derivative that exhibits Type H3 hysteresis: A*PEI profile correlates with the so-called Type II(a) isotherm, while SA*PEI corresponds more to Type II(b) isotherm. H3 hysteresis loops are usually associated with aggregates of platy particles [86].

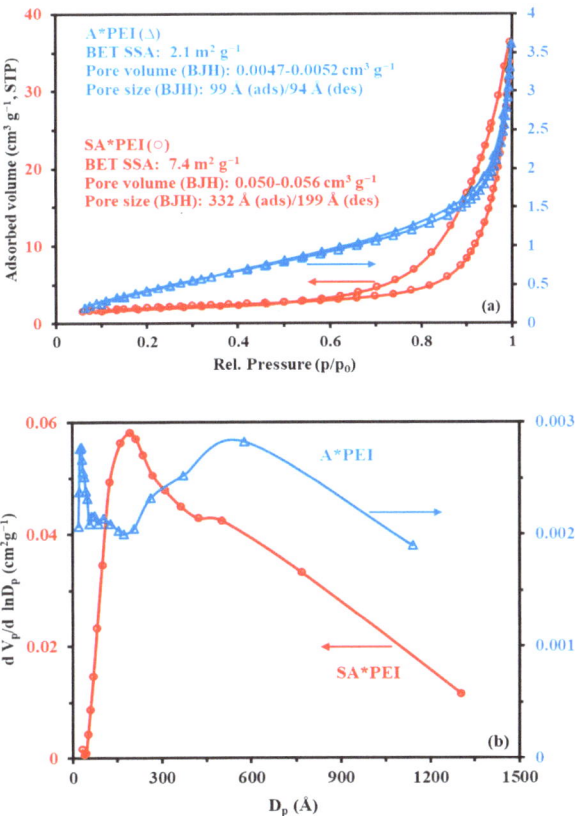

Figure A3. Textural properties of A*PEI and SA*APEI beads: (a) N_2 adsorption/desorption isotherms, (b) pore size distribution.

In the current case, the sorbent is based on a hydrogel structure resulting from the interaction of alginate fraction from algal biomass and polyethylenimine; the sulfonation of amine groups involves an expansion of the polymer structure with increased flexibility (which may explain the larger hysteresis loop). The porous volume increases tenfold after sulfonation; this is consistent with the larger SSA. Figure A3b shows the distribution of pore width for the two sorbents. Not considering the broad section of the curves that correspond to the macroporous regions (above 500 Å), A*PEI shows a principal pore size fraction (which maximum is close to 34 Å); the pore width is calculated (BJH method) close to 99–94 Å (restricted variation between measurements from adsorption and desorption branches). In the case of sulfonated sorbent, a broader maximum is observed around 214 Å; the difference in the values obtained from adsorption and desorption branches (332–199 Å) is substantially larger than for A*PEI. From these observations, it is possible anticipating that SA*PEI may have more favorable mass transfer characteristics than A*PEI.

Appendix A.3. Thermal Degradation Properties

The investigation of the thermal degradation of sorbents is reported in TGA and DrTG profiles (Figure 1); Figure A4 shows the inset and outset temperatures with detailed weight losses (WLs) for TGA curve. In the TGA curves, three transitions can be identified for A*PEI against four transitions in SA*PEI.

The first transition corresponds to water release from sorbent at temperature below 177–211 °C, representing 20.2% and 10.8% WLs for A*PEI and SA*PEI, respectively.

In the second stage of the degradation, the WLs represent 46.9% (177–404 °C) and 31.0% (211–389 °C), respectively. This step may be associated with the depolymerization of PEI and carbohydrate ring and the degradation of carboxyl groups.

The last transition for A*PEI (404–890 °C) leads to a complementary WL of 26%. This phase corresponds to the degradation of PEI (into CO, CO_2, and CH_4) [87], and the final degradation of the char (produced in the preceding step).

This degradation profile is roughly consistent with the profile reported by Godiya et al. [88]; however, the curve is shifted toward higher temperatures of degradation.

Contrary to A*PEI, the end of the degradation profile of SA*PEI shows two successive waves: (a) 389–604 °C (WL: 27%), and (b) 604–890 °C (WL: 14%). The first transition corresponds to sulfonate degradation (reported above 320 for sulfonate polystyrene/silica composite, [89]). The last section corresponds to the degradation of the char. The sulfonate derivative increases the thermal resistance of the composite. The residue represents 17.2% while for the precursor the residue was less than 7%.

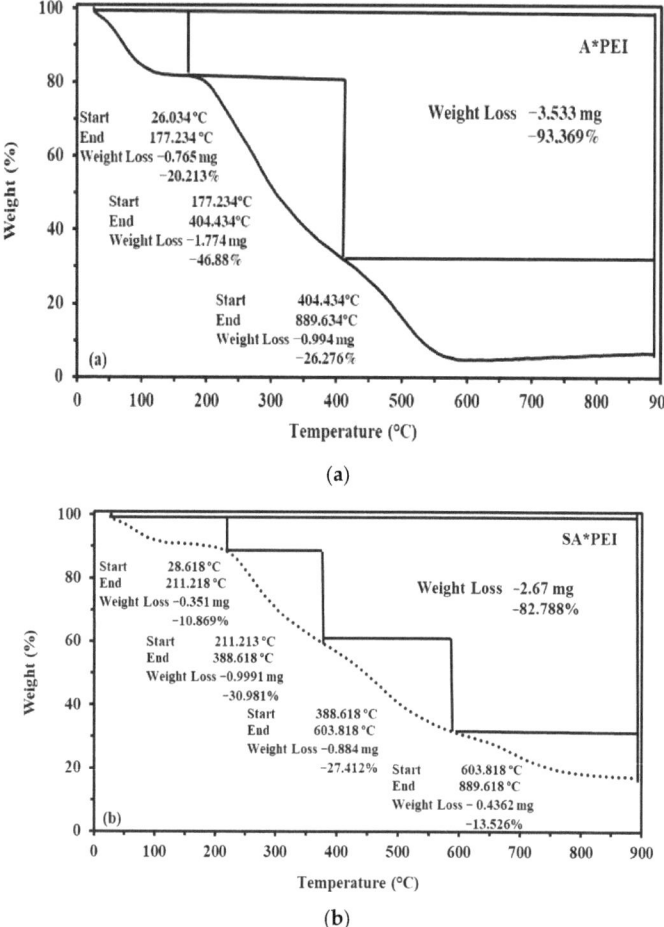

Figure A4. TGA analysis of A*PEI (a) and SA*PEI beads (b).

DrTG curves confirm these trends (Figure 1b) and the global higher stability of the sulfonated derivative. The main observations can be summarized: Hardly marked inflection at the water release phase, shift from 293.2 to 308.4 °C for the second step in the degradation, the shift of the peak at 498.4 °C (in A*PEI) toward lower temperature with functionalization to 468.8 °C, accompanied by a shoulder at 537 °C, and shift of sharp weakly intense peak at 580.85 °C, which is shifted toward higher temperature (i.e., 681.1 °C).

Appendix A.4. FTIR Spectroscopy

The sorbents are constituted of algal biomass (including alginate, meaning carboxylic acid), polyethyleneimine (PEI, meaning primary, secondary, and tertiary amines), and sulfonic acid (after functionalization) (Figure 2). The proper complexity of algal biomass that contains many constituents [90,91], such as proteins, minerals, and carbohydrates (alginate, mannitol, laminarin, cellulose, or polyphenols) makes the interpretation of FTIR spectra very complex.

Rodriguez et al. [91] identified alginate in brown algae through: Guluronic acid residues with bands at 1290, 1080, 1025, and 787 cm^{-1}, mannuronic acid residues at 1320, 1030, 1019, 878, and 808 cm^{-1}, and the ν(C-O) of uronic acids at 948 cm^{-1}. The sulfate esters of fucoidan (representing a weak fraction of carbohydrates) may be identified by bands at 1260–1195 cm^{-1} (ν(S=O)). In the case of the interactions of PEI with Mesquite Gum (MG, which also bears carboxylic groups), Pinilla-Torres et al. [92] reported a series of bands representing the typical bands associated with PEI and carboxylic functions from the MG:

- 3420 cm^{-1}: ν(N-H) primary amine
- 3280 cm^{-1}: ν(N-H), primary amine and Amide A band
- 1648 cm^{-1}: ν(C=O)/ν(C-N), Amide I band
- 1551 cm^{-1}: ν(N-H)$_{primary}$ in-plane δ(N-H)/ν(C-N)/ν(C-C), Amide II band
- 1241 cm^{-1}: ν(C-N)/δ(N-H), Amide III band
- 1074 cm^{-1}: ν(C-N)

Table A1 summarizes the main peaks appearing on the spectra of the sorbents before and after Sr(II) sorption, and after the fifth cycle of desorption (to evaluate the potential degradation of the sorbents). The sorbents show a broad band at 3430–3420 cm^{-1}, which is assigned to the overlapping of ν(N-H) and ν(O-H), while the bands at 2950–2930 cm^{-1} correspond to ν(C-H). These bands are not significantly influenced by A*PEI functionalization nor interactions with Sr(II) (metal sorption and desorption).

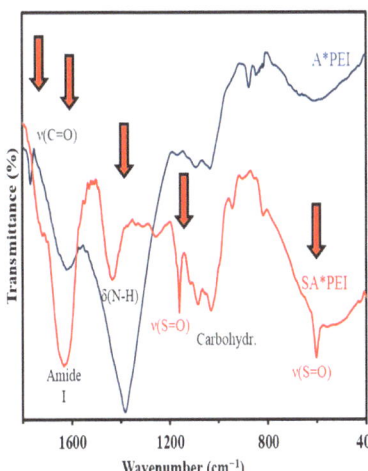

Figure A5. Main modifications of FTIR spectrum brought by the sulfonation of A*PEI.

The main effects of sulfonation on the FTIR spectrum of A*PEI are identified in Figure A5 in the region of amine/amide bands, and carboxylic groups: The bands are shifted and/or width-changed. In addition, typical bands of sulfonic groups are identified at 1159 cm^{-1}, for example.

Table A1. Assignment of main FTIR bands for A*PEI and SA*PEI before and after Sr(II) sorption, and after the fifth desorption (sorbent recycling).

Vibration	A*PEI	A*PEI + Sr(II)	A*PEI after 5th Desorption	SA*PEI	SA*PEI + Sr(II)	SA*PEI after 5th Desorption	Ref.
ν(N-H) + ν(O-H)	3429	3449–3428	3447–3429	3418	3443–3429	3422	[93]
ν(C-H) (aliphatic)	2934	2939	2934	2947	2932	2953	[93]
ν(C=O)	1767	1721	1765				[93]
δ(N-H)$_{prim.}$	1622	1634	1620	1632	1618	1622	[93]
O-H, δ(N-H)$_{2nd}$, δ(C-H)	1383	1429	1422	1435	1429	1435	[93]
ν(S=O)						1339–1315	
ν(C-O)				1256		1256	
ν$_s$(C-N), ν$_{as}$(S=O)				1159	1171	1165	[94,95]
ν(C-O-C) carbohydr.	1092	1083	1086	1084	1086	1084	[96]
δ(O-H), ν$_s$(S=O), ν$_{sk}$(C-O), ν(C-N)	1034	1031	1031	1030	1032	1030	[95–98]
δ(O-H)		943	945	941		941	[93]
ν(S=O) and δ(C-H)	847, 876	818	816, 852, 876	818, 852	816, 851, 878	818, 854	[98,99]
Sr-N, Sr-O bonds, ν(S=O)				602	573	555	[99,100]

In Figure A6, the mains changes brought by Sr(II) sorption are identified. For A*PEI (Figure A6a), the most significant changes concern:

- the region 1750–1700 cm^{-1}, assigned to ν(C=O) for carboxylic groups: Shift toward lower wavenumber and formation of a triplet of bands,
- the region 1660–1580 cm^{-1}, assigned to amide bands (overlapped with amine groups): Stronger signal with reduced width,
- the intense and broad band, resulting from the overlapping of different vibrations (O-H, δ(N-H)$_{2nd}$, δ(C-H)): Shift toward higher wavenumber and width reduction, and
- the region 950–800 cm^{-1}, assigned to δ(O-H), δ(C-H) signals: Variations in the intensity of the relevant signal.

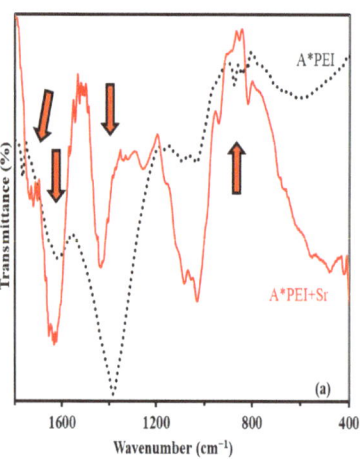
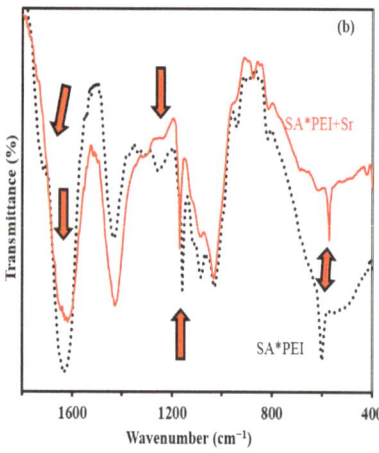

Figure A6. Main modifications of FTIR spectra of A*PEI (**a**) and SA*PEI (**b**) brought by Sr(II) sorption.

Different changes in FTIR spectra can be observed for SA*PEI (Figure A6b). The main differences are observed:

(a) the region at ≈1702 cm^{-1} (weak shoulder), associated with residual ν(C=O) of carboxylic groups, that disappears (or is shifted toward lower wavenumber around 1652 cm^{-1}, where it contributed to the widening of the band at 1632 cm^{-1},
(b) the region at ≈1632 cm^{-1}, attributed to δ(N-H) (associated with amide I band): Widening,
(c) the region at ≈1256 cm^{-1}, assigned to ν(C-O): Intensity reduction,
(d) the band at 1159 cm^{-1}, corresponding mainly to ν_{as}(S=O): Shift toward higher wavenumber,
(e) the band at 602 cm^{-1}, assigned to ν(S=O): Shifted toward lower wavenumber (and/or replace with a signal associated with Sr-N and Sr-O bond).

After regeneration (at the end of the fifth sorption/desorption cycle), the FTIR spectra maintain significant changes (Figure A7). Hence, for A*PEI, the most representative differences concern:

(a) the band at 1620 cm^{-1}: Width reduction,
(b) the band at 1383 cm^{-1}: Width reduction and shift toward higher wavenumber, and
(c) the region at 950–800 cm^{-1}: (weak) intensity reductions and shifts of local peaks.

In the case of SA*PEI, the sorbent appears relatively well restored and the FTIR spectra are close between pristine sorbent and regenerated material:

(a) the shoulder at 1702 cm^{-1}: Significantly reduced,
(b) the band at 1159 cm^{-1}: Shift toward higher wavenumber (though less than after Sr(II) sorption, and
(c) the band at 602 cm^{-1} is again shifted toward the lower wavenumber (more extensive than after Sr(II) sorption).

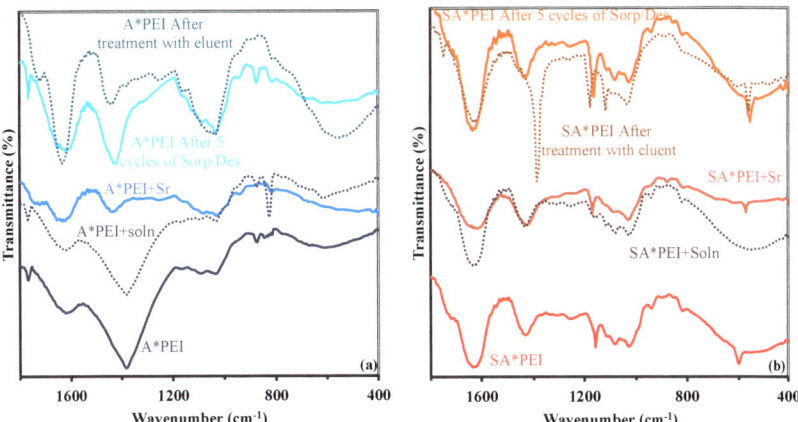

Figure A7. Effect of "environmental parameters" on the FTIR spectra of A*PEI (**a**) and SA*PEI (**b**) (sorbent after contact with solution at the pH of sorption, and sorbent after contact with the eluent).

The comparison of the spectra suggests that the sorption of Sr(II) onto A*PEI involves amine and carboxyl groups; in the case of SA*PEI, the modification of the chemical environment of sulfonic appearing in the FTIR spectrum means that sulfonate groups are the main binding functional groups (with the contribution of amine and carboxylic groups). The regeneration of the sorbents (desorption with 0.3 M HCl solution, after the fifth desorption step) does not completely restore the material in their neat form: The functional groups (amine/amide, carboxyl, sulfonate) are partially regenerated.

This final analysis depends on the FTIR spectra of the materials conditioned at the pH used for Sr(II) sorption (Figure A7) and analyzed after contact with 0.3 M HCl solution (with strontium) to isolate the relevant effects of sorbent conditioning and proper metal-sorbent interactions (and desorption). Globally, the contact of the sorbents with the metal-free solutions at the pH selected for strontium sorption does not show significant differences with pristine sorbents. Therefore, the changes in the spectra observed after Sr(II) binding can be specifically attributed to the interactions of the metal with identified functional groups.

The contact of A*PEI with 0.3 M HCl solution weakly affects the FTIR spectrum: The most significant changes concern the bands at 1383 cm^{-1} (δ(N-H)), 1632 cm^{-1} (δ(N-H)), and 1713 cm^{-1} (shifted from 1767 cm^{-1}, ν(C=O)); the modification of these bands was also observed after Sr(II) sorption. These bands are affected by the protonation of reactive groups, and/or their interaction with Sr(II). It is noteworthy that the regeneration of the sorbent restores the band at 1765 cm^{-1} contrary to the 0.3 M HCl-conditioned sorbent (Figure A7). In the case of SA*PEI, some specific bands appear (or are shifted) after being in contact with 0.3 M HCl solution: The band at 1748 cm^{-1}, at 1383 cm^{-1} (strong shift of the band at 1435 cm^{-1}), and 1128 cm^{-1}. The recycled sorbent (after 5 cycles) shows a spectrum globally closer to the pristine sorbent than the protonated sorbent collected after contact with 0.3 M HCl solution (Figure A8). In addition, the FTIR spectrum of SA*PEI is less affected by sorption and desorption cycles than A*PEI; consistently with higher stability of sorption performances (Table 3).

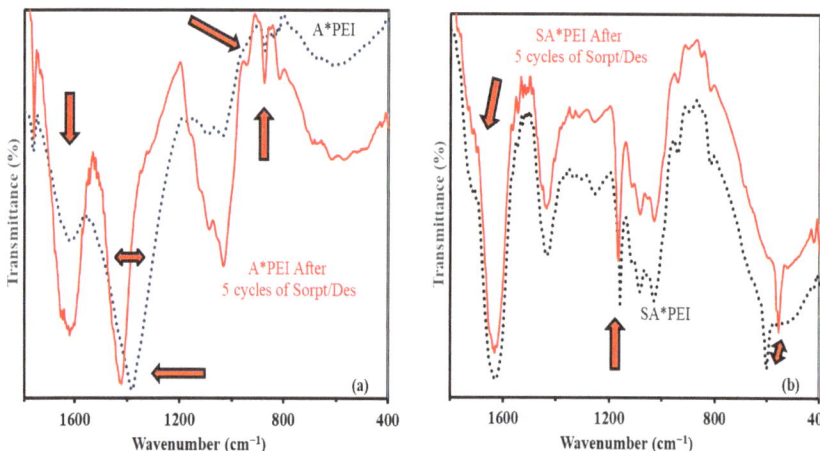

Figure A8. Main modifications of FTIR spectra of A*PEI (**a**) and SA*PEI (**b**) after the 5th cycle of sorption/desorption.

Appendix A.5. Elemental Analysis and pH$_{PZC}$

The elemental analysis confirms the successful sulfonation of raw material (1.04 mmol S g^{-1}; consistent with the increase in O content); N content is not affected by chemical modification ($\approx$2.14 mmol N g^{-1}). The grafting of acid reactive groups (sulfonic acid) induces a strong decrease of the pH$_{PZC}$ from 7.73–7.84 to 4.53–4.64. This is consistent with the well-known character of sulfonic-based resins as strong cation exchangers. The concentration of background salt (0.1 M vs. 1 M NaCl) hardly changes pH$_{PZC}$ values (Figure A9).

Table A2. Elemental analysis of A*PEI and SA*APEI.

Sorbent	C (%)	H (%)	O (%)	O (mmol g^{-1})	N (%)	N (mmol g^{-1})	S (%)	S (mmol g^{-1})
A*PEI	35.97	11.95	36.13	22.58	3.02	2.156	0.19	0.0593
SA*PEI	34.84	13.84	43.61	27.26	2.97	2.121	3.32	1.036

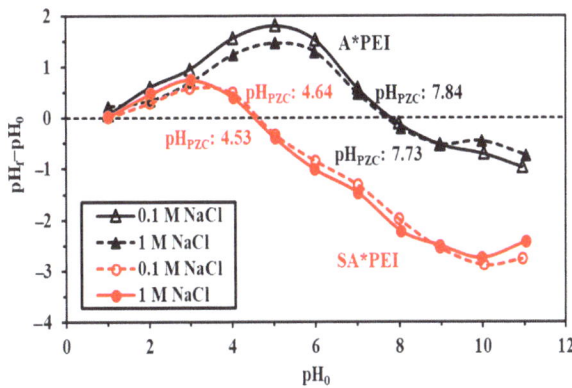

Figure A9. pH$_{PZC}$ determination using pH-drift method for A*PEI and SA*APEI (sorbent dose, SD: 2 g L^{-1}; background salt: NaCl (at concentrations 0.1 M and 1 M); time: 48 h).

Appendix B. Sorption Properties

Appendix B.1. Modeling of Uptake Kinetics and Sorption Isotherms

Table A3. (**a**). Reminder on equations used for modeling uptake kinetics; (**b**). Reminder on equations used for modeling sorption isotherms.

(a)			
Model	Equation	Parameters	Ref.
PFORE	$q(t) = q_{eq,1}(1 - e^{-k_1 t})$	$q_{eq,1}$ (mmol g^{-1}): Sorption capacity at equilibrium k_1 (min^{-1}): Apparent rate constant of PFORE	[101]
PSORE	$q(t) = \dfrac{q_{eq,2}^2 k_2 t}{1 + k_2 q_{eq,2} t}$	$q_{eq,2}$ (mmol g^{-1}): Sorption capacity at equilibrium k_2 (g mmol^{-1} min^{-1}): Apparent rate constant of PSORE	[101]
RIDE	$\dfrac{q(t)}{q_{eq}} = 1 - \sum\limits_{n=1}^{\infty} \dfrac{6\alpha(\alpha+1) \exp\left(\frac{-D_e q_n^2}{r^2} t\right)}{9 + 9\alpha + q_n^2 \alpha^2}$ With q_n being the non-zero roots of $\tan q_n = \dfrac{3 q_n}{3 + \alpha q_n^2}$ and $\dfrac{m}{V}\dfrac{q}{C_0} = \dfrac{1}{1+\alpha}$	D_e (m^2 min^{-1}): Effective diffusivity coefficient	[102]

(b)			
Model	Equation	Parameters	Ref.
Langmuir	$q_{eq} = \dfrac{q_{m,L} C_{eq}}{1 + b_L C_{eq}}$	$q_{m,L}$ (mmol g^{-1}): Sorption capacity at saturation of monolayer b_L (L mmol^{-1}): Affinity coefficient	[103]
Freundlich	$q_{eq} = k_F C_{eq}^{1/n_F}$	k_F (mmol g^{-1})/(mmol L^{-1})^{n_F} and n_F: Empirical parameters of Freundlich equation	[104]
Sips	$q_{eq} = \dfrac{q_{m,S} b_S C_{eq}^{1/n_S}}{1 + b_S C_{eq}^{1/n_S}}$	$q_{m,S}$ (mmol g^{-1}), b_S (mmol L^{-1})^{n_S}, and n_S: Empirical parameters of Sips equation (based on Langmuir and Freundlich equations)	[105]
Temkin	$q_{eq} = \dfrac{RT}{b_T} \ln(A_T C_{eq})$	A_T (L mmol^{-1}): equilibrium binding capacity; b_T: Temkin constant related to sorption heat (J kg^{-1} mol^{-2})	[71,106]

(m (g): mass of sorbent; V (L): Volume of solution; C_0 (mmol L^{-1}): initial concentration of the solution).

Akaike Information Criterion, AIC [107]:

$$\text{AIC} = N \ln\left(\frac{\sum_{i=0}^{N}\left(y_{i,\text{exp.}} - y_{i,\text{model}}\right)^2}{N}\right) + 2N_p + \frac{2N_p(N_p+1)}{N-N_p-1}$$

where N is the number of experimental points, N_p the number of model parameters, $y_{i,\text{exp.}}$ and $y_{i,\text{model}}$ the experimental and calculated values of the tested variable.

Appendix B.2. Effect of pH on Sr(II) Sorption

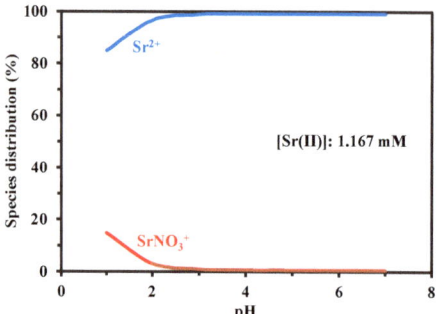

Figure A10. pH effect on Sr(II) speciation (under experimental conditions from pH study; calculated using Visual Minteq, [108]).

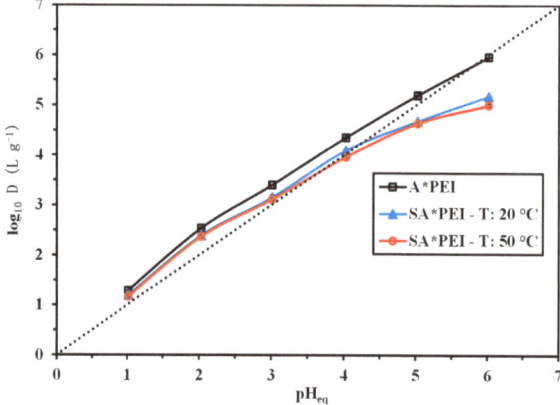

Figure A11. pH effect on Sr(II) sorption using A*PEI and SA*PEI beads—pH$_{eq}$ vs. pH$_0$ (C$_0$: 1.17 mmol Sr L^{-1}; Sorbent dosage, SD: 0.666 g L^{-1}; T: 20 ± 1 °C (except for SA*PEI test at T: 50 ± 1 °C); time: 48 h).

The distribution ratio is defined, at equilibrium, as the ratio between sorption capacity and residual metal concentration in the solution (D = q_{eq}/C_{eq}, L g^{-1}). Figure A12 shows the log$_{10}$ plots of D vs. equilibrium pH. The slope of this plot may be used for evaluating the stoichiometry of proton exchange in ion-exchange processes (i.e., +2 for the exchange of divalent cation with protons, +1 for monovalent cation; in the case of strong acid resin, or +1 for weak acid resin) [109]. From Figure 3, it is possible anticipating substantial differences between A*PEI and SA*PEI. For the reference material, the distribution coefficient continuously increases, and the slope is close to 0.163. On the opposite hand, for SA*PEI, whatever the temperature, two linear segments can be identified (with a breakpoint close to pH 4): in acidic solutions, the slopes range between 0.39 and 0.43, while near neutral pH, the slopes evolve between −0.36 and −0.40. Despite the good quality of the fits (R^2 values), the slopes are not consistent with the divalent charge of strontium. The contribution of several types of reactive groups and/or alternative binding mechanisms could explain the impossibility of appropriately fitting the slopes with the stoichiometric exchange ratio. Notably, the absolute values of the slopes of the ascending and descending segments are very close (+0.431/−0.401 at 50 °C, +0.390/−0.360 at 20 °C); meaning that the numbers of protons exchanged (bound/released) per strontium are the same on both side of the extrema.

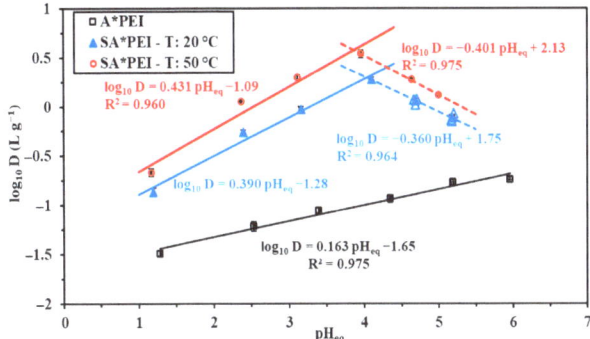

Figure A12. pH effect on Sr(II) sorption using A*PEI and SA*PEI beads—distribution ratio vs. pH_{eq} (C_0: 1.17 mmol Sr L^{-1}; SD: 0.666 g L^{-1}; T: 20 ± 1 °C (except for SA*PEI test at T: 50 ± 1 °C); time: 48 h).

Appendix B.3. Uptake Kinetics

The calculated sorption capacities at equilibrium slightly overestimate experimental values:

For A*PEI: 0.198 ± 0.003 mmol g^{-1} vs. 0.179 ± 0.010 mmol g^{-1}.
For SA*PEI: 0.974 ± 0.009 mmol g^{-1} vs. 0.926 ± 0.002 mmol g^{-1}.
The increase in temperature slightly increases the apparent rate coefficient:
A*PEI at 20 °C (5.02 ± 0.23 × 10^{-2} min^{-1}) < SA*PEI at 50 °C (5.80 ± 0.07 × 10^{-2} min^{-1}).
The fitted sorption capacities also slightly overestimate experimental values when varying temperature:
For SA*PEI at 20 °C: 0.974 ± 0.009 mmol g^{-1} vs. 0.926 ± 0.002 mmol g^{-1}.
For SA*PEI at 20 °C: 1.227 ± 0.034 mmol g^{-1} vs. 1.170 ± 0.029 mmol g^{-1}.

Table A4. Sr(II) uptake kinetics using A*PEI and SA*APEI—Parameters of PFORE, PSORE, and RIDE models (individual replicates).

Sorbent		A*PEI			SA*PEI					
Temperature		T: 20 ± 1 °C			T: 20 ± 1 °C			T: 50 ± 1 °C		
Model	Parameter	#1	#2	#3	#1	#2	#3	#1	#2	#3
Experim.	q_{eq}	0.181	0.177	0.180	0.939	0.921	0.917	1.18	1.20	1.13
PFORE	$q_{eq,1}$	0.194	0.200	0.199	0.986	0.968	0.967	1.24	1.26	1.18
	$k_1 \times 10^2$	2.20	1.89	2.06	5.14	5.21	4.70	5.72	5.88	5.80
	R^2	0.976	0.949	0.971	0.975	0.974	0.973	0.977	0.976	0.977
	AIC	−151	−137	−149	−97	−97	−96	−92	−92	−92

Table A4. Cont.

Sorbent		A*PEI			SA*PEI					
Temperature		T: 20 ± 1 °C			T: 20 ± 1 °C			T: 50 ± 1 °C		
Model	Parameter	#1	#2	#3	#1	#2	#3	#1	#2	#3
PSORE	$q_{eq,2}$	0.257	0.277	0.268	1.16	1.14	1.15	1.44	1.46	1.38
	$k_2 \times 10^2$	7.43	5.40	6.45	5.03	5.22	4.56	4.72	4.83	5.01
	R^2	0.962	0.929	0.955	0.936	0.935	0.937	0.937	0.937	0.937
	AIC	−146	−134	−145	−86	−86	−86	−80	−80	−80
RIDE	$D_e \times 10^9$	7.06	6.54	6.93	8.30	8.54	7.48	7.22	7.14	7.26
	R^2	0.956	0.920	0.947	0.939	0.939	0.938	0.936	0.935	0.936
	AIC	−140	−129	−138	−81	−82	−81	−75	−74	−75

Appendix B.4. Sorption Isotherms

Table A5. (a). Sr(II) sorption isotherms using A*PEI (individual replicates); (b). Sr(II) sorption isotherms using SA*PEI, at T: 20 °C (individual replicates); (c). Sr(II) sorption isotherms using SA*PEI, at T: 50 °C (individual replicates).

(a)				
		Series #		
Model	Parameter	1	2	3
Experim.	$q_{m,exp.}$	0.584	0.570	0.607
Langmuir	$q_{eq,L}$	1.140	0.864	0.988
	b_L	0.195	0.366	0.286
	R^2	0.987	0.993	0.983
	AIC	−75	−83	−74
Freundlich	k_F	0.199	0.233	0.230
	n_F	1.54	1.77	1.70
	R^2	0.991	0.993	0.991
	AIC	−82	−85	−81
Temkin	A_T	8.32	9.51	10.4
	b_T	18388	18348	18550
	R^2	0.893	0.943	0.908
	AIC	−54	−62	−56

(b)				
		Series #		
Model	Parameter	1	2	3
Experim.	$q_{m,exp.}$	1.86	1.84	1.91
Langmuir	$q_{eq,L}$	2.02	2.02	2.13
	b_L	1.93	2.15	1.75
	R^2	0.993	0.995	0.994
	AIC	−55	−58	−57
Freundlich	k_F	2.90	1.20	1.19
	n_F	1.67	2.92	2.77
	R^2	0.908	0.971	0.975
	AIC	28	−42	−43
Sips	$q_{eq,S}$	2.34	2.24	2.41
	b_S	1.21	1.47	1.21
	n_S	1.34	1.26	1.27
	R^2	0.995	0.996	0.995
	AIC	−58	−61	−58
Temkin	A_T	47.1	49.9	42.3
	b_T	7187	7114	6785
	R^2	0.985	0.988	0.982
	AIC	−49	−51	−47

Table A5. Cont.

		(c)		
		Series #		
Model	Parameter	1	2	3
Experim.	$q_{m,exp.}$	2.40	2.28	2.34
Langmuir	$q_{eq,L}$	2.38	2.27	2.33
	b_L	4.90	4.68	4.70
	R^2	0.968	0.969	0.988
	AIC	−33	−34	−46
Freundlich	k_F	1.70	1.63	1.66
	n_F	3.66	3.64	3.55
	R^2	0.963	0.968	0.960
	AIC	−33	−36	−33
Sips	$q_{eq,S}$	2.97	2.96	2.61
	b_S	1.69	1.48	2.51
	n_S	1.69	1.80	1.37
	R^2	0.985	0.986	0.995
	AIC	−40	−42	−52
Temkin	A_T	165.7	205.3	121.8
	b_T	7404	8081	7144
	R^2	0.987	0.985	0.995
	AIC	−45	−45	−56

Units: q, mmol g^{-1}; b, L mmol^{-1}; n, dimensionless; k_F, mmol$^{1-1/nF}$ L$^{-1/nF}$ g^{-1}; A_T, L mmol^{-1}; b_T, J mmol^{-1}.

Table A6. Comparison of Sr(II) sorption properties with alternative sorbents (for T ≈ 20 °C; Nat.: natural pH or non-documented).

Sorbent	pH	Time (min)	$q_{m,exp}$ (mmol g^{-1})	$q_{m,L}$ (mmol g^{-1})	b_L (L mmol^{-1})	Ref.
Dowex 50W8 sulfonic resin	3.7	60	1.43	1.43	206	[36]
Alginate microsphere	6	1440	1.20	1.27	8.24	[58]
Resorcinol-formaldehyde resin	7	1440	-	1.14	-	[34]
Sulfonated polyaniline sorbent	Nat.	40	1.01	1.05	4.73	[110]
Amidoximated algal/PEI beads	6	90	2.16	2.36	2.01	[38]
Crab carapace	Nat.	240	0.038	0.045	11.4	[111]
SrTreat®	Nat.	60	0.104	0.109	265	[111]
Kurion-TS™	Nat.	60	0.128	0.230	492	[111]
Mixed-bed resin (T-46/A-33)	7	30	-	0.109	0.084	[35]
Functionalized silica beads	8	60	1.38	1.57	1.41	[74]
Magnetic composite sulfonated sorbent	10	180	0.539	-	-	[30]
Salvadora persica biomass	7	60	-	0.474	0.237	[27]
Fly ash-based zeolite	5.4	720	0.681	0.749	0.756	[112]
Photinia serrulata leaf	>4	30	0.120	0.138	11.0	[113]
S. cerevisiae-Fe$_3$O$_4$ composite	6	960	-	0.234	1.33	[29]
Bacillus pumilus SWU7–1	7	7200	3.14	3.42	2.89	[28]
Modified montmorillonite	7	30	-	0.028	40.2	[114]
Granular manganese oxide	-	4800	1.2	1.9	0.1	[73]
Zr-metal-organic framework	Nat.	5	-	0.871	2.51	[115]
Functionalized graphene oxide	2	60	1.37	1.44	3.03	[75]
SLS/polyacrylonitrile	11.5	800	0.4	0.376	10.8	[64]
PVA/graphene oxide aerogel	7	480	0.228	0.229	239	[116]
Graphene oxide	5	20	0.97	1.50	0.850	[76]
ZrSn(IV) phosphate nanocomp.	8	120	-	0.202	3.94	[117]
A*PEI	5	120	0.607	0.977	0.278	This study
SA*PEI	5	40	1.91	2.05	1.94	This study

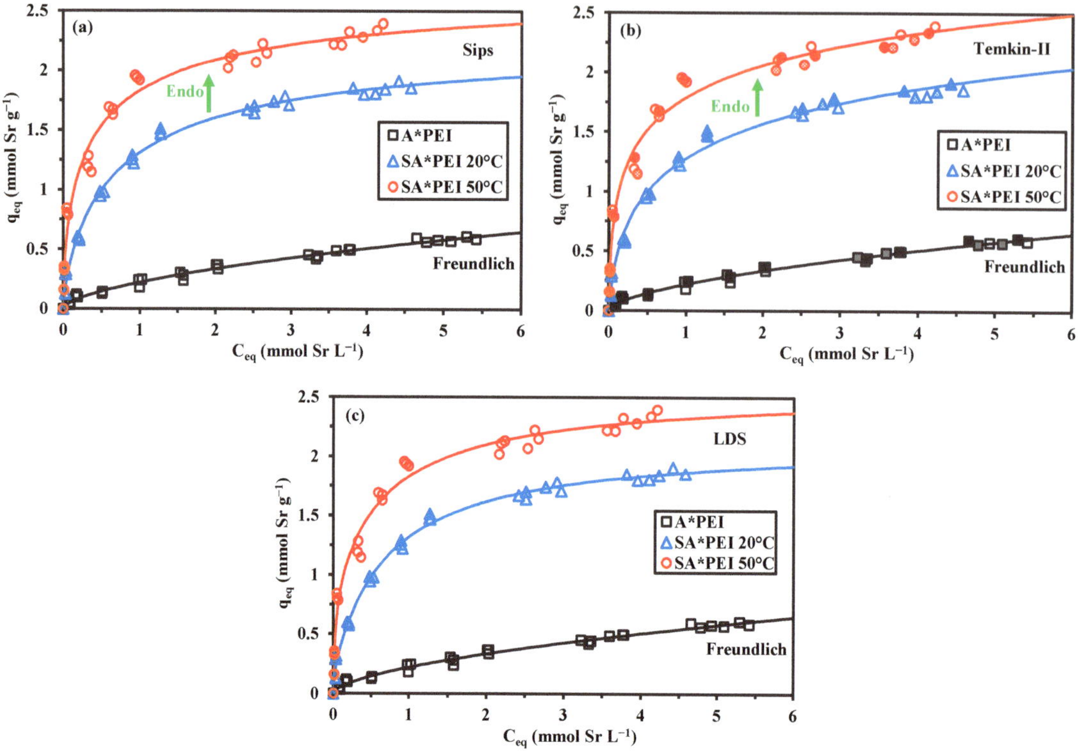

Figure A13. Sr(II) sorption isotherms on A*PEI and SA*PEI beads—Modeling with Freundlich equation for A*PEI and Sips equation (**a**), Temkin-II equation (**b**), and Langmuir dual site (LDS) (**c**) for SA*PEI (C_0: 0.12–5.82 mmol Sr L^{-1}; SD: 0.666 g L^{-1}; T: 20 ± 1 °C (for A*PEI and SA*APEI) and at T: 50 ± 1 °C (for SA*PEI, c); pH_0: 5; triplicated series; fitting on cumulated series).

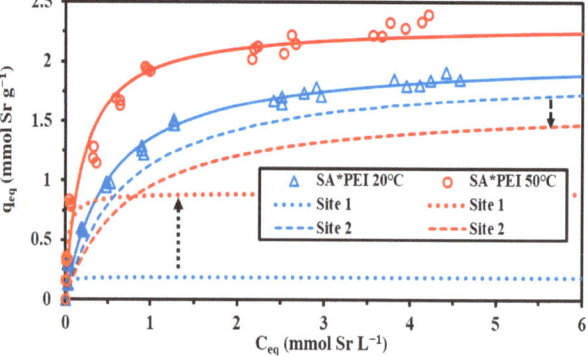

Figure A14. Contributions of Sites 1 and 2 in the Sr(II) sorption isotherms for SA*PEI at T: 20 and 50 °C.

Appendix B.5. Sorption Selectivity

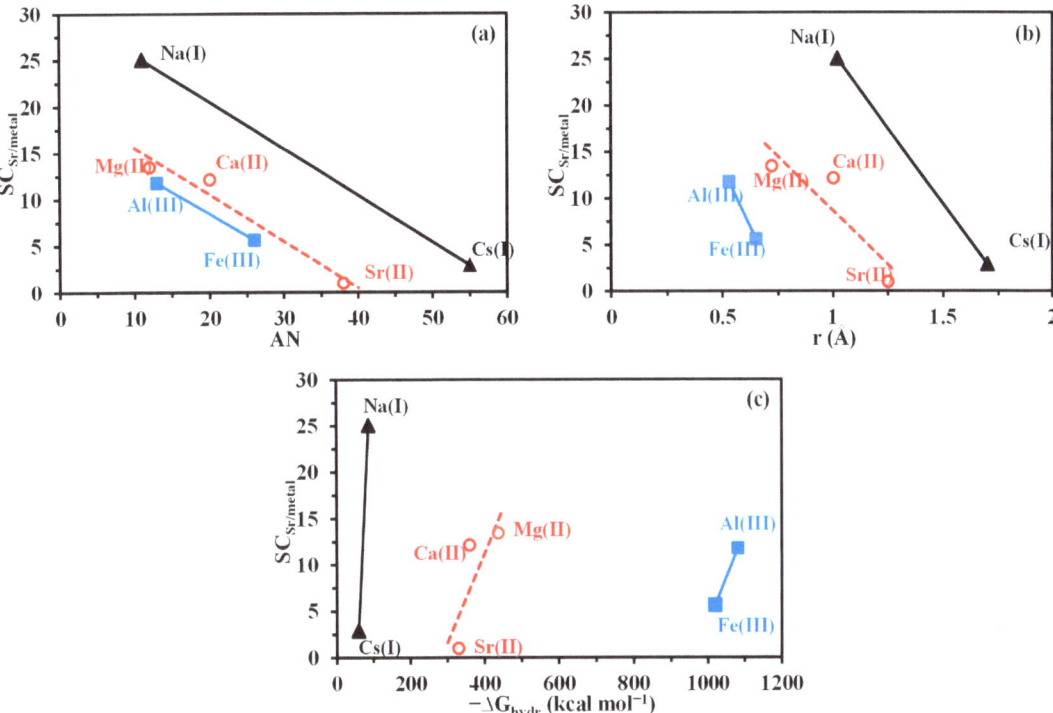

Figure A15. Correlation of $SC_{Sr/metal}$ with Atomic Number (AN) (**a**), ionic radius (**b**) and enthalpy of hydration (**c**) (grouped by ionic charge: +1, +2, and +3).

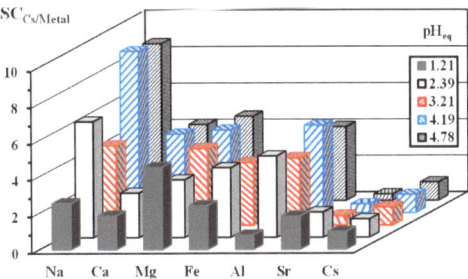

Figure A16. Effect of pH on the selectivity coefficient $SC_{Cs/Metal}$ for multi-component equimolar solutions (C_0: 1 mmol L^{-1}; SD: 1.6 g L^{-1}; T: 20 ± 1 °C; time: 24 h).

The distribution ratios (i.e., D, L g^{-1}, not shown) increase with pH for Na(I), contrary to Sr(II) (which showed a maximum for pH 4), and decrease with increasing Na,Cl concentration (from 0.0093 to 0.0029 L g^{-1}). At pH 4, the D ratio for Sr(II) is 215 to 465 times higher than the value for Na(I). This is consistent with competitor studies (Sections 2.2.5 and 2.2.6) and with the remarkable preference of SA*PEI for Sr(II) against Na(I); this is a good indication of the interest of the sorbent for application in seawater (see Section 2.3).

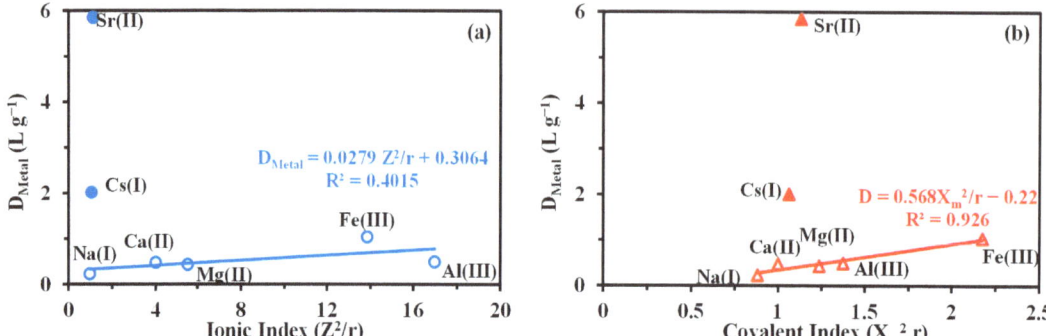

Figure A17. Distribution ratios vs. Ionic Index (Z^2/r) (**a**) and covalent index ($X_m^2 r$) (**b**).

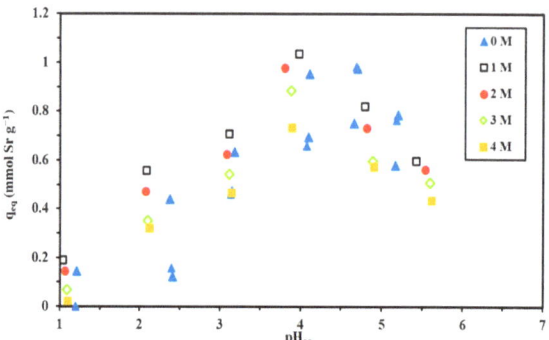

Figure A18. Influence of NaCl concentration on Sr(II) sorption (C_0: 1.17 mmol Sr L^{-1}; SD: 1 g L^{-1}; T: 20 ± 1 °C; time: 20 h; Curve 0 M correspond to Figure 2 and SD: 0.666 g L^{-1}; time 48 h; average value/std. dev. for 3 replicates).

Appendix B.6. Sr(II) Desorption

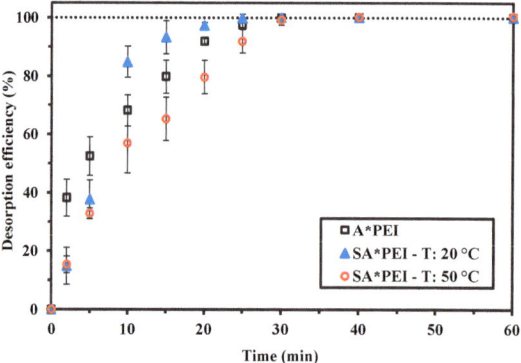

Figure A19. Desorption kinetics for Sr(II)-loaded sorbents (collected from uptake kinetics) using 0.3 M HCl solutions (SD: 2.67 g L^{-1}; T: 20 ± 1 °C; v: 210 rpm).

Appendix B.7. Application to Seawater Samples

The Red Sea is characterized by a higher Sr(II) content than the Mediterranean sea (Table A6). Actually, this may be explained by the geologic environment of the local

Egyptian coast. Strontium deposits occur in the Quseir-Marsa Alam area in the sedimentary rocks of the late Miocene. The concentration of strontium reaches up to 1% (as $SrSO_4$ mineral). The Quseir-Marsa Alam area is regularly exposed to violent and short rains during the winter season, forming torrents that lead to the transfer of many blocks and/or fragments of rocks (containing $SrSO_4$) to the Red Sea, in addition to natural drainage of strontium-containing solutions [118,119].

Table A7. Composition of seawater samples (major metal ions and selected traces).

Metal Ion	Units	Mediterranean Sea	Red Sea
Na(I)	g L^{-1}	13.03	14.10
K(I)	mg L^{-1}	554.8	489.5
Mg(II)	mg L^{-1}	1424	1504
Ca(II)	mg L^{-1}	555.3	568.4
Sr(II)	mg L^{-1}	4.218	5.968
B(III)	mg L^{-1}	3.856	4.119
U(VI)	µg L^{-1}	9.8	10.9

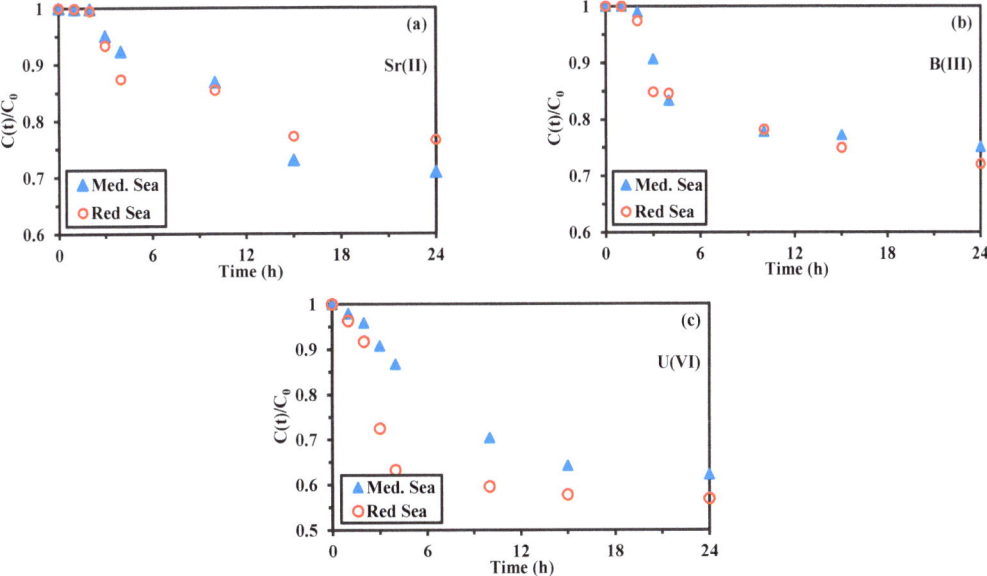

Figure A20. Uptake kinetics for the recovery of trace metal ions (Sr(II) (**a**), B(III) (**b**), and U(VI) (**c**)) from seawater samples (natural pH: ≈7.5; SD: 0.2 g L^{-1}; v: 210 rpm).

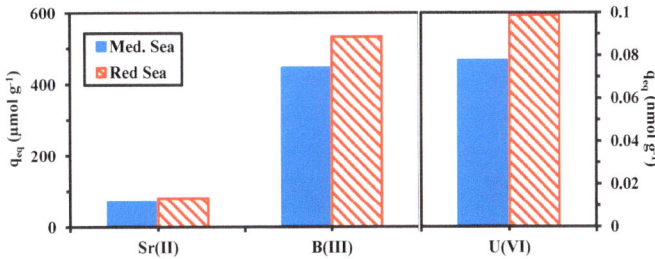

Figure A21. Recovery of trace metal ions (Sr(II), B(III), and U(VI)) from seawater samples (natural pH: ≈7.5; SD: 0.2 g L^{-1}; time: 24 h; v: 210 rpm).

Appendix C. Synthesis of Sorbents

*Appendix C.1. Manufacturing of A*PEI Beads*

The synthesis of algal-PEI beads was already described [120]. Algal/PEI beads were prepared through three steps (see Scheme A1). After grinding (≈250 µm) and sieving the algal biomass (*L. digitata*), the algae (30 g) was dispersed into 800 mL of Na_2CO_3 solution (1%, *w/w*), this suspension was maintained under agitation at 50 °C for 24 h. This step allows the partial extraction of alginate from algae. Five mL of PEI (50%, *w/w*) was added to the mixture (second step) before dropping the mixture into 2 L of $CaCl_2$ solution (1%, *w/w*) for ionotropic gelation of extracted alginate (third step) (by the interaction of calcium with carboxylate groups). This gelation bath was completed with glutaraldehyde (GA, 5 mL, 50% *w/w*) for linkage between amine groups in PEI and for creating an interpenetrating network (alginate carboxylic groups/Ca(II) and PE amine groups/GA). These reactions contribute to the stability of the material. The beads were freeze-dried for two days (−52 °C, 0.1 mbar) for work organization reasons; however, alternative tests showed that shorter operating time (i.e., overnight, ≈15 h) is sufficient for achieving the freeze-drying. Scheme A1 reports the structure and the principle of A*PEI synthesis.

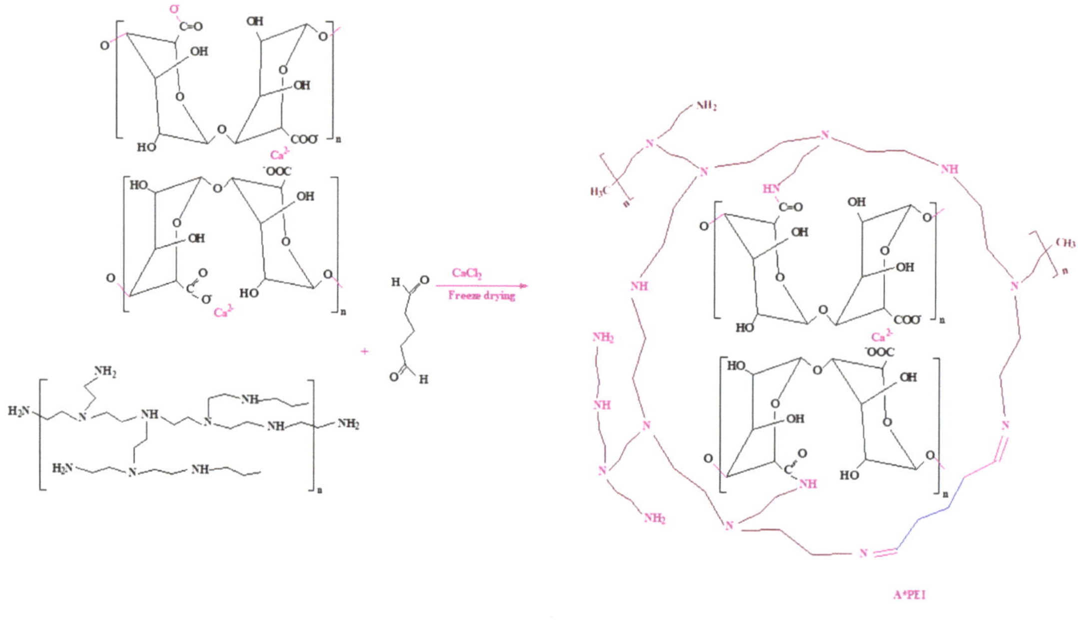

Scheme A1. Synthesis and structure of A*PEI.

*Appendix C.2. Functionalization of A*PEI (Synthesis of SA*PEI)*

The one-pot sulfonation of A*PEI was processed in a closed reactor (adapted for processing redox reactions). Four g of dry A*PEI beads were dispersed into demineralized water (20 mL) in the reactor. Potassium persulfate (0.2 g) and sodium metasulfite (0.1 g) were added (and dissolved in the aqueous suspension). The sulfonating agents (i.e., 2-propene-1-sulfonic acid (1.0 g) and hydroxylamine-O-sulfonic acid (0.8 g)) were first dissolved in water (10 mL), before adding dropwise into the reactor. The suspension was refluxed (at 90 °C) for 9 h. The beads were filtered off, washed with water, and acetone before being dried at 50 °C for 12 h. Persulfate and bisulfide redox initiators orientate the bonding of C=C with >NH and –NH_2 groups [121]. The synthesis is summarized in Scheme A2.

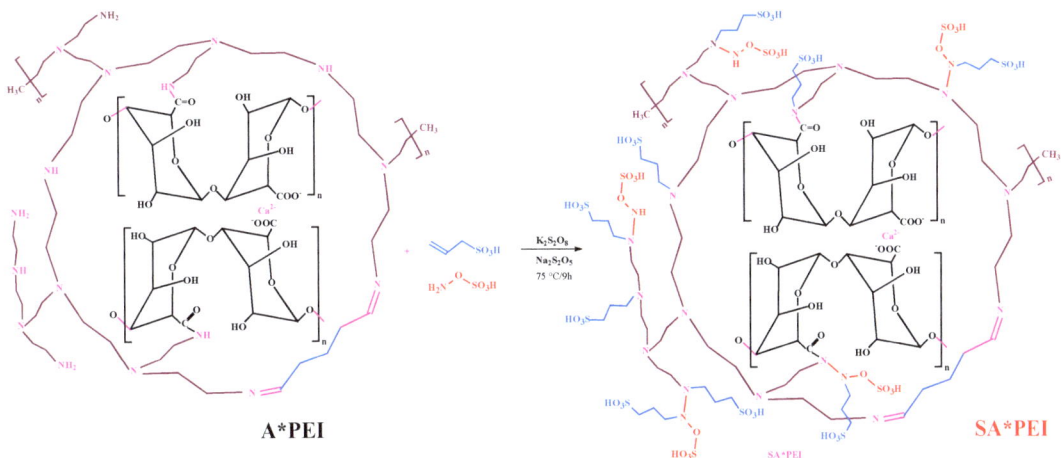

Scheme A2. Synthesis and structure of SA*PEI.

References

1. RSC. Periodic Table. Available online: https://www.rsc.org/periodic-table/ (accessed on 5 October 2021).
2. Mangano, J.J.; Sternglass, E.J.; Gould, J.M.; Sherman, J.D.; Brown, J.; McDonnell, W. Strontium-90 in newborns and childhood disease. *Arch. Environ. Health* **2000**, *55*, 240–244. [CrossRef] [PubMed]
3. Gould, J.M.; Sternglass, E.J.; Sherman, J.D.; Brown, J.; McDonnell, W.; Mangano, J.J. Strontium-90 in deciduous teeth as a factor in early childhood cancer. *Int. J. Health Serv.* **2000**, *30*, 515–539. [CrossRef] [PubMed]
4. Mangano, J.J.; Sherman, J.D. Elevated in vivo strontium-90 from nuclear weapons test fallout among cancer decedents: A case-control study of deciduous teeth. *Int. J. Health Serv.* **2011**, *41*, 137–158. [CrossRef] [PubMed]
5. Shimura, H.; Itoh, K.; Sugiyama, A.; Ichijo, S.; Ichijo, M.; Furuya, F.; Nakamura, Y.; Kitahara, K.; Kobayashi, K.; Yukawa, Y.; et al. Absorption of radionuclides from the Fukushima nuclear accident by a novel algal strain. *PLoS ONE* **2012**, *7*, e44200. [CrossRef] [PubMed]
6. Aamodt, N.O. The potential for disease initiation by inhaled beta-emitting nuclear particles. *Med. Hypotheses* **2018**, *116*, 124–131. [CrossRef] [PubMed]
7. Adigun, O.A.; Oninla, V.O.; Babarinde, N.A.; Oyedotun, K.O.; Manyala, N. Characterization of sugarcane leaf-biomass and investigation of its efficiency in removing nickel(II), chromium(III) and cobalt(II) ions from polluted water. *Surf. Interfaces* **2020**, *20*, 100621. [CrossRef]
8. Shooto, N.D. Removal of toxic hexavalent chromium (Cr(VI)) and divalent lead (Pb(II)) ions from aqueous solution by modified rhizomes of *Acorus Calamus*. *Surf. Interfaces* **2020**, *20*, 100624. [CrossRef]
9. Jayakumar, V.; Govindaradjane, S.; Rajasimman, M. Efficient adsorptive removal of zinc by green marine macro alga *Caulerpa scalpelliformis*—Characterization, optimization, modeling, isotherm, kinetic, thermodynamic, desorption and regeneration studies. *Surf. Interfaces* **2021**, *22*, 100798. [CrossRef]
10. Ayouch, I.; Kassem, I.; Kassab, Z.; Barrak, I.; Barhoun, A.; Jacquemin, J.; Draoui, K.; El Achaby, M. Crosslinked carboxymethyl cellulose-hydroxyethyl cellulose hydrogel films for adsorption of cadmium and methylene blue from aqueous solutions. *Surf. Interfaces* **2021**, *24*, 101124. [CrossRef]
11. Dronov, M.; Koza, T.; Schwiers, A.; Schmidt, T.C.; Schram, J. Strontium carbonate precipitation as a sample preparation technique for isotope ratio analysis of Sr in mineral water and wine by quadrupole-based inductively coupled plasma mass spectrometry. *Rapid Commun. Mass Spectrom.* **2018**, *32*, 149–158. [CrossRef]
12. Thanh, L.H.V.; Liu, J.C. Flotation separation of strontium via phosphate precipitation. *Water Sci. Technol.* **2017**, *75*, 2520–2526. [CrossRef] [PubMed]
13. Guo, Y.; Nhung, N.T.H.; Dai, X.; He, C.; Wang, Y.; Wei, Y.; Fujita, T. Strontium ion removal from artificial seawater using a combination of adsorption with biochar and precipitation by blowing CO_2 nanobubble with neutralization. *Front. Bioeng. Biotechnol.* **2022**, *10*, 819407. [CrossRef] [PubMed]
14. Abdollahi, T.; Towfighi, J.; Rezaei-Vahidian, H. Sorption of cesium and strontium ions by natural zeolite and management of produced secondary waste. *Environ. Technol. Innov.* **2020**, *17*, 100592. [CrossRef]
15. Ogata, F.; Kobayashi, Y.; Uematsu, Y.; Nakamura, T.; Kawasaki, N. Zeolite produced from fly ash by thermal treatment in alkaline solution and Its capability to adsorb Cs(I) and Sr(II) in aqueous solution. *Yakugaku Zasshi-J. Pharm. Soc. Jpn.* **2020**, *140*, 729–737. [CrossRef] [PubMed]

16. Moamen, O.A.A.; Hassan, H.S.; Zaher, W.F. Taguchi L-16 optimization approach for simultaneous removal of Cs^+ and Sr^{2+} ions by a novel scavenger. *Ecotoxicol. Environ. Saf.* **2020**, *189*, 110013. [CrossRef]
17. Karmaker, S.C.; Eljamal, O.; Saha, B.B. Response surface methodology for strontium removal process optimization from contaminated water using zeolite nanocomposites. *Environ. Sci. Pollut. Res.* **2021**, *28*, 56535–56551. [CrossRef] [PubMed]
18. Prajitno, M.Y.; Harbottle, D.; Hondow, N.; Zhang, H.; Hunter, T.N. The effect of pre-activation and milling on improving natural clinoptilolite for ion exchange of cesium and strontium. *J. Environ. Chem. Eng.* **2020**, *8*, 102991. [CrossRef]
19. Zhang, X.; Liu, Y. Ultrafast removal of radioactive strontium ions from contaminated water by nanostructured layered sodium vanadosilicate with high adsorption capacity and selectivity. *J. Hazard. Mater.* **2020**, *398*, 122907. [CrossRef]
20. Jiao, Z.; Meng, Y.; He, C.; Yin, X.; Wang, X.; Wei, Y. One-pot synthesis of silicon-based zirconium phosphate for the enhanced adsorption of Sr (II) from the contaminated wastewater. *Microporous Mesoporous Mater.* **2021**, *318*, 111016. [CrossRef]
21. Amesh, P.; Venkatesan, K.A.; Suneesh, A.S.; Maheswari, U. Tuning the ion exchange behavior of cesium and strontium on sodium iron titanate. *Sep. Purif. Technol.* **2021**, *267*, 118678. [CrossRef]
22. Park, B.; Ghoreishian, S.M.; Kim, Y.; Park, B.J.; Kang, S.-M.; Huh, Y.S. Dual-functional micro-adsorbents: Application for simultaneous adsorption of cesium and strontium. *Chemosphere* **2021**, *263*, 128266. [CrossRef] [PubMed]
23. Li, T.T.; He, F.; Dai, Y.D. Prussian blue analog caged in chitosan surface-decorated carbon nanotubes for removal cesium and strontium. *J. Radioanal. Nucl. Chem.* **2016**, *310*, 1139–1145. [CrossRef]
24. Ali, M.M.S.; Sami, N.M.; El-Sayed, A.A. Removal of Cs^+, Sr^{2+} and Co^{2+} by activated charcoal modified with Prussian blue nanoparticle (PBNP) from aqueous media: Kinetics and equilibrium studies. *J. Radioanal. Nucl. Chem.* **2020**, *324*, 189–201. [CrossRef]
25. El-Bahy, S.M.; Fadel, D.A.; El-Bahy, Z.M.; Metwally, A.M. Rapid and highly efficient cesium removal by newly synthesized carbomer encapsulated potassium copper hexacyanoferrate composite. *J. Environ. Chem. Eng.* **2018**, *6*, 1875–1885. [CrossRef]
26. Vincent, C.; Barre, Y.; Vincent, T.; Taulemesse, J.M.; Robitzer, M.; Guibal, E. Chitin-Prussian blue sponges for Cs(I) recovery: From synthesis to application in the treatment of accidental dumping of metal-bearing solutions. *J. Hazard. Mater.* **2015**, *287*, 171–179. [CrossRef]
27. Hassan, S.S.M.; Kamel, A.H.; Youssef, M.A.; Aboterika, A.H.A.; Awwad, N.S. Removal of barium and strontium from wastewater and radioactive wastes using a green bioadsorbent, Salvadora persica (Miswak). *Desalin. Water Treat.* **2020**, *192*, 306–314. [CrossRef]
28. Dai, Q.; Zhang, T.; Zhao, Y.; Li, Q.; Dong, F.; Jiang, C. Potentiality of living Bacillus pumilus SWU7-1 in biosorption of strontium radionuclide. *Chemosphere* **2020**, *260*, 127559. [CrossRef] [PubMed]
29. Feng, J.; Zhao, X.; Zhou, H.; Qiu, L.; Dai, Y.; Luo, H.; Otero, M. Removal of strontium by high-performance adsorbents Saccharomyces cerevisiae-Fe_3O_4 bio-microcomposites. *J. Radioanal. Nucl. Chem.* **2020**, *326*, 525–535. [CrossRef]
30. El-Saied, H.A.; El-Din, A.M.S.; Masry, B.A.; Ibrahim, A.M. A promising superabsorbent nanocomposite based on grafting biopolymer/nanomagnetite for capture of ^{134}Cs, ^{85}Sr and ^{60}Co radionuclides. *J. Polym. Environ.* **2020**, *28*, 1749–1765. [CrossRef]
31. Abou-Lilah, R.A.; Rizk, H.E.; Elshorbagy, M.A.; Gamal, R.; Ali, A.M.; Badawy, N.A. Efficiency of bentonite in removing cesium, strontium, cobalt and uranium ions from aqueous solution: Encapsulation with alginate for column application. *Int. J. Environ. Anal. Chem.* **2020**, *102*, 2913–2936. [CrossRef]
32. Maqbool, M.; Sadaf, S.; Bhatti, H.N.; Rehmat, S.; Kausar, A.; Alissa, S.A.; Iqbal, M. Sodium alginate and polypyrrole composites with algal dead biomass for the adsorption of Congo red dye: Kinetics, thermodynamics and desorption studies. *Surf. Interfaces* **2021**, *25*, 101183. [CrossRef]
33. Vorster, C.; van der Walt, T.N.; Coetzee, P.P. Ion exchange separation of strontium and rubidium on Dowex 50W-X8, using the complexation properties of EDTA and DCTA. *Anal. Bioanal.Chem.* **2008**, *392*, 287–296. [CrossRef] [PubMed]
34. Nur, T.; Loganathan, P.; Kandasamy, J.; Vigneswaran, S. Removal of strontium from aqueous solutions and synthetic seawater using resorcinol formaldehyde polycondensate resin. *Desalination* **2017**, *420*, 283–291. [CrossRef]
35. Kumar, R.; Malodia, P.; Kachwaha, M.; Verma, S. Adsorptive and kinetic studies of resin for removal of Cs^+ and Sr^{2+} from aqueous solution. *J. Water Chem. Technol.* **2019**, *41*, 292–298. [CrossRef]
36. Hafizi, M.; Abolghasemi, H.; Moradi, M.; Milani, S.A. Strontium adsorption from sulfuric acid solution by Dowex 50W-X resins. *Chin. J. Chem. Eng.* **2011**, *19*, 267–272. [CrossRef]
37. Dragan, E.S.; Humelnicu, D.; Ignat, M.; Varganici, C.D. Superadsorbents for strontium and cesium removal enriched in amidoxime by a homo-IPN strategy connected with porous silica texture. *ACS Appl. Mater. Interfaces* **2020**, *12*, 44622–44638. [CrossRef] [PubMed]
38. Wei, Y.; Salih, K.A.M.; Lu, S.; Hamza, M.F.; Fujita, T.; Vincent, T.; Guibal, E. Amidoxime functionalization of algal/polyethyleneimine beads for the sorption of Sr(II) from aqueous solutions. *Molecules* **2019**, *24*, 3893. [CrossRef] [PubMed]
39. Nada, A.M.A.; El-Wakil, N.A.; Hassan, M.L.; Adel, A.M. Differential adsorption of heavy metal ions by cotton stalk cation-exchangers containing multiple functional groups. *J. Appl. Polym. Sci.* **2006**, *101*, 4124–4132. [CrossRef]
40. Erenturk, S.A.; Haciyakupoglu, S.; Senkal, B.F. Investigation of interaction behaviours of cesium and strontium ions with engineering barrier material to prevent leakage to environmental. *J. Environ. Radioact.* **2020**, *213*, 106101. [CrossRef]

41. Kwak, N.S.; Yang, J.R.; Hwang, C.W.; Hwang, T.S. The effect of a molecular weight and an amount of PEGDA (poly(ethylene glycol)diacrylate) on a preparation of sodium methallyl sulfonate-co-PEGDA microspheres and sorption behavior of Co(II). *Chem. Eng. J.* **2013**, *223*, 216–223. [CrossRef]
42. Kyzas, G.Z.; Kostoglou, M.; Lazaridis, N.K.; Lambropoulou, D.A.; Bikiaris, D.N. Environmental friendly technology for the removal of pharmaceutical contaminants from wastewaters using modified chitosan adsorbents. *Chem. Eng. J.* **2013**, *222*, 248–258. [CrossRef]
43. Vasudevan, T.; Pandey, A.K.; Das, S.; Pujari, P.K. Poly(ethylene glycol methacrylate phosphate-co-2-acrylamido-2-methyl-1-propane sulfonate) pore-filled substrates for heavy metal ions sorption. *Chem. Eng. J.* **2014**, *236*, 9–16. [CrossRef]
44. Yang, J.J.; Dong, Y.H.; Li, J.; Liu, Z.J.; Min, F.L.; Li, Y.Y. Removal of Co(II) from aqueous solutions by sulfonated magnetic multi-walled carbon nanotubes. *Korean J. Chem. Eng.* **2015**, *32*, 2247–2256. [CrossRef]
45. Borai, E.H.; Hamed, M.G.; El-kamash, A.M.; Siyam, T.; El-Sayed, G.O. Synthesis, characterization and application of a modified acrylamide–styrene sulfonate resin and a composite for sorption of some rare earth elements. *New J. Chem.* **2015**, *39*, 7409–7420. [CrossRef]
46. Parlak, E.; Arar, O. Removal of copper (Cu^{2+}) from water by sulfonated cellulose. *J. Disp. Sci. Technol.* **2018**, *39*, 1403–1408. [CrossRef]
47. Arar, O. Co-precipitative preparation of a sulfonated cellulose-magnetite hybrid sorbent for the removal of Cu^{2+} ions. *Anal. Sci.* **2020**, *36*, 81–86. [CrossRef]
48. Miller, D.D.; Siriwardane, R.; McIntyre, D. Anion structural effects on interaction of rare earth element ions with Dowex 50W X8 cation exchange resin. *J. Rare Earths* **2018**, *36*, 879–890. [CrossRef]
49. Hamza, M.F.; Lu, S.M.; Salih, K.A.M.; Mira, H.; Dhmees, A.S.; Fujita, T.; Wei, Y.Z.; Vincent, T.; Guibal, E. As(V) sorption from aqueous solutions using quaternized algal/polyethyleneimine composite beads. *Sci. Total Environ.* **2020**, *719*, 137396. [CrossRef]
50. Hamza, M.F.; Mubark, A.E.; Wei, Y.; Vincent, T.; Guibal, E. Quaternization of composite algal/PEI beads for enhanced uranium sorption-application to ore acidic leachate. *Gels* **2020**, *6*, 6020012. [CrossRef] [PubMed]
51. Hamza, M.F.; Wei, Y.; Guibal, E. Quaternization of algal/PEI beads (a new sorbent): Characterization and application to scandium recovery from aqueous solutions. *Chem. Eng. J.* **2020**, *383*, 123210. [CrossRef]
52. Wei, Y.; Salih, K.A.M.; Hamza, M.F.; Fujita, T.; Rodríguez-Castellón, E.; Guibal, E. Synthesis of a new phosphonate-based sorbent and characterization of its interactions with lanthanum (III) and terbium (III). *Polymers* **2021**, *13*, 1513. [CrossRef] [PubMed]
53. Wei, Y.; Salih, K.A.M.; Rabie, K.; Elwakeel, K.Z.; Zayed, Y.E.; Hamza, M.F.; Guibal, E. Development of phosphoryl-functionalized algal-PEI beads for the sorption of Nd(III) and Mo(VI) from aqueous solutions—Application for rare earth recovery from acid leachates. *Chem. Eng. J.* **2021**, *412*, 127399. [CrossRef]
54. Zhang, Y.; Hamza, M.F.; Vincent, T.; Roux, J.-C.; Faur, C.; Guibal, E. Tuning the sorption properties of amidoxime-functionalized algal/polyethyleneimine beads for La(III) and Dy(III) using EDTA: Impact of metal speciation on selective separation. *Chem. Eng. J.* **2021**, *431*, 133214. [CrossRef]
55. Hamza, M.F.; Salih, K.A.M.; Abdel-Rahman, A.A.H.; Zayed, Y.E.; Wei, Y.; Liang, J.; Guibal, E. Sulfonic-functionalized algal/PEI beads for scandium, cerium and holmium sorption from aqueous solutions (synthetic and industrial samples). *Chem. Eng. J.* **2021**, *403*, 126399. [CrossRef]
56. Demadis, K.D.; Paspalaki, M.; Theodorou, J. Controlled release of bis(phosphonate) pharmaceuticals from cationic biodegradable polymeric matrices. *Ind. Eng. Chem. Res.* **2011**, *50*, 5873–5876. [CrossRef]
57. Haug, A. Dissociation of alginic acid. *Acta Chem. Scand.* **1961**, *15*, 950–952. [CrossRef]
58. Hong, H.J.; Ryu, J.; Park, I.S.; Ryu, T.; Chung, K.S.; Kim, B.G. Investigation of the strontium (Sr(II)) adsorption of an alginate microsphere as a low-cost adsorbent for removal and recovery from seawater. *J. Environ. Manag.* **2016**, *165*, 263–270. [CrossRef]
59. Dong, H.T.; Du, H.B.; Wickramasinghe, S.R.; Qian, X.H. The effects of chemical substitution and polymerization on the pK_a values of sulfonic acids. *J. Phys. Chem. B* **2009**, *113*, 14094–14101. [CrossRef]
60. Hamza, M.F.; Mahfouz, M.G.; Abdel-Rahman, A.A.H. Adsorption of uranium (VI) ions on hydrazinyl amine and 1,3,4-thiadiazol-2(3 H)-thion chelating resins. *J. Disp. Sci. Technol.* **2012**, *33*, 1544–1551. [CrossRef]
61. Lu, S.J.; Duan, M.L.; Lin, S.B. Synthesis of superabsorbent starch graft-poly(potassium acrylate-co-acrylamide) and its properties. *J. Appl. Polym. Sci.* **2003**, *88*, 1536–1542. [CrossRef]
62. Pourjavadi, A.; Barzegar, S.; Zeidabadi, F. Synthesis and properties of biodegradable hydrogels of kappa-carrageenan grafted acrylic acid-co-2-acrylamido-2-methylpropanesulfonic acid as candidates for drug delivery systems. *React. Funct. Polym.* **2007**, *67*, 644–654. [CrossRef]
63. Hubbe, M.A.; Azizian, S.; Douven, S. Implications of apparent pseudo-second-order adsorption kinetics onto cellulosic materials: A review. *BioResources* **2019**, *14*, 7582–7626. [CrossRef]
64. Mahmoud, M.R.; Hassan, R.S.; Rashad, G.M. One-pot synthesis of sodium lauryl sulfate-loaded polyacrylonitrile solid-phase extractor for investigating the adsorption behavior of radioactive strontium(II) from aqueous solutions. *Appl. Radiat. Isot.* **2020**, *163*, 109198. [CrossRef]
65. Shin, J.; Lee, Y.-G.; Kwak, J.; Kim, S.; Lee, S.-H.; Park, Y.; Lee, S.-D.; Chon, K. Adsorption of radioactive strontium by pristine and magnetic biochars derived from spent coffee grounds. *J. Environ. Chem. Eng.* **2021**, *9*, 105119. [CrossRef]
66. Bezhin, N.A.; Dovhyi, I.I.; Tokar, E.A.; Tananaev, I.G. Physical and chemical regularities of cesium and strontium recovery from the seawater by sorbents of various types. *J. Radioanal. Nucl. Chem.* **2021**, *330*, 1101–1111. [CrossRef]

67. Morig, C.R.; Gopala Rao, M. Diffusion in ion exchange resins: Sodium ion-strontium ion system. *Chem. Eng. Sci.* **1965**, *20*, 889–893. [CrossRef]
68. Yamaguchi, T.; Sakamoto, Y.; Senoo, M. Consideration on effective eiffusivity of strontium in granite. *J. Nucl. Sci. Technol.* **1993**, *30*, 796–803. [CrossRef]
69. Severin, A.V.; Gopin, A.V.; Vasiliev, A.N.; Enikeev, K.I. Diffusion and sorption of radium and strontium in a layer of porous sorbent based on hydroxyapatite. *Radiochemistry* **2021**, *63*, 51–55. [CrossRef]
70. Wang, F.; Wang, L.J.; Li, J.S.; Sun, X.Y.; Zhang, L. Synthesis of modified D401 chelating resin and its adsorption properties for Pb^{2+}. *J. Cent. South Univ. Technol.* **2009**, *16*, 575–580. [CrossRef]
71. Chu, K.H. Revisiting the Temkin isotherm: Dimensional inconsistency and approximate forms. *Ind. Eng. Chem. Res.* **2021**, *60*, 13140–13147. [CrossRef]
72. Maruyama, H.; Seki, H. Adsorption modeling by two sites Langmuir type isotherm for adsorption of bisphenol-A and diethyl phthalate onto activated carbon in single and binary system. *Sep. Sci. Technol.* **2022**, *57*, 1535–1542. [CrossRef]
73. Matskevich, A.I.; Tokar, E.A.; Ivanov, N.P.; Sokolnitskaya, T.A.; Parot'kina, Y.A.; Dran'kov, A.N.; Silant'ev, V.E.; Egorin, A.M. Study on the adsorption of strontium on granular manganese oxide. *J. Radioanal. Nucl. Chem.* **2021**, *327*, 1005–1017. [CrossRef]
74. Wei, Y.; Rakhatkyzy, M.; Salih, K.A.M.; Wang, K.; Hamza, M.F.; Guibal, E. Controlled bi-functionalization of silica microbeads through grafting of amidoxime/methacrylic acid for Sr(II) enhanced sorption. *Chem. Eng. J.* **2020**, *402*, 125220. [CrossRef]
75. Alamdarlo, F.V.; Solookinejad, G.; Zahakifar, F.; Jalal, M.R.; Jabbari, M. Study of kinetic, thermodynamic, and isotherm of Sr adsorption from aqueous solutions on graphene oxide (GO) and (aminomethyl)phosphonic acid-graphene oxide (AMPA-GO). *J. Radioanal. Nucl. Chem.* **2021**, *329*, 1033–1043. [CrossRef]
76. Abu-Nada, A.; Abdala, A.; McKay, G. Isotherm and kinetic modeling of strontium adsorption on graphene oxide. *Nanomaterials* **2021**, *11*, 2780. [CrossRef] [PubMed]
77. Bonner, O.D.; Smith, L.L. A selectivity scale for some divalent cations on Dowex 50. *J. Phys. Chem.* **1957**, *61*, 326–329. [CrossRef]
78. Iyer, S.T.; Nandan, D.; Venkataramani, B. Alkaline earth metal ion-proton-exchange equilibria on Nafion-117 and Dowex 50W X8 in aqueous solutions at 298+/−1K. *React. Funct. Polym.* **1996**, *29*, 51–57. [CrossRef]
79. Pehlivan, E.; Altun, T. The study of various parameters affecting the ion exchange of Cu^{2+}, Zn^{2+}, Ni^{2+}, Cd^{2+}, and Pb^{2+} from aqueous solution on Dowex 50W synthetic resin. *J. Hazard. Mater.* **2006**, *134*, 149–156. [CrossRef] [PubMed]
80. Gupta, A.R. Molecular level interpretation of ion-exchange selectivity. *Indian J. Chem. Sect A* **1990**, *29*, 409–417.
81. Yin, J.; Yang, S.; He, W.; Zhao, T.; Li, C.; Hua, D. Biogene-derived aerogels for simultaneously selective adsorption of uranium(VI) and strontium(II) by co-imprinting method. *Sep. Purif. Technol.* **2021**, *271*, 118849. [CrossRef]
82. Kirishima, A.; Sasaki, T.; Sato, N. Solution chemistry study of radioactive Sr on Fukushima Daiichi NPS site. *J. Nucl. Sci. Technol.* **2015**, *52*, 152–161. [CrossRef]
83. Amphlett, J.T.M.; Choi, S.; Parry, S.A.; Moon, E.M.; Sharrad, C.A.; Ogden, M.D. Insights on uranium uptake mechanisms by ion exchange resins with chelating functionalities: Chelation vs. anion exchange. *Chem. Eng. J.* **2020**, *392*, 123712. [CrossRef]
84. Yazdani, M.R.; Virolainen, E.; Conley, K.; Vahala, R. Chitosan-zinc(II) complexes as a bio-sorbent for the adsorptive abatement of phosphate: Mechanism of complexation and assessment of adsorption performance. *Polymers* **2018**, *10*, 25. [CrossRef] [PubMed]
85. Rouquerol, F.; Rouquerol, J.; Sing, K.S.W.; Maurin, G.; Llewellyn, P. (Eds.) 1—Introduction. In *Adsorption by Powders and Porous Solids*, 2nd ed.; Academic Press: Oxford, UK, 2014; pp. 1–24.
86. Sing, K.S.W.; Rouquerol, F.; Rouquerol, J.; Llewellyn, P. 8—Assessment of mesoporosity. In *Adsorption by Powders and Porous Solids*, 2nd ed.; Rouquerol, F., Rouquerol, J., Sing, K.S.W., Llewellyn, P., Maurin, G., Eds.; Academic Press: Oxford, UK, 2014; pp. 269–302.
87. Sato, D.M.; Guerrini, L.M.; de Oliveira, M.P.; de Oliveira Hein, L.R.; Botelho, E.C. Production and characterization of polyetherimide mats by an electrospinning process. *Mater. Res. Express* **2018**, *5*, 115302. [CrossRef]
88. Godiya, C.B.; Liang, M.; Sayed, S.M.; Li, D.; Lu, X. Novel alginate/polyethyleneimine hydrogel adsorbent for cascaded removal and utilization of Cu^{2+} and Pb^{2+} ions. *J. Environ. Manag.* **2019**, *232*, 829–841. [CrossRef]
89. Sui, C.; Preece, J.A.; Zhang, Z. Novel polystyrene sulfonate-silica microspheres as a carrier of a water soluble inorganic salt (KCl) for its sustained release, via a dual-release mechanism. *RSC Adv.* **2017**, *7*, 478–481. [CrossRef]
90. Schiener, P.; Black, K.D.; Stanley, M.S.; Green, D.H. The seasonal variation in the chemical composition of the kelp species *Laminaria digitata, Laminaria hyperborea, Saccharina latissima* and *Alaria esculenta*. *J. Appl. Phycol.* **2015**, *27*, 363–373. [CrossRef]
91. Rodrigues, D.; Freitas, A.C.; Pereira, L.; Rocha-Santos, T.A.P.; Vasconcelos, M.W.; Roriz, M.; Rodríguez-Alcalá, L.M.; Gomes, A.M.P.; Duarte, A.C. Chemical composition of red, brown and green macroalgae from Buarcos bay in Central West Coast of Portugal. *Food Chem.* **2015**, *183*, 197–207. [CrossRef]
92. Pinilla-Torres, A.M.; Carrion-Garcia, P.Y.; Sanchez-Dominguez, C.N.; Gallardo-Blanco, H.; Sanchez-Dominguez, M. Modification of branched polyethyleneimine using Mesquite Gum for its improved hemocompatibility. *Polymers* **2021**, *13*, 2766. [CrossRef]
93. Zaaeri, F.; Khoobi, M.; Rouini, M.; Javar, H.A. pH-responsive polymer in a core-shell magnetic structure as an efficient carrier for delivery of doxorubicin to tumor cells. *Int. J. Polym. Mater. Polym. Biomater.* **2018**, *67*, 967–977. [CrossRef]
94. Zhengbo, H.; Zhu, W.; Song, H.; Chen, P.; Yao, S. The adsorption behavior and mechanism investigation of Cr(VI) ions removal by poly(2-(dimethylamino)ethyl methacrylate)/poly(ethyleneimine) gels. *J. Serb. Chem. Soc.* **2015**, *80*, 889–902.
95. Xu, Z.P.; Braterman, P.S. High affinity of dodecylbenzene sulfonate for layered double hydroxide and resulting morphological changes. *J. Mater. Chem.* **2003**, *13*, 268–273. [CrossRef]

96. Sakugawa, K.; Ikeda, A.; Takemura, A.; Ono, H. Simplified method for estimation of composition of alginates by FTIR. *J. Appl. Polym. Sci.* **2004**, *93*, 1372–1377. [CrossRef]
97. Lawrie, G.; Keen, I.; Drew, B.; Chandler-Temple, A.; Rintoul, L.; Fredericks, P.; Grondahl, L. Interactions between alginate and chitosan biopolymers characterized using FTIR and XPS. *Biomacromolecules* **2007**, *8*, 2533–2541. [CrossRef]
98. Dangi, Y.R.; Bediako, J.K.; Lin, X.; Choi, J.-W.; Lim, C.-R.; Song, M.-H.; Han, M.; Yun, Y.-S. Polyethyleneimine impregnated alginate capsule as a high capacity sorbent for the recovery of monovalent and trivalent gold. *Sci. Rep.* **2021**, *11*, 17836. [CrossRef]
99. Saxena, N.; Pal, N.; Ojha, K.; Dey, S.; Mandal, A. Synthesis, characterization, physical and thermodynamic properties of a novel anionic surfactant derived from *Sapindus laurifolius*. *RSC Adv.* **2018**, *8*, 24485–24499. [CrossRef]
100. Fernando, I.R.; Daskalakis, N.; Demadis, K.D.; Mezei, G. Cation effect on the inorganic-organic layered structure of pyrazole-4-sulfonate networks and inhibitory effects on copper corrosion. *New J. Chem.* **2010**, *34*, 221–235. [CrossRef]
101. Ho, Y.S.; McKay, G. Pseudo-second order model for sorption processes. *Process Biochem.* **1999**, *34*, 451–465. [CrossRef]
102. Crank, J. *The Mathematics of Diffusion*, 2nd ed.; Oxford University Press: Oxford, UK, 1975; p. 414.
103. Langmuir, I. The adsorption of gases on plane surfaces of glass, mica and platinum. *J. Am. Chem. Soc.* **1918**, *40*, 1361–1402. [CrossRef]
104. Freundlich, H.M.F. Uber die adsorption in lasungen. *Z. Phys. Chem.* **1906**, *57*, 385–470.
105. Tien, C. *Adsorption Calculations and Modeling*; Butterworth-Heinemann: Newton, MA, USA, 1994; p. 243.
106. Kegl, T.; Kosak, A.; Lobnik, A.; Novak, Z.; Kralj, A.K.; Ban, I. Adsorption of rare earth metals from wastewater by nanomaterials: A review. *J. Hazard. Mater.* **2020**, *386*, 121632.
107. Falyouna, O.; Eljamal, O.; Maamoun, I.; Tahara, A.; Sugihara, Y. Magnetic zeolite synthesis for efficient removal of cesium in a lab-scale continuous treatment system. *J. Colloid Interface Sci.* **2020**, *571*, 66–79. [CrossRef]
108. Puigdomenech, I. *MEDUSA (Make Equilibrium Diagrams Using Sophisticated Algorithms)*; 32 Bit Version; Royal Institute of Technology: Stockholm, Sweden, 2010.
109. AFSSA. *Guidelines on the Assessment of Ion Exchangers Used for the Treatment of Water Intended for Human Consumption*; AFSSA: Maisons-Alfort, France, 2009; p. 60.
110. Lu, T.T.; Zhu, Y.F.; Wang, W.B.; Qi, Y.X.; Wang, A.Q. Polyaniline-functionalized porous adsorbent for Sr^{2+} adsorption. *J. Radioanal. Nucl. Chem.* **2018**, *317*, 907–917. [CrossRef]
111. Rae, I.B.; Pap, S.; Svobodova, D.; Gibb, S.W. Comparison of sustainable biosorbents and ion-exchange resins to remove Sr^{2+} from simulant nuclear wastewater: Batch, dynamic and mechanism studies. *Sci. Total Environ.* **2019**, *650*, 2411–2422. [CrossRef]
112. Lihareva, N.; Petrov, O.; Dimowa, L.; Tzvetanova, Y.; Piroeva, I.; Üblekov, F.; Nikolov, A. Ion exchange of Cs^+ and Sr^{2+} by natural clinoptilolite from bi-cationic solutions and XRD control of their structural positioning. *J. Radioanal. Nucl. Chem.* **2020**, *323*, 1093–1102. [CrossRef]
113. Li, Y.; Luo, X.; Bai, X.; Lv, W.; Liao, Y. Adsorption of strontium onto adaxial and abaxial cuticle of *Photinia serrulata* leaf. *Int. J. Environ. Res. Public Health* **2020**, *17*, 1061. [CrossRef]
114. Kumar, R.; Verma, S.; Harwani, G.; Patidar, D. Adsorptive and kinetic studies of the extraction of toxic metal ion from contaminated water using modified montmorillonites. *J. Water Chem. Technol.* **2021**, *43*, 321–329. [CrossRef]
115. Ren, L.; Zhao, X.; Liu, B.; Huang, H. Synergistic effect of carboxyl and sulfate groups for effective removal of radioactive strontium ion in a Zr-metal-organic framework. *Water Sci. Technol.* **2021**, *83*, 2001–2011. [CrossRef]
116. Huo, J.-B.; Yu, G.; Wang, J. Adsorptive removal of Sr(II) from aqueous solution by polyvinyl alcohol/graphene oxide aerogel. *Chemosphere* **2021**, *278*, 130492. [CrossRef]
117. Abass, M.R.; Maree, R.M.; Sami, N.M. Adsorptive features of cesium and strontium ions on zirconium tin(IV) phosphate nanocomposite from aqueous solutions. *Int. J. Environ. Anal. Chem.* **2022**, 2016728. [CrossRef]
118. Mahmoud, S.A.E.A.; Mosleh, N.M.; Mansour, G.M.R. The Zug El Bohar hydrothermal uranium-base metal deposits, Southwest El Quseir, Red Sea Coast, Egypt. *Arab J. Nucl. Sci. Appl.* **2018**, *51*, 228–247. [CrossRef]
119. Mahmoud, S.A.E.A.; Gehad, M.R.M. The hydrothermal uranium and some other metal deposits of the extensional faulting during the advanced opening of the Red Sea, Central Eastern Desert, Egypt. *Beni-Suef Univ. J. Basic Appl. Sci.* **2018**, *7*, 752–766. [CrossRef]
120. Hamza, M.F.; Salih, K.A.M.; Zhou, K.; Wei, Y.; Abu Khoziem, H.A.; Alotaibi, S.H.; Guibal, E. Effect of bi-functionalization of algal/polyethyleneimine composite beads on the enhancement of tungstate sorption: Application to metal recovery from ore leachate. *Sep. Purif. Technol.* **2022**, *290*, 120893. [CrossRef]
121. Wallace, R.G. Hydroxylamine-O-sulfonic acid—A versatile synthetic reagent. *Aldrichimica Acta* **1980**, *13*, 3–11.

Article

Synthesis and Characterization of Novel Thiosalicylate-based Solid-Supported Ionic Liquid for Removal of Pb(II) Ions from Aqueous Solution

Nur Anis Liyana Kamaruddin [1,*], Mohd Faisal Taha [2] and Cecilia Devi Wilfred [2]

[1] Centre of Research in Ionic Liquids, Universiti Teknologi PETRONAS, Seri Iskandar, Perak 32610, Malaysia
[2] Fundamental and Applied Sciences Department, Universiti Teknologi PETRONAS, Seri Iskandar, Perak 32610, Malaysia
* Correspondence: nur_21000857@utp.edu.my; Tel.: +60-1129019793

Abstract: The main objectives of this study are to synthesize a new solid-supported ionic liquid (SSIL) that has a covalent bond between the solid support, i.e., activated silica gel, with thiosalicylate-based ionic liquid and to evaluate the performance of this new SSIL as an extractant, labelled as Si-TS-SSIL, and to remove Pb(II) ions from an aqueous solution. In this study, 1-methyl-3-(3-trimethoxysilylpropyl) imidazolium thiosalicylate ([MTMSPI][TS]) ionic liquid was synthesized and the formation of [MTMSPI][TS] was confirmed through structural analysis using NMR, FTIR, IC, TGA, and Karl Fischer Titration. The [MTMSPI][TS] ionic liquid was then chemically immobilized on activated silica gel to produce a new thiosalicylate-based solid-supported ionic liquid (Si-TS-SSIL). The formation of these covalent bonds on Si-TS-SSIL was confirmed by solid-state NMR analysis. Meanwhile, BET analysis was performed to study the surface area of the activated silica gel and the prepared Si-TS-SSIL (before and after washing with solvent) with the purpose to show that all physically immobilized [MTMSPI][TS] has been washed off from Si-TS-SSIL, leaving only chemically immobilized [MTMSPI][TS] on Si-TS-SSIL before proceeding with removal study. The removal study of Pb(II) ions from an aqueous solution was carried out using Si-TS-SSIL as an extractant, whereby the amount of Pb(II) ions removed was determined by AAS. In this removal study, the experiments were carried out at a fixed agitation speed (400 rpm) and fixed amount of Si-TS-SSIL (0.25 g), with different contact times ranging from 2 to 250 min at room temperature. The maximum removal capacity was found to be 8.37 mg/g. The kinetics study was well fitted with the pseudo-second order model. Meanwhile, for the isotherm study, the removal process of Pb(II) ions was well described by the Freundlich isotherm model, as this model exhibited a higher correlation coefficient (R^2), i.e., 0.99, as compared to the Langmuir isotherm model.

Keywords: ionic liquids; solid-supported ionic liquids; thiosalicylate; Pb(II) removal

Citation: Kamaruddin, N.A.L.; Taha, M.F.; Wilfred, C.D. Synthesis and Characterization of Novel Thiosalicylate-Based Solid-Supported Ionic Liquid for Removal of Pb(II) Ions from Aqueous Solution. *Molecules* 2023, 28, 830. https://doi.org/10.3390/molecules28020830

Academic Editors: Yongchang Sun and Dimitrios Giannakoudakis

Received: 8 December 2022
Revised: 1 January 2023
Accepted: 10 January 2023
Published: 13 January 2023

Copyright: © 2023 by the authors. Licensee MDPI, Basel, Switzerland. This article is an open access article distributed under the terms and conditions of the Creative Commons Attribution (CC BY) license (https://creativecommons.org/licenses/by/4.0/).

1. Introduction

Fast evolution in industrial enterprise and poor wastewater treatment has led to the discharge of a tremendous amount of waste, which contains hazardous chemicals and expels pollutants into the environment. Thus, it is critically important to remove or reduce the amounts of heavy metals in wastewater to an allowable safe limit before discharging them into the natural environment [1]. One of the common hazardous metal ions present in wastewater, such as in the petroleum industry, is the Pb(II) ion. According to Akpoveta et al., in the petroleum industry, Pb(II) ions are found because of the chemicals used during the refining process, metals absorption coming from pipelines, and vessels and tanks, together with the natural metals, existing in sandstone during the crude extraction process [2]. Pb(II) ions can be absorbed and accumulated in the human body and cause serious health effects such as cancer, causing damage to the kidneys, liver, heart, brain, bones, and neurological system in humans through inhaling and swallowing polluted food and water [3]. They represent serious threats to the human population and the fauna and flora of the receiving water bodies, as the low amounts of this heavy metal are highly toxic. Hence, it is very crucial to remove Pb(II) ions from wastewater and the environment.

Several commercial methods are being used in wastewater treatment to remove metal ions, namely, reverse osmosis, ion exchange, chemical precipitation, irradiation, biosorption, and extraction [4]. Reverse osmosis can remove many types of molecules and ions from solutions, including bacteria, and is used in both industrial processes. Reverse osmosis involves a diffusive mechanism, so the separation efficiency is dependent on a solute concentration, pressure, and water flux rate [5]. Ion exchange can attract soluble ions from the liquid phase to the solid phase, which is the most widely used method in the water treatment industry. As a cost-effective method, the ion exchange process normally involves low-cost materials and convenient operations. However, it is very sensitive to the pH of an aqueous solution and can be used only at low concentrated metal solutions [6]. Chemical precipitation is widely used because of its simple operation [7]. However, it requires a large amount of chemicals to reduce metals to an acceptable level for discharge. Other drawbacks are huge sludge production, slow metal precipitation, poor settling, the aggregation of metal precipitates, and the long-term environmental impacts of sludge disposal. Another method that has gained attention during recent years for heavy metal removal is the usage of plant biomass as the extractant through the biosorption process, owing to its good performance and being a low-cost material [8]. Although biological methods are inexpensive, environmentally friendly techniques, they need large areas and proper maintenance and operation [5].

Liquid–liquid extraction or solvent extraction is one of the frequently used methods in several industries such as electronic and battery industries to remove metal ions from their wastewater. Liquid–liquid extraction may be an excellent option for wastewater treatment as it offers several advantages such as high extractability and high selectivity for liquid separation [9]. However, there are a few drawbacks to liquid–liquid extraction, such as emulsion formation, the usage of huge volatile organic solvents that are carcinogenic and non-biodegradable and, hence, the formation of large amounts of pollutants, making it costly, time-consuming, and ecologically unfriendly [10]. The problems associated with the usage of volatile organic solvents in liquid–liquid extraction have motivated many researchers to find alternative metal extractants that would be efficient, cost effective, and environmentally friendly. One of the alternatives that is being extensively studied by researchers to replace organic solvents as metal extractants in liquid–liquid extraction is ionic liquids (ILs) [11].

Ionic liquid mainly consists of an organic cation with an organic or inorganic anion, which is a liquid organic salt that is bulky in chemical structure [12]. ILs are salts and generally exist in liquid form at temperatures below 100 °C. The physical and chemical properties of ILs can be altered as desired by modifying the cation and anion combinations [13]. Because of their excellent tunability, ILs have great potential to form different combinations in the designing of task-specific fluids. This feature would give access to more applications of ILs, such as replacing common volatile organic solvents with ILs in the liquid–liquid extraction method, which allows for the development of more effective separation processes. The substitution of volatile organic solvents with ILs also prevents organic solvent losses by evaporation; hence, resulting in material loss and reducing environmental impact [14]. Furthermore, ILs have unique properties which make them different from ordinary salts, such as being non-volatile, non-flammable solvents that have a negligible vapor pressure [15].

Aside from the fact that ILs are considered as greener solvents as compared to volatile organic solvents, they also have limitations such as being expensive, having high viscosity, high cost separation, and being faced with the difficulty to maintain extracted analytes in their phase (mass transfer problem) [16]. To cope with these problems, a method based on immobilization or impregnation of ionic liquid on a solid support as a solid phase extractor, known as a solid-supported ionic liquid (SSIL), is being studied extensively by many researchers [17–19]. This immobilization method producing SSIL creates a thin layer of ionic liquid on a solid support which in turn decreases the amount of ionic liquid required to extract metal ions compared to using ionic liquid in liquid–liquid extraction [20].

The removal studies have shown that SSIL performed better in removing metal ions from aqueous solution as compared to well-known solid acids such as silica, alumina, and zeolites. This is caused by the high surface area for contact between the ionic liquid (contain functional group(s) having high affinity towards metal ions) immobilized on a solid support in SSIL with the targeted metal ions. In addition to offering different functional groups in ionic liquid immobilized on its solid support, the acidity of SSIL can be tunable depending on the choice of solid support [21]. In fact, SSIL is easy to apply for a large scale of operation. Moreover, SSIL can be successfully employed in both batch and flow processes [20].

In this piece of work, an attempt was made to chemically immobilize 1-methyl-3-(3-trimethoxysilylpropyl) imidazolium thiosalicylate ([MTMSPI][TS]) ionic liquid on activated silica gel to produce a new solid-supported ionic liquid known as thiosalicylate-based solid-supported ionic liquid (Si-TS-SSIL). The [MTMSPI][TS] ionic liquid containing the thiosalicylate functional group was chosen to be chemically immobilized on activated silica gel in SSIL (as an extraction agent for Pb(II) ions from aqueous solution) because of the high affinity of thiosalicylate towards metal ions [21]. Meanwhile, Pb(II) ions have been selected as this metal ion is commonly found in industrial wastewater [11]. The kinetics studies for the removal of Pb(II) ions was carried out at different contact times (2–250 min). As for the adsorption isotherms, this study was conducted to elaborate the insight of the adsorption process with different initial Pb(II) concentrations ranging from 10–200 ppm. For the first time, the newly synthesized Si-TS-SSIL extractant in the removal of Pb(II) ions in an aqueous solution and wastewater were performed. It is an advanced creation that could potentially provide both low-cost and efficiency compared to the sole usage of ILs as metal extractants. Therefore, this current study has great potential to contribute to wastewater and environmental cleanliness. In addition, this adsorption method is easy to apply and has a short test period.

2. Results and Discussions

2.1. Characterization

The ^{1}H-NMR spectra of the synthesized [MTMSPI][TS] ionic liquid is depicted in Figure 1. As shown by Figure 1, chemical shifts between 6.8 to 7.6 ppm could clearly be assigned to aromatic hydrogen atoms of the thiosalicylate anion [22]. The ^{1}H-NMR of thiosalicylate anion in [MTMSPI][TS] is summarized as: δ 6.89 (2H, CH), δ 7.13 (1H, CH), δ 7.6 (1H, CH). Meanwhile, the ^{1}H-NMR of 1-methyl-3-(3-trimethoxysilylpropyl) imidazolium cation in [MTMSPI][TS] is summarized as: δ 0.55 (2H, CH_2), δ 1.91 (2H, CH_2), δ 3.66 (9H, CH_3), δ 3.94 (3H, CH_3), δ 4.27 (2H, CH_2), δ 7.84 (2H, CH), δ 9.78 (1H, CH). The chemical shift for the NMR solvent, i.e., dimethyl sulfoxide (DMSO), is 2.5 ppm.

The FTIR spectra of the synthesized [MTMSPI][TS] ionic liquid was recorded by Thermo Scientific spectrometer using the attenuated total reflectance (ATR) method and is shown in Figure 2. In activated silica gel, the silanol group was observed by the presence of broad -OH stretch at 3425 cm^{-1}. The siloxane group (Si-O-Si) asymmetric stretching was appeared at 1081 cm^{-1} and the corresponding symmetric stretching was observed at 793 cm^{-1} [23].

In the [MTMSPI][TS] ionic liquid, the absorption band at 2942 cm^{-1} was indicated to the stretching mode of -CH_2 groups, which were related to trimethoxysilylpropyl. Si-O-Si tensile vibration was observed at 1034 cm^{-1} [24]. Another absorption band appeared at 1571 cm^{-1} because of C-N stretching [25]. Additionally, a weak peak detected around 2232 cm^{-1} belongs to the S-H group of the aromatic compound in thiosalicylate ion of [MTMSPI][TS] [26].

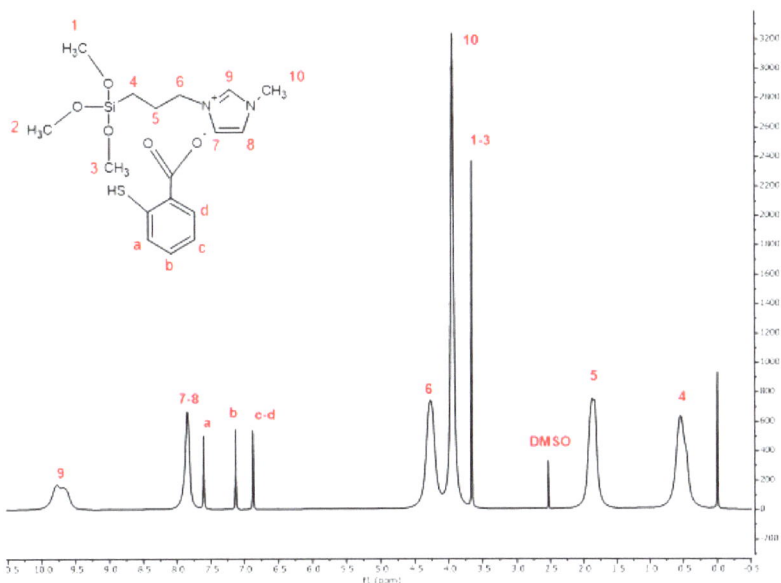

Figure 1. ^{1}H-NMR spectra of the newly synthesized 1-methyl-3-(3-trimethoxysilylpropyl) imidazolium thiosalicylate ([MTMSPI][TS]) ionic liquid.

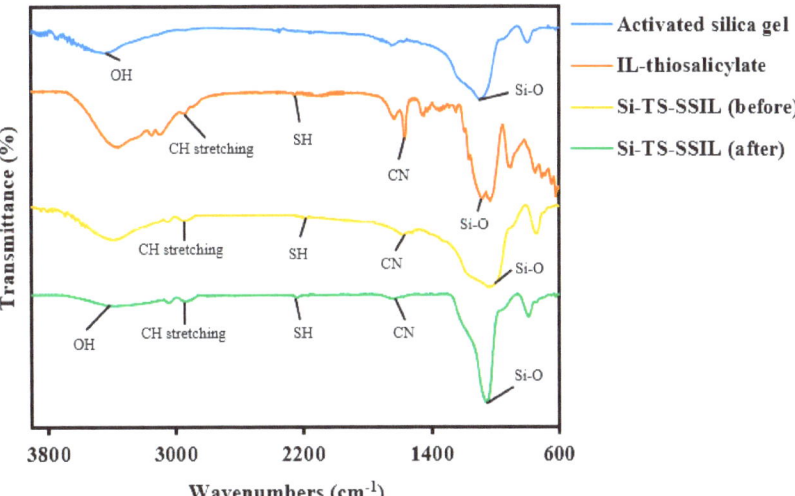

Figure 2. FTIR spectra of the newly synthesized 1-methyl-3-(3-trimethoxysilylpropyl) imidazolium thiosalicylate ([MTMSPI][TS]) ionic liquid, activated silica gel, and Si-TS-SSIL before and after the adsorption process.

For Si-TS-SSIL, which contains the combination of activated silica gel and [MTMSPI][TS] ionic liquid, stretching vibrations of the Si-O-Si groups were observed at 1030 cm^{-1} [23]. The peaks observed in the range of 1550–1650 cm^{-1} were related to the C-N stretching of the imidazolium ring. Meanwhile, the O-H vibration observed in the silica gel particle will be decreased when it is modified with ionic liquid. It can be seen in the figure below where the lack of peak in the range of 3300–3500 cm^{-1} is significant. This indicates that the imidazolium cation is in interaction with the activated silica gel [27]. The presence of the expectable functional groups in the prepared materials at the respective position on the FTIR spectrum indicates the successful anchorage of the organic ligands and alkyl silanes onto the silica framework.

Meanwhile, the moisture content of the [MTMSPI][TS] ionic liquid was below 10.00 ppm. Since sodium chloride (NaCl) dissolved in the solvent is the by-product for synthesizing [MTMSPI][TS], the chloride content was determined as part of the purity study. As shown in Table 1, the chloride content in [MTMSPI][TS] was found to be 9.08 ppm. With regards to chloride content, the purity of synthesized [MTMSPI][TS] was calculated to be 99.91%.

Table 1. Chloride content and the purity of the newly synthesized [MTMSPI][TS] ionic liquid.

Chloride Content (ppm)	Concentration (%)	Purity (%)
9.08	0.0908	99.91

The thermogravimetric analysis of the synthesized [MTMSPI][TS] was studied in the range of 100 °C to 700 °C to observe its thermal stability. As shown in Figure 3, three characteristic decomposition stages were observed. The first TGA curve shows a mass loss of about 9.09% up to 196 °C because of the removal of adsorbed water molecules [28]. The second weight loss occurred from 224 °C to 401 °C, which was assigned to the degradation of thiol group [29]. Further weight loss (8.19%) that was noticed from 417 °C to 532 °C could be associated with the degradation of the remaining organic molecules [30]. From the TGA analysis, the newly synthesized [MTMSPI][TS] ionic liquid exhibited high thermal stability.

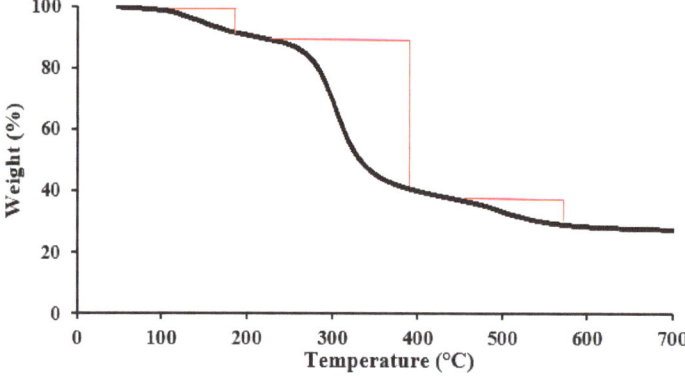

Figure 3. TGA curve of [MTMSPI][TS] ionic liquid.

The newly synthesized solid-supported ionic liquid with a thiol functional group, Si-TS-SSIL, was analyzed using solid-state NMR with the purpose to confirm the formation of covalent bonds (Si-O bonds) between [MTMSPI][TS] and activated silica gel in Si-TS-SSIL, i.e., the covalent between the cation of [MTMPSI][TS] and the silanol group (Si-OH) on activated silica gel. The Si-OH group dominates the surface properties of activated silica gel, and this group corresponds to the cation of [MTMSPI][TS], 1-methyl-3-(3-trimethoxysilylpropyl) imidazolium, to perform chemical immobilization. In the solid-state NMR analysis, the solid-state of silica cross-polarization magic angle spinning (^{29}Si CP-MAS) NMR spectroscopy was utilized to show that [MTMSPI][TS] was covalently bonded to activated silica gel in Si-TS-SSIL. As shown in Figure 4a, the ^{29}Si MAS NMR spectrum of pure activated silica gel revealed the presence of three signals with resolved peaks at −91, −100, and −109 ppm. These peaks are assigned to silicon atoms in the silanediol groups (Q^2), silanol groups (Q^3) and silicon-oxygen tetrahedra (Q^4) of the SiO_2, respectively [31]. After chemical modification of activated silica gel surface with the [MTMSPI][TS] ionic liquid producing Si-TS-SSIL, an increase in the intensity growth was observed for signal Q^4 followed by a significant reduction in the signal Q^2 and Q^3 [32]. Additionally, the appearance of new peaks was found, which appeared at −57 and −66 ppm of the T^2 and T^3 units as shown in Figure 4b. Both peaks are attributed to the silanisation of the silica particles surface, which produced the covalent bonds of Si-O, thus, confirmed that [MTMSPI][TS] covalently bonded to the activate silica gel surface in Si-TS-SSIL [33].

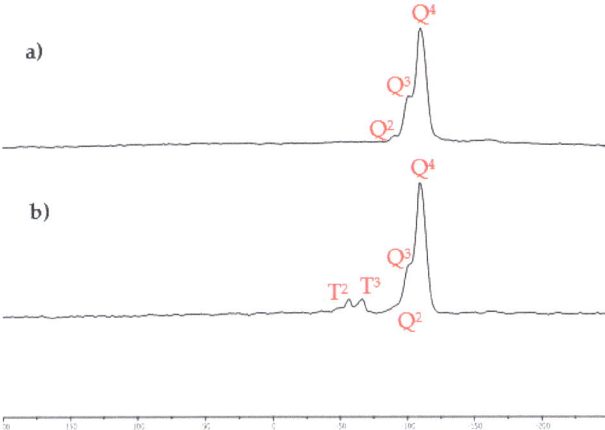

Figure 4. ^{29}Si NMR spectra of (**a**) pure silica and (**b**) Si-TS-SSIL (modified activated silica gel with [MTMSPI][TS] ionic liquid).

The surface area, pore volume, and pore diameter of activated silica gel and Si-TS-SSIL (before and after washing with dichloromethane (solvent)) were determined using BET. Based on Table 2, the activated silica gel and Si-TS-SSIL could be classified into the category of mesoporous as their pore size is in the range of 2–10 nm [34]. The surface area, pore volume and pore diameter of Si-TS-SSIL, for both samples of before and after washing with dichloromethane, were reduced as compared to activated silica gel. These results could be explained by the fact that the immobilization, either physically or chemically, of [MTMSPI][TS] on activated silica gel would reduce the surface area, pore volume, and pore diameter of Si-TS-SSIL. After washing with dichloromethane, as expected, the surface area, pore volume, and pore diameter of Si-TS-SSIL were increased compared to the one before washing with dichloromethane. These achieved results might be caused by the removal of [MTMSPI][TS], which was physically immobilized on the activated silica gel in Si-TS-SSIL by dichloromethane, which in turn left only chemically immobilized [MTMSPI][TS]. In this study, it is important to remove physically immobilized [MTMSPI][TS] on activated silica gel in Si-TS-SSIL to prevent the leaching of [MTMSPI][TS] during the removal study.

Table 2. The results of BET analyses for activated silica gel and Si-TS-SSIL (before and after washing with dichloromethane).

Sample	BET Surface Area (m^2/g)	Pore Volume (cm^3/g)	Pore Diameter (nm)
Activated silica gel	275.8132	0.7021	11.5176
Si-TS-SSIL (before washing with dichloromethane)	225.6244	0.5758	9.9599
Si-TS-SSIL (after washing with dichloromethane)	232.6600	0.5793	10.2075

2.2. Adsorption Studies

2.2.1. Effect of [MTMSPI][TS]: Activated Silica gel Mass Ratio in Si-TS-SSIL on the Removal Efficiency

In this piece of work, an attempt was first made to analyze the performance of three control samples of extractant ([MTMSPI][TS], activated silica gel, and Si-TS-SSIL) to remove Pb(II) ions from the aqueous solution with the following constant parameters: initial Pb(II) ions concentration, with mixing time and extractant dosage being 200 mg L^{-1}, 30 min, and 0.25 g, respectively. The results obtained are shown in Table 3. As expected, Si-TS-SSIL exhibited better removal efficiency compared to activated silica gel, but lower performance compared to [MTMSPI][TS]. With this confirmation on the results of removal efficiency by Si-TS-SSIL, the removal study proceeded using four samples of Si-TS-SSIL as an extractant.

Table 3. Removal capacity of activated silica gel, [MTMSPI][TS], and Si-TS-SSIL control samples.

Extractant	Removal Capacity (mg/g)
Activated silica gel	1.370
[MTMSPI][TS]	5.801
Si-TS-SSIL	5.285

Four samples of Si-TS-SSIL that have a different mass ratio of [MTMSPI][TS] and activated silica gel, i.e., [MTMSPI][TS](in g):activated silica gel (in g), were analyzed on the perspective of removal efficiency. The purpose of this analysis is to study the dosage effect of [MTMSPI][TS] in Si-TS-SSIL in removing Pb(II) ions from the aqueous solution. Figure 5 depicts the relationship between the Si-TS-SSIL's removal capacity and the [MTMSPI][TS]:activated silica gel mass ratio in Si-TS-SSIL to remove Pb(II) ions from the aqueous solution. From this figure, it can be seen that the increasing amount of [MTMSPI][TS] resulted in an increase of removal efficiency until the system reached equilibrium, when the [MTMSPI][TS]:activated silica gel mass ratio was 0.2:1, whereby there was no significant difference in removal efficiency beyond the mass ratio of 0.2:1. Thus, further experiments of removal study were conducted based on 0.2:1 ([MTMSPI][TS]:activated silica gel) mass ratio as it was the optimum mass ratio for the removal process of Pb(II) ions from an aqueous solution using Si-TS-SSIL as an extractant.

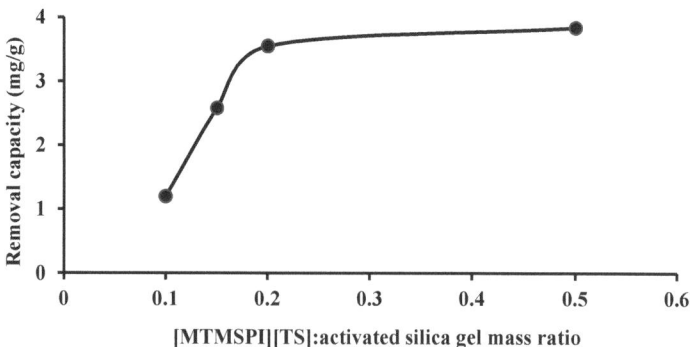

Figure 5. The effect of different [MTMSPI][TS]:activated silica gel mass ratios on the removal process of Pb(II) ions onto Si-TS-SSIL.

2.2.2. Effect of Contact Time

In the batch removal process, contact time is one of the most important factors as it defines the time required for the extractant to reach a dynamic equilibrium stage. Temperature (25 °C), Si-TS-SSIL dosage (0.25 g), initial concentration Pb(II) ions (200 mg L^{-1}), and agitation speed (400 rpm) were all held constant at this step, except for the contact time. Figure 6 depicts the effect of contact time on Pb(II) ions removal efficiency using an [MTM-

SPI][TS]:activated silica gel mass ratio of 0.2:1 in Si-TS-SSIL. As can be seen in Figure 6, the removal rate grew rapidly at first and was followed by a subsequent slow uptake. The optimum removal efficiency of Pb(II) ions was established in approximately 120 min. These results of the removal process could be explained, wherein when the number of accessible sites (functional group, i.e., thiol group, and number of pores) is substantially greater than the number of metal species to be removed, the removal process appears to proceed rapidly. As the contact time increased, the amount of Pb(II) ions removed also increased up to a certain contact time, whereby the removal phase was reached at a steady state. Thus, in this removal study, the optimum contact time was decided at 120 min for the removal of Pb(II) metal ions from the aqueous solution with 87% removal efficiency.

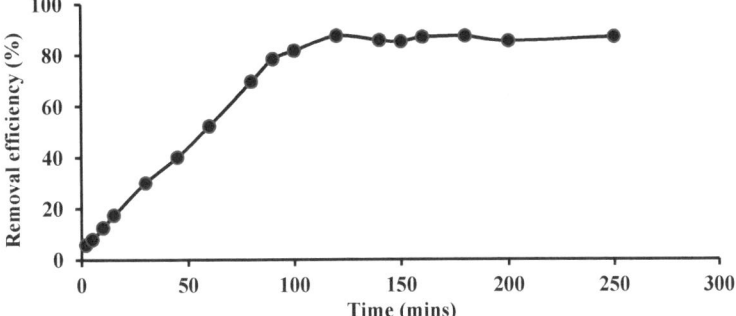

Figure 6. Effect of contact time on Pb(II) ions removal efficiency using Si-TS-SSIL.

2.2.3. Effect of pH

The pH of the aqueous solution from which Pb(II) ions were removed significantly affected the extraction process because of the active sites and charge distribution on the Si-TS-SSIL surface, solubility, ionization, and speciation of Pb(II) ions in the solution. In this study, the effect of pH on Pb(II) metal removal in an aqueous solution was investigated between pH 3 to 9 using an Si-TS-SSIL extractant. The removal efficiency of the Si-TS-SSIL extractant increased with an increase in the pH with highest removal efficiency observed at pH 6, as shown in Figure 7. Above this value, the amount of Pb(II) metal ions removed decreased due to possible precipitation of Pb(II) hydroxide. In an acidic condition, the removal efficiency is low, which could be caused by the presence of protonated functional groups containing lone pairs [35].

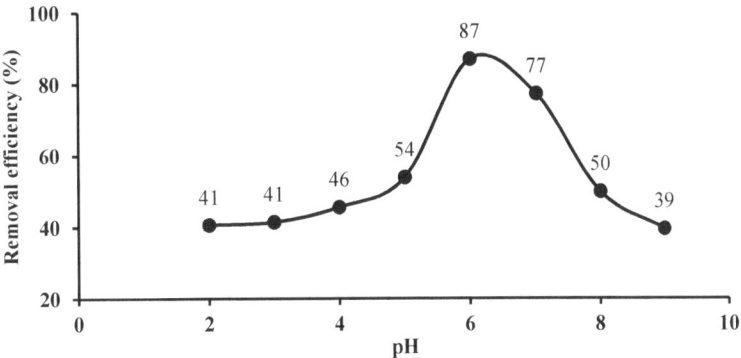

Figure 7. Effect of pH on the removal of Pb(II) ions from aqueous solution using Si-TS-SSIL.

2.3. Adsorption Kinetics

One of the essential characteristics that determines the extraction efficiency is the kinetic of the removal process, which describes the rate of removal of Pb(II) ions. The results of the kinetic analysis were used to establish the optimum mechanism for Pb(II) ions removal. Table 4 provides the summary of all the relative values and the statistical analysis for all models, with the most significant values highlighted. Pseudo-first order and pseudo-second order were applied in this kinetic study to understand the dynamics extraction of Pb(II) ions onto an Si-TS-SSIL extractant. The linearized forms of pseudo-first order and pseudo-second order kinetic models are presented in the Equations (1) and (2), respectively [36].

$$\log(q_e - q_t) = \log q_e - \left(\frac{k_1}{2.303}\right)t \quad (1)$$

$$\frac{t}{q_t} = \frac{1}{k_2 q_e^2} + \left(\frac{1}{q_e}\right)t \quad (2)$$

where q_t and q_e are the amount of the Pb(II) ions removed (mg g^{-1}) at any time and the amount of Pb(II) ions removed (mg g^{-1}) at equilibrium, respectively. The pseudo-first order rate constant (min^{-1}) is represented as k_1 whereas the rate constant for pseudo-second order (g mg^{-1} min^{-1}) is k_2. The rate constant and removal capacities for both models were calculated from the slope and intercept of the graphs (Figures 8 and 9).

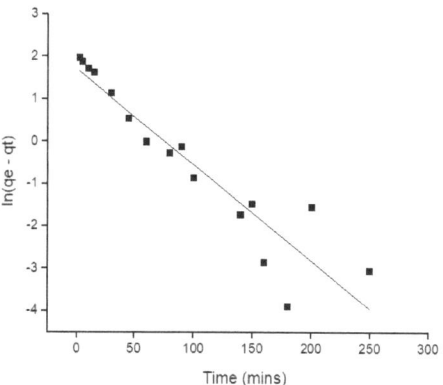

Figure 8. Pseudo-first order kinetic plots for the removal of Pb(II) ions onto Si-TS-SSIL.

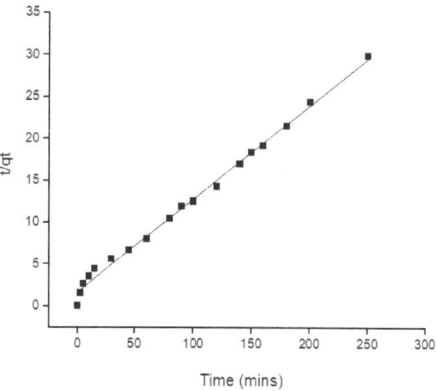

Figure 9. Pseudo-second order kinetic plots for the removal of Pb(II) ions onto Si-TS-SSIL.

Table 4. Kinetics model for the removal of Pb(II) ions by the Si-TS-SSIL extractant.

Extractant	Pseudo-First Order			Pseudo-Second Order		
Si-TS-SSIL	$q_{e\,exp}$ (mg/g)	$q_{e\,cal}$ (mg/g)	R^2	$q_{e\,exp}$ (mg/g)	$q_{e\,cal}$ (mg/g)	R^2
	8.3970	5.5182	0.8705	8.3970	8.9718	0.9942

To identify the best fit of kinetic models for the removal of Pb(II) ions, the values of the correlation coefficient (R^2) of the linear plots and the calculated removal capacities ($q_{e\,calc}$) against the experimental removal capacities ($q_{e\,exp}$) were compared between the pseudo-first and second order kinetic models. From this comparison, the value of R^2 for pseudo-second order reaction (0.9942) was higher compared to the pseudo-first order (0.8705). Moreover, the calculated removal capacities ($q_{e\,calc}$) for the pseudo-second order, which is 8.9718 mg g^{-1}, was found to be very close to the experimental values (8.397 mg g^{-1}). Meanwhile, the $q_{e\,calc}$ for pseudo-first order was 5.5182 mg g^{-1}, which was significantly different from the experimental removal capacity. Thus, this study suggested that that the pseudo-second order reaction better represents the uptake of Pb(II) ions onto an Si-TS-SSIL extractant from an aqueous solution. These results also show that the removal process was controlled by the chemisorption process [22].

2.4. Adsorption Isotherm

The experimental data of isotherms study has been described using a variety of adsorption models. The most used models are the Langmuir and Freundlich models. Both models were employed in this study to analyse the best equilibrium interaction. The Langmuir isotherm theory assumes that a monolayer adsorbate is formed on a homogeneous adsorbent surface. The Langmuir isotherm predicts monolayer adsorption on a homogenous surface with a limited amount of adsorption sites, with no intermolecular interactions occurring between the adsorbed molecules [37]. The Langmuir isotherm equation relates the amount of adsorbate adsorbed on the adsorbent to the equilibrium concentration, which is shown as follows in Equation (3):

$$\frac{C_e}{q_e} = \frac{C_e}{q_m} + \left(\frac{1}{K_L \cdot q_m}\right) \quad (3)$$

where q_e is the amount of Pb(II) ions removed by the Si-TS-SSIL extractant, K_L and q_m are the Langmuir constants, representing the energy constant (L mg^{-1}) and the maximum removal capacity (mg g^{-1}), respectively. From the graph in Figure 11, it was found that the correlation coefficient, R^2 of the Langmuir isotherm, was 0.9102. The Langmuir equation can also be used to determine a dimensionless equilibrium parameter called the separation factor, R_L which is mathematically defined in Equation (4).

$$R_L = \frac{1}{(1 + K_L \cdot C_o)} \quad (4)$$

where K_L and C_o are Langmuir constant (L mg^{-1}) and initial concentration of Pb(II) ions (mg L^{-1}), respectively. The linear removal process is represented by $R_L = 1$, whereas the irreversible removal process is represented by $R_L = 0$. Favourable removal is represented by $0 < R_L < 1$, whereas unfavorable removal is represented by $R_L > 1$ [38]. In this case, the R_L value was calculated and was found to be 0.3808. This result indicated that the removal of Pb(II) ions onto the Si-TS-SSIL extractant is a favourable removal process.

The Freundlich equation is expressed by Equation (5).

$$\ln q_e = \frac{1}{n}\left(\frac{\ln}{Ce}\right) + \ln K_F \quad (5)$$

where q_e is the Pb(II) ions uptake capacity (mg g^{-1}), C_e is the residual concentration of Pb(II) ions at equilibrium (mg L^{-1}), and n and K_F are the Freundlich constants, representing the removal intensity and removal capacity (mg g^{-1}), respectively. A greater n value ($n > 1$) suggests better removal performance, whereas $n < 1$ indicates poor removal performance [25]. In this study, as shown in Table 5, the value of n indicates a favourable removal process. The most prevalent condition is $n > 1$, which can be caused by a distribution of surface sites or any other circumstance that causes a decrease in adsorbent–adsorbate interaction as surface density rises. From Figures 10 and 11, the isotherm data fitted well with the Freundlich model with correlation coefficient (R^2) of 0.9961 compared to the Langmuir isotherm model.

Table 5. Parameters of Langmuir and Freundlich isotherms for the removal of Pb(II) ions by Si-TS-SSIL extractant.

Extractant	Langmuir Isotherm			Freundlich Isotherm		
	K_L (L/mg)	q_m (mg/g)	R^2	K_F (mg/g)	n	R^2
Si-TS-SSIL	0.1626	9.3729	0.9102	0.8592	1.8149	0.9961

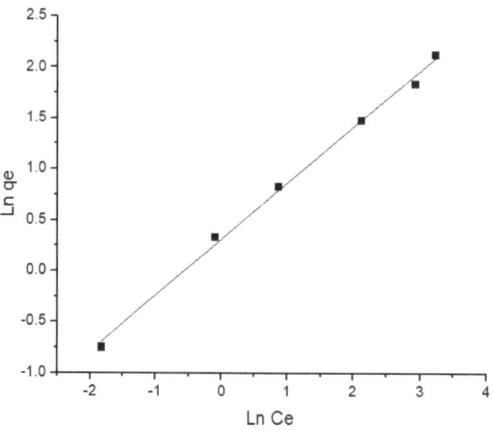

Figure 10. Freundlich isotherm for the removal of Pb(II) ions from aqueous solution by the Si-TS-SSIL.

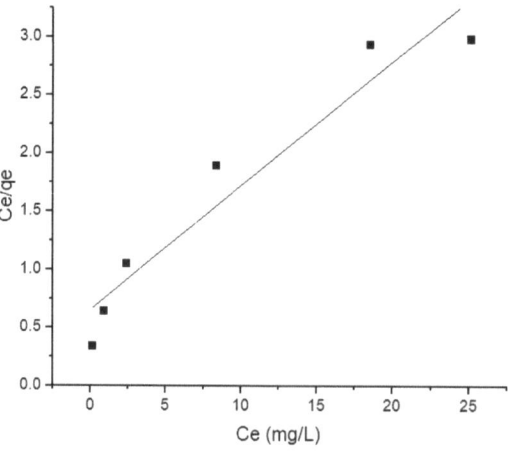

Figure 11. Langmuir isotherm for the removal of Pb(II) ions from aqueous solution by Si-TS-SSIL.

2.5. Adsorption Mechanism

For the adsorption mechanism using Si-TS-SSIL, the electron donors coming from the hydroxyl group, and the oxygen and sulphur of the thiosalicylate functional group in Si-TS-SSIL could be responsible for the chemisorption of Pb(II) ions. In addition to this, the organic and inorganic constituents' presence in the Si-TS-SSIL extractant offered more active sites to bind with Pb(II) ions in the aqueous solution [39]. Additionally, Pb(II) ions would have been trapped in the pores of Si-TS-SSIL, which could contribute to the removal efficiency. Therefore, greater removal efficiency of Pb(II) ions by Si-TS-SSIL could be caused by the thiol functional group existing on Si-TS-SSIL extractant surface for binding.

2.6. Adsorption from Wastewater

The adsorption performance of the Si-TS-SSIL extractant was evaluated by the experiments carried out using crude oil effluent containing 10 mg L^{-1} of Pb(II) ions. In this test, 0.25 g of Si-TS-SSIL extractant was dispersed into 12 mL of the wastewater samples, and the mixture was shaken for 120 min. It was found from the obtained results that Pb(II) was removed by 99%. Table 6 below shows the comparison of the removal efficiency of Pb(II) ions using Si-TS-SSIL from an aqueous solution and wastewater. From these results, Si-TS-SSIL can potentially be used as a metal extractant for the treatment of wastewater containing Pb(II) ions.

Table 6. Comparison of removal efficiency for Pb(II) removal using Si-TS-SSIL extractant from aqueous solution and wastewater containing 10 mg L^{-1} of Pb(II) ions.

Extractant	Source	Removal Efficiency (%)
Si-TS-SSIL	Aqueous solution	99.33
	Wastewater	80.6

3. Materials and Methods

3.1. Chemicals

Silica gel, sodium hydroxide pellets (98% purity), hydrochloric acid fuming 37%, diethyl ether, methanol, toluene, dichloromethane, and acetonitrile were acquired from Merck, NJ, USA. Meanwhile, 1-methylimidazole, 3-chloropropyltrimethoxysilane (CPTMS), thiosalicylic acid (97% purity), and Pb(II) nitrate salt (99.99% purity) were purchased from Sigma-Aldrich, St. Louis, MO, USA.

3.2. Instrumentation

A nuclear magnetic resonance (NMR) instrument (Model 1200 Series Avance III, Bruker, Billerica, MA, USA) was used to obtain the ^{1}H-NMR spectra of newly synthesized [MTMSPI][TS] ionic liquid, and to perform solid-state NMR analysis to confirm the formation of the covalent bond formed between [MTMSPI][TS] and activated silica gel in a new thiosalicylate-based solid-supported ionic liquid (Si-TS-SSIL). The functional groups of the [MTMSPI][TS] ionic liquid were determined using Fourier-transform infrared spectrophotometer (FTIR) instrument (Model Frontier, Thermo Fisher Scientific, Waltham, MA, USA). A Karl Fischer titrator (Model KF V30, Mettler Toledo, Columbus, OH, USA) was used to analyze the water content in the [MTMSPI][TS] ionic liquid. A thermal analyzer (Model STA 6000, Perkin Elmer, Waltham, MA, USA) was used to record the thermal stability whereas the halide content in [MTMSPI][TS] ionic liquid was analyzed using an ion chromatography (IC) instrument (Model Compact 930, Metrohm, Herisau, Switzerland). The surface area of Si-TS-SSIL was measured using a Brunauer-Emmett-Teller (BET) instrument (Tristar 3020, Micrometritics, Norcross, GA, USA). The concentration of Pb(II) ions in an aqueous solution was determined by an atomic absorption spectrometer (AAS) from Agilent (Model 240 FS, Agilent Technologies, Santa Clara, CA, USA).

3.3. Procedures

3.3.1. Synthesis of [MTMSPI][TS] Ionic Liquid

1-methyl-3-(3-trimethoxysilylpropyl) imidazolium thiosalicylate ([MTMSPI][TS]) ionic liquid was synthesized in two steps. The first step is the synthesis of the reagent required to produce [MTMSPI][TS], i.e., the 1-methyl-3-(3-trimethoxysilylpropyl) imidazolium chloride ([MTMSPI][Cl]) ionic liquid. A-4.13 g (25 mmol) of 1-methylimidazole was added to a stirred solution of 3-chloropropyltrimethoxysilane (25 mmol) to form [MTMSPI][Cl] ionic liquid. This mixture was refluxed under 80 °C and stirred at 400 rpm for 72 h in the silicone oil bath. The product, i.e., [MTMSPI][Cl], was decanted with diethyl ether three times to remove impurities and unreacted starting materials [40] and the remaining solvent was extracted by using a rotary evaporator.

Thereafter, in the second step, sodium thiosalicylate (25 mmol) was added to the obtained [MTMSPI][Cl] and stirred for 48 h at room temperature to complete the anion exchange process to obtain the new [MTMSPI][TS] ionic liquid with the by-product of NaCl. The by-product salt, NaCl, was filtered after the addition of acetonitrile and the solvent was then removed by rotary evaporator. The [MTMSPI][TS] ionic liquid that was obtained was found to be a clear and viscous yellow liquid and this newly synthesized ionic liquid was dried under high vacuum. The synthesis route for the formation of [MTMSPI][TS] ionic liquid is shown in Figure 12. The newly synthesized [MTMSPI][TS] ionic liquid was analyzed for structural analysis using NMR and FTIR. The purity of this newly synthesized ionic liquid was also analyzed in terms of halide content and moisture content using IC and the Karl Fischer titrator, respectively. Meanwhile, the thermal analysis including the decomposition temperature of [MTMSPI][TS] was studied using TGA.

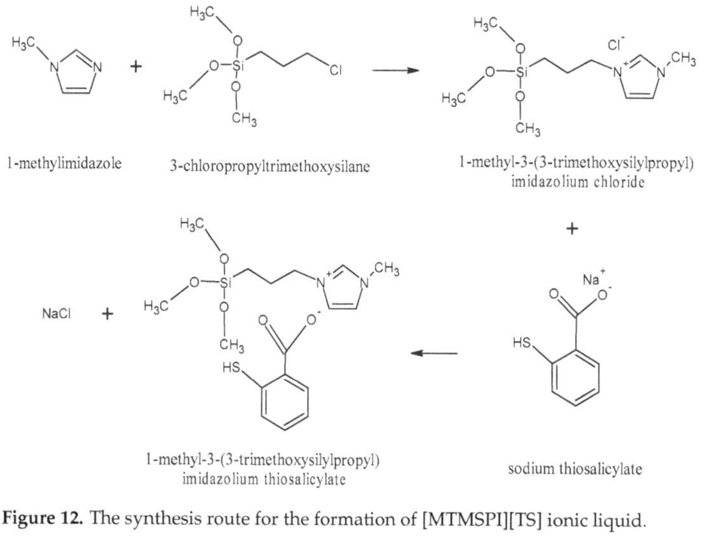

Figure 12. The synthesis route for the formation of [MTMSPI][TS] ionic liquid.

3.3.2. Preparation of Solid-Supported Ionic Liquid Containing the Thiosalicylate Functional Group (Si-TS-SSIL)

Si-TS-SSIL was prepared through covalently immobilized [MTMSPI][TS] onto activated silica gel. This preparation involved two steps. The first step was to activate the silica gel through acid activation to obtain activated silica gel surfaces, which is to activate surface silanol groups. These silanol groups would make covalent bonds with the cation of [MTMSPI][TS] to form a new solid-supported ionic liquid solid (Si-TS-SSIL). In this step, activated silica gel (10 g) was immersed in 100 mL of 6 M HCl. The mixture was then refluxed and stirred for 8 h. The suspension was filtered and washed several times with distilled water and ethanol to remove any HCl residue. It was then dried in a vacuum oven for about five hours at the temperature of 70 °C to remove any excess moisture to obtain the activated silica gel.

In the second step, 1 g of the synthesized [MTMSPI][TS] was dissolved in methanol (20 mL) and toluene (20 mL), followed with the addition of 6 g of activated silica gel. This mixture was left stirred and refluxed at 100 °C for 48 h and Si-TS-SSIL was vacuum-filtered and washed with dichloromethane to ensure the complete elimination of solvent. Si-TS-SSIL was then dried in a vacuum oven at 120 °C overnight. The synthesis route to prepare Si-TS-SSIL is shown in Figure 13.

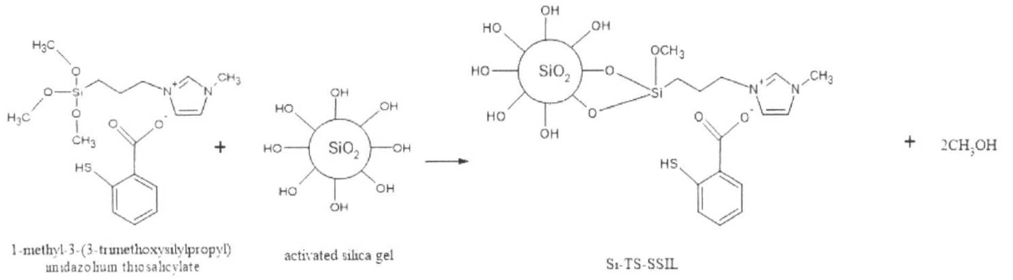

Figure 13. Synthesis route to produce Si-TS-SSIL extractant.

Based on the second step, an attempt was also made to prepare Si-TS-SSIL with a different mass ratio of [MTMSPI][TS] and activated silica gel, i.e., [MTMSPI][TS](in g):activated silica gel (in g). The purpose of this attempt is to study the effect of different amounts of [MTMSPI][TS] covalently bonded onto activated silica gel in Si-TS-SSIL on the removal efficiency of Pb(II) ions from an aqueous solution. The mass ratios of ([MTMSPI][TS](in g): activated silica gel (in g)) studied were 0.1:1, 0.12:1, 0.2:1, and 0.5:1.

3.3.3. Removal Study

An attempt was first made to compare the removal efficiency between three extractant, i.e., [MTMSPI][TS] ionic liquid, activated silica gel, and Si-TS-SSIL, to remove Pb(II) ions from an aqueous solution. In this removal study, 200 mg L^{-1} of aqueous Pb(II) ions solution was first prepared by mixing the appropriate amount of Pb(II) nitrate salt in distilled water. The removal experiments were conducted by mixing 0.25 g of [MTMSPI][TS] with 12 mL of 200 mg L^{-1} aqueous Pb(II) ions solution in propylene bottles. The propylene bottles were shaken in the orbital shaker with a constant agitation speed (400 rpm) for 30 min at room temperature. The mixture was then separated through centrifugation at 4000 rpm for 15 min to obtain a clear aqueous solution. This clear aqueous solution was analyzed for concentration of Pb(II) ions using AAS. The same experimental procedures were applied to study the performance of activated silica and Si-TS-SSIL to remove Pb(II) ions from the aqueous solution.

pH studies were carried out at various pH values ranging from 2 to 9. 0.25 g of the Si-TS-SSIL extractant, which was mixed with 12 mL of 200 mg L^{-1} aqueous Pb(II) ions solution in propylene bottles. The mixture was shaken at 400 rpm at room temperature. The final concentration of Pb(II) ions was analyzed using AAS and the removal efficiency was then calculated.

Based on the results of the effect of the [MTMSPI][TS] dosage in Si-TS-SSIL pH on the removal efficiency, the experiments for the removal of Pb(II) ions from the aqueous solution were proceeded further using Si-TS-SSIL (mass ratio of [MTMSPI][TS](in g):activated silica gel (in g) = 0.2:1) at pH 6 with the same initial concentration of aqueous Pb(II) ions solution (200 mg L^{-1}), but with different contact times ranging from 2 min to 250 min to determine the equilibrium time for the purpose of adsorption isotherm study. The results of the removal efficiency showed that the equilibrium time, t_{eq}, was 120 min. Thus, to obtain the data for adsorption isotherm study, the experiments for the removal of Pb(II) ions from the aqueous solution were carried out using different initial concentrations of aqueous

Pb(II) ions solution, but with the contact time, i.e., 120 min, at room temperature. In this adsorption isotherm study, different initial concentrations of aqueous Pb(II) ions solution were 10, 30, 50, 100, 150, and 200 mg L^{-1}.

In the experiments for the removal of Pb(II) ions from aqueous solution, the removal efficiency and removal capacity of extractant were calculated using Equations (6) and (7), respectively [41].

$$\text{Removal efficiency (\%)} = \frac{C_o - C_e}{C_o} \times 100 \quad (6)$$

$$\text{Removal capacity } (q_e) = \frac{(C_o - C_e)}{m} \times V \quad (7)$$

where q_e (removal capacity) represents the amount of Pb(II) ions removed at equilibrium (mg g^{-1}), and C_o and C_e (mg L^{-1}) are the initial and equilibrium concentrations of Pb(II) ions in the aqueous solution, whereas V (L) and m (g) represent the volume of the aqueous lead solution and mass of the extractant, respectively.

3.4. Adsorption Isotherm and Kinetics Models

The adsorption isotherms for Pb(II) ions were obtained via removal for 120 min, which was sufficient to reach equilibrium. The effect of the initial concentration of Pb(II) ions (10–200 mg L^{-1}) was investigated. The adsorption isotherm and kinetics models were used to determine the ideal fitting in this removal study. Adsorption isotherm models such as Langmuir and Freundlich are used to evaluate and compare the removal capacities of the extractants for the removal of Pb(II) ions in an aqueous solution, whereas for the kinetics studies, the pseudo first and second order models were applied. All models were used to process numerous experimental equilibrium data and to verify which model represents the best fit for the obtained data. The parameters obtained from the Langmuir and Freundlich models provide crucial information about the surface properties, the affinity of the adsorbent, and the extraction mechanism [42].

4. Conclusions

In this work, the newly ionic liquid, [MTMSPI][TS], and the functionalized solid supported ionic liquid, Si-TS-SSIL, were successfully synthesized. The formation of an [MTMSPI][TS] ionic liquid has been confirmed with the results from the characterization analyses such as NMR, FTIR, IC, moisture content, and TGA. Meanwhile, the formation of a chemical bond between [MTMSPI][TS] and activated silica gel in the newly synthesized Si-TS-SSIL, i.e., Si-O bond, was confirmed by solid-state NMR analysis. The results of solid-state NMR analysis have shown that an attempt to chemically immobilize [MTMSPI][TS] on activated silica gel to produce Si-TS-SSIL was successful.

The optimum removal efficiency of Pb(II) ions from the aqueous solution was achieved at a contact time of 120 min with the highest removal efficiency of 87% using Si-TS-SSIL that has an [MTMSPI][TS]:activated silica gel mass ratio of 0.2:1. The equilibrium data was fitted with the Langmuir and Freundlich isotherm models whereby the best correlation was obtained by the Freundlich model. The uptake of Pb(II) ions using the Si-TS-SSIL extractant was well described by the kinetics study, which pointed towards the pseudo-second order kinetic model, suggesting the removal process was controlled by the chemisorption process.

The novelty of this study is the synthesis of the [MTMSPI][TS] ionic liquid and the chemical immobilization of the [MTMSPI][TS] ionic liquid onto the surface of a solid support, i.e., activated silica gel, producing Si-TS-SSIL extractant for the removal of Pb(II) ions from an aqueous solution. This work is crucial to ensure that the newly synthesized extractant (Si-TS-SSIL) is effective in removing Pb(II) ions metal ions from aqueous solutions, which would be applicable for other types of metal ions, as well as provide the precise characterization of the newly synthesized [MTMSPI][TS] ionic liquid.

Author Contributions: Conceptualization, N.A.L.K. and M.F.T.; methodology, N.A.L.K. and M.F.T.; formal analysis, N.A.L.K.; writing—original draft preparation, N.A.L.K.; writing—review and editing, M.F.T. and C.D.W.; supervision, M.F.T. and C.D.W.; funding acquisition, M.F.T. and C.D.W. All authors have read and agreed to the published version of the manuscript.

Funding: This research was funded by Yayasan Universiti Teknologi PETRONAS Fundamental Re-search Grant (YUTP-FRG), 015LC0-267.

Institutional Review Board Statement: Not applicable.

Informed Consent Statement: Not applicable.

Data Availability Statement: All relevant data are contained in the present manuscript.

Acknowledgments: The authors acknowledge the use of the facilities within the UTP Centralized Analytical Laboratory (CAL).

Conflicts of Interest: The authors declare no conflict of interest.

References

1. Liu, W.; Zheng, J.; Ou, X.; Liu, X.; Song, Y.; Tian, C.; Rong, W.; Shi, Z.; Dang, Z.; Lin, Z. Effective Extraction of Cr(VI) from Hazardous Gypsum Sludge via Controlling the Phase Transformation and Chromium Species. *Environ. Sci. Technol.* **2018**, *52*, 13336–13342. [CrossRef] [PubMed]
2. Akpoveta, O.V.; Osakwe, S.A. Determination of Heavy Metal Contents in Refined Petroleum Products. *IOSR J. Appl. Chem.* **2014**, *7*, 01–02. [CrossRef]
3. Kinuthia, G.K.; Ngure, V.; Beti, D.; Lugalia, R.; Wangila, A.; Kamau, L. Levels of heavy metals in wastewater and soil samples from open drainage channels in Nairobi, Kenya: Community health implication. *Sci. Rep.* **2020**, *10*, 8434. [CrossRef] [PubMed]
4. Vo, T.S.; Hossain, M.M.; Jeong, H.M.; Kim, K. Heavy metal removal applications using adsorptive membranes. *Nano Converg.* **2020**, *7*, 36. [CrossRef]
5. Gunatilake, S.K. Methods of Removing Heavy Metals from Industrial Wastewater. *Methods* **2015**, *1*, 12–18.
6. Xiao, W.; Pan, D.; Niu, Z.; Fan, Y.; Wu, S.; Wu, W. Opportunities and challenges of high-pressure ion exchange chromatography for nuclide separation and enrichment. *Chinese Chem. Lett.* **2022**, *33*, 3413–3421. [CrossRef]
7. Al-Qodah, Z.; Yahya, M.A.; Al-Shannag, M. On the performance of bioadsorption processes for heavy metal ions removal by low-cost agricultural and natural by-products bioadsorbent: A review. *Desalin. Water Treat.* **2017**, *85*, 339–357. [CrossRef]
8. Naseem, K.; Farooqi, Z.H.; Begum, R.; Ur Rehman, M.Z.; Shahbaz, A.; Farooq, U.; Ali, M.; Ur Rahman, H.M.A.; Irfan, A.; Al-Sehemi, A.G. Removal of Cadmium (II) from Aqueous Medium Using Vigna radiata Leave Biomass: Equilibrium Isotherms, Kinetics and Thermodynamics. *Zeitschrift fur Phys. Chemie* **2019**, *233*, 669–690. [CrossRef]
9. Hadj-kali, M.K.; El Blidi, L.; Mulyono, S.; Wazeer, I.; Ali, E. Deep Eutectic Solvents for the Separation of Toluene/1-Hexene via Liquid—Liquid Extraction. *Separations* **2022**, *9*, 369. [CrossRef]
10. Phakoukaki, Y.V.; O'Shaughnessy, P.; Angeli, P. Intensified liquid-liquid extraction of biomolecules using ionic liquids in small channels. *Sep. Purif. Technol.* **2022**, *282*, 120063. [CrossRef]
11. Chowdhury, I.R.; Chowdhury, S.; Mazumder, M.A.J.; Al-Ahmed, A. *Removal of Lead Ions (Pb2+) from Water and Wastewater: A Review on the Low-Cost Adsorbents*; Springer International Publishing: Cham, Switzerland, 2022; Volume 12, p. 8. ISBN 0123456789.
12. Wang, Y.L.; Li, B.; Sarman, S.; Mocci, F.; Lu, Z.Y.; Yuan, J.; Laaksonen, A.; Fayer, M.D. Microstructural and Dynamical Heterogeneities in Ionic Liquids. *Chem. Rev.* **2020**, *120*, 5798–5877. [CrossRef] [PubMed]
13. Angeli, P.; Ortega, E.G.; Tsaoulidis, D.; Earle, M. Intensified liquid-liquid extraction technologies in small channels: A review. *Johnson Matthey Technol. Rev.* **2019**, *63*, 299–310. [CrossRef]
14. Ventura, S.P.M.; Silva, F.A.E.; Quental, M.V.; Mondal, D.; Freire, M.G.; Coutinho, J.A.P. Ionic-Liquid-Mediated Extraction and Separation Processes for Bioactive Compounds: Past, Present, and Future Trends. *Chem. Rev.* **2017**, *117*, 6984–7052. [CrossRef] [PubMed]
15. Tang, B.; Bi, W.; Tian, M.; Row, K.H. Application of ionic liquid for extraction and separation of bioactive compounds from plants. *J. Chromatogr. B Anal. Technol. Biomed. Life Sci.* **2012**, *904*, 1–21. [CrossRef]
16. Ola, P.D.; Matsumoto, M. Metal Extraction with Ionic Liquids-Based Aqueous Two-Phase System. *Recent Adv. Ion. Liq.* **2018**, *8*, 146–158. [CrossRef]
17. Friess, K.; Izák, P.; Kárászová, M.; Pasichnyk, M.; Lanč, M.; Nikolaeva, D.; Luis, P.; Jansen, J.C. A review on ionic liquid gas separation membranes. *Membranes* **2021**, *11*, 97. [CrossRef]
18. Lanaridi, O.; Sahoo, A.R.; Limbeck, A.; Naghdi, S.; Eder, D.; Eitenberger, E.; Csendes, Z.; Schnürch, M.; Bica-Schröder, K. Toward the Recovery of Platinum Group Metals from a Spent Automotive Catalyst with Supported Ionic Liquid Phases. *ACS Sustain. Chem. Eng.* **2021**, *9*, 375–386. [CrossRef]
19. Wang, J.; Luo, J.; Feng, S.; Li, H.; Wan, Y.; Zhang, X. Recent development of ionic liquid membranes. *Green Energy Environ.* **2016**, *1*, 43–61. [CrossRef]

20. Wolny, A.; Chrobok, A. Silica-Based Supported Ionic Liquid-like Phases as Heterogeneous Catalysts. *Molecules* **2022**, *27*, 5900. [CrossRef]
21. Gawande, M.; Hosseinpour, R.; Luque, R. Silica Sulfuric Acid and Related Solid-supported Catalysts as Versatile Materials for Greener Organic Synthesis. *Curr. Org. Synth.* **2014**, *11*, 526–544. [CrossRef]
22. Regina, K.; Bernhard, K. Novel thiosalicylate-based ionic liquids for heavy metal extractions. *J. Hazard. Mater.* **2016**, *314*, 164–171. [CrossRef]
23. Ellerbrock, R.; Stein, M.; Schaller, J. Comparing amorphous silica, short-range-ordered silicates and silicic acid species by FTIR. *Sci. Rep.* **2022**, *12*, 1–8. [CrossRef]
24. Fadavipoor, E.; Badri, R.; Kiasat, A.; Sanaeishoar, H. CuO supported 1-methyl-3-(3-(trimethoxysilyl) propyl) imidazolium chloride (MTMSP-Im/Cl) nanoparticles as an efficient simple heterogeneous catalysts for synthesis of β-azido alcohols. *J. Iran. Chem. Soc.* **2019**, *16*, 1451–1458. [CrossRef]
25. Bilgiç, A.; Çimen, A. Removal of chromium(vi) from polluted wastewater by chemical modification of silica gel with 4-acetyl-3-hydroxyaniline. *RSC Adv.* **2019**, *9*, 37403–37414. [CrossRef]
26. Kogelnig, D.; Stojanovic, A.; Galanski, M.; Groessl, M.; Jirsa, F.; Krachler, R.; Keppler, B.K. Greener synthesis of new ammonium ionic liquids and their potential as extracting agents. *Tetrahedron Lett.* **2008**, *49*, 2782–2785. [CrossRef]
27. Konshina, D.N.; Lupanova, I.A.; Konshin, V. V Imidazolium-Modified Silica Gel for Highly Selective Preconcentration of Ag (I) from the Nitric Acid Medium. *Chemistry* **2022**, *4*, 1702–1713. [CrossRef]
28. Cestari, A.R.; Airoldi, C. A new elemental analysis method based on thermogravimetric data and applied to alkoxysilane immobilized on silicas. *J. Therm. Anal.* **1995**, *44*, 79–87. [CrossRef]
29. Chen, H.; Zhu, S.; Zhou, R.; Wu, X.; Zhang, W.; Han, X.; Wang, J. Thermal Degradation Behavior of Thiol-ene Composites Loaded with a Novel Silicone Flame Retardant. *Polymers* **2022**, *14*, 20. [CrossRef]
30. Akl, M.A.A.; Kenawy, I.M.M.; Lasheen, R.R. Organically modified silica gel and flame atomic absorption spectrometry: Employment for separation and preconcentration of nine trace heavy metals for their determination in natural aqueous systems. *Microchem. J.* **2004**, *78*, 143–156. [CrossRef]
31. Protsak, I.S.; Morozov, Y.M.; Dong, W.; Le, Z.; Zhang, D.; Henderson, I.M. A 29Si, 1H, and 13C Solid-State NMR Study on the Surface Species of Various Depolymerized Organosiloxanes at Silica Surface. *Nanoscale Res. Lett.* **2019**, *14*, 160. [CrossRef]
32. Duan, Y.; Chen, Y.; Lei, M.; Hou, C.; Li, X.; Chen, S.; Fang, K.; Wang, T. Hybrid silica material as a mixed-mode sorbent for solid-phase extraction of hydrophobic and hydrophilic illegal additives from food samples. *J. Chromatogr. A* **2022**, *1672*, 463049. [CrossRef]
33. Chrobok, A.; Baj, S.; Pudło, W.; Jarzebski, A. Supported hydrogensulfate ionic liquid catalysis in Baeyer-Villiger reaction. *Appl. Catal. A Gen.* **2009**, *366*, 22–28. [CrossRef]
34. Sharifah, T.; Tengku, M.; Yarmo, M.A.; Hakim, A.; Najiha, M.; Tahari, A.B.U.; Hin, T.Y.U.N. Immobilization and Characterizations of Imidazolium-Based Ionic Liquid on Silica for CO 2 Adsorption/Desorption Studies. *Mater. Sci. Forum* **2016**, *840*, 404–409. [CrossRef]
35. Saleem, S.; Nauman, A.; Saqib, S.; Mujahid, A.; Hanif, M. Extraction of Pb (II) from water samples by ionic liquid-modified silica sorbents Extraction of Pb (II) from water samples by ionic liquid-modified silica sorbents. *Desalination Water Treat.* **2014**, *52*, 7915–7924. [CrossRef]
36. Lutfullah; Rashid, M.; Haseen, U.; Rahman, N. An advanced Cr(III) selective nano-composite cation exchanger: Synthesis, characterization and sorption characteristics. *J. Ind. Eng. Chem.* **2014**, *20*, 809–817. [CrossRef]
37. Panda, H.; Tiadi, N.; Mohanty, M.; Mohanty, C.R. Studies on adsorption behavior of an industrial waste for removal of chromium from aqueous solution. *South African J. Chem. Eng.* **2017**, *23*, 132–138. [CrossRef]
38. Adina, N.; Lupa, L.; Mihaela, C.; Negrea, P. Characterization of Strontium Adsorption from Aqueous Solutions Using Inorganic Materials Impregnated with Ionic Liquid Characterization of strontium adsorption from aqueous solutions using inorganic materials impregnated with ionic liquid. *Int. J. Chem. Eng. Appl.* **2013**, *4*, 326. [CrossRef]
39. Rahman, N.; Nasir, M. Facile synthesis of thiosalicylic acid functionalized silica gel for effective removal of Cr(III): Equilibrium modeling, kinetic and thermodynamic studies. *Environ. Nanotechnol. Monit. Manag.* **2020**, *14*, 100353. [CrossRef]
40. Safari, J.; Zarnegar, Z. A highly efficient magnetic solid acid catalyst for synthesis of 2,4,5-trisubstituted imidazoles under ultrasound irradiation. *Ultrason. Sonochem.* **2013**, *20*, 740–746. [CrossRef]
41. Mohammad, N.; Atassi, Y. Enhancement of removal efficiency of heavy metal ions by polyaniline deposition on electrospun polyacrylonitrile membranes. *Water Sci. Eng.* **2021**, *14*, 129–138. [CrossRef]
42. Tian, M.; Fang, L.; Yan, X.; Xiao, W.; Row, K.H. Determination of Heavy Metal Ions and Organic Pollutants in Water Samples Using Ionic Liquids and Ionic Liquid-Modified Sorbents. *J. Anal. Methods Chem.* **2019**, *2019*, 1948965. [CrossRef]

Disclaimer/Publisher's Note: The statements, opinions and data contained in all publications are solely those of the individual author(s) and contributor(s) and not of MDPI and/or the editor(s). MDPI and/or the editor(s) disclaim responsibility for any injury to people or property resulting from any ideas, methods, instructions or products referred to in the content.

Article

Treatment of Organic and Sulfate/Sulfide Contaminated Wastewater and Bioelectricity Generation by Sulfate-Reducing Bioreactor Coupling with Sulfide-Oxidizing Fuel Cell

Thi Quynh Hoa Kieu [1,2,*], Thi Yen Nguyen [1] and Chi Linh Do [3]

1. Institute of Biotechnology, Vietnam Academy of Science and Technology, 18 Hoang Quoc Viet, Cau Giay, Hanoi 100000, Vietnam
2. Faculty of Biotechnology, Graduate University of Science and Technology, Vietnam Academy of Science and Technology, 18 Hoang Quoc Viet Str., Cau Giay, Hanoi 100000, Vietnam
3. Institute of Material Sciences, Vietnam Academy of Science and Technology, 18 Hoang Quoc Viet, Cau Giay, Hanoi 100000, Vietnam
* Correspondence: ktquynhhoa@ibt.ac.vn or kieuthiquynhhoa@gmail.com

Abstract: A wastewater treatment system has been established based on sulfate-reducing and sulfide—oxidizing processes for treating organic wastewater containing high sulfate/sulfide. The influence of COD/SO_4^{2-} ratio and hydraulic retention time (HRT) on removal efficiencies of sulfate, COD, sulfide and electricity generation was investigated. The continuous operation of the treatment system was carried out for 63 days with the optimum COD/SO_4^{2-} ratio and HRT. The result showed that the COD and sulfate removal efficiencies were stable, reaching 94.8 ± 0.6 and 93.0 ± 1.3% during the operation. A power density level of 18.0 ± 1.6 mW/m² was obtained with a sulfide removal efficiency of 93.0 ± 1.2%. However, the sulfide removal efficiency and power density decreased gradually after 45 days. The results from scanning electron microscopy (SEM) with an energy dispersive X-ray (EDX) show that sulfur accumulated on the anode, which could explain the decline in sulfide oxidation and electricity generation. This study provides a promising treatment system to scale up for its actual applications in this type of wastewater.

Keywords: electricity generation; sulfate-reducing bacteria (SRB); microbial fuel cell; sulfate reduction; sulfide oxidization

1. Introduction

Organic carbon and sulfate are widespread environmental contaminants resulting from human activities such as tanning processes, chemical manufacturing, landfills, food processing, swine, and the petrochemical industry [1–3], Under anaerobic conditions, sulfate-reducing bacteria (SRB) utilize organic compounds as carbon and energy and sulfate as the terminal electron acceptor for sulfide production (H_2S, HS^-, S^{2-}) and bicarbonate, according to (Equation (1)) below [4–6]:

$$2 CH_2O + SO_4^{2-} \rightarrow H_2S + 2 HCO_3^- \quad (1)$$

Therefore, besides organic compounds and sulfate, sulfide is also ubiquitous in these types of wastewater. The organic and sulfide/sulfate contaminated wastewater is a typical corrosive, odorous pollutant and toxic to human health and living organisms, especially in anoxic sulfate-rich environment. This wastewater is a hazardous substance that must be removed from wastewater before discharge into the environment. Sulfide can cause inhibition of the cytochrome oxidase enzyme system resulting in a lack of oxygen use in the cells. Anaerobic metabolism causes the accumulation of lactic acid leading to an acid-base imbalance. The nervous system and cardiac tissues are particularly vulnerable to the disruption of oxidative metabolism and death is often the result of respiratory arrest.

Hydrogen sulfide also irritates skin, eyes, mucous membranes, and the respiratory tract. Pulmonary effects may not be apparent for up to 72 h after exposure [7–9].

Conventional methods for removing sulfate/sulfide from wastewater include chemical oxidization by chloride (Cl^-), potassium permanganate ($KMnO_4$) and hydrogen peroxide (H_2O_2), chemical removal by metal salts [9,10], increasing redox potential to control by sulfide formation by air injection and biological oxidation by sulfide-oxidizing bacteria (SOB) can be used to prevent the formation of sulfide [11]. Although these methods can effectively remove sulfide, they share a common limitation of high energy and chemical consumption, which would result in high operating costs and enormous sludge. Moreover, these methods cannot remove organic carbon and sulfate/sulfide pollutants simultaneously.

In comparison with chemical methods, sulfide removal by sulfide-oxidizing bacteria (SOB) has the advantages of cost-effectiveness and minimization of chemical sludge [12]. However, SOB is commonly autotrophic so inorganic electron acceptors such as nitrate are needed for sulfide oxidizing. In wastewaters lacking nitrate, simultaneous removal of sulfide is not achievable; while the addition of electron acceptor (nitrate) is not a cost-effective and environmentally friendly option [13,14]. Therefore, it is necessary to seek novel methods to simultaneously remove sulfate/sulfide and organic carbon compounds from wastewater [15].

To solve the problems, microbial fuel cell (MFC), a novel method, which has been considered a promising method in reducing operating costs, energy, and toxic by-products compared with the traditional treatments is proposed in this study [16–18]. The use of microbial fuel cells (MFCs) for the treatment of organic carbon wastewater containing sulfate/sulfide attracts great attention nowadays due to the capability of bioelectricity generation and simultaneous removal of sulfate/sulfide and organic compounds based on the dissimilative microbial sulfate-reduction process [15]. Moreover, this technology would generate much less sludge than a conventional activated sludge process [19,20].

In the sulfate-reduction process, sulfate-reducing bacteria were selected as catalysts in wastewater treatment systems. The sulfide produced biologically based on the organic carbon-oxidizing and sulfate-reducing processes by SRB plays a key role of electron mediator in MFCs, which transfers electrons to the anode electrode to produce electricity and the additional amount of synthetic endogenous mediators, toxic and expensive compounds, is not necessary. Then, the released protons in the anodic chamber migrate through a proton—selective membrane into the cathode chamber [21,22]. In the last years, the number of studies on the application of MFC has increased [23].

MFCs have also been employed for effective organic compounds removal with bioelectricity recovery [24–28], while MFCs studies for sulfate/sulfide [2,29–32], especially simultaneous organics removal which is often presented together with sulfate/sulfide in wastewaters are still limited [33,34].

However, in previous studies, the anodic chamber of MFC was designed to oxidize organic compounds and convert sulfate to sulfide before sulfide oxidation occurs at the anode. This MFC configuration supports the co-existence of sulfate-reducing and sulfide-oxidizing processes in an anodic chamber. However, the main drawback of this configuration is that the MFC trended to have a long hydraulic retention time with a small extent of mixing, which could also have affected the sulfide oxidation and electricity generation of the MFC [31,34]. Therefore, the integration of sulfate-reducing bioreactor and sulfide-oxidizing fuel cell was designed for the improvement of the pollutant removal efficiency in this study.

The aim of this study was to (i) investigate the performance of a wastewater treatment system in the treatment of organic wastewater containing high sulfate under continuous operation and (ii) evaluate the influence of COD/SO_4^{2-} ratios and hydraulic retention time (HRTs) on the performance of the treatment system. The success of this study will help minimize environmental pollution and human health protection.

2. Results

2.1. Effect of Different Ratios of COD/SO_4^{2-} on Removal Efficiencies of Sulfate and COD in SRRB

Operating factors could affect the performance of sulfate-reducing bioreactor (SRRB) and sulfide-oxidizing fuel cell (SOFC) thus typical factors such as COD/SO_4^{2-} ratio and HRT were investigated. Nine different initial ratios of COD/SO_4^{2-} (0.5, 1, 1.5, 2, 2.5, 3, 4, 5, and 6) were conducted with sulfate concentration of 1300 mgL^{-1} and HRT fixed at 72 h in SRBR. It was found that both sulfate and COD removals seemed negatively correlated with the initial COD/SO_4^{2-} ratio, due to its inhibition to SRB.

2.1.1. Effect of Different COD/SO_4^{2-} Ratios on Sulfate Removal Efficiency in SRBR

The COD/SO_4^{2-} ratio was an important parameter related to electron flow in anaerobic fermentation. Results (Figure 1) indicated that when the ratios of COD/SO_4^{2-} increased, sulfide production was improved. The lowest sulfide production of 99 ± 7 mg L^{-1} with a sulfate reduction efficiency of 37.1% was obtained with a COD/SO_4^{2-} ratio of 0.5. When the ratios were increased to 1, 1.5, 2, 2.5, and 3, the sulfide production improved up to 153 ± 4, 213 ± 4, 320 ± 8, 344 ± 5, 364 ± 3 mg S L^{-1} with sulfate reduction of 55.8, 65, 93.5, 98.2 and 99.5%, respectively. The strong sulfate-reducing activity was observed at a COD/SO_4^{2-} ratio of 3 with a maximum of sulfide production (364 ± 3 mg L^{-1}) and sulfate reduction (99.5%). The results showed that at COD/SO_4^{2-} ratios higher than 3, sulfate removal efficiency decreased gradually. Sulfate was converted by 83.5, 69.4, and 47.8% of initial concentrations to sulfide of 263 ± 4, 233 ± 5, and 128 ± 4 mg L^{-1} under the feed COD/SO_4^{2-} ratios of 4, 5, and 6, respectively.

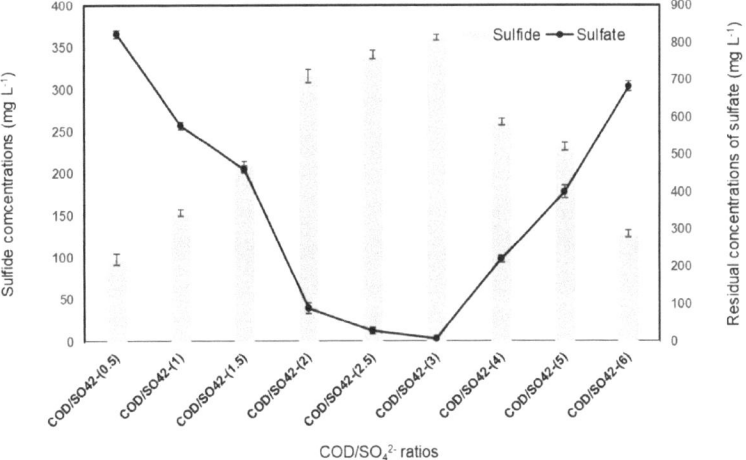

Figure 1. Effect of different ratios of COD/SO_4^{2-} on sulfate reduction and -sulfide production in SRBR with time. Error bars denote standard deviations.

2.1.2. Effect of Different COD/SO_4^{2-} Ratios on COD Removal Efficiency in SRBR

Figure 2 shows the residual concentrations of COD at various COD/SO_4^{2-} ratios (1.5; 2; 2.5, and 3). These results indicate that when the COD/SO_4^{2-} ratio increased to more than 1.5, sulfate reduction and sulfide production were improved (Figure 1), whereas COD was incompletely oxidized (Figure 2).

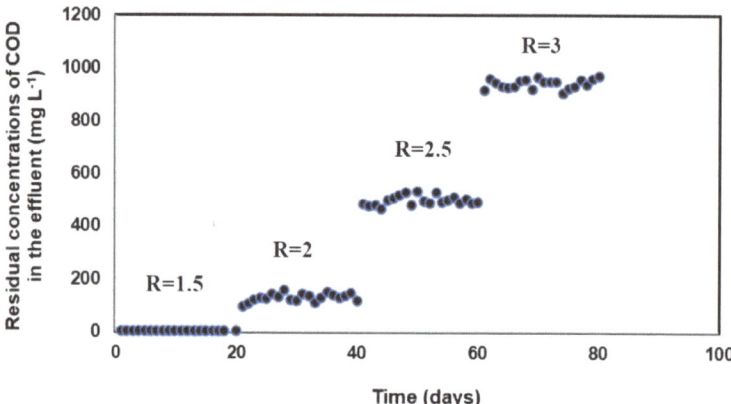

Figure 2. Residual concentrations of COD at various COD/SO$_4^{2-}$ ratios in SRBR with time (R = COD/SO$_4^{2-}$ ratio).

At the COD/SO$_4^{2-}$ ratio of ≤1.5, the COD removal efficiency was 99.7 ± 0.3%, and COD was not detected in the effluent. When the initial COD/SO$_4^{2-}$ ratio became 2 by adjusting the lactate concentration in the influent, the COD removal efficiency decreased to 95 ± 0.6%. At an initial COD/SO$_4^{2-}$ ratio of 2.5 and 3, the COD removal efficiency dropped further to 84.6 ± 0.5 and 75.8 ± 0.5%. COD removal was close to 100% at feed COD/SO$_4^{2-}$ ratios between 0.5 and 1.5. At higher ratios (2; 2.5 and 3), COD was detected in the effluent with residual COD concentration of 130 ± 15.6; 499 ± 17.7; and 942 ± 18 mg L^{-1}, respectively (Figure 2). This can be attributed to the sulfate-limiting conditions (Figure 1). The incomplete oxidation of COD was compensated by adding an excess of lactate to obtain the high sulfide production at the feed COD/SO$_4^{2-}$ ratio of 1.5, 2, 2.5, and 3, with sulfate removal efficiencies of 65, 93.5, 98.2, and 99.5%, respectively (Figure 1). Based on the efficiencies of sulfate and COD removal, the COD/SO$_4^{2-}$ ratio of 2 was selected for further determination of a suitable HRT in SOFC.

2.2. Effect of HRTs on Sulfide Removal and Electricity Generation in SOFC

Different levels of HRT (12, 18, and 24 h) were examined with the average initial sulfide concentrations of 316 ± 5.8 mg L^{-1}. The power density was calculated by voltage and current every 2 h in SOFC during operation. With prolonged HRT, the removal of sulfide was enhanced. Microbes could have sufficient opportunity to contact and react with sulfide when HRT increased.

Figures 3 and 4 show the results of power density and sulfide removal in SOFC. At an HRT of 12 h, the highest power density was observed, reaching 47.1 ± 0.9 mW/m^2. Meanwhile, the sulfide concentration in the effluent was 130 ± 4.9 mg L^{-1}, corresponding to sulfide removal efficiency of 58.8 ± 1.5%. As the HRT increased, the sulfide removal efficiency increased, while power density dropped. At the HRT of 18 and 24 h, the power densities of 34.5 ± 1 and 18.9 ± 1.1 mW/m^2 were obtained, respectively, with the stabilized voltage was 0.02 V. The sulfide concentrations of HRT 18 and 24 h in the effluent of SOFC were 83 ± 6.3 and 20 ± 3.8 mg L^{-1}, corresponding to sulfide removal efficiencies of 73.7 ± 1.7 and 93.7 ± 1.2%, respectively (Figure 4). The maximum sulfide removal efficiency achieved was as high as 93.7 ± 1.2% with HRT of 24 h.

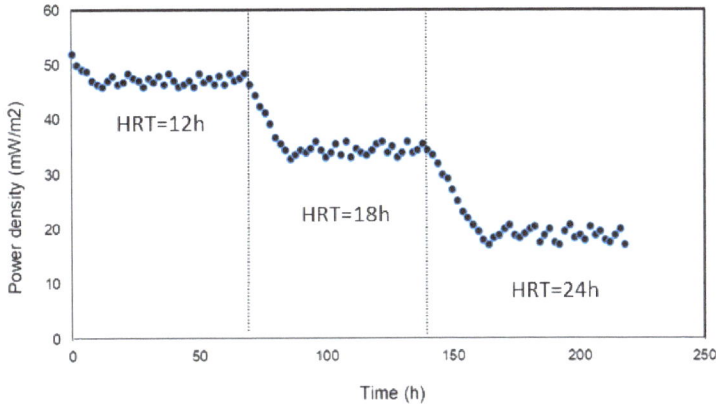

Figure 3. Power density in the SOFC at different HRTs (12, 18 and 24 h) with time.

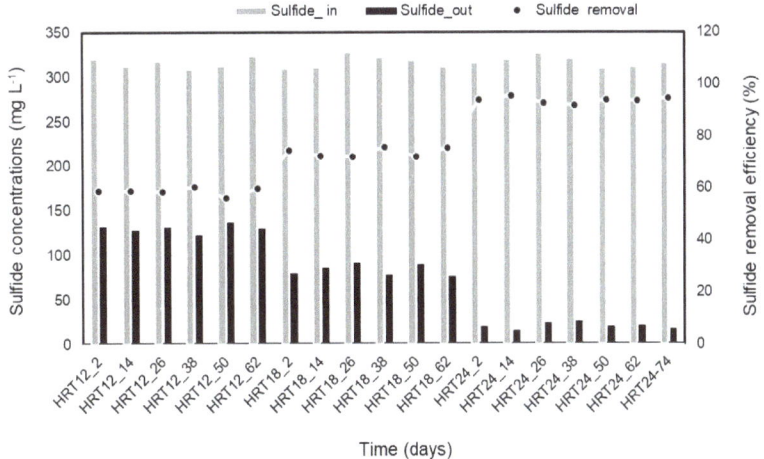

Figure 4. Sulfide removal efficiency in the SOFC at different HRTs (12, 18 and 24 h) with time.

2.3. The Removal of COD, Sulfate, Sulfide and Electricity Generation in Wastewater Treatment System

Based on the previous experiments, continuous operation of the wastewater treatment system (SRBR integrated with SOFC) was set up with the optimum COD/SO$_4^{2-}$ ratio of 2 and sulfate concentration of 1300 mg L^{-1}. The experiment was carried out for 63 days. The synthetic wastewater (see Section 4.2.1) was fed continuously at the bottom of the SRBR by a peristaltic pump with HRT of 72 h. The effluent of SRBR was supplied using a peristaltic pump into the anode of SOFC at HRT of 24 h.

* Removal efficiency of COD and sulfate in the SRBR:

Figures 5 and 6 display the daily performance outcomes of the SRBR including the removal efficiencies and residual concentrations of COD and sulfate under the selected conditions. The obtained results showed that the concentrations of COD and sulfate decreased significantly after 7 days. The lag period observed in the initial phase was attributable to the slow initiation of activities of SRB. The COD and sulfate removal efficiencies were stable during the continuous operation from day 8 to 63, with an average COD removal efficiency of 94.8 ± 0.6% (Figure 5). Sulfate was converted by 93 ± 1.3% of the initial concentration to sulfide (Figure 6).

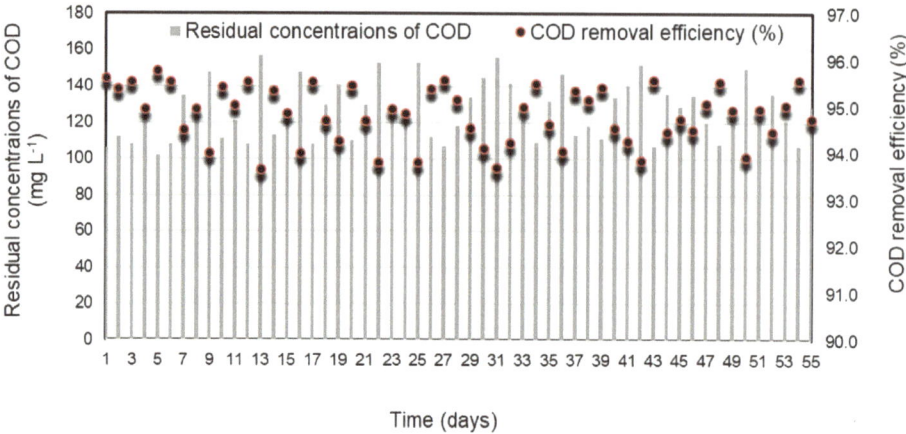

Figure 5. Removal efficiency and residual concentrations of COD in SRBR with time.

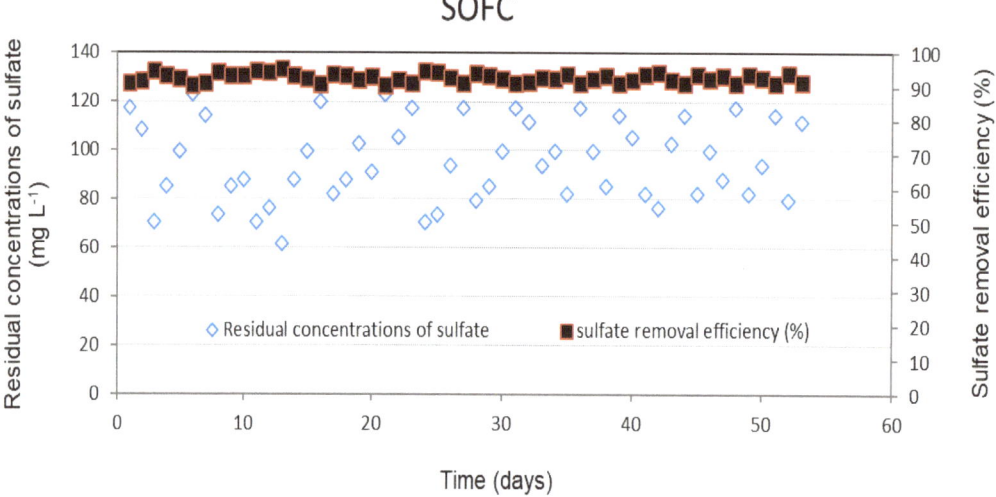

Figure 6. Removal efficiency and residual concentrations of sulfate in SRBR with time.

* Sulfide removal and electricity generation in SOFC

In this study, the anode chamber was filled with the effluent of the SRBR at HRT of 24 h during 63-day operation with the average sulfide concentrations of 316 ± 5.8 mg L^{-1}. Concentrations of sulfide in influent and effluent of the SOFC are shown in Figure 7. Within 38-day operation from day 8 to 45, sulfide was removed stably from the anode chamber with an amount of $93 \pm 1.2\%$. The average sulfide concentration in the effluent of SOFC was 21 ± 3.8 mg L^{-1}. This means that the sulfide present in the anode solution was oxidized, releasing electrons to the anode when the SOFC operation began, resulting in electricity production.

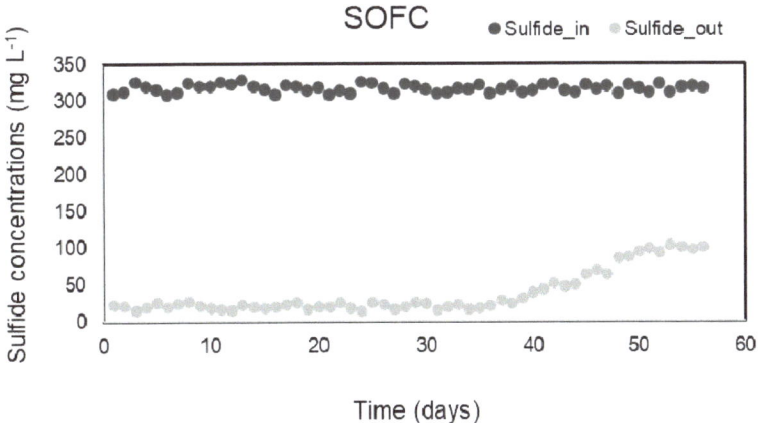

Figure 7. Sulfide concentrations in the influent and effluent of SOFC with time.

Electricity was generated continuously from the sulfide oxidation process of SOFC during 63-day operation. The power density reached a maximum power density (18.2 ± 1.6 mW/m^2) after the first week and remained stable for the next 35 days of SOFC operation (days 8 to 43). During the early stage of the SOFC operation, sulfide oxidation in the SOFC appeared to correspond to electricity generation. Figures 7 and 8 showed the relative positive correlation between bioelectricity generation and initial sulfide concentration, due to the reason that the anode potential decreased with the increase in sulfide as it possesses lower redox potential.

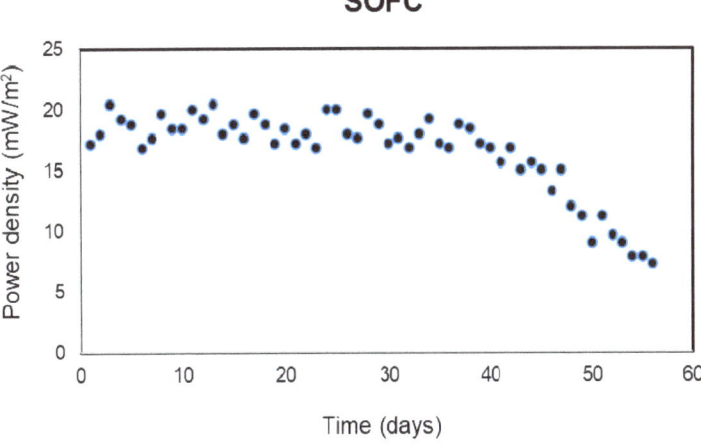

Figure 8. Power density during continuous operation of SOFC with time.

However, from day 43 to 45 onwards, the sulfide removal efficiency and power density dropped gradually to 68% and 7.2 mW/m^2, respectively, at the end of the operation (on day-63) (Figures 7 and 8).

The results of anode surface analysis by SEM-EDX are shown in Figure 9. As presented in Figure 9, many solid deposits were formed on the anode surface after SOFC operation. The characterization of the solids using EDX revealed that their major component was elemental sulfur. These indicate that sulfide was oxidized at the anode surface, resulting in the formation of insoluble elemental sulfur.

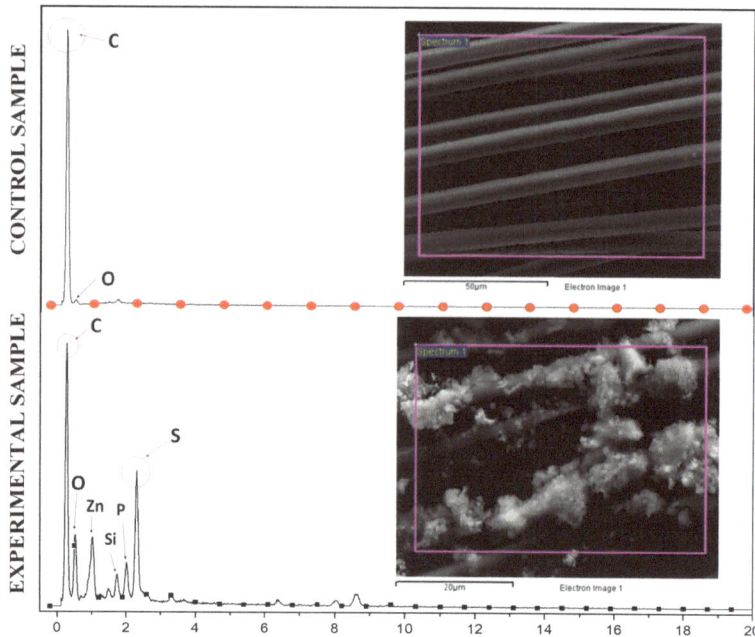

Figure 9. SEM images and EDX analysis of the anode surface before (**Above**) and after (**Below**) SOFC operation.

3. Discussion

The performance of the wastewater treatment system in this study revealed that a SOFC can be integrated effectively with a SRBR. To minimize environmental pollution, the SOFC was operated without the addition of synthetic electron transport mediators during the operation. The produced sulfide from the sulfate-reducing process in SRBR, a biological compound, was used as an electron mediator, which transfers electrons to the anode electrode [35,36]. Sulfide can be removed in the SOFC when it is oxidized through electrochemical reactions at the anode. Electrons and protons were produced during the process of oxidation of sulfide. The resulting electrons are delivered to the anode and transported to the cathode through an external circuit, producing electricity. Then, the released protons in the anodic chamber migrate through a proton exchange membrane (PEM) into the cathode chamber [36,37].

Oxygen, ferricyanide, nitrate, persulfate and permanganate are widely used as electron acceptors in the cathode due to their high oxidation potential in the cathode. In this study, the protons are taken up and consumed by ferricyanide and oxygen. Both ferricyanide and oxygen in the presence of electrons donated from the cathode surface react with protons and are reduced to ferrocyanide and water, as shown in Equations (2) and (3) [2,38,39]:

$$O_2 + 4e^- + 4H^+ \rightarrow 2H_2O \tag{2}$$

$$Fe(CN)_6^{3+} + e^- \rightarrow Fe(CN)_6^{4+} \tag{3}$$

Using a sulfide mediator to produce electricity was observed in different studies. Ref. [35] reported that dissolved sulfide can be converted to elemental sulfur by MFCs. The MFCs were connected with an up-flow anaerobic sludge reactor, providing removal of up to 98% and 46% of the sulfide and acetate, respectively. Ref. [40] studied the electricity generation potential and demonstrated the anodic potential was controlled by the sulfide concentrations in the chamber when treating sulfate-laden wastewaters. These investigations demonstrated that simultaneous sulfide and organic carbon removals with electricity generation can be achieved in MFCs. Ref. [29] studied sulfide removal by MFCs prior to irrigation water reuse. Ref. [33] demonstrated that removal efficiencies of 49.7 and 70% of the initial sulfide concentrations (150 and 60 mg L^{-1}) and power density of 1.2 mW/m^2 were achieved in the continuous MFC.

The ratio of COD/SO$_4^{2-}$ and HRT influences sulfate-reducing and COD oxidizing capability of SRB [41–43]. In this study, they were recognized as the most two important factors for sulfate and COD removal in SRBR as well as sulfide removal and bioelectricity generation in SOFC. The obtained results showed that at initial COD/SO$_4^{2-}$ ratios of 2–3 most of the sulfate in the influent was converted to sulfide with sulfate reduction of 93.5–99.5%. The finding was similar to results obtained by Ref. [44], who reported sulfate removal efficiencies over 91% at COD/SO$_4^{2-}$ ratios equal to or higher than 2.5 for sulfate concentrations up to 1960 mg L^{-1}.

On the other hand, the obtained data (Figure 1) showed that the lower (0.5, 1, 1.5) or higher COD/SO$_4^{2-}$ ratios (4, 5, 6) might result in low sulfate removal efficiency in the SRBR (37.1 to 83.5%). The lowest sulfide production with sulfate reduction efficiency of 37.1 and 47.8% were observed with COD/SO$_4^{2-}$ ratios of 0.5 and 6, respectively. COD removal was close to 100% at COD/SO$_4^{2-}$ ratios $\leq$ 1.5. At higher ratios ($\geq$2), COD was detected in the influence of the SRBR. Low sulfate reduction at COD/SO$_4^{2-}$ ratios of 0.5, 1, and 1.5 might be attributed to the inhibition of the anaerobic process due to the lack of carbon source [45]. However, at high COD/SO$_4^{2-}$ ratios (4, 5, and 6) and in sulfate-limiting conditions, methane-producing bacteria (MPB) might be dominant in the competition with SRB. Ref. [46] demonstrated that the highest sulfate removal efficiency was obtained when lactate or acetate was used as carbon and electron sources at COD/SO$_4^{2-}$ ratios between 1.5 and 2.25. Ref. [47] also suggested that a negative effect on the SRB activity can be observed at COD/SO$_4^{2-}$ ratios of more than 2.7 because competition for nutrients can occur between SRB and MPB. Ref. [43] reported that COD/sulfate ratio and HRT influence sulfate loadings and were recognized as the most two important factors for sulfate removal and bioelectricity generation. In their study, the maximum electricity generation and sulfate removal (83.9%), with the fixed COD/sulfate ratio of 4 and the influent COD of 2400 mg L^{-1}, were established at an HRT of 60 h. However, as mentioned above, COD was detected in the influence of the SRBR at COD/SO$_4^{2-}$ ratio equal to or higher than 2. Therefore, based on the efficiencies of sulfate and COD removal, the COD/SO$_4^{2-}$ of 2 was selected for the 63-day -continuous operation.

Besides the COD/SO$_4^{2-}$ ratio, HRT is also one of the most important factors during the operation of biological processes, as it determines the contact duration between pollutants and microbes. In this study, the maximum sulfide removal efficiency achieved was as high as 93.7 $\pm$ 1.2% with HRT of 24 h. However, the highest power density was observed, reaching 47.1 $\pm$ 0.9 mW/m^2 at HRT of 12 h. This might explain that the higher sulfide removal efficiency at the increased HRT resulted from the slower flow with the longer retention time providing sulfide ions more opportunities to undergo electrochemical reactions on the anode. In terms of the power density performance, however, a slower flow for a higher HRT is undesirable due to the lack of the supply of fresh sulfide ions from SRBR. In such cases, the anode loses the opportunity to encounter electron-rich fresh sulfides, thereby undergoing mass transport polarization. Electricity generation displayed an inverse phenomenon as electron donors (sulfide) became insufficient. The abundant consumption of sulfide also raised the anode potentials which did not benefit bioelectricity generation.

MFCs often faced trade-off in aspects of electricity generation and pollutant removal [48]. Based on the consent MFC technology should be applied as a waste/wastewater treatment unit rather than a renewable energy source. In this study, considering the sulfide removal efficiency, which is one of the most important values in a wastewater treatment system, the optimum HRT was determined to be 24 h. The continuous operation of the treatment system was carried out for 63 days with the optimum COD/SO$_4^{2-}$ ratio and

HRT. The results indicated that the COD and sulfate removal efficiencies were stable in SRRB, reaching 94.8 ± 0.6% and 93 ± 1.3% during the operation. Electricity was generated continuously and stably (18 ± 1.6 mW/m^2) from the sulfide oxidation process of SOFC for the first 45 days with 93 ± 1.2% of the sulfide being oxidized from the anode chamber. However, the sulfide removal efficiency and power density decreased gradually after 45 days. At the end of the operation, sulfide removal efficiency and power density were 68% and 7.2 W/m^2, respectively.

The results obtained in our study showed that the generated bioelectricity is proportional to the sulfide concentration, the higher concentration of sulfide obtained, the more electricity is generated. The results are consistent with the report from [40,49]. They demonstrated that the bioelectricity produced by the electrodes was dependent on the concentration of the sulfide, which on the other hand indicated the sulfide oxidation process.

The decline in sulfide oxidation and electricity generation after 45 days of operation can be attributed to the deposition of elemental sulfur, which hinders the effective mass transport of fresh sulfide ions to the anode electrode. The concentration of sulfate did not increase during the sulfide removal, suggesting that under this condition, sulfide was oxidized to sulfur, not sulfate. This was confirmed through an anode surface analysis with SEM and EDX at the end of the SOFC operation. The accumulation of sulfur on the anode may have decreased the electrical conductivity of the anode, thereby increasing the overpotential of the anode in the SOFC over time. This explanation is in agreement with previous studies [50,51]. Ref. [50] observed losses in the current output of MFCs and the reduction in removal of sulfide or sulfate from wastewater due to the deposition of elemental sulfur on the electrode surface. Furthermore, exopolysaccharides produced from SRB hamper electron transfer between bacterial cells and the electrode and thus reduce the voltage. Ref. [52] who investigated the performance of MFC treating organic wastewater containing high sulfate showed that the ohmic loss or internal resistance of the MFC increased over time from day 13 to day 54. The anode replacement on day 81 resulted in a significant reduction in the ohmic loss, suggesting the important role of the anode in the internal resistance of the MFC.

To remove effectively pollutants from wastewater and generate stable electricity in long-term operation, the integrated treatment system should be (i) separated from sulfate-reducing and sulfide-oxidizing processes, and (ii) operated in continuous flow mode. In such systems, substrate and other nutrients will be continuously supplied to the SRB. Moreover, the continuous flow mode could also remove the by-product elemental sulfur from the anode compared with MFCs operated in batch mode. Furthermore, the presence of sulfate in the anode media has negative effects on sulfide oxidation and electricity generation in MFCs. Ref. [31] reported that low concentrations of sulfate (≤ 1470 mg L^{-1}) benefited the MFC efficiency, while higher sulfate presence blocked the sulfide oxidization and electricity generation.

4. Materials and Methods

4.1. Inoculum and Culture Medium

A consortium of SRB was enriched from anaerobic sludge rich in sulfide from a crude oil tanker, Vung Tau, Vietnam, and used as the inoculum. This culture was cultivated under anaerobic conditions using Postgate's medium B [53] with a slight modification. Modified Postgate's B medium contained (in g/L): KH$_2$PO$_4$ 0.5; NH$_4$Cl 1.0; Na$_2$SO$_4$ 1.0; MgSO$_4$·7H$_2$O 2.0; Sodium lactate 3.2; Yeast extract 0.5; FeSO$_4$·7H$_2$O 0.01; ascorbic acid 0.1; thioglycolic acid 0.1. The pH was adjusted to 7.0 ± 0.2 using HCl 1 M or NaOH 1 M.

All procedures during the preparation of the medium and cultivation were performed according to the modified Hungate's method for anaerobes [54]. To enrich the SRB number, the cultivation step was repeated three times before inoculating into the SRBR. The enrichment culture was obtained as follows: The culture was seeded with 10% (v/v) inoculum and incubated at 30 °C in a Hungate tube or glass culture bottle. The medium was sparged with pure nitrogen gas to maintain anaerobic conditions before inoculation. Every week

10% of the volume of the culture in the bottle (or Hungate tube) was replaced by fresh medium. After 3-time cultivations, a culture containing a high density of SRB was achieved. The cell density of the enrichment culture was approximately 1×10^8 cells mL^{-1} in all experiments.

4.2. Design of Wastewater Treatment System

The system consisted of two identical components: (1) a SRBR to reduce sulfate to sulfide; (2) a SOFC to oxidize subsequently sulfide to element sulfur (S°)/sulfate (SO$_4^{2-}$). The integrated use of SRBR and SOFC can remove organic matter (COD), sulfate and sulfide simultaneously with electricity generation. The experimental apparatus is illustrated in Figure 10.

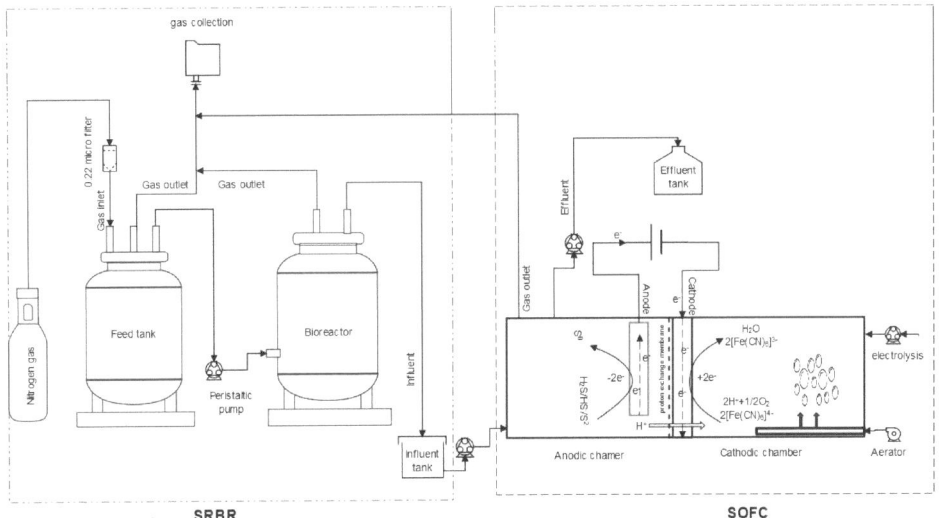

Figure 10. Configuration of a sulfate-reducing bioreactor (SRBR) integrated with a sulfide-oxidizing fuel cell (SOFC).

4.2.1. Sulfate-Reducing Bioreactor (SRBR)

The schematic diagram of the feeding tank and sulfate-reducing bioreactor are present in Figure 10.

* Feeding tank

Synthetic wastewater (see below) was prepared aseptically to avoid contamination and then fed continuously at the bottom of the bioreactors by a peristaltic pump (Ismatec SA, Zuerich, Switzerland) with a volumetric flow of 125 mL d^{-1} (HRT of 72 h).

To maintain the anaerobic condition, feeding tank was purged with filter sterilized nitrogen gas (0.22 µm). Gas produced during the treatment process was trapped by 4% (w/v) NaOH solution.

* Synthetic wastewater composition

The composition of the synthetic wastewater (g L^{-1}) consisted of KH$_2$PO$_4$, 0.5; NH$_4$Cl; Na$_2$SO$_4$, 1.9; MgSO$_4$·7H$_2$O, 0.06. Synthetic wastewater was fed with different COD/SO$_4^{2-}$ ratios. The sulfate concentration (1300 mg L^{-1}) was maintained constant during the whole experiment and the sodium lactate concentration (in COD) was varied in the medium (1070, 2140, 3210, 4280, 5350, 6420, 8560, 10,700 and 12,840 mg L^{-1}) to obtained feed COD/SO$_4^{2-}$ ratios of 0.5, 1.0, 1.5, 2.0, 2.5, 3, 4, 5 and 6, respectively. The synthetic wastewater was not supplemented with Fe^{2+} and reducing agents such as yeast extract, ascorbic acid, sodium thioglycolate and Na$_2$S to prevent the precipitation of FeS. This avoids clogging pipes and membranes, and loss of sulfide in the treatment system. The pH was adjusted to 7.0 ± 0.2 using HCl 1 M or NaOH 1 M.

* SRBR and operating conditions

For sulfate reduction and removal of organic matter (COD), the experiments were carried out in a continuous anaerobic SRBR in an up-flow mode. This reactor was fabricated from glass, having a total volume of 350 mL and a working volume of 250 mL. The SRBR was soaked in a 3 M HNO$_3$ solution for 72 h and rinsed with de-ionized water before use to avoid contamination.

The SRBR was first inoculated with 10% (v/v) of the enriched SRB consortium containing 1×10^8 cells mL^{-1} using synthetic wastewater containing sulfate and sodium lactate as electron acceptor and donor, respectively. The synthetic wastewater was continuously injected into the bottom of the SRBR by a peristaltic pump. It then flowed upward. After inoculation with enriched SRB consortium, the reactor was purged by N$_2$ gas to provide the anaerobic condition. To investigate the roles of SRB in the removal of organic (COD) and sulfate in SRBR, the reactor was operated continuously with a HRT of 72 h for 63 days at ambient room temperature (25 ± 2 °C).

4.2.2. Sulfide-Oxidizing Fuel Cell (SOFC)

* Design and fabrication of SOFC

Double-chambered SOFC design was used to investigate the removal of sulfide and electricity generation in SOFC. The SOFC consisted of two identical chambers (working volume of 80 mL each): (i) anode chamber, where electrochemical oxidation of sulfide on an anode surface derives electrons and protons and (ii) cathode chamber, where the oxygen and ferricyanide are the terminal electrons acceptors and react with the released protons. The design of this SOFC precludes the possibility of oxygen ingress in the anodic chamber because the anode and cathode chambers are divided by proton-exchange membrane (PEM). It is permeable only for H$^+$ cations.

Both chambers were made of mica acrylic plates to prevent any corrosion from sulfide. These plates were fabricated by using a CNC machine. After machining, the plates were glued with acrylic resin to form the SOFC chambers. Before the experiment, all the electrodes were first immersed in 1 M NaOH then 1 M HCl for one-hour each to remove microbial residues on the electrodes surface.

* Preparation of SOFC electrode

An electrocatalyst ink was prepared by mixing Pt/C 40% wt. catalyst powder (Johnson Mathey—Wayne, PA, USA), Nafion solution 5% and iso-propanol. Afterward, the mixture was ultrasonicated and stirred by magnetic to ensure that the catalyst powder was uniformly dispersed in the ink. The Pt/C catalyst layer was prepared by brushing the catalyst ink on a carbon cloth with an active area of around 10 cm^2. After each brushing, the sample was dried in air and this process was repeated until reaching the desired Pt loading of 1 mg/cm^2. Finally, the catalyst layer was dried at 130 °C for 30 min. The cathodic electrode was fabricated by hot pressing the catalyst layer on a Nafion 117 membrane. The hot-pressing conditions were executed at a temperature of 135 °C, duration time of 180 s and pressure of 21 kg/cm^2. For the anode, a plain carbon cloth (1071 HBC, Pinon Driver,—USA) was utilized with a working area of about 5 cm^2.

* SOFC operation conditions

To investigate the feasibility of sulfide removal and the electricity generation capacity of the SOFC, the effluent of SRBR was fed continuously into the anode chamber of SOFC by a peristaltic pump with HRT of 24 h. The SOFC was operated for 63 days at ambient room temperature (25 ± 2 °C).

The cathode chamber was filled with ferricyanide solution (50 mM K$_3$Fe (CN)$_6$) and air was supplied through a diffuser connected to an aerator. Ferricyanide solution is often used in the cathodic medium in SOFC to substitute oxygen as a cathodic electron acceptor due to low overpotential. To maintain sufficient deoxidation by oxygen in the cathode chamber, fresh ferricyanide solution was replenished in the cathodic chamber every week.

4.2.3. Determination of Suitable Conditions for SRBR and SOFC

* Determination of COD/SO_4^{2-} ratios for sulfate removal in SRBR

To select a COD/SO_4^{2-} ratio that simultaneous removal of COD, sulfate and production of sulfide effectively in SRBR, the continuous experiments were operated with HRT of 72 h for 180 days at various COD/sulfate ratios (0.5, 1, 1.5, 2, 2.5, 3, 4, 5, and 6). The period of each COD/sulfate ratio lasted for 20 days. After 20-day operation of COD/SO_4^{2-} ratio of 0.5, other ratios (1, 1.5, 2, 2.5, 3, 4, 5, and 6) were respectively started up and having similar performance. During the optimization of the COD/SO_4^{2-} ratio, sulfate concentration was maintained constant (1300 mg L^{-1}) while the COD/sulfate ratios were increased in a stepwise manner from 0.5 to 6 by increasing the amount of sodium lactate (in COD). The sulfate removal efficiency was estimated by measuring the concentration of sulfide and sulfate in the effluent of SRBR with all ratios of COD/SO_4^{2-} after reaching a dynamic equilibrium (after 7 days). The COD removal efficiency was only monitored at COD/SO_4^{2-} ratios of 1.5, 2, 2.5 and 3 that strong sulfate-reducing activity was observed.

* Determination of suitable HRTs for sulfide removal and electricity generation in SOFC

For the effect of HRTs on the sulfide removal and the electricity generation capacity of SOFC, a feed COD/SO_4^{2-} ratio of 2 and a sulfate concentration of 1300 mg L^{-1} were used. The effluent of the SRBR was supplied using a peristaltic pump into the SOFC system at various HRTs in the range of 12, 18, and 24 h. Sulfide concentrations in the influent and effluent of the anode chamber in SOFC were measured at 12 h intervals during 3 days of each HRT. The power density was examined every 2 h in SOFC.

* Performance of wastewater treatment system

Based on the selected COD/SO_4^{2-} ratio of 2 and optimum HRT from the test in SRBR (72 h) and SOFC (24 h), respectively, continuous operation of the wastewater treatment system was carried out for 63 days. The SOFC was operated without the addition of external substrates or electron transport mediators during the experiment. After reaching a dynamic equilibrium (7 days), the influent and effluents from SRBR and SOFC were sampled for investigating removal efficiencies of sulfate, sulfide, and COD every day. The power density was monitored every 2 h to check for changes in the electricity generation efficiency during the operation.

4.3. Analytical Methods

Influent and effluent samples of SRBR and SOFC were collected over time for COD, sulfate, and sulfide measurement. Before each analysis, the samples were filtered through a 0.45 μm nitrocellulose membrane syringe filter. Sulfate was measured using the turbidimetric method based on the addition of barium chloride to form a colloidal suspension of barium sulfate at 420 nm [55]. Sulfide (H_2S, HS^-, and S^{2-}) was measured at 480 nm according to Cord–Ruwish method based on CuS precipitation [56]. The organic substrate utilization was estimated by measuring the chemical oxygen demand (COD). The concentrations of COD were monitored in a concentrated sulfuric acid based on digestion with potassium dichromate for 2 h at 150 °C [55]. Power density (P, mW/m^2) was counted according to P = IU/A. Where I (A) is current, U (mV) is voltage, and A (m^2) is the surface area of the cathode.

The surfaces of the clean anode and the used anode were analyzed using scanning electron microscopy (SEM) (HITACHI, S—4800, Tokyo, Japan) equipped with an energy dispersive X-ray (EDX) detector (HORIBA, model 7593—H, Kyoto, Japan). For SEM examination, samples were first immersed in glutaraldehyde (2.5%, 60 min) and then washed with phosphate buffer (0.1 M, pH 7.0, 3 times). Finally, the samples were treated with critical point drying to dehydrate the biological tissues and coated with Pt.

5. Conclusions

In this study, the integrated treatment system consisting of an anaerobic SRBR and a SOFC has been successfully applied to treat organic wastewater containing high concentrations of sulfate/sulfide. Sulfide produced in the sulfate reduction process by SRB acts not only as an endogenous electron mediator but also as an electron donor to oxidize sulfide into non-toxic sulfur and recoverable by precipitation. High sulfate and COD removal were attained in SRBR at feed COD/SO_4^{2-} ratios of 2 and sulfate concentration of 1300 mg L^{-1} under continuous operation at HRT of 72 h. The COD and sulfate removal efficiencies of SRBR were 94.8 ± 0.6 and 93 ± 1.3%, respectively, during the operation. The maximum sulfide removal efficiency achieved was as high as 93.7 ± 1.2% and power density reached 18.2 ± 1.6 mW/m^2 with HRT of 24 h and initial sulfide of 316 ± 5.8 mg L^{-1}. However, the sulfide removal efficiency and power density dropped gradually after 45 days of operation. At the end of the operation, sulfide removal efficiency and power density were 68% and 7.2 mW/m^2, respectively. This might explain why the accumulation of sulfur on the anode may have decreased the electrical conductivity of the anode, thereby increasing the overpotential of the anode in the SOFC over time. The results presented in this study clearly revealed the feasibility of using an integrated treatment system to control the removal of pollutants from wastewater and electricity generation.

However, a major limitation of the method is the decrease in electrochemical activity over time due to the deposition of elemental sulfur. The precipitated sulfur forms a barrier towards the further oxidation of sulfides over long periods of time. Therefore, an efficient method to re-activate the electrode and recover sulfur from the electrode surface needs to be developed in the future.

Author Contributions: Methodology, T.Q.H.K. and C.L.D.; Validation, T.Y.N.; Formal analysis, T.Y.N. and C.L.D.; Investigation, T.Q.H.K.; Data curation, T.Y.N. and C.L.D.; Writing—original draft, T.Y.N. and C.L.D.; Writing—review & editing, T.Q.H.K.; Supervision, T.Q.H.K. All authors have read and agreed to the published version of the manuscript.

Funding: This research was funded by National Foundation for Science and Technology Development (NAFOSTED) under grant number 106.04-2017.314.

Institutional Review Board Statement: Not applicable.

Informed Consent Statement: Not applicable.

Data Availability Statement: Not applicable.

Conflicts of Interest: The authors declare no conflict of interest.

Sample Availability: Not applicable.

References

1. Cai, J.; Quaisar, M.; Ding, A.; Wang, K.; Sun, Y.; Wang, R. Microbial fuel cells simultaneously treating sulfide and nitrate under different influent sulfide to nitrate molar ratios. *Energy Fuels* **2020**, *34*, 3858–3866.
2. Chen, Z.; Zhang, S.H.; Zhong, L. Simultaneous sulfide removal, nitrogen removal and electricity generation in a coupled microbial fuel cell system. *Bioresour. Technol.* **2019**, *291*, 121888. [CrossRef] [PubMed]
3. Silva, A.M.; Lima, R.M.F.; Leao, V.A. Water treatment with limestone for sulfate removal. *J. Hazard. Mater.* **2012**, *221–222*, 45–55.
4. Angelov, A.; Bratkova, S.; Loukanov, A. Microbial fuel cell based on electroactive sulfate-reducing biofilm. *Energy Convers. Manag.* **2013**, *67*, 283–286. [CrossRef]
5. Guerrero-Rangel, N.; Rodriguez-de la Garza, J.A.; Garza-Garcia, Y.; Rios-Gonzalez, L.J.; Sosa-Santillan, G.J. Comparative study of three cathodic electron acceptors on performance of mediatorless microbial fuel cell. *Int. J. Electr. Power. Eng.* **2010**, *4*, 27–31.
6. Neculita, C.M.; Zagury, G.J.; Busiere, B. Passive treatment of acid mine drainage in bioreactors using sulfate-reducing bacteria: Critical review and research needs. *J. Environ. Qual.* **2007**, *36*, 1–16.
7. Batterman, S.; Grant-Alfieri, A.; Seo, S.-H. Low level exposure to hydrogen sulfide: A review of emissions, community exposure, health effects, and exposure guidelines. *Crit. Rev. Toxicol.* **2023**, *53*, 244–295.
8. Lloyd, D. Hydrogen sulfide: Clandestine microbial messenger. *Trends Microbiol.* **2006**, *14*, 456–462.
9. Tang, K.; Baskaran, V.; Nemati, M. Bacteria of the sulphur cycle: An overview of microbiology, biokinetics and their role in petroleum and mining industries. *Biochem. Eng. J.* **2009**, *44*, 73–94.

10. Zhang, L.; de Schryver, P.; de Gusseme, B.; de Myynck, W.; Boon, N.; Vertraete, W. Chemical and biological technologies for hydrogen sulfide emission control in sewer systems: A review. *Water Res.* **2008**, *42*, 1–12.
11. Dou, L.; Zhang, M.; Pan, L. Sulfide removal characteristics, pathways and potential application of a novel chemolithotrophic sulfide-oxidizing strain. *Marinobacter sp. SDSWS8*. *Environ. Res.* **2022**, *212*, 113176.
12. Lee, E.Y.; Cho, K.S.; Ryu, H.W. Simultaneous removal of H_2S and NH_3 in biofilter inoculates with *Acidithiobacillus thiooxidans* TAS. *J. Biosci. Bioeng.* **2005**, *99*, 25–31. [CrossRef] [PubMed]
13. Chen, C.; Ho, K.I.; Liu, F.C.; Ho, M.N.; Wang, A.J.; Ren, N.Q.; Lee, D.J. Autotrophic and heterotrophic denitrification by a newly isolated strain *Pseudomonas* sp. C27. *Bioresour. Technol.* **2013**, *145*, 351–356. [CrossRef] [PubMed]
14. Sun, M.; Tong, Z.H.; Sheng, G.P.; Chen, Y.Z.; Zhang, F.; Mu, Z.X.; Wang, H.L.; Zeng, R.J.; Liu, X.W.; Yu, H.Q.; et al. Microbial communities involved in electricity generation from sulfide oxidation in a microbial fuel cell. *Biosens. Bioelectron.* **2010**, *26*, 470–476. [CrossRef] [PubMed]
15. Song, T.S.; Wang, D.B.; Wang, H.; Li, X.; Liang, Y.; Xie, J. Cobalt oxide/nanocarbon hybrid materials as alternative cathode catalyst for oxygen reduction in microbial fuel cell. *Int. J. Hydrogen Energy* **2015**, *40*, 3868–3874. [CrossRef]
16. Barua, P.K.; Deka, D. Electricity generation from biowaste based microbial fuel cells. *Int. J. Energy Inf. Commun.* **2010**, *1*, 77–92.
17. Koók, L.; Nemestóthy, N.; Bélafi-Bakó, K.; Bakonyi, P. Investigating the specific role of external load on the performance versus stability trade-off in microbial fuel cells. *Bioresour. Technol.* **2020**, *309*, 123313. [CrossRef] [PubMed]
18. Zhang, B.; Zhao, H.; Shi, C.; Zhou, S.; Ni, J. Simultaneous reduction of nitrate and oxidation of by-products using electrochemical method. *J. Hazard. Mater.* **2009**, *171*, 724–730.
19. Holzman, D.C. Microbe power. *Environ. Health Persp.* **2005**, *113*, A754–A757. [CrossRef]
20. Jang, J.K.; Pham, T.H.; Chang, I.S.; Khang, K.H.; Moon, H.; Cho, K.S.; and Kim, B.H. Construction and operation of a novel mediator- and membrane-less microbial fuel cell. *Process Biochem.* **2004**, *39*, 1007–1012. [CrossRef]
21. Malik, S.; Kishore, S.; Dhasmana, A.; Kumari, P.; Mitra, T.; Chaudhary, V.; Kumari, R.; Bora, J.; Ranjan, A.; Minkina, T.; et al. A perspective review on microbial fuel cells in treatment and product recovery from wastewater. *Water* **2023**, *15*, 316.
22. Van Den Brand, T.P.; Roest, K.; Chen, G.H.; Brdjanovic, D.; van Loosdrecht, M.C.M. Potential for beneficial application of sulfate reducing bacteria in sulfate containing domestic wastewater treatment. *World J. Microbiol. Biotechnol.* **2015**, *31*, 1675–1681. [CrossRef] [PubMed]
23. Roy, H.; Rahman, T.U.; Tasnim, N.; Arju, J.; Rafid, M.M.; Islam, M.R.; Pervez, M.N.; Cai, Y.; Naddeo, V.; Islam, M.S. Microbial Fuel cell construction features and application for sustainable wastewater treatment. *Membranes* **2023**, *13*, 490. [PubMed]
24. Kim, H.; Kim, B.; Yu, J. Power generation response to readily biodegradable COD in single-chamber microbial fuel cells. *Bioresour. Techno.* **2015**, *186*, 136–140. [CrossRef]
25. Moharir, P.V.; Tembhurkar, A.R. Effect of recirculation on bioelectricity generation using microbial fuel cell with hfood waste leachate as substrate. *Int. J. Hydrogen Energy* **2018**, *43*, 10061–10069. [CrossRef]
26. Subha, C.; Kavitha, S.; Abishekar, S.; Tamilarasan, K.; Arulazhagan, P.; Rajesh Banu, J. Bioelectricity generation and effect studies from organic rich chocolaterie wastewater using continous upflow anaerobic microbial fuel cell. *Fuel* **2019**, *251*, 224–232. [CrossRef]
27. Xin, X.; Hong, J.; Liu, Y. Insights into microbial community profiles associated with electric energy production in microbial fuel cells fed with food waste hydrolysate. *Sci. Total Environ.* **2019**, *670*, 50–58. [CrossRef]
28. Yang, Z.; Pei, H.; Hou, Q.; Jiang, L.; Zhang, L.; Nie, C. Algal biofilm-assisted micoabial fuel cell to enhance domestic wastewater treatment nutrient, organics removal and bioenergy production. *Chem. Eng. J.* **2018**, *332*, 277–285. [CrossRef]
29. Abourached, C.; English, M.J.; Liu, H. Wastewater treatment by Microbial fuel cell (MFC) prior irrigation water reuse. *J. Clean. Prod.* **2016**, *137*, 144–149. [CrossRef]
30. Gacitúa, M.A.; Muñoz, E.; González, B. Bioelectrochemical sulphate reduction on batch reactors: Effect of inoculum-type and applied potential on sulfphate consumption and pH. *Bioelectrochemistry* **2018**, *199*, 26–32. [CrossRef]
31. Gao, C.; Wang, A.; Zhao, Y. Contribution of sulfate-reducing bacteria to the electricity generation in microbial fuel cells. *Adv. Mater. Res.* **2014**, *1008–1009*, 285–289. [CrossRef]
32. Zhang, K. Simultaneous sulfide removal and hydrogen production in a microbial electrolysis cell. *Int. J. Electrochem. Sci.* **2017**, *12*, 10553–10566.
33. Liu, H.; Zhang, B.; Liu, Y.; Wang, Z.; Hao, L. Continuous bioelectricity generation with simultaneous sulfide and organics removals in an anaerobic baffled stacking microbial fuel cell. *Inter. J. Hydrogen Energy* **2015**, *40*, 8128–8136. [CrossRef]
34. Sangcharoen, A.; Niyom, W.; Suwannasilp, B.B. A microbial fuel cell treating organic wastewater containing high sulfate under continuous operation: Performance and microbial community. *Process Biochem.* **2015**, *50*, 1648–1655. [CrossRef]
35. Rabaey, K.; Boon, N.; Hofte, M.; Verstraete, W. Microbial phenazine production enhances electron transfer in biofuel cells. *Environ. Sci. Technol.* **2006**, *39*, 3401–3408. [CrossRef]
36. Venkatramanan, V.; Shah, S.; Prasad, R. A critical review on microbial fuel cells technology: Perspectives on wastewater treatment. *Open Biotechnol. J.* **2021**, *15*, 131–141. [CrossRef]
37. Hong, S.W.; Chang, I.S.; Choi, Y.S.; Kim, B.H.; Chung, T.H. Response from freshwater sediment during electricity generation using microbial fuel cells. *Bioprocess Biosyst. Eng.* **2009**, *32*, 389–395. [CrossRef]
38. Kong, X.; Sun, Y.; Yuan, Z.; Li, D.; Li, L.; Li, Y. Effect of cathode electron-receiver on the performance of microbial fuel cells. *Int. J. Hydrog. Energy* **2010**, *35*, 7224–7227. [CrossRef]

39. You, S.J.; Ren, N.Q.; Zhao, Q.L.; Kiely, P.D.; Wang, J.Y.; Yang, F.L.; Fu, L.; Peng, L. Improving phosphate buffer-free cathode performance of microbial fuel cell base on biological nitrification. *Biosens. Bioelectron.* **2009**, *24*, 3698–3701. [CrossRef]
40. Zhao, F.; Rahunen, N.; Varcoe, J.R.; Roberts, A.J.; Avignone-Rossa, C.; Thumser, A.E. Factors affecting the performance of microbial fuel cells for sulfur pollutants removal. *Biosens. Bioelectron.* **2009**, *24*, 1931–1936. [CrossRef]
41. Moon, C.; Singh, R.; Veeravalli, S.S.; Shanmugam, S.R.; Chaganti, S.R.; Lalman, J.A.; Heath, D.D. Effect of COD:SO_4^{2-} ratio, HRT and Linoleic acid concentration on mesophilic sulfate reduction: Reactor performance and microbial population dynamics. *Water* **2015**, *7*, 2275–2292. [CrossRef]
42. Rossi, R.; Cario, B.P.; Santoro, C.; Yang, W.; Saikaly, P.E.; Logan, B.E. Evaluation of electrode and solution area-based resistance enables quantitative comparisons of factors impacting microbial fuel cell performance. *Environ. Sci. Technol.* **2019**, *53*, 3977–3986. [CrossRef] [PubMed]
43. Zhang, B.; Zhang, J.; Yang, Q.; Feng, C.; Zhu, Y.; Ye, Z.; Ni, J. Investigation and optimization of the novel UASB-MFC integrated system for sulfate removal and bioelectricity generation using the response surface methodology (RSM). *Biores. Technol.* **2012**, *124*, 1–7. [CrossRef]
44. Damianovic, M.H.R.Z.; Foresti, E. Anaerobic Degradation of synthetic wastewaters at different levels of sulfate and COD/Sulfate ratios in horizontal-flow anaerobic reactors (HAIB). *Environ. Engine Sci.* **2007**, *24*, 384–393. [CrossRef]
45. Mohan, S.V.; Rao, N.C.; Prasad, K.K.; Sarma, P.N. Bioaugmentaion of anaerobic sequencing batch biofilm reactor (AnSBBR) with immobilized sulphate reducing bacteria (SRB) for the treatment of sulphate bearing chemical wastewater. *Process Biochem.* **2005**, *40*, 2849–2857. [CrossRef]
46. El Bayoumy, M.A.; Bewtra, J.K.; Ali, H.I.; Biswas, N. Removal of heavy metals and COD by SRB in UAFF reactor. *J. Environ. Eng.* **1999**, *125*, 532–539. [CrossRef]
47. Choi, E.; Rim, J.M. Competition and inhibition of sulfate reducers and methane producers in anaerobic treatment. *Water Sci. Technol.* **1991**, *23*, 1259–1264. [CrossRef]
48. Zhang, J.; Zhang, B.; Tian, C.; Ye, Z.; Liu, Y.; Lei, Z. Simultaneous sulfide removal and electricity generation with corn stover biomass as co-substrate in microbial fuel cells. *Bioresour. Technol.* **2013**, *138*, 198–203. [CrossRef]
49. Ieropoulos, I.; Greenman, J.; Melhuish, C.; Hart, J. Comparative study of three types of microbial fuel cell. *Enzyme Technol.* **2005**, *37*, 238–245. [CrossRef]
50. Dutta, P.K.; Rabaey, K.; Yuan, Z.; Keller, J. Spontaneous electrochemical removal of aqueous sulfide. *Water Res.* **2008**, *42*, 4965–4975. [CrossRef]
51. Wan, Y.; Zhang, D.; Wang, Y. Electron transfer from sulfate-reducing bacteria biofilm promoted by reduced graphene sheets. *Chin. J. Ocean Limnol.* **2012**, *30*, 12. [CrossRef]
52. Miran, W.; Jang, J.; Nawaz, M.; Shahzad, A.; Jeong, S.; Jeon, E.; Lee, D.S. Mixed sulfate-reducing bacteria-enriched microbial fuel cells for the treatment of wastewater containing copper. *Chemosphere* **2017**, *189*, 134–142. [CrossRef] [PubMed]
53. Postgate, J.R. *The Sulfate-Reducing Bacteria*, 2nd ed.; Cambridge University Press: Cambridge, UK, 1984.
54. Miller, T.L.; Wolin, M.J. A serum bottle modification of the Hungate technique for cultivating obligate anaerobes. *Appl. Microbiol.* **1974**, *27*, 985. [CrossRef] [PubMed]
55. APHA. *Standard Methods for the Examination of Water and Wastewater*, 22nd ed.; American Public Health Association: Washington, DC, USA, 2012.
56. Cord-Ruwisch, R. A quick method for the determination of dissolved and precipitated sulfides in cultures of sulfate-reducing bacteria. *J. Microbiol. Meth.* **1985**, *4*, 33–36. [CrossRef]

Disclaimer/Publisher's Note: The statements, opinions and data contained in all publications are solely those of the individual author(s) and contributor(s) and not of MDPI and/or the editor(s). MDPI and/or the editor(s) disclaim responsibility for any injury to people or property resulting from any ideas, methods, instructions or products referred to in the content.

Article

An Investigation of a Natural Biosorbent for Removing Methylene Blue Dye from Aqueous Solution

Basma G. Alhogbi *[ID] and Ghadeer S. Al Balawi

Department of Chemistry, Faculty of Science, King Abdulaziz University, Jeddah 21589, Saudi Arabia
* Correspondence: balhogbi@kau.edu.sa

Abstract: The current study reports the use of zeolite prepared from a kaolin composite via physical mixing with different ratios from fiber of palm tree (Zeo-FPT) as a sustainable solid sorbent for the removal of methylene blue (MB) dye from aqueous solutions. The prepared biosorbent was fully characterized using XRD, TGA, SEM, and FTIR. The impacts of various analytical parameters, for example, contact time, dosage, MB dye concentration, and the pH of the solution, on the dye adsorption process were determined. After a contact time of 40 min, the capacity to remove MB dye was 0.438 mg g^{-1} at a Zeo-FPT composition ratio of 1F:1Z. At pH 8, Zeo-FPT (1F:1Z) had a removal efficiency of 87% at a sorbent dosage of 0.5 g for a concentration of MB dye in an aqueous phase of 10 mg L^{-1}. The experimental data were also analyzed using the kinetic and adsorption isotherm models. The retention process fitted well with the pseudo-second-order model (R^2 0.998), where the $Q_{e,calc}$ of 0.353 mg g^{-1} was in acceptable agreement with the $Q_{e,exp}$ of 0.438 mg g^{-1}. The data also fitted well with the Freundlich isotherm model, as indicated by the correlation coefficient value (R^2 0.969). The Zeo-FPT attained a high percentage (99%) in the removal of MB dye from environmental water samples (tap water, bottled water, and well water). Thus, it can be concluded that the proposed zeolite composite with fiber of palm tree (Zeo-FPT) is a suitable, environmentally friendly, and low-cost adsorbent for removing dyes from wastewater.

Keywords: environmentally friendly product; fiber of palm tree; kaolin; zeolite; dye retention; adsorption

1. Introduction

Recently, from a global perspective, environmental protection awareness has increased. Following the industrial revolution and the rapid development of new technologies, a great deal of waste has been produced that is harmful to the environment and to public health. Water bodies are most affected, resulting in the water becoming unsuitable for consumption. Dyes are used in many industries, such as textiles, food, and cosmetics, and are also used for medical purposes, including as antiseptic agents, making them common water pollutants [1]. During the dyeing process, many chemicals are discharged into the water, affecting living organisms, damaging the soil and plants, and poisoning drinking water. For example, the dye methylene blue (MB) is a heterocyclic aromatic compound, referred to as 3, 7 bis (dimethyl amino) phenothiazine-5-ium chloride by the International Union of Pure and Applied Chemistry (IUPAC). It has been defined as a biologically active substance and is a cationic thiazine dye that, when oxidized, has a deep blue color, but when reduced to leucomethylene blue, it becomes colorless [2,3]. At different doses, MB dye is toxic to animals, aquatic ecosystems, and humans; for example, at high doses, it may cause chest pain, dyspnea, restlessness, tremors, and damage to the digestive and respiratory systems of humans [4,5].

Recently, much attention has been directed toward developing low-cost and widely applicable technologies to meet the significant challenge of the treatment of wastewater contaminated with dyes. The elements of the dyes vary with respect to their chromosphere groups and substituents, thus affecting characteristics such as polarity and solubility. The most widely used methods for removing dyes from wastewater include coagulation [6], ozonation [7], chlorination [8], electrochemical processes [9], chemical precipitation [10], membrane methods [11], biological treatments [12] and adsorption [13]. The removal of MB dye from wastewater is extremely important, due to the serious environmental damage that it can cause [14,15]. The most common method of dye removal is adsorption as there are a large number of adsorbents available in the field. Adsorption is considered a surface phenomenon with several mechanisms for the inorganic and organic removal of pollutants. Adsorption is a mass transfer operation by which a substance is conveyed from a liquid stage onto a solid surface and becomes bound by chemical and/or physical interactions. The massive surface area causes high surface reactivity and adsorption capacity [16]. Adsorption procedures are preferred because of their advantages in terms of low cost, reliability, ease of process, and, most crucially, the possibility of acquiring the adsorbent from a wide range of natural, artificial, and waste materials, such as lignocellulosic biomass [16].

Presently, the search for low-cost and effective solid-phase extractors that are applicable for the removal of dye from wastewater is ongoing [12]. The efficiency of this process can also be successfully increased by optimizing the conditions through the use of various surface modifications [12]. In the literature, hydroxyapatite modified by magnetic particles is utilized as an effective, stable, biodegradable adsorbent that more easily and rapidly performs the process of removing heavy metal ions, inorganic elements, and organic dyes that constitute environmental pollution [10]. Materials classified as natural adsorbents include clays, charcoal, zeolites, ores, and biomass [13,16]. These natural elements are generally inexpensive and present considerable potential for alteration, with the aim of eventually improving their adsorption abilities [14]. Every adsorbent has its own features in terms of pore structure, porosity, and the essence of its adsorbing surfaces. Some of the waste elements used include coconut shell, fruit waste, bark, scrap tires, rice husk, sawdust, fertilizer waste, petroleum waste, sugar industry waste, palm fiber, fly ash, seafood processing waste, chitosan, peat moss, red mud, clays, and zeolites [1,13,17].

Adsorption technology includes an effective range of techniques for removing most color pigments from wastewater using modified synthesized zeolite to enhance the adsorption capacity via cation exchange [18]. The adsorption of organic dyes using clay has been shown to be a cost-effective approach for the purification of wastewater [13,16]. Many researchers have studied synthetic zeolite, which is promising and relatively inexpensive, and is an ideal scavenger of water pollutants such as organic dyes and heavy metals by means of adsorption and ion exchange. Moreover, zeolite is already used in several applications due to its surface area and high ion exchange capacity [19–21]. Zeolites are microporous, crystalline hydrated aluminosilicates with symmetrically stacked AlO_4 and SiO_4 tetrahedral forms, which result in an open and stable three-dimensional honeycomb structure with a negative charge within the pores that can be neutralized by positively charged ions (cations) such as sodium. The structural and chemical formula of the crystallographic unit cell zeolite with oxides is $M_{x/n} [(SiO_2)_y (AlO_2)_x] m\, H_2O$. In this formula, M is exchangeable metal cations, n is the valence of the cation (M), x and y are the total number of tetrahedra per unit cell, and m is the number of water molecules per unit cell [22]. Zeolites are distinguished by the mobile actions inside their channels and cages [23].

Currently, date palm trees are generally found in tropical and subtropical regions of Saudi Arabia and Asia, and they have been essential for the daily routine of citizens inhabiting the Arabian Peninsula for the last 7000 years [24]. A normal date palm tree produces 35 kg of fiber of palm tree (FPT) annually, which is a burnable waste that consists of dried leaves, spathes, sheaths and petioles [25]. This waste is burned on ranches, causing ecological contamination and the demise of important soil microorganisms [26]. Due to their highly lignocellulosic composition and low ash content, date palm tree components have been intensively assessed recently with respect to their suitability for assembling enacted carbon at minimal expense [25]. In contrast to commercial activated carbon, the activated carbon obtained from date palm trees has prominent textural qualities that provide it with a notably greater adsorption limit for water remediation purposes [27]. The important biomass components of fiber of palm tree (FPT) include hemicellulose (18%), cellulose (46%), ash (10.54%), and lignin (20%) [28]. Moreover, lignocellulosic fibers are renewable, biodegradable, recyclable, and supportable materials [29]. Therefore, fiber of palm tree (FPT) can be used as an adsorbent material for wastewater management. In addition, it has been shown that palm fiber can be altered to fully activate carbon for MB dye removal [14,30].

Adsorption processes using zeolite and biosorbents such as fiber of palm tree (FPT) are becoming the most typical and low-cost solution for removing organic pollutants from wastewater such as effluent bodies polluted by industry [15,23,31]. Thus, this study is focused on (i) testing the efficiency of zeolite prepared from local Saudi kaolin and composite with fiber of palm tree (FPT) physical mixing as a sustainable sorbent for obtaining a more cost-effective material for use as an adsorbent in wastewater treatment for MB dye; (ii) Biosorbent characterization of the prepared biosorbent and finally; (iii) calculating the adsorption efficiencies of various zeolite–fiber ratios as a function of contact time, adsorption pH, sorbent dosages, and different concentrations. The data that can be obtained from the kinetic and isotherm models will help to establish the most probable retention mechanism of MB dye by Zeo-FPT sorbent in the test aqueous solution. Moreover, the trace spectrophotometric determination of MB dye in a variety of environmental waters was attempted.

2. Results and Discussion

2.1. Adsorbent Characterization

The prepared biosorbent sample with mixed physical composite (Zeo-FPT) was fully characterized using different techniques, i.e., XRD, TGA, SEM, and FTIR. The lattice constant of synthesized zeolite A is illustrated in Table 1. The XRD pattern of the percentage crystal of the zeolite A structure was used to calculate the intensities of the peaks at 2θ values of 7.18, 10.17, 12.46, 21.68, 24.06, 27.14, 29.97, 34.21, 52.89, and 54.43, which are presented in Figure 1a. The Scherrer formula was successfully used to calculate the crystallite size of the zeolite A, which was found to be 254 Å (25.4 nm) [20,22].

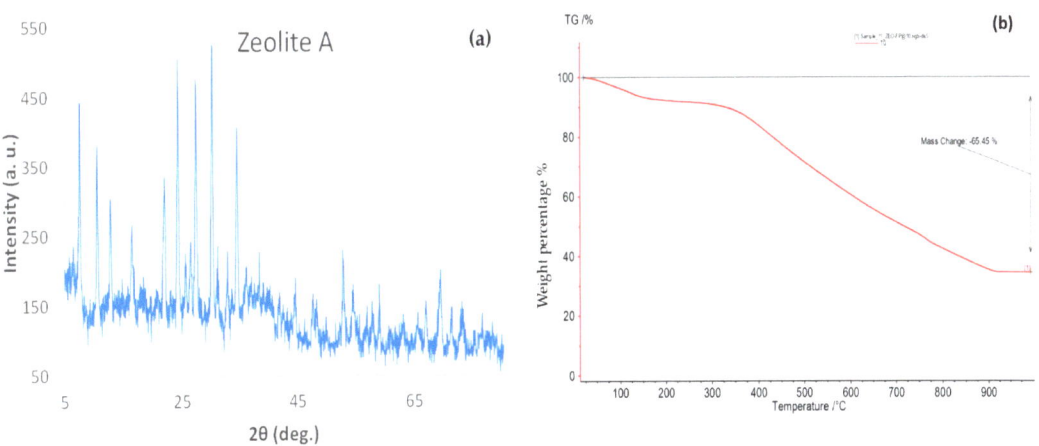

Figure 1. XRD diffractogram of zeolite A (**a**) and TGA thermographs analysis of Zeo-FPT composite (**b**).

Table 1. Lattice constant of zeolite A.

Sample	a	b	c	α	β	γ	Formula
Zeolite A, reference	12.144	12.144	12.144	90	90	90	$(Li, Na)_2\ Al_2\ Si_{1.85}\ O_{7.7} \times H_2$

Thermogravimetric (TGA) analysis was used to evaluate the mass sample transformation and thermal stability at different temperatures. The TGA of the Zeo-FPT showed three phases, as presented in Figure 1b. The first phase appeared at an approximate range of 100–150 °C, accompanied with a loss of weight of 8.9% due to dehydration of water molecules. The second degradation phase happened around 200–300 °C, which may have been due to the dehydroxylation of the Zeo-FPT [32]. The final phase occurred at a wider range of 300–600 °C and can be attributed to decomposition of the carbonic natural material in the fiber (lignin) [33]. The percentage decomposition of mass loss was 65.45% at the end of all thermal phases. This indicates that Zeo-FPT showed an appropriate thermal stability up to high temperatures.

The SEM images of synthesized pure zeolite A (Zeo.), FPT-AC, Zeo-FPT, and Zeo-FPT-MB are shown in Figure 2. These images reveal that the surface of zeolite A was characterized by a rhombohedral crystal (pseudocubic) morphology. FPT-AC and Zeo-FPT presented irregular aggregates, rough surfaces, porous cavities on the surface and different particle sizes. There was partial darkness in the blue color after the MB dye was adsorbed on the pseudocubic crystals of the Zeo-FPT-MB, indicating the irregular and abundant adsorption of dye on the sorbent surface [18].

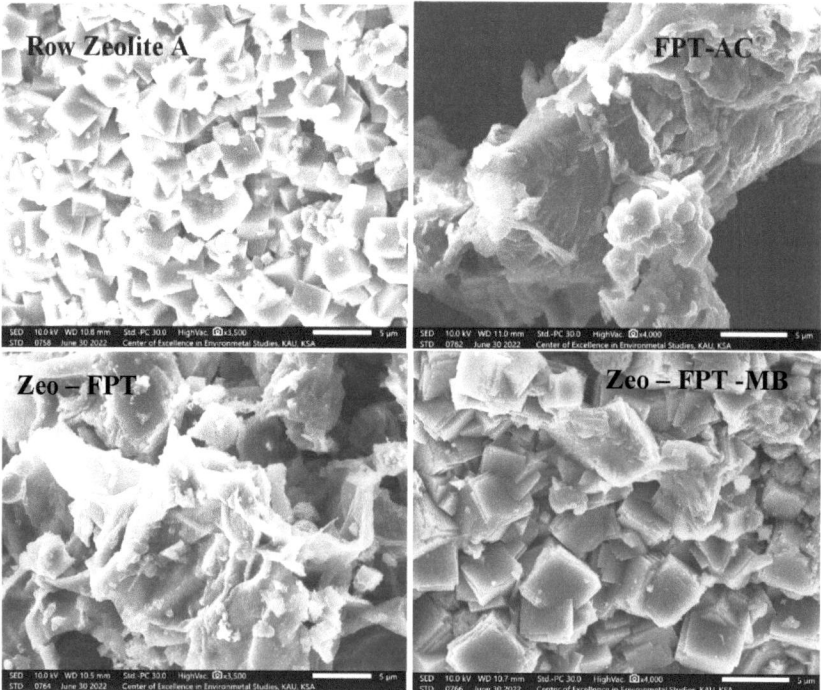

Figure 2. The SEM images of synthesized pure zeolite A, FPT-AC, Zeo-FPT composite before sorption, and Zeo-FPT-MB after sorption.

The FTIR spectroscopy results shown in Figure 3 and Table 2 were used to examine the chemical structures of the Zeo-FPT absorbent before and after the removal of the MB dye.

In contrast to the spectrum of Zeo, Zeo-FPT exhibited an absorption band intensity that decreased at 902 cm^{-1}, which may be proof of the validity of the combination of zeolite and FPT. The adsorption of the MB dye onto Zeo-FPT showed that the spectrum was controlled by the substitution of a broad band around 1008–902 cm^{-1}, probably due to the MB dye interacting with the functional groups of Si–O–T (T = Si or Al) present on the Zeo-FPT surface [34,35].

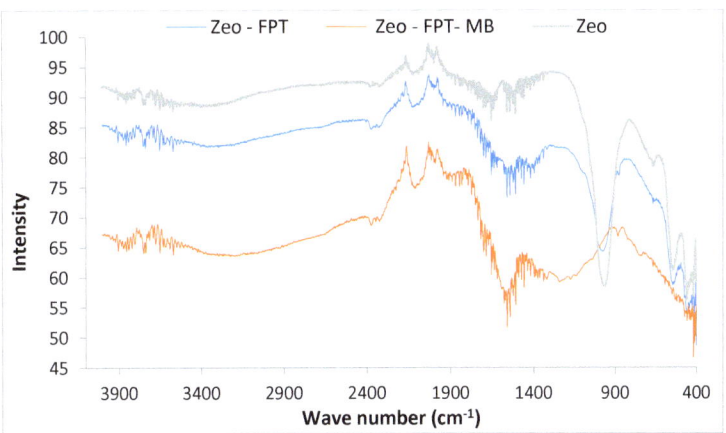

Figure 3. FTIR spectra of Zeo, the Zeo-FPT composite, and Zeo-FPT-MB dye after adsorption.

Table 2. FTIR spectroscopy absorption bands of Zeo, FPT, and the Zeo-FPT composite.

Wave Number cm^{-1}	Stretching	Bending
518–478	stretching of Zeo-FPT vibrations of the bonds inside the TO$_4$ (T = Al, Si) tetrahedrons	weak bands of Zeo-FPT bending vibrations of O–T–O
670–650	weak bands of symmetric stretching vibration of the T–O bonds (Si–O–Si bending)	
1000–952	Zeo and Zeo-FPT, the internal asymmetric stretching vibration of Si–O–T (T = Si or Al)	C=C bending; strong monosubstituted alkene
1000	strong, broad CO–O–CO stretching anhydride	
1044	the internal asymmetric stretching vibration of Si–O–T (T = Si or Al) in zeolite	
1500		bending vibration bands of strong hydroxyl groups
2165–2030	triple bond region (2000–2500) presence of alkynes C=C=N stretching ketenimine C=C=O stretching ketene	
3200–2700 cm^{-1}	weak broad O–H stretching	weak bending vibration bands of hydroxyl groups (H–O–H)
3740	stretching vibrations of the OH groups; vibration found for water vapor	

2.2. Optimization of the Dye Retention by the Established Extractor

The effects of different parameters on Zeo-FPT adsorption efficiency were investigated. The effects of different contact times of 0.5 g of different ratios of fiber (F), zeolite (Z), and the composite (Zeo-FPT) (0F:1Z, 1F:0Z, 1F:1Z, 1F:2Z, and 2F;1Z) with 25 mL of 10 mg L^{-1} MB dye solution were combined with different shaking durations (15, 30, 45, 60, 75, 90, 105, and 120 min) at 300 rpm at room temperature. The results shown in Figure 4a indicate that in the first 15 to 90 min, the removal capacity (Q) of 0F:1Z for the MB dye was around 0.291 mg g^{-1}, while that of 1F:0Z was 0.356 mg g^{-1} at 120 min, and the removal capacity increased to 0.348 mg g^{-1} for 0F:1Z and 0.368 mg g^{-1} for 1F:0Z. Therefore, with increasing contact time, the removal capacity increased, and adsorption reached equilibrium. This indicates that there were free adsorbent sites and an intense interaction between the MB dye and the adsorbent sites involved. The results shown in Figure 4a for 1F:2Z indicate that there was a gradual decline in the adsorption capacity (Q), which began at 0.4 mg g^{-1} to 0.3 mg g^{-1} at 120 min, after which there was an increase, and the final Q value was 0.402 mg g^{-1}. The results for 2F:1Z show that there was a gradual decline in the adsorption capacity (Q), which began around 0.35 mg g^{-1} to 0.2 mg g^{-1} at 120 min, after which there was an increase, and the final Q value was 0.358 mg g^{-1}. Moreover, the results for 1F:1Z show that there was a gradual decline in the adsorption capacity (Q), which began at 0.47 mg g^{-1} to 0.36 mg g^{-1} at 120 min. Hence, an increase in the contact time resulted no significant change in the removal capacity, which means there was sufficient time for the adsorbate to adsorb into the layer of the adsorbate. The highest MB dye removal capacity was displayed by 1F:1Z—0.438 mg g^{-1} at 40 min. Dosa et al. (2022) [20] investigated the use of Clinoptilolite, a natural zeolite, in the adsorption of MB dye, and found that the efficiency of the removal of the MB dye was nearly 96% at 210 min. Thus, the ratio of 1F:1Z of Zeo-FPT was selected to carry out further investigations of the different factors, such as the pH of the solution, the MB concentration, and the adsorbent dosage.

The variations in the efficiency of Zeo-FPT in removing MB dye were tested at different pH levels. The solution was composed of 0.2 g 1F:1Z and 25 mL of 50 mg L^{-1} MB dye, and was introduced at pH values of 4, 6, 8, and 10 for 40 min. The results regarding the pH effect are presented in Figure 4b. The lowest percentage removal was 75.35% at pH 4, and in a slightly acidic medium, at pH 6, there was a minor increase to 84.48%. Finally, at pH 8 and 10, the activity remained stationary at 86.8%, which represents the highest removal percentage. The lower adsorption of MB dye in acidic pH solution was probably due to the fact that, in an acidic pH suspension, the surface may become positively charged, thus causing H$^+$ ions to effectively compete with MB dye cations for the active sites of Zeo-FPT. This may cause a decrease in the amount of dye adsorbed [36]. However, the cationic MB dye at different ranges of pH showed a high percentage removal 84.48%, which makes the Zeo-FPT a significant sorbent for wastewater remediation. A similar result was found when removing MB dye from textile wastewater using zeolite-x and kaolin adsorbents; the removal percentages were 97.77% at pH 4 and 86.86% at pH 6, respectively [6].

The effect of various concentrations of MB dye solution obtained by mixing 0.2 g of 1F:1Z from Zeo–FPT with 25 mL of (5, 20, 30, 40, 50 or 70 mg L^{-1}) for 40 min is shown in Figure 4c. There was a rapid rise in the adsorption capacity (Q) from 0.07 to 1.12 mg g^{-1} for the 5 mg L^{-1} to 50 mg L^{-1} MB dye concentration. The maximum adsorption capacity (Q) was 1.26 mg g^{-1}, which was attained by the 70 mg L^{-1} MB dye concentration; this was due to the increase in the driving force of the concentration gradient as a result of the increase in the initial dye concentration. Li et al. (2015) [37] found a similar pattern when investigating the ability of zeolite synthesized from electrolytic manganese to remove MB dye from wastewater, showing that the amount of adsorbed MB increased with an increasing MB dye concentration.

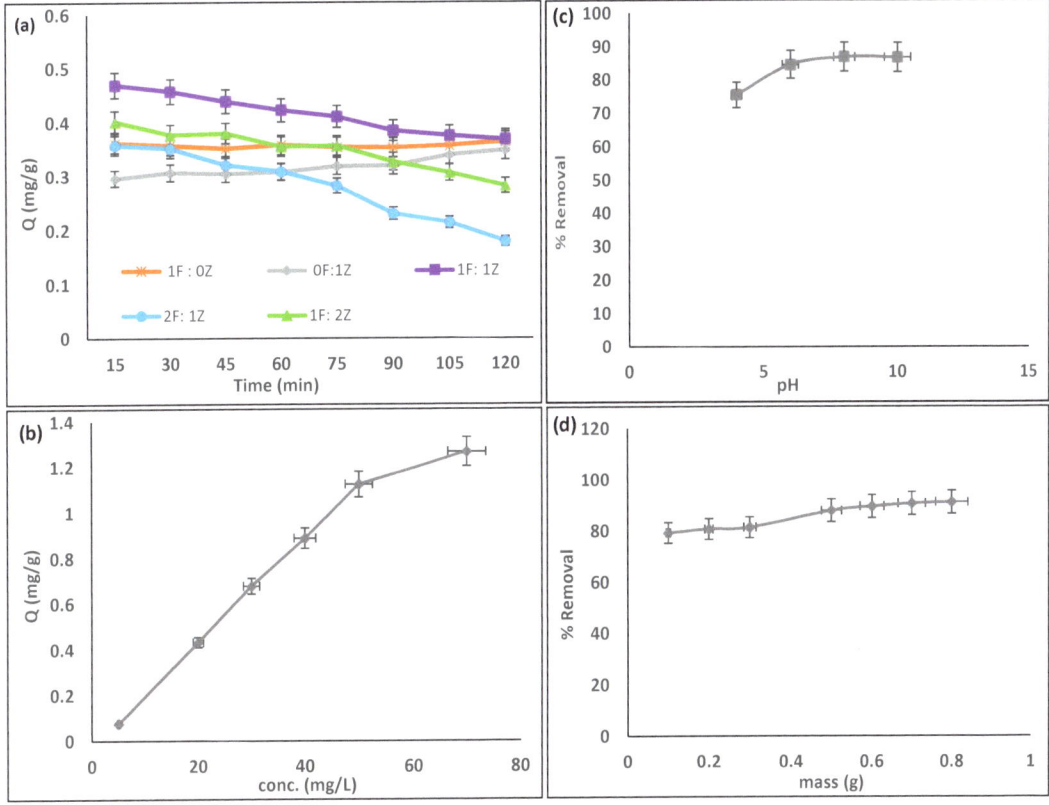

Figure 4. The effects of different factors on the MB dye removal capacity of Zeo-FPT: (**a**) contact time at different ratios of Zeo-FPT; (**b**) solution concentration of Zeo-FPT (1F:1Z); (**c**) the removal percentage at different pH values; and (**d**) the mass of Zeo-FPT (1F:1Z).

Further evaluations of the effect of Zeo-FPT (1F:1Z) on the MB dye adsorption process were carried out. The relationship between the sorbent dosages and the percentage removal was investigated. Figure 4d presents the different masses of Zeo-FPT (0.1, 0.2, 0.3, 0.5, 0.6, 0.7, and 0.8 g) in 25 mL of 10 mg L^{-1} MB dye solution after 40 min at room temperature. It was observed that the percentage removal at 0.1 g was 79.3%, followed by a gradual increase at 0.3 g to 81.4%, and another increase at 0.8 g to stability at 91.2%. The percentage removal demonstrates an increasing trend with increasing adsorbent dosages. This is because, at higher Zeo-FPT masses, there was a greater availability of active sites, and they were surrounded by higher amounts of MB dye-adsorbent molecules [36].

2.3. Sorption Kinetics

The relationship between the concentration of Zeo-FPT (1F:1Z) and adsorption time was determined using pseudo-first- and pseudo-second-order kinetics. The results in Figure 5a,b show that the correlation coefficient R^2 was 0.748 in the pseudo-first-order kinetics compared with R^2 0.998 in the pseudo-second-order kinetics. The values of the adsorption kinetics parameters in Table 3 show that the sorption capacity of the pseudo-second-order model was $Q_{e,calc}$ 0.353 mg g^{-1} for the Zeo-FPT at equilibrium, which was estimated to be similar to the experimental data of $Q_{e,exp}$ 0.438 mg g^{-1}. The constant rate of adsorption K_2 was -0.787 (g mg^{-1} min^{-1}). This indicates that the adsorption rate-limiting step followed the pseudo-second-order kinetics model, and the adsorption rate of Zeo-FPT was dependent on the MB dye concentration. The capability of the synthesized zeolite 4A adsorbent from Ethiopian kaolin was investigated regarding the removal of MB dye. The data fit best to pseudo-second-order kinetics [13]. Alhogbi et al. (2021) [30] studied activated carbon (AC) from palm tree fiber (PTF) waste for the removal of Congo red (CR), an anionic dye, and Rhodamine B (RhB), a cationic dye, from wastewater; the results show that the adsorption mechanism followed a pseudo-second-order model. Huang et al. (2019) [38] studied the removal efficiency of MB dye onto modified zeolite via magnetic graphene oxide (Cu-Z-GO-M) composites using two ratios (1:2 and 1:1), and found that the adsorption behavior followed pseudo-second-order kinetics.

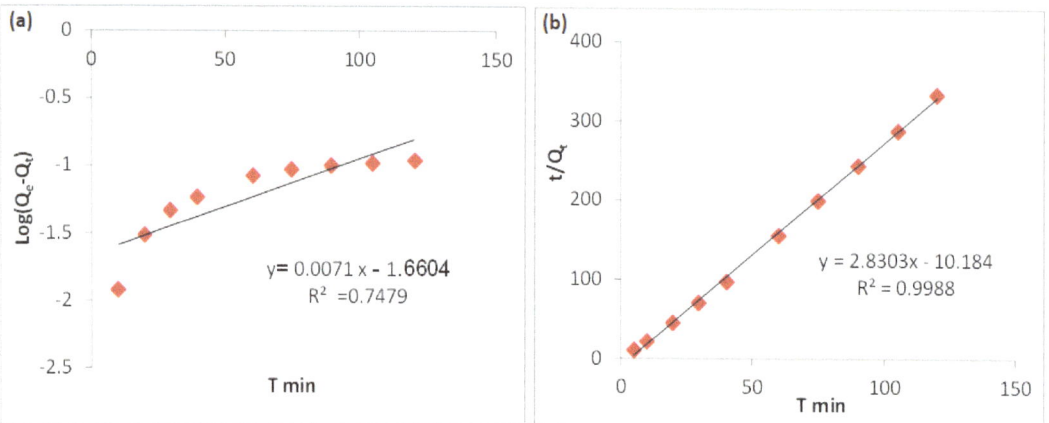

Figure 5. Adsorption kinetics: (**a**) pseudo-first-order model and (**b**) pseudo-second-order model for the removal of MB using Zeo-FPT.

Table 3. Adsorption kinetics parameters calculated for the pseudo-first- and pseudo-second-order kinetics for the removal of MB using Zeo-FPT.

Pseudo-First-Order		Pseudo-Second-Order	
$Q_{e,calc}$ (mg g^{-1})	0.220	$Q_{e,calc}$ (mg g^{-1})	0.353
		$Q_{e,exp}$ (mg g^{-1})	0.438
R^2	0.748	R^2	0.998
K_1 (min^{-1})	0.016	K_2 (g mg^{-1} min^{-1})	0.787

2.4. Adsorption Isotherm

The adsorption isotherm shows the correlation between the adsorbate adsorbed by the adsorbent (Q_e) and that remaining in solution after reaching equilibrium (C_e). The adsorption isotherm models are shown in Figure 6a–c; the value of the correlation coefficient of R^2 in the Langmuir model was 0.547. The adsorption mechanism followed the Freundlich model with a correlation coefficient of R^2 0.969 (linear); this indicates that the MB dye was adsorbed in multiple layers and was heterogeneous on the surface of Zeo-FPT (1F:1Z). Based on the results presented in Table 4, $1/n = 0.88$, which indicates that the sorption was in favorable [38]. The data applied to the Temkin isotherm model indicate that the adsorption of MB dye onto Zeo-FPT (1F:1Z) was inferior, with the R^2 of 0.732 and the adsorption heat of 9.36 kJ mol^{-1}. Domingues et al. (2020) [27] studied the use of *Zygia cauliflora* seeds for water purification, and found that the adsorption mechanism was best described by the Freundlich isotherm model.

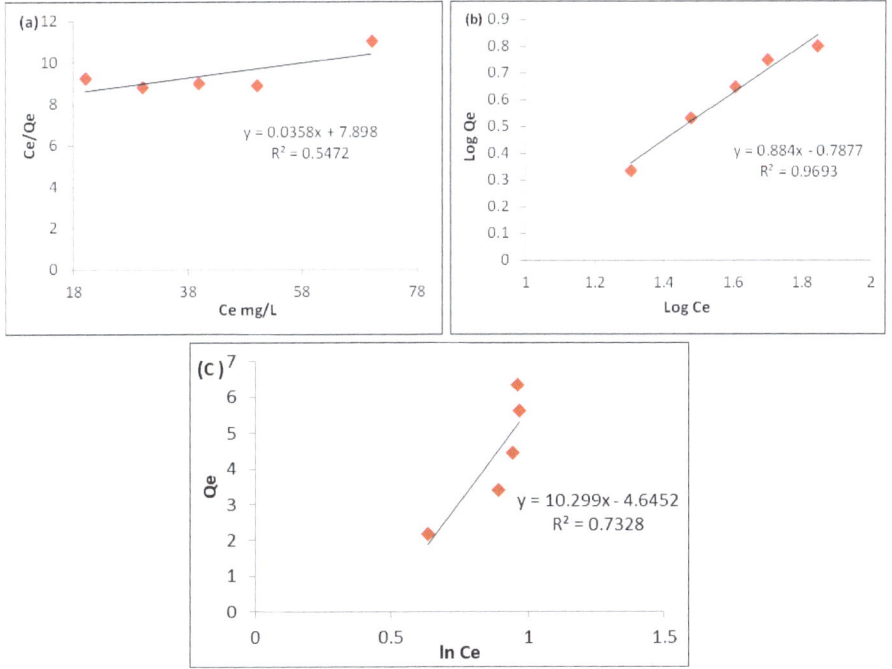

Figure 6. Adsorption isotherms (Langmuir model (**a**) Freundlich model (**b**) and Temkin model (**c**)) for removal of MB dye using Zeo-FPT.

Table 4. Adsorption isotherm parameters calculated for the removal of MB using Zeo-FPT.

Isotherm Model	Parameter	Value
Langmuir	Q_{max} (mg g^{-1})	27.933
	K_L (L mg^{-1})	4.533×10^{-3}
	R^2	0.547
Freundlich	K_f (mg$^{1-1/n}$ L$^{1/n}$ g^{-1})	0.163
	$1/n$	0.884
	R^2	0.969
Temkin	K_T (L g^{-1})	1.57
	R^2	0.732
	B (kJ mol^{-1})	9.36

Table 5 contains a list of results from recent research conducted on the removal of MB dye from wastewater using natural solid sorbents. Each study had a different method of eliminating MB dye from the aqueous solution. However, in this study, the removal capacity of MB dye onto Zeo-FPT ($Q_{e,exp}$ 0.438 mg g^{-1}) at 25 °C within 40 min proved that Zeo-FPT (1F:1Z) has potential for use in the removal of MB dye from an aqueous solution with minimal time and an adequate removal capacity.

Table 5. A comparison between the sorption characteristics of the established Zeo-FPT (1:1) biosorbent and some other biosorbents used for MB dye removal.

Biosorbent	Q_{max} mgg^{-1}	Adsorption Time (min)	Kinetic/Isotherm Models	Reference
Zeolite modified by magnetic graphene oxide (Cu-Z-GO-M) (1:1)	99.2	600	(PSO)/(FRH)	[38]
Sugarcane waste-activated carbon modified with natural zeolite (AC500/NZ)	51	-	(PSO)/(TMK)	[21]
Zeolite 4A from Ethiopian kaolin	44.35	179.82	(PSO)/(LNR)	[13]
Zeolite-X from Ethiopian kaolin	0.61	180	(PSO)/(FRH)	[6]
Clove leaves activated with sodium hydroxide (CL-NaOH)	9.80	60	(PSO)/(LNR)	[39]
Rice straw biochars	131.58	-	(PSO)/(LNR)	[40]
(Zeo-FPT) (1:1) biosorbent	0.438	40	(PSO)/(FRH)	this study

Freundlich (FRH), Langmuir (LNR), Temkin (TMK), pseudo-first-order (PFO) and pseudo-second-order (PSO) models.

2.5. Environmental Samples

Different environmental samples such as tap water, well water, and bottled water were collected to study the efficiency of the Zeo-FPT (1F:1Z) composite. Environmental sampling was performed to establish whether different sources of water could be purified to the maximum extent through adsorption by Zeo-FPT. The results are shown in Table 6. The removal percentages of tap water, well water, and bottled water were 99.04%, 99.38%, and 97.86%, respectively. The adsorption capacities (Q) of tap water, well water, and bottled water were 1.238 mg g^{-1}, 1.242 mg g^{-1}, and 1.22 mg g^{-1}, respectively. The obtained results are outstanding; they show a high removal percentage and removal capacity of MB dye from tap water, bottled water, and well water. This confirms that Zeo-FPT has high potential utility as a solid adsorbent for the removal of MB dye from environmental water samples.

Table 6. The removal percentage and adsorption capacity at fixed parameters for environmental samples.

Water	Time min	m (g)	C (mg L^{-1})	V (L)	%	Q (mg g^{-1})
Tap	70	0.2	10	0.025	99.040	1.238
Well	70	0.2	10	0.025	99.382	1.242
Bottle	70	0.2	10	0.025	97.856	1.223

3. Experimental

3.1. Chemicals and Materials

The two main raw materials used in this investigation, i.e., kaolin clay and fiber of palm tree (FPT), were supplied from the northern part of Saudi Arabia by the Saudi Geological Survey, and the FPT was collected from Jeddah city, in the Western Region of Saudi Arabia. The MB dye chosen as the target compound has the molecular formula $C_{16}H_{18}ClN_3S$ and a molecular weight of 319.09 g mol^{-1}. A high purity methylene blue dye (Sigma–Aldrich, Darmstadt, Germany) was used, and sodium hydroxide (NaOH, Riedel-deHaen Laboratories Supplies, Wycombe, UK) was used to control the pH.

3.2. Instrumentation

The synthesized samples of zeolite (Z), fiber of palm tree active carbon (FPT-AC) and composite (Zeo-FPT) before and after adsorption of MB dye were evaluated and characterized based on their morphology and phase analysis. The crystalline stage and crystallite size of the zeolite (Z) sample were examined by X-ray powder diffraction (XRD) on a Rigaku XRD system with a RINT 2000 broad-angle goniometer with a power of 40 kV × 30 mA and Cu Kα radiation (λ = 0.15478 nm). The 2θ intensity information was determined at temperatures ranging from 10 to 80 °C. The thermal stability of the developed materials was analyzed using the thermogravimetric analyzer (TGA) under a N_2 atmosphere at 10–1000 °C. A scanning electron microscope (SEM) is an essential apparatus for characterizing mesoporous and microporous molecular sieve elements. The obtained micrographs show the particle morphology of the synthesized materials and the presence of any amorphous stages in the selected samples [16]. The samples were analyzed using a field emission JEOL JSM-7600F SEM. The composition of solids and liquids was revealed by Fourier transform infrared spectroscopy (FTIR). Briefly, FTIR analysis evaluates a sample's absorption of infrared light at various wavelengths to ascertain the material's molecular composition and structure by exposing it to infrared (IR) radiation.

3.3. Preparation of the Biosorbent

Zeolite A was prepared from Saudi kaolin using the hydrothermal treatment method mentioned in [31,41]. A certain amount of Saudi kaolin was calcinated at 600 °C for 4 h, then left in an oven for 24 h. Calcined kaolin was mixed with NaOH dissolved in distilled water, stirred, and left at room temperature for 24 h. The synthesized zeolite was kept in a water bath for 6 h at 60 °C, after which it was filtered and washed with distilled water until the pH reached 7–8, then left to completely dry [41]. The fiber of palm tree (FPT) was washed several times with distilled water to remove surface impurities and dust, and was left to completely dry at room temperature, then was ground and sifted. The carbonized fiber was placed in a muffle furnace at 400 °C for a duration of 120 min using a porcelain crucible with lid to lower oxygen release, as shown in Scheme 1a.

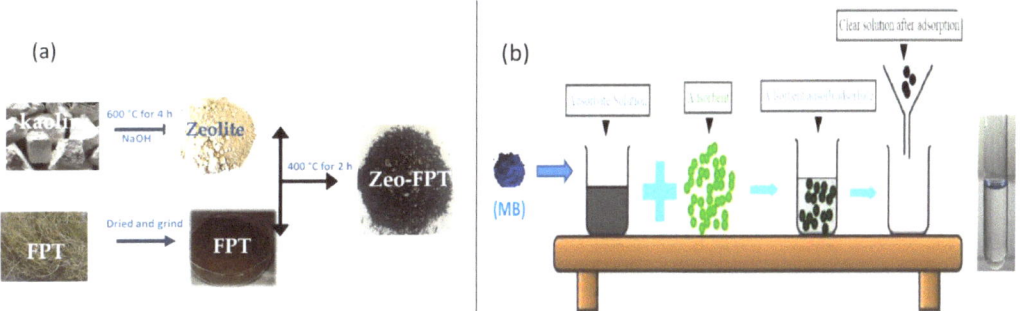

Scheme 1. Schematic illustrations of the required steps for the production of the biosorbent Zeo-FPT (**a**) and batch experiment (**b**).

3.4. Recommended Batch Experiments

The batch experiment shown in Scheme 1b was used to investigate the adsorption performance of prepared samples with different ratios of a fiber (F) and zeolite (Z) composite (Zeo-FPT) (0F:1Z, 1F:0Z, 1F:1Z, 1F:2Z and 2F;1Z). The effect of contact time was determined with 25 mL of 10 mg L^{-1} MB dye solution added to 0.5 g of sorbent and shaken for different periods of time (15, 30, 45, 60, 75, 90, 105, and 120 min) at 300 rpm at room temperature. Then, the samples were filtered using No. 2 Whatman filter paper. The filtered samples were analyzed to determine the MB dye concentration using UV–vis spectrophotometry at a maximum absorbance of λ_{max} = 664 nm to study the efficiency of the Zeo-FPT composite as a sorbent material based on different factors, such as the effect of the solution's pH (4, 6, 8, and 10), different MB dye concentrations (5, 20, 30, 40, 50, 70, and 100 mg L^{-1}), and different sorbent dosages (0.1, 0.2, 0.3, 0.5, 0.6, 0.7, and 0.8 g). It is important to note that the homogeneity and stability of the composite derived from the physical mixing of Zeo-FPT under investigation were demonstrated by the fact that all the experimental data reported are based on the average of three replicates and had an error of less than 5%. The percentage removal (%) and the adsorption capacity (Q) of MB dye retention on the

Zeo-FPT sorbent from the aqueous solution were calculated by employing equations that have been applied in previous publications [30].

3.5. Sorption Kinetic Study

In the treatment of aqueous effluents, adsorption kinetic studies are crucial because they provide useful information about the adsorption process mechanism. Many kinetic models have been developed to determine fundamental kinetic adsorption constants. The pseudo-first- and second-order models are the most used kinetic models for describing the adsorption process based on chemical reaction kinetics. In this study, the Zeo-FPT had the capacity to adsorb MB dye and was soluble in water [42].

3.5.1. The Pseudo-First-Order Model

The linear form of the pseudo-first-order model is given by Equation (1):

$$\log(Q_e - Q_t) = \log Q_e - \frac{k_1}{2.303}t \qquad (1)$$

where Q_e is the adsorption capacity at equilibrium (mg g^{-1}), Q_t is the amount of adsorbate adsorbed at time t (mg g^{-1}), k_1 is the pseudo-first-order rate constant (min^{-1}), and t is the contact time (min). The values of the adsorption rate constant k_1, equilibrium adsorption capacity, and the correlation coefficient R^2 were determined from the plot of log $(Q_e - Q_t)$ versus t at different concentrations.

3.5.2. The Pseudo-Second-Order Model

The linearized form of the second-order kinetic model to calculate the k_2 where the pseudo-second-order rate constant (g mg^{-1} min^{-1}) is from the given Equation (2):

$$\frac{t}{Q_t} = \frac{1}{k_2 Q_e^2} + \frac{1}{Q_e}t \qquad (2)$$

3.6. Adsorption Isotherm Model

The adsorption isotherm, at a constant temperature, provides useful data on the mechanism and surface properties of the adsorbate to adsorbed by the adsorbent and that remaining in solution after reaching equilibrium. Several equations can be used to describe the adsorbate's adsorption equilibrium on an adsorbent; the most well known are the Langmuir and Freundlich isotherm models [43].

3.6.1. Langmuir Isotherm Model

The Langmuir isotherm model is generally useful under the following conditions:
- An adsorbate monolayer is created on the adsorbent surface when it becomes saturated;
- Adsorbates are adsorbed at a specified number of sites;
- All sites are energetically equivalent;
- All sites hold one adsorbate species;
- No interactions occur between the species of the adsorbate.

The Langmuir model is given by Equation (3):

$$\frac{C_e}{Q_e} = \frac{1}{K_l Q_m} + \left(\frac{1}{Q_m}\right) C_e \qquad (3)$$

In this equation, C_e represents the supernatant concentration of the adsorbate (mg L^{-1}). K_l represents the equilibrium Langmuir constant associated with the affinity of the binding sites (L mg^{-1}), and Q_{max} represents the maximum capacity of adsorption (mg g^{-1}). While applying the linearized Langmuir equation by plotting C_e/Q_e against C_e, Q_{max} and K_l were obtained from the slope and the intercept, respectively.

3.6.2. Freundlich Isotherm Model

The Freundlich isotherm model presents an exponential form which assumes that an adsorbate concentration on the surface of the adsorbent increases as the concentration of the adsorbate increases. Therefore, an infinite amount of adsorption can take place when using this model. The model also states that adsorption can occur through several heterogeneous layers. The Freundlich isotherm model is expressed by Equation (4):

$$\log Q_e = \left(\frac{1}{n}\right) \log C_e + \log K_F \quad (4)$$

where K_F represents the constant of the Freundlich equilibrium (mg g^{-1}(mg L^{-1})$^{-1/n}$) while $1/n$ represents the Freundlich indicator (dimensionless), where sorption is irreversible if $1/n = 0$, sorption is unfavorable if $1/n > 1$, and sorption is favorable if $(0 < 1/n < 1)$ [34].

3.6.3. Temkin Isotherm Model

The Temkin model presumes that the adsorbate and adsorbent interaction causes the heat of the adsorption of analytes in the surface layer to decrease linearly with increasing surface coverage. The Timken isotherm model can be applied via Equations (5) and (6):

$$Q_e = \frac{RT}{b} Ln K_T + \frac{RT}{b} ln C_e \quad (5)$$

$$B = \frac{RT}{b} \quad (6)$$

Where K_T corresponds to the Temkin isotherm constant (L g^{-1}), R is the gas constant, and b is the Temkin constant given by plotting Q_e vs. $ln\ C_e$ attained from the slope and K_T from the intercept. B is a constant associated with the heat of analyte retention (kJ mol^{-1}).

3.7. Study of Environmental Samples

Environmental water samples (tap water, bottled water, and well water) were filtered to remove impurities. A spiked solution was prepared with 10 mL of 10 mg L^{-1} MB dye solution added to 90 mL of each of the three water samples (tap water, well water, and bottled water). The adsorption was studied for optimum parameters with 0.2 g of (1F:1Z) Zeo-FPT added to 25 mL of the spiked 10 mg L^{-1} MB dye solution at pH 8, with a reaction time of 70 min at room temperature. The samples were filtered using No. 2 Whatman filter paper and analyzed using a UV–vis spectrophotometer.

4. Conclusions

In conclusion, fiber of palm tree (FPT) and Saudi kaolin waste were successfully used as starting materials for the preparation of effective solid composites of zeolite and FPT (Zeo-FPT) in a diverse ratio. The newly created (Zeo-FPT) composite exhibited a remarkable efficiency for the removal of MB dye from aqueous solutions. The synthesized zeolite A was characterized as having a shape of rhombohedral crystal, and the MB dye was adsorbed on the surface of Zeo-FPT. The results show that the percentage removal of MB dye in an equal (1F:1Z) ratio of Zeo-FPT was 84.99%, and the removal capacity was 0.438 mg g^{-1} after 40 min at pH 8, with a sorbent mass of 0.5 g and a concentration of 10 mg L^{-1}, which are the optimal parameters. The results demonstrate that the kinetics models adsorption mechanism was described well by a pseudo-second-order model with R^2 0.998, and the obtained value of the sorption capacity ($Q_{e,calc}$ 0.353 mg g^{-1}) was in agreement with the experimental capacity ($Q_{e,exp}$ 0.438 mg g^{-1}). The adsorption isotherm strongly agreed with that of the Freundlich model, while the R^2 was 0.969, verifying that the adsorption process was heterogeneous, with multiple layers adsorbed onto the surface of the Zeo-FPT. The data of the environmental water samples indicate that Zeo-FPT has great potential utility as an adsorbent for removing MB dye. We conclude that the combination of low cost and easy synthesis using natural raw materials makes Zeo-FPT an attractive adsorbent—as well as being environmentally friendly and cost-effective—for the removal of MB dyes from wastewater. This study suggests encourages further investigations into the possibility of using a low-cost sorbent, such as kaolin, and fiber of palm tree in the preparation of sorbents, and enhances the economic utility of these prepared materials.

Author Contributions: B.G.A.—validation, data curation, and writing—review and editing; G.S.A.B.—formal analysis, writing draft. All authors have read and agreed to the published version of the manuscript.

Funding: This research received no external funding.

Institutional Review Board Statement: Not applicable.

Informed Consent Statement: Not applicable.

Data Availability Statement: Not applicable.

Acknowledgments: The authors are gratefully acknowledged King Abdulaziz University for facilitating the research project.

Conflicts of Interest: The authors declare no conflict of interest.

Sample Availability: Samples of the compound are not available from the authors.

References

1. Hassan, M.M.; Carr, C.M. Biomass-derived porous carbonaceous materials and their composites as adsorbents for cationic and anionic dyes: A review. *Chemosphere* **2021**, *265*, 129087. [CrossRef]
2. Tani, A.; Thomson, A.J.; Butt, J.N. Methylene blue as an electrochemical discriminator of single- and double-stranded oligonucleotides immobilised on gold substrates. *Analyst* **2001**, *126*, 1756–1759. [CrossRef]
3. Bożęcka, A.M.; Orlof-Naturalna, M.M.; Kopeć, M. Methods of Dyes Removal from Aqueous Environment. *J. Ecol. Eng.* **2021**, *22*, 111–118. [CrossRef]
4. Allafchian, A.; Mousavi, Z.S.; Hosseini, S.S. Application of cress seed musilage magnetic nanocomposites for removal of methylene blue dye from water. *Int. J. Biol. Macromol.* **2019**, *136*, 199–208. [CrossRef]
5. Hassan, A.; Abdel-Mohsen, A.; Fouda, M.M. Comparative study of calcium alginate, activated carbon, and their composite beads on methylene blue adsorption. *Carbohydr. Polym.* **2014**, *102*, 192–198. [CrossRef]
6. Mulushewa, Z.; Dinbore, W.T.; Ayele, Y. Removal of methylene blue from textile waste water using kaolin and zeolite-x synthesized from Ethiopian kaolin. *Environ. Anal. Health Toxicol.* **2021**, *36*, e2021007. [CrossRef]
7. Quan, X.; Luo, D.; Wu, J.; Li, R.; Cheng, W.; Ge, S. Ozonation of acid red 18 wastewater using $O_3/Ca(OH)_2$ system in a micro bubble gas-liquid reactor. *J. Environ. Chem. Eng.* **2017**, *5*, 283–291. [CrossRef]
8. Musz-Pomorska, A.; Widomski, M.K. Variant analysis of the chlorination efficiency of water in a selected water supply network. *Proc. ECOpole'19 Conf.* **2020**, *14*, 19–28. [CrossRef]
9. Tahir, K.; Miran, W.; Jang, J.; Maile, N.; Shahzad, A.; Moztahida, M.; Ghani, A.A.; Kim, B.; Jeon, H.; Lee, D.S. MXene-coated biochar as potential biocathode for improved microbial electrosynthesis system. *Sci. Total Environ.* **2021**, *773*, 145677. [CrossRef] [PubMed]
10. Biedrzycka, A.; Skwarek, E.; Hanna, U.M. Hydroxyapatite with magnetic core: Synthesis methods, properties, adsorption and medical applications. *Adv. Colloid Interface Sci.* **2021**, *291*, 102401. [CrossRef] [PubMed]
11. Babu, J.; Murthy, Z. Treatment of textile dyes containing wastewaters with PES/PVA thin film composite nanofiltration membranes. *Sep. Purif. Technol.* **2017**, *183*, 66–72. [CrossRef]
12. Sarayu, K.; Sandhya, S. Current Technologies for Biological Treatment of Textile Wastewater—A Review. *Appl. Biochem. Biotechnol.* **2012**, *167*, 645–661. [CrossRef]
13. Belachew, N.; Hinsene, H. Preparation of Zeolite 4A for Adsorptive Removal of Methylene Blue: Optimization, Kinetics, Isotherm, and Mechanism Study. *Silicon* **2022**, *14*, 1629–1641. [CrossRef]
14. Neolaka, Y.A.; Riwu, A.A.; Aigbe, U.O.; Ukhurebor, K.E.; Onyancha, R.B.; Darmokoesoemo, H.; Kusuma, H.S. Potential of activated carbon from various sources as a low-cost adsorbent to remove heavy metals and synthetic dyes. *Results Chem.* **2023**, *5*, 100711. [CrossRef]
15. Xue, H.; Gao, X.; Seliem, M.K.; Mobarak, M.; Dong, R.; Wang, X.; Fu, K.; Li, Q.; Li, Z. Efficient adsorption of anionic azo dyes on porous heterostructured MXene/biomass activated carbon composites: Experiments, characterization, and theoretical analysis via advanced statistical physics models. *Chem. Eng. J.* **2023**, *451*, 138735. [CrossRef]
16. Zhou, Y.; Lu, J.; Zhou, Y.; Liu, Y. Recent advances for dyes removal using novel adsorbents: A review. *Environ. Pollut.* **2019**, *252*, 352–365. [CrossRef]

17. El Bardiji, N.; Ziat, K.; Naji, A.; Saidi, M. Fractal-Like Kinetics of Adsorption Applied to the Solid/Solution Interface. *ACS Omega* **2020**, *5*, 5105–5115. [CrossRef]
18. Łach, M.; Grela, A.; Pławecka, K.; Guigou, M.D.; Mikuła, J.; Komar, N.; Bajda, T.; Korniejenko, K. Surface Modification of Synthetic Zeolites with Ca and HDTMA Compounds with Determination of Their Phytoavailability and Comparison of CEC and AEC Parameters. *Materials* **2022**, *15*, 4083. [CrossRef] [PubMed]
19. Neolaka, Y.A.; Lawa, Y.; Riwu, M.; Darmokoesoemo, H.; Setyawati, H.; Naat, J.; Widyaningrum, B.A.; Aigbe, U.O.; Ukhurebor, K.E.; Onyancha, R.B.; et al. Synthesis of Zinc(II)-natural zeolite mordenite type as a drug carrier for ibuprofen: Drug release kinetic modeling and cytotoxicity study. *Results Chem.* **2022**, *4*, 100578. [CrossRef]
20. Dosa, M.; Grifasi, N.; Galletti, C.; Fino, D.; Piumetti, M. Natural Zeolite Clinoptilolite Application in Wastewater Treatment: Methylene Blue, Zinc and Cadmium Abatement Tests and Kinetic Studies. *Materials* **2022**, *15*, 8191. [CrossRef]
21. Mohamed, F.; Shaban, M.; Zaki, S.K.; Abd-Elsamie, M.S.; Sayed, R.; Zayed, M.; Khalid, N.; Saad, S.; Omar, S.; Ahmed, A.M.; et al. Activated carbon derived from sugarcane and modified with natural zeolite for efficient adsorption of methylene blue dye: Experimentally and theoretically approaches. *Sci. Rep.* **2022**, *12*, 18031. [CrossRef] [PubMed]
22. Van Dang, L.; Nguyen, T.T.M.; Van Do, D.; Le, S.T.; Pham, T.D.; Le, A.T.M. Study on the Synthesis of Chabazite Zeolites via Interzeolite Conversion of Faujasites. *J. Anal. Methods Chem.* **2021**, *2021*, 5554568. [CrossRef]
23. Kadja, G.T.; Culsum, N.T.; Mardiana, S.; Azhari, N.J.; Fajar, A.T. Recent advances in the enhanced sensing performance of zeolite-based materials. *Mater. Today Commun.* **2022**, *33*, 104331. [CrossRef]
24. Allbed, A.; Kumar, L.; Shabani, F. Climate change impacts on date palm cultivation in Saudi Arabia. *J. Agric. Sci.* **2017**, *155*, 1203–1218. [CrossRef]
25. Chowdhury, S.; Pan, S.; Balasubramanian, R.; Das, P. Date palm based activated carbon for the efficient removal of organic dyes from aqueous environment. *Sustain. Agric. Rev.* **2019**, *34*, 247–263. [CrossRef]
26. Elseify, L.A.; Midani, M.; Shihata, L.A.; El-Mously, H. Review on cellulosic fibers extracted from date palms (*Phoenix dactylifera* L.) and their applications. *Cellulose* **2019**, *26*, 2209–2232. [CrossRef]
27. Domingues, J.A.; Consolin Filho, N.; Souza LA, G.D.; Medeiros, F.V.D.S. Coagulation activity of the seed extract from *Zygia cauliflora* (WILLD.) KILLIP Applied in Water Treatment. *Rev. Ambiente Água* **2020**, *15*, 1–12. [CrossRef]
28. Shafiq, M.; Alazba, A.A.; Amin, M.T. Removal of heavy metals from wastewater using date palm as a biosorbent: A comparative review. *Sains Malays.* **2018**, *47*, 35–49. [CrossRef]
29. Asim, M.; Jawaid, M.; Khan, A.; Asiri, A.M.; Malik, M.A. Effects of Date Palm fibres loading on mechanical, and thermal properties of Date Palm reinforced phenolic composites. *J. Mater. Res. Technol.* **2020**, *9*, 3614–3621. [CrossRef]
30. Alhogbi, B.G.; Altayeb, S.; Bahaidarah, E.A.; Zawrah, M.F. Removal of Anionic and Cationic Dyes from Wastewater Using Activated Carbon from Fiber of palm tree Waste. *Processes* **2021**, *9*, 416. [CrossRef]
31. El Maksod, I.A.; Elzaharany, E.A.; Kosa, S.A.; Hegazy, E.Z. Simulation program for zeolite A and X with an active carbon composite as an effective adsorbent for organic and inorganic pollutants. *Microporous Mesoporous Mater.* **2016**, *224*, 89–94. [CrossRef]
32. Neto, J.S.S.; de Queiroz, H.F.M.; Aguiar, R.A.A.; Banea, M.D. A Review on the Thermal Characterisation of Natural and Hybrid Fiber Composites. *Polymers* **2021**, *13*, 4425. [CrossRef] [PubMed]
33. Mosai, A.K.; Chimuka, L.; Cukrowska, E.M.; Kotzé, I.A.; Tutu, H. The Recovery of Rare Earth Elements (REEs) from Aqueous Solutions Using Natural Zeolite and Bentonite. *Water Air Soil Pollut.* **2019**, *230*, 188. [CrossRef]
34. Dhiman, V.; Kondal, N. ZnO nanoadsorbents: A potent material for removal of heavy metal ions from wastewater. *Colloid Interface Sci. Commun.* **2021**, *41*, 100380. [CrossRef]
35. Dehghani, M.H.; Dehghan, A.; Alidadi, H.; Dolatabadi, M.; Mehrabpour, M.; Converti, A. Removal of methylene blue dye from aqueous solutions by a new chitosan/zeolite composite from shrimp waste: Kinetic and equilibrium study. *Korean J. Chem. Eng.* **2017**, *34*, 1699–1707. [CrossRef]
36. Gao, W.; Zhao, S.; Wu, H.; Deligeer, W.; Asuha, S. Direct acid activation of kaolinite and its effects on the adsorption of methylene blue. *Appl. Clay Sci.* **2016**, *126*, 98–106. [CrossRef]
37. Li, C.; Zhong, H.; Wang, S.; Xue, J.; Zhang, Z. Removal of basic dye (methylene blue) from aqueous solution using zeolite synthesized from electrolytic manganese residue. *J. Ind. Eng. Chem.* **2015**, *23*, 344–352. [CrossRef]
38. Huang, T.; Yan, M.; He, K.; Huang, Z.; Zeng, G.; Chen, A.; Peng, M.; Li, H.; Yuan, L.; Chen, G. Efficient removal of methylene blue from aqueous solutions using magnetic graphene oxide modified zeolite. *J. Colloid Interface Sci.* **2019**, *543*, 43–51. [CrossRef]
39. Kusuma, H.S.; Aigbe, U.O.; Ukhurebor, K.E.; Onyancha, R.B.; Okundaye, B.; Simbi, I.; Ama, O.M.; Darmokoesoemo, H.; Widyaningrum, B.A.; Osibote, O.A.; et al. Biosorption of Methylene blue using clove leaves waste modified with sodium hydroxide. *Results Chem.* **2023**, *5*, 100778. [CrossRef]
40. Wang, K.; Peng, N.; Sun, J.; Lu, G.; Chen, M.; Deng, F.; Dou, R.; Nie, L.; Zhong, Y. Synthesis of silica-composited biochars from alkali-fused fly ash and agricultural wastes for enhanced adsorption of methylene blue. *Sci. Total Environ.* **2020**, *729*, 139055. [CrossRef]
41. Ibrahim, H.; Jamil, T.S.; Hegazy, E. Application of zeolite prepared from Egyptian kaolin for the removal of heavy metals: II. Isotherm models. *J. Hazard. Mater.* **2010**, *182*, 842–847. [CrossRef] [PubMed]

42. Sandollah, N.A.S.; Ghazali, S.A.I.S.M.; Ibrahim, W.N.W.; Rusmin, R. Adsorption-Desorption Profile of Methylene Blue Dye on Raw and Acid Activated Kaolinite. *Indones. J. Chem.* **2020**, *20*, 755–765. [CrossRef]
43. Ruthven, D.M. *Principles of Adsorption and Adsorption Processes*; John Wiley & Sons: New York, NY, USA, 1984; ISBN 978-0-471-86606-0.

Disclaimer/Publisher's Note: The statements, opinions and data contained in all publications are solely those of the individual author(s) and contributor(s) and not of MDPI and/or the editor(s). MDPI and/or the editor(s) disclaim responsibility for any injury to people or property resulting from any ideas, methods, instructions or products referred to in the content.

Hydrothermal Synthesis of a Technical Lignin-Based Nanotube for the Efficient and Selective Removal of Cr(VI) from Aqueous Solution

Qiongyao Wang [1,†], Yongchang Sun [1,*,†], Mingge Hao [1], Fangxin Yu [1] and Juanni He [2,*]

[1] Key Laboratory of Subsurface Hydrology and Ecological Effects in Arid Region, Ministry of Education, School of Water and Environment, Chang'an University, Xi'an 710054, China; wqy03218520@163.com (Q.W.); haomingge_813@163.com (M.H.); y526328876@foxmail.com (F.Y.)
[2] Huijin Technology Holding Group Corporation Limited, Xi'an 710000, China
* Correspondence: sunyongchang60@163.com (Y.S.); jnhe2007@163.com (J.H.); Tel.: +86-29-82339952 (Y.S.)
† These authors contributed equally to this work.

Abstract: Aminated lignin (AL) was obtained by modifying technical lignin (TL) with the Mannich reaction, and aminated lignin-based titanate nanotubes (AL-TiNTs) were successfully prepared based on the AL by a facile hydrothermal synthesis method. The characterization of AL-TiNTs showed that a Ti–O bond was introduced into the AL, and the layered and nanotubular structure was formed in the fabrication of the nanotubes. Results showed that the specific surface area increased significantly from 5.9 m^2/g (TL) to 188.51 m^2/g (AL-TiNTs), indicating the successful modification of TL. The AL-TiNTs quickly adsorbed 86.22% of Cr(VI) in 10 min, with 99.80% removal efficiency after equilibration. Under visible light, AL-TiNTs adsorbed and reduced Cr(VI) in one step, the Cr(III) production rate was 29.76%, and the amount of total chromium (Cr) removal by AL-TiNTs was 90.0 mg/g. AL-TiNTs showed excellent adsorption capacities of Zn^{2+} (63.78 mg/g), Cd^{2+} (59.20 mg/g), and Cu^{2+} (66.35 mg/g). After four cycles, the adsorption capacity of AL-TiNTs still exceeded 40 mg/g. AL-TiNTs showed a high Cr(VI) removal efficiency of 95.86% in simulated wastewater, suggesting a promising practical application in heavy metal removal from wastewater.

Keywords: lignin; nanotubes; adsorption; photocatalysis; Cr(VI)

1. Introduction

Water pollution with heavy metals is commonly more sustainable than organic pollution such as pesticides and dyes, and severe heavy metal pollution could result in devastating effects on humans [1,2]. Particularly, in industrial processes such as metallurgy, galvanization, and tannery, chromium (Cr) is widely used [3,4]. Generally, there are two stable forms of Cr: Cr(III) and Cr(VI) [5]. The high toxicity, carcinogenicity, extreme solubility, and high mobility of Cr(VI), compared to Cr(III), make soil and water pollution treatment extremely difficult [6]. Generally, treatment methods for Cr(VI) pollution include biological reduction, chemical reduction, adsorption, photocatalysis, and electrochemical technologies [7,8]. Among them, adsorption is regarded as a high-performance, low-cost, and green technology in practice [9,10].

Recently, nanotube-based materials have attracted extensive attention as efficient heavy metal adsorption materials [11,12]. For example, a carbon nanotube (CNT)-doped hydrotalcite-hydroxyapatite (HT-HAp) material was prepared to adsorb Cr(VI) from tannery wastewater. Under the action of ion exchange and electrostatic attraction, 76.97% of Cr(VI) was adsorbed by HT-HAp [13]. Guo et al. loaded zirconium oxide (ZrO$_2$) onto halloysite nanotubes (ZrO$_2$/HNTs) to enhance the average pore size and specific surface area of halloysite nanotubes (HNTs), which resulted in an improved adsorption performance of HNTs on As(III) (36.08 mg/g) [14]. The surface of the synthesized magnetic nanoparticles-phosphate-titanium nanotubes (MNP-PN-TNT) possessed rich hydroxyl groups, which gave MNP-PN-TNT a removal performance of 35 mg/g of Cr(VI) by ion exchange under acidic conditions [15]. Although these nanotube materials exhibited a certain amount of heavy metal adsorption capacity, the high preparation costs significantly hindered their applications [16]. Consequently, it is urgent to develop a cheap and effective method to prepare nanotube materials to remove heavy metals from wastewater [17].

Lignin, as the second most naturally occurring amorphous aromatic molecular polymer in nature after cellulose, is regarded as a key component of the lignocellulosic biomass [18,19]. Technical lignin (TL) is mainly derived from by-products of the paper and pulp industry, and the annual amount of TL produced is estimated to be 225 million tons by 2030 [20]. However, only 5% of TL is utilized at a high-value level, and a large amount of TL is burned as low-value fuel or disposed of as waste [21]. Low utilization is still a challenge for TL [22]. To increase the TL utilization efficiency, chitin was introduced to lignin to prepare chitin/lignin hybrid material for heavy metal removal. It was found that chitin/lignin exhibited superb adsorption capacity of Cu^{2+} (75.70 mg/g), Zn^{2+} (82.41 mg/g), Ni^{2+} (70.41 mg/g), and Pb^{2+} (91.74 mg/g), which was probably attributed to the abundant functional groups of chitin and lignin, such as phenols, hydroxyl, carboxyl, and ethers [23]. In addition, with functional groups and a three-dimensional structure, lignin-based carbon nanotubes (L-CNTs) were prepared by introducing lignin into the carbon nanotubes, which exhibited excellent adsorption performance of L-CNTs for Pb^{2+} (235 mg/g) and the advantages of strong water dispersion and environmental friendliness [24]. The results indicate that lignin is an effective material in heavy metal removal and has a high potential in water pollution treatment.

However, poor water solubility and low surface area remain a challenge for lignin utilization [25,26]. Structural modifications of lignin were reported to enhance the properties of lignin [27]. The main methods for lignin modification are acid hydrolysis, sulfonation, ionic liquid modification, alkylation, steam activation, carboxylation, ball milling, amination, etc. [19,28]. Amination is a prospective method because the amine groups introduced into the lignin ionize and generate a positive charge under acidic conditions. The Mannich reaction for lignin modification is an excellent method because the aminated product is highly dispersed, which improves the lignin adsorption capacity for heavy metals by four to five times [29,30].

To our knowledge, titanate nanotubes have been extensively studied as semiconductor materials with high specific surface area and reusable properties [31]. However, studies combining the lignin-rich functional groups with the tubular structure of titanate nanotubes to effectively remove Cr(VI) have not been reported. In this study, TL was modified by activation using the Mannich reaction to improve its disadvantages of easy agglomeration and low adsorption capacity. AL-TiNTs were successfully prepared by a simple hydrothermal method for the first time (Scheme 1). The purpose of this study was to explore (1) the physicochemical properties of AL-TiNTs; (2) the adsorption performance of AL-TiNTs under different conditions; (3) the reduction performance of AL-TiNTs on Cr(VI) in the photocatalytic process; (4) the selectivity of AL-TiNT adsorption in mixed solutions and the wide absorbability on other heavy metal ions; and (5) the removal mechanism of AL-TiNTs for Cr(VI) and the regenerative properties of AL-TiNTs.

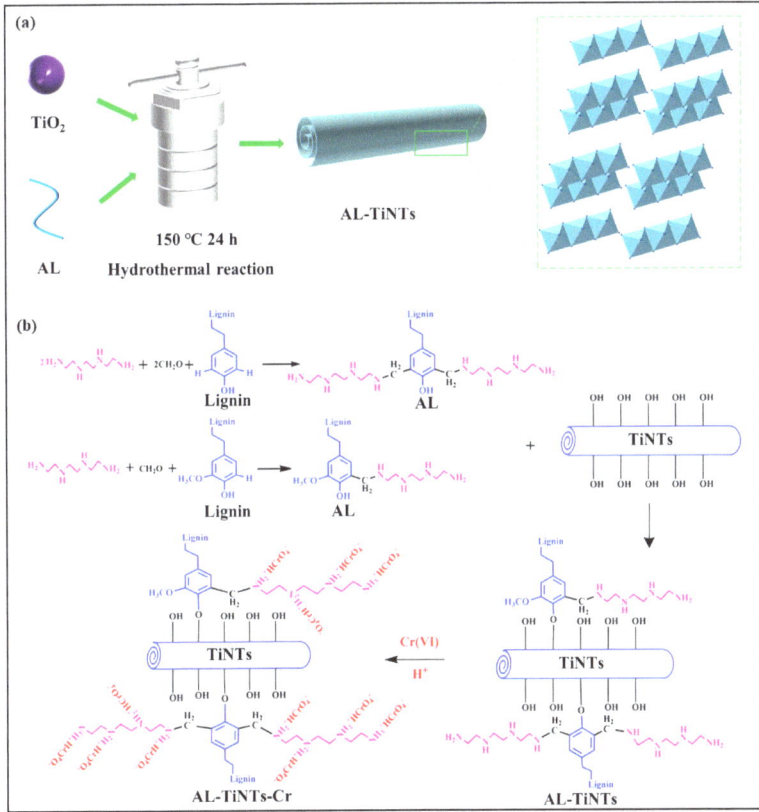

Scheme 1. The synthesis process of AL-TiNTs: (**a**) schematic diagram; (**b**) structural diagram.

2. Results and Discussion

2.1. Characterization of AL-TiNTs

2.1.1. SEM

In Figure 1, the SEM images of TL, AL, and AL-TiNTs are shown. When the magnification was 5 kx, the pure TL particles were aggregated into a blocky structure (Figure 1a), the modified AL exhibited a more compact structure with some pores formed by depressions on the surface (Figure 1b), and the AL-TiNTs presented a fluffy structure with an irregular spherical shape formed by the surface projections and more pores (Figure 1c). When the magnification reached 50 kx, the TL surface was smooth with fewer pores (Figure 1d), while the AL surface was rough and presented a few pores (Figure 1e). On the one hand, this could be due to the modified amino grafting on the pure TL, which changed its surface structure [32]; on the other hand, the washing and freeze-drying process might leave pores [30]. In particular, the AL-TiNTs presented a randomly entangled nanotube-like structure, along with a nanotube cross-shaped porous network structure (Figure 1f), which was related to the Ti–O–Ti bond breaking and the formation of sodium titanate during the hydrothermal process [33]. Results indicated that the hydrothermal and modification synthesis of AL-TiNTs not only improved the disadvantages of lignin which was prone to agglomeration but also formed the nanotubular structure, which greatly improved the specific surface area of AL-TiNTs and thus promoted the adsorption and photocatalytic reduction properties of AL-TiNTs [34].

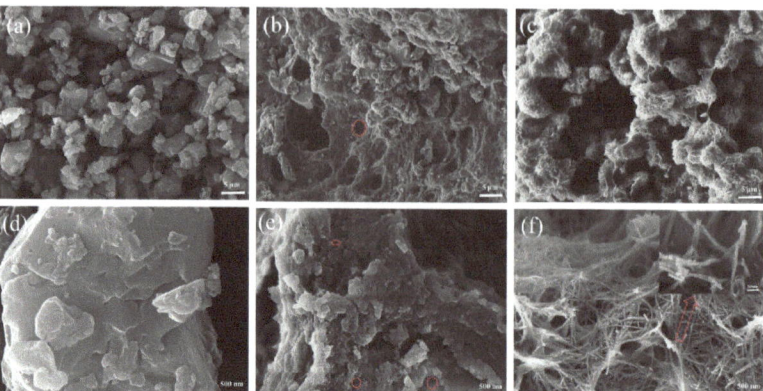

Figure 1. SEM pictures: (**a**) TL with 5 kx; (**b**) AL with 5 kx; (**c**) AL-TiNTs with 5 kx; (**d**) TL with 50 kx; (**e**) AL with 50 kx; (**f**) AL-TiNTs with 50 kx and 200 kx.

2.1.2. EDS

The EDS mapping of the AL-TiNTs is presented in Figure 2. The presence of the element N in Figure 2c was evidence of the presence of the amino group. The high degree of overlap between the Ti (Figure 2a) and O (Figure 2d) elements indicated the possible presence of Ti-O bonds in the AL-TiNTs, and it was assumed that the synthesis of $Na_2Ti_3O_7$ took place. Meanwhile, the dense distribution of Ti (Figure 2a), C (Figure 2b), and N (Figure 2c) elements proved the involvement of AL in the dispersion of titanate nanotubes, further proving the successful synthesis of AL-TiNTs.

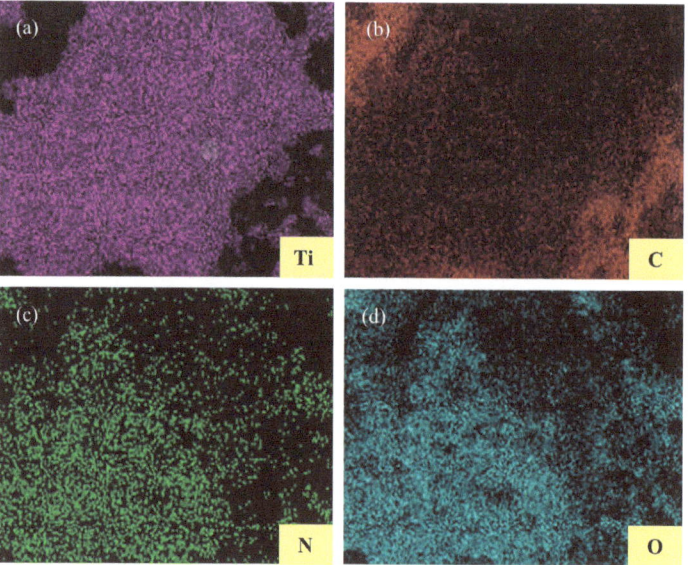

Figure 2. EDS mapping of AL-TiNTs: (**a**) Ti; (**b**) C; (**c**) N; and (**d**) O.

2.1.3. BET

Figure 3a presents a type II isotherm of a very small H3 hysteresis loop, which was in accordance with the previously reported conclusion that lignin is predominantly a microporous structure and has less content of mesoporous structure [35]. Compared to the TL, although AL presented a relatively large H3 hysteresis loop (Figure S1), it was still a type II isotherm, suggesting that the amination modification was not significant in affecting the lignin pore structure and that the microporous structure was still predominant in AL. Conversely, Figure 3b presents the typical appearance of type IV isotherms and H3 hysteresis loops, showing their mesoporous structures with diameters of 2–50 nm [36]. The PSD curves calculated by the BJH method are depicted in the insets of Figure 3a,b. Compared to Figure 3a, the PSD curves in Figure 3b exhibit dramatic peaks, and the pore diameter at 3–4 nm corresponds to the pores contained within the AL-TiNT tubes [5]. The 9–12 nm corresponds to spaces between the aggregated tubes of AL-TiNTs, which was in accordance with the SEM characterization results [37]. The BET results for TL, AL, and AL-TiNTs are presented in Table S1. Compared to TL (5.9037 m^2/g) and AL (6.7762 m^2/g), the AL-TiNTs offered a larger specific surface area (188.51 m^2/g), owing to the special tubular structure and resulting in a higher adsorption capacity [16].

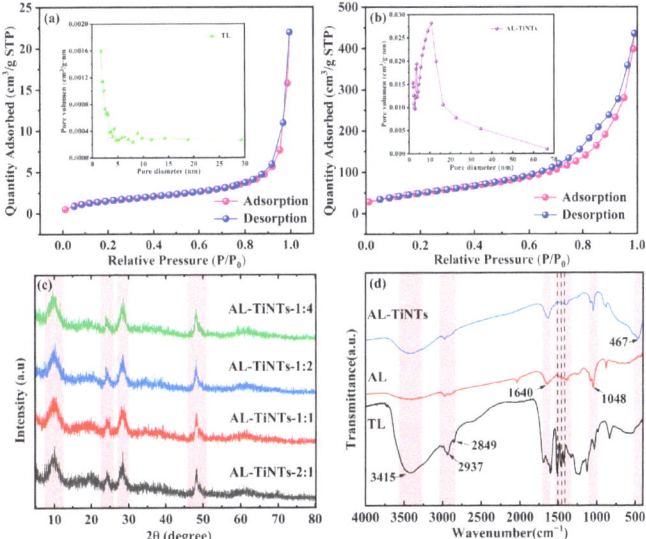

Figure 3. Adsorption–desorption isotherms (**a**) TL and (**b**) AL-TiNTs; (**c**) XRD patterns of the AL-TiNTs; (**d**) FT-IR spectra of AL-TiNTs, AL, and TL.

2.1.4. XRD

The composites prepared with different AL to nano-TiO$_2$ ratios displayed similar XRD patterns (Figure 3c). The AL-TiNTs synthesized in different ratios presented characteristic peaks with diffraction angles of 48°, 28°, and 24.5°, which were the characteristic peaks of titanate nanotubes and belonged to the (020), (211), and (110) planes of TiNTs, respectively [38]. The characteristic peak at 10° corresponded to one of the (001) crystal planes of Na$_2$Ti$_3$O$_7$ [39], and the peaks at 48°, 28°, and 24.5° indicated that the composite contains sodium titanate compounds and hydrogen titanate compounds [39]. The broad peak at 10° corresponded to a layer spacing of approximately 0.7–0.85 nm between layers of titanate [40]. In addition, the lack of anatase diffraction peaks demonstrated that all of the TiO$_2$ was converted to titanate nanotubes [41].

2.1.5. FT-IR

It was obvious that the absorption peaks of modified lignin were weakened, which was because of the involvement of the O–H bond during the Mannich reaction, resulting in the reduction of the O–H group (Figure 3d). However, following the synthesis of the AL-TiNTs, the increased absorption peak at 3415 cm^{-1} suggested the formation of a new

O–H group, which could be the stretching vibration of O–H in the Ti–OH bond on the nanotubes [2]. High peaks at 2849 and 2937 cm^{-1} occurred due to the aromatic methoxy C–H vibration. Aromatic skeletal vibrational peaks at 1422, 1461, 1511, and 1600 cm^{-1} were found in TL, AL, and AL-TiNT fractions, which were typical peaks of lignin [42,43]. The C–N vibrational peak at 1048 cm^{-1} and the N–H in-plane bending vibrational peak at 1640 cm^{-1} were found in the AL and AL-TiNTs, which indicated the successful introduction of amino groups on lignin [44]. Meanwhile, the O–Ti and Ti–O–Ti bond at 467 cm^{-1} was found in AL-TiNTs [45], indicating the successful synthesis of sodium titanate $Na_2Ti_3O_7$, which was in accordance with the SEM and XRD analyses [46].

2.2. Adsorption

2.2.1. Mass Ratios of AL/Nano-TiO$_2$

The AL/nano-TiO$_2$ mass ratio played a remarkable influence on Cr(VI) removal when the AL/nano-TiO$_2$ ratios were set as 0.25, 0.5, 1, and 2. Pure AL and nano-TiO$_2$ were also considered (Figure 4a). When the AL/nano-TiO$_2$ mass ratio was 0.25, only 78.14% of Cr (VI) was removed. As the AL/nano-TiO$_2$ mass ratio improved, the content of amino functional groups gradually increased, more adsorption active sites were exposed, and the removal efficiency increased gradually. When the mass ratio of AL/nano-TiO$_2$ was 1:1, the removal efficiency could reach 99.80%, which was found to be the optimum mass ratio of AL/nano-TiO$_2$.

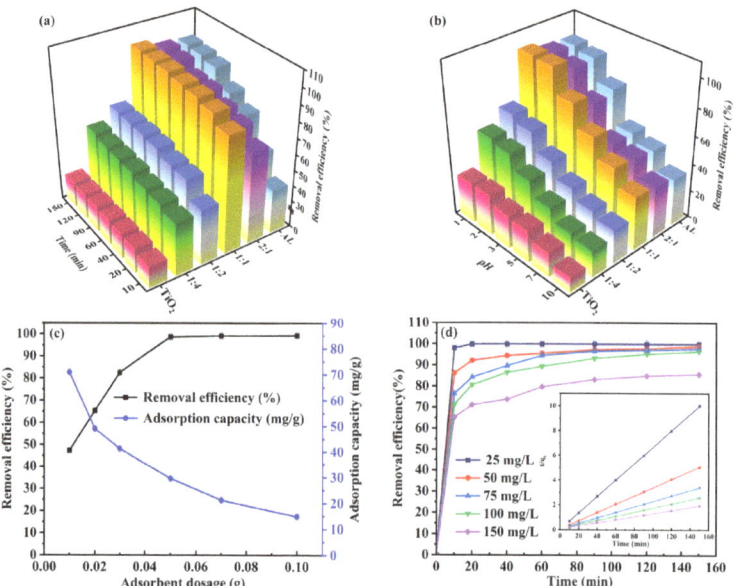

Figure 4. Adsorption performance of AL-TiNTs: (**a**) the mass ratio of AL/nano-TiO$_2$; (**b**) pH; (**c**) AL-TiNT dosage; and (**d**) initial concentration of Cr (VI) and kinetic analysis (AL-TiNT dosage: 0.05 g; pH = 2; temperature: 25 °C).

2.2.2. pH

Cr(VI) existed in three main forms in aqueous solution, mainly in CrO_4^{2-} at pH > 6.0 and in H_2CrO_4 and $HCrO_4^-$ at pH < 6.0 [47]. It was found that the acidic environments were beneficial in improving the AL-TiNT adsorption capacity (Figure 4b). AL-TiNTs showed a clear downward trend of the Cr(VI) removal efficiency as the pH increased. When the solution became neutral, the removal efficiency increased slightly. Results indicated that under strongly acidic conditions, the AL-TiNTs exhibited better adsorption performance and could remove 99.80% of Cr(VI) at pH 2. The reactions during adsorption were as follows [48]:

$$-NH_3^+ + HCrO_4^- \rightleftharpoons -NH_3^+ HCrO_4^- \quad (1)$$

$$2-NH_3^+ + CrO_4^{2-} \rightleftharpoons (-NH_3^+)_2 CrO_4^{2-} \quad (2)$$

$$2-NH_3^+ + Cr_2O_7^{2-} \rightleftharpoons (-NH_3^+)_2 Cr_2O_7^{2-} \quad (3)$$

2.2.3. AL-TiNT Dosage

Figure 4c exhibits the effect of the AL-TiNT dosage (0.01, 0.02, 0.03, 0.05, 0.07, and 0.10 g) on the adsorption performance. When the adding amount of AL-TiNTs increased, the Cr(VI) removal efficiency gradually increased, while the AL-TiNT adsorption capacity decreased. There was a slope of the removal efficiency curve that tended to slow down between 0.02 and 0.10 g and a 100% removal of Cr(VI) by AL-TiNTs at 0.10 g. In contrast, the adsorption capacity of AL-TiNTs at the lower dosing of 0.01 g reached 70.87 mg/g, which was higher than the adsorption capacity of 0.1 g AL-TiNTs. The AL-TiNT adsorption capacity decreased sharply with increasing the dosage of AL-TiNTs from 0.01 to 0.03 g. This could be explained as follows: as the amount of the AL-TiNTs increased, the AL-TiNTs were exposed to more adsorption sites, resulting in more Cr(VI) being adsorbed on AL-TiNTs [49]. Therefore, the optimum dosage of AL-TiNTs was 0.05 g based on the removal efficiency and cost.

2.2.4. Temperature

The highest Cr(VI) removal efficiency was found to be 100% at 35 °C, while it was only 96.51% at 20 °C (Figure S2). The performance of AL-TiNTs in removing Cr(VI) decreased with decreasing temperatures, which may be due to the decreased vibrational frequency of Cr(VI) at lower temperature [50].

2.2.5. Cr(VI) Concentration

Figure 4d presents the effect of the initial Cr(VI) concentration (C_0). As the results indicated, with the increase in C_0, the removal efficiency declined. The equilibrium removal efficiency of AL-TiNTs for Cr(VI) was 85.72% at the C_0 of 150 mg/L, which was comparatively lower. However, the AL-TiNT adsorption capacity was relatively higher reaching 77.15 mg/g. This could be explained by that when the dosage of AL-TiNTs was constant, the total number of active adsorption sites was also limited. At the lower Cr(VI) concentrations, the AL-TiNTs displayed a relatively greater number of adsorption sites where Cr(VI) could be adsorbed more efficiently. Adsorption sites of AL-TiNTs were continuously occupied with the increasing amount of pollutant, and thus, the excess amount of Cr(VI) could not be adsorbed.

2.3. Photodegradation Study

The photodegradation of the AL-TiNTs was investigated under light (Figure 5a,b). At low Cr(VI) concentration (50 mg/L), the AL-TiNTs were able to adsorb almost 100% of Cr(VI). Therefore, to study the reductive efficiency of AL-TiNTs under visible light, Cr(VI) concentration with 150 mg/L was used. The removal efficiency of Cr(VI) by the AL-TiNTs of AL/nano-TiO_2 ratios of 1:1 and 2:1 could reach 90.48% and 94.48% after the adsorption and photocatalytic reaction, respectively (Figure 5a). However, the AL-TiNT materials with the AL/nano-TiO_2 mass ratio of 1:1 achieved a conversion of Cr(III) of up to 29.76% at 40 min under light exposure (Figure 5b). Therefore, the reduction of Cr(VI) provided more evidence of the better performance of AL-TiNTs with the AL/nano-TiO_2 ratio of 1:1. The reason for the existence of low amounts of Cr(III) before the photocatalytic reaction was that under strongly acidic conditions, a high concentration of Cr(VI) could react with some hydrochloric acid to produce low amounts of Cr(III) (Figure 5b). After the photocatalysis, the Cr(III) concentration gradually increased. The Cr(III) in solution might have entered the titanate nanotubes and limited the complexation of electrons and holes, increasing the AL-TiNT photocatalytic activity and leading to a higher rate of Cr(III) production [35]. At 40 min, the production rate of Cr(III) decreased because of the photocatalysis, which could result from the gradual photocatalysis equilibrium at this time, and more Cr(III) could be adsorbed by the AL-TiNTs [51]. Finally, the removal amount of AL-TiNTs for total Cr was 90.00 mg/g. The reduction of AL-TiNTs for Cr(VI) following photoelectron production is displayed in Text S1 [52].

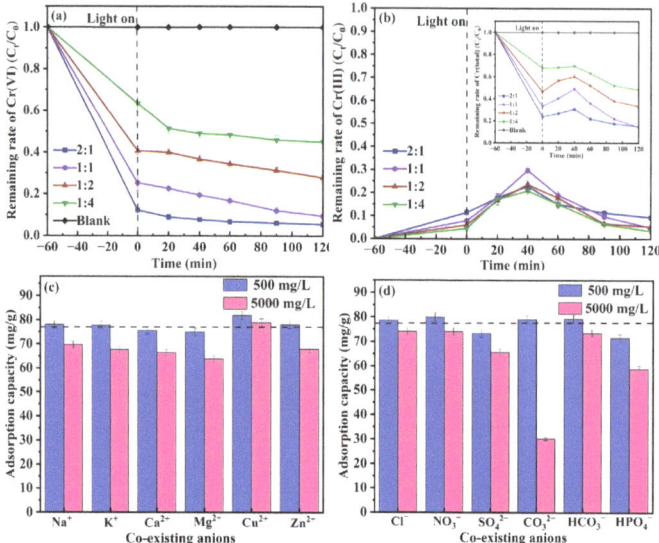

Figure 5. The effect of photocatalysis and co-existing anions on the performance of AL-TiNTs: (**a**) variation of Cr(VI); (**b**) variation of Cr(III) and total Cr; (**c**) cations; and (**d**) anions of different ionic strengths (AL-TiNTs: 0.05 g; pH = 2; temperature: 25 °C; xenon lamp current: 8.5 A).

The effects of different light intensities on AL-TiNTs were investigated by varying the current of the xenon lamp light source to regulate the light intensity (Figures S3 and S4). When the operating current was 5 A, the photocatalytic performance of different AL/nano-TiO$_2$ mass ratios of AL-TiNTs was poor, among which the photocatalytic reduction of AL-TiNTs with an AL/nano-TiO$_2$ mass ratio of 2:1 was the worst, which was only 5.54% (Figure S4a). When the operating current was adjusted to 12 A, the photocatalytic effect of AL-TiNTs was significantly improved, and the Cr(VI) reduction efficiency of AL-TiNTs with the AL/nano-TiO$_2$ mass ratio of 1:4 could reach up to 23.53%, which was twice as much as that of its Cr(VI) reduction efficiency when the current of the xenon lamp source was 5 A (Figure S4b). It could be seen that light intensity was crucial for the photocatalytic activity of AL-TiNTs.

2.4. Co-Existing Anions

When the co-existing ion concentrations were 5 and 50 mg/L, the influence of co-existing ions on Cr(VI) removal can be neglected (Figure S5), indicating that AL-TiNTs were highly resistant to complex solutions. It was found that Cu^{2+} and Zn^{2+} had a positive impact on the Cr(VI) removal by AL-TiNTs when each of the co-existing cation concentrations was 500 mg/L (Figure 5c). However, the adsorption performance of AL-TiNTs was slightly inhibited by the high concentration of Ca^{2+} and Mg^{2+} (5000 mg/L). On the one hand, this may be due to the high ion concentration which shielded the electrostatic attraction. On the other hand, the high concentration of the co-existing cation significantly hindered the movement of Cr(VI) to the AL-TiNTs [53]. The six co-existing anions exhibited a slight influence on the AL-TiNT adsorption performance. However, at the high co-existing anion concentration of 5000 mg/L, the six anions indicated different degrees of inhibition of the removal capacity of AL-TiNTs (Figure 5d). This could be explained by a reduction in the AL-TiNT active sites due to the occupation by an excess amount of anions. In addition, CO_3^{2-} inhibited Cr(VI) removal by AL-TiNTs, probably because the addition of CO_3^{2-} could increase the pH of the solution, which was not conducive to the protonation of amino groups on the AL-TiNTs, and the electrostatic attraction was not liable to occur [29]. The inhibitory effect of SO_4^{2-} may be attributed to the nature of SO_4^{2-} with its greater charge and higher radius ratio (z/r) and its stronger interaction with the adsorption site [54].

2.5. Theoretical Study
2.5.1. Adsorption Kinetics

The kinetic studies were performed at different Cr(VI) concentrations at 298 K (Figures 4d and 6, Text S2). The pseudo-second-order kinetic model (Figure 4d) and the largest R^2 values (0.9991–0.9999) for the kinetic model (Table S2) suggested that the secondary kinetic model was highly relevant to the Cr(VI) adsorption process. Compared to other kinetic models (Figure 6a–d), the theoretical equilibrium adsorption capacities of the proposed secondary kinetic model showed a high degree of similarity to the actual equilibrium adsorption capacities, demonstrating that the kinetic model exactly described the adsorption process of the AL-TiNTs [55]. It could be concluded that processes such as electron transfer and the sharing of active sites between Cr(VI) and the AL-TiNTs occurred at the microscopic level with chemisorption playing a crucial role. Both the second and third curves did not pass through the origin of the coordinates in the fitted curves of the intraparticle diffusion model (Figure 6d), and the fit was poor. This indicated that intraparticle diffusion would not be the single limiting factor and would not be as dominant as chemical interactions.

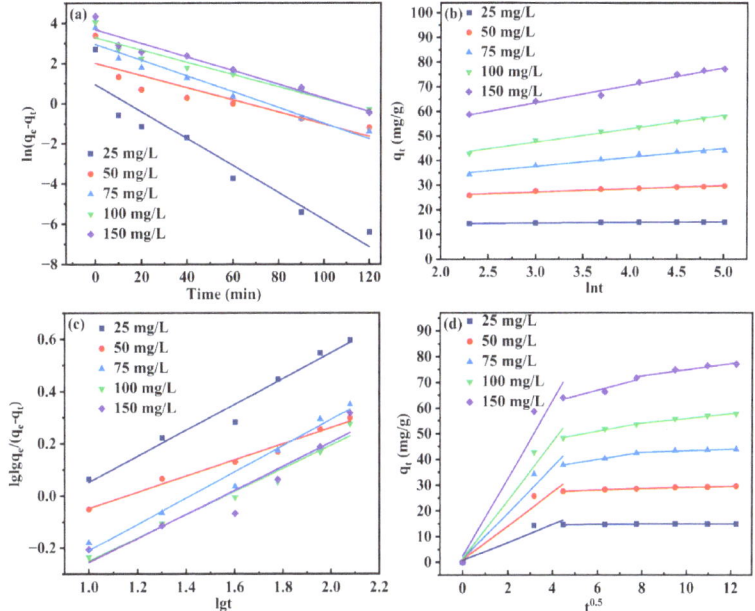

Figure 6. (a) Pseudo-first-order kinetic; (b) Elovich; (c) Bingham; and (d) intraparticle diffusion models (AL-TiNTs: 0.05 g; pH = 2; temperature: 25 °C).

2.5.2. Adsorption Isotherms

Adsorption was analyzed using four different isotherm models such as Langmuir (Figure S6a), Freundlich (Figure S6b), Temkin (Figure S6c), and D-R isotherm models (Figure S6d), and the details are presented in Text S3. The highest R^2 value (0.9912–0.9972) for the Langmuir model in Table 1 suggested that the Langmuir model described the adsorption processes involved in this study to a considerable extent, suggesting a consistent adsorption process for single-molecular-layer adsorption [35]. At different temperatures, the R_L < 1 indicated that AL-TiNTs readily adsorb Cr(VI) [27]. The R_L value diminished as the concentration increased, indicating that the adsorption process of AL-TiNTs was irreversible as Cr(VI) concentration increased, and there are intense interactions between the AL-TiNTs and the adsorbate molecules Cr(VI) [56]. In the Freundlich isotherm fit results, $0 < 1/n < 1$ indicated that the adsorption process involved in this study was easy to carry out, which led to the same conclusion as that obtained from the Langmuir isotherm fit [57]. Furthermore, in the D-R model, the high E value (>8 kJ/mol) indicated that there was a predominantly chemisorption process for the AL-TiNTs.

Table 1. Fitting results for different sorption isotherm models.

Isotherm Model	Parameter	Temperature (K)			
		293 K	298 K	303 K	308 K
Langmuir	q_m (mg/g)	88.97	89.85	88.18	86.13
	K_L (L/mg)	0.24	0.43	0.61	1.12
	R^2	0.9923	0.9972	0.9948	0.9912
Freundlich	K_F ((mg/g)(L/mg)$^{1/n}$)	30.39	37.83	42.65	49.86
	$1/n$	0.30	0.28	0.24	0.21
	R^2	0.9843	0.9839	0.9905	0.9774
Temkin	K_T (L/mol)	3.34	7.41	15.30	49.06
	B_1 (J/mol)	17.66	16.63	14.82	12.80
	R^2	0.9856	0.9893	0.9891	0.9731
Dubinin–Radushkevich	q_m (mol/g)	0.0042	0.0041	0.0037	0.0033
	K_D (mol^2/g^2)	0.0029	0.0025	0.0020	0.0016
	R^2	0.9866	0.9879	0.9917	0.9804
	E (kJ/mol)	13.09	14.29	15.69	17.68

2.5.3. Thermodynamics

A thermodynamic approach to reveal the orientation and complexity of the reaction process is shown in Figure S7 and Text S4. The consequences are given in Table S3. A negative value for ΔG_0 indicated that the reaction was spontaneous. It was confirmed by $\Delta H_0 > 0$ that the AL-TiNT adsorption process was heat-absorbing. $\Delta S_0 > 0$ implied an improvement in the stoichiometry of the AL-TiNT/solution interface [50].

2.6. Selective Adsorption

The selective adsorption performance of AL-TiNTs on heavy metals was investigated (Figure 7a). The methods for the analysis of Zn^{2+}, Cd^{2+}, and Cu^{2+} are presented in Text S5. Under acidic conditions (pH = 2), AL-TiNTs exhibited excellent adsorption performance for Cr(VI), while poorer adsorption performance for Zn^{2+}, Cd^{2+}, and Cu^{2+} was observed. These results were also found in the binary and quaternary systems. The reason may be that Cr(VI) appeared in the solution as the negatively charged anions ($Cr_2O_7^{2-}$, CrO_4^{2-}, and $HCrO_4^{-}$), which were more easily adsorbed to the protonated amino groups on AL-TiNTs by the electrostatic attraction. It could be demonstrated that under acidic conditions, titanate nanotubes (TiNTs) were positively charged, while Zn^{2+}, Cd^{2+}, and Cu^{2+} were also positively charged metal ions, which could be electrostatically repelled by AL-TiNTs [58]. Ultimately, these heavy metal ions could only weakly physically adsorb through the pores of the AL-TiNTs. The adsorption performance of AL-TiNTs on Zn^{2+}, Cd^{2+}, and Cu^{2+} under neutral conditions without pH adjustment is displayed in Figure 7b. Interestingly, a significant improvement in the adsorption capacities of Zn^{2+} (63.78 mg/g), Cd^{2+} (59.20 mg/g), and Cu^{2+} (66.35 mg/g) by AL-TiNTs was observed. It revealed that under acidic conditions, AL-TiNTs performed exceptional selective adsorption of Cr(VI), and under neutral conditions, AL-TiNTs performed broad-spectrum adsorption of heavy metals, which was beneficial to the wide application of AL-TiNTs in heavy-metal-containing wastewater treatment.

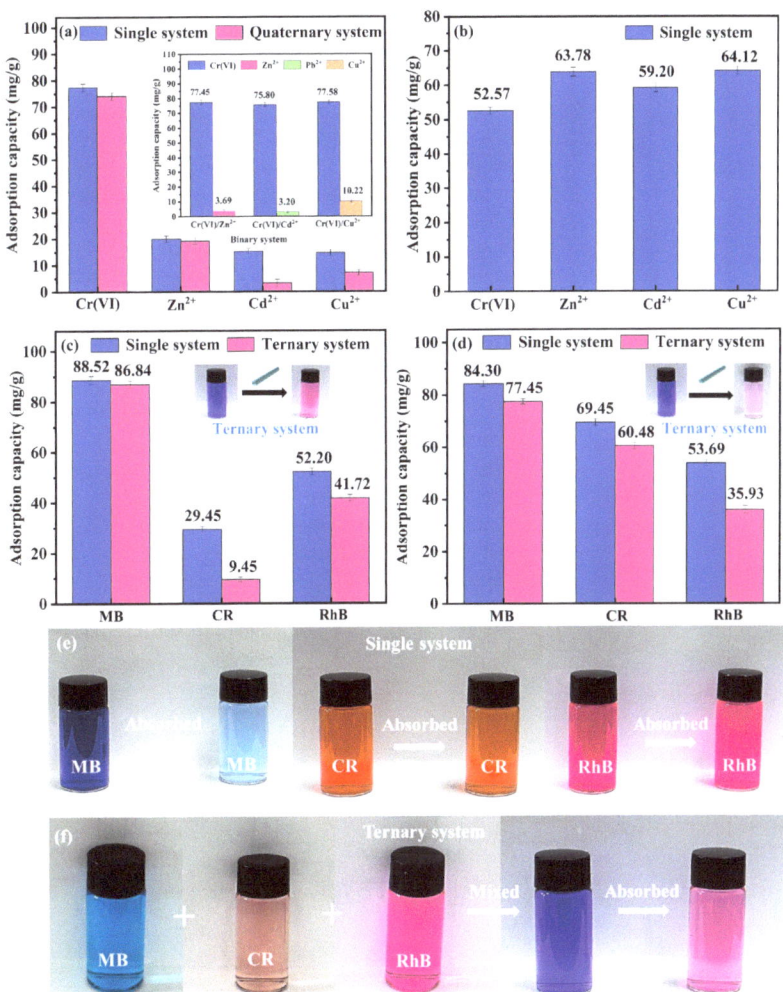

Figure 7. Adsorption of AL-TiNTs on heavy metal in single, binary, and quaternary systems (**a**) pH = 2 and (**b**) pH = 7; adsorption of AL-TiNTs on dyes in single and ternary systems (**c**) pH = 7 and (**d**) pH = 5.5; color change of dyes before and after adsorption by AL-TiNT (**e**) single system and (**f**) ternary system (AL-TiNTs: 0.05 g; temperature: 25 °C).

AL-TiNTs were added to the solutions containing methylene blue (MB), Congo red (CR), and rhodamine B (RhB). The selective adsorption performance of AL-TiNTs on dyes was investigated in the mixed ternary system at different pH conditions, respectively (Figure 7c,d). The methods for the testing of MB, CR, and RhB are described in Table S4. Results revealed that under neutral conditions, the adsorption performance of AL-TiNTs on dyes was in the order of MB > RhB > CR. Considering that the discoloration range of CR was 3.5–5.2, the adsorption of AL-TiNTs for CR was further investigated by adjusting the pH to 5.5 (Figure 7d). There was a remarkable change in the adsorption performance of AL-TiNTs for CR (69.45 mg/g). However, in the mixed ternary system, AL-TiNTs still prefer to adsorb MB. Under neutral conditions without pH adjustment, MB exhibited the most obvious color change from dark blue to light blue, while CR and RhB showed very little color change in the single system, (Figure 7e). In the mixed ternary system, the color changed from blue-purple to light pink because of the selective adsorption of MB (86.86 mg/g) by the AL-TiNTs, and only 41.72 and 9.45 mg/g were found for RhB and CR, respectively (Figure 7f). The excellent selective adsorption ability of AL-TiNTs for MB may be explained by the π-π stacking and electrostatic interactions between MB molecules and the aromatic rings on AL-TiNTs [59,60]. On the contrary, the spatial site resistance caused by the long side chains of the RhB molecule would inhibit its π-π stacking and electrostatic attraction with the AL-TiNTs, making it difficult for RhB to be removed by AL-TiNTs. The low AL-TiNT adsorption capacity for CR was attributed to the fact that both TiNTs and CR were negatively charged under neutral conditions, which may produce a strong electrostatic repulsion [61]. The results indicated that AL-TiNTs exhibited strong selective adsorption of MB in a wide pH range, which was beneficial for the application of AL-TiNTs in dye-containing wastewater treatment.

2.7. Mechanism Study

Figure 8 explains the removal mechanism of AL-TiNTs for Cr(VI), which involved the following processes:

(1). The amino group in AL grafted onto titanate nanotubes was protonated and adsorbed Cr(VI) by electrostatic attraction under acidic conditions;
(2). Under visible light, the AL-TiNTs generated electrons, and the Cr(VI) adsorbed on the AL-TiNTs was reduced to Cr(III);
(3). The positively charged Cr(III) reached the electronegative surface of the titanate nanotubes by electrostatic attraction and was eventually absorbed by the titanate nanotubes through ion exchange with Na^+ and H^+ between the nanotube layers or through complexation and co-precipitation [4].

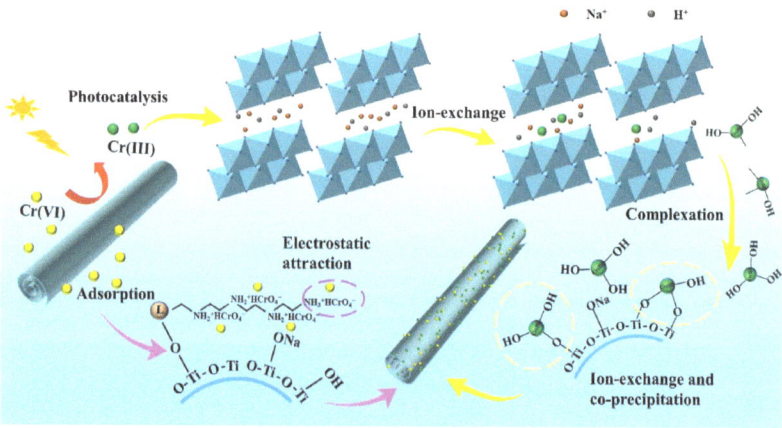

Figure 8. The mechanism of Cr(VI) removal by AL-TiNTs.

2.8. Regenerative and Applicative Study

The durability and stability of the material are important indicators for assessing the performance of the AL-TiNT materials, as well as for evaluating their application prospects. It was appropriate to use NaOH solution as the desorption reagent for the reasons presented in Text S6. The effect of desorption solution pH on the desorption efficiency of Cr(VI) was explored (Figure S8). The result indicated that the desorption efficiency was essentially stable at 81.32% when the pH value was greater than 10. Continuing to increase the pH value, the desorption efficiency was hardly improved. The adsorption capacity of TL and AL-TiNTs diminished with the number of applications (Figure 9a). On the one hand, the structure of the AL-TiNTs may have changed after several adsorption–desorption cycles. On the other hand, the re-obtained AL-TiNTs after the adsorption of Cr(VI) (AL-TiNTs-Cr) are more stable compared to the original AL-TiNTs, and many active sites may be blocked. In addition, in the first-time application, the adsorption capacity for Cr(VI) was seven to eight times higher by AL-TiNTs than that by TL. After four cycles, the adsorption capacity of AL-TiNTs was 49.50 mg/g, which was much higher than the adsorption capacity of amino-modified titanate nanotubes TNTs-RNH$_2$ after three cycles (23.9 mg/g) [37]. Under the condition of the adsorbent undergoing alternating acidic and basic conditions, the AL-TiNTs still retained excellent stability.

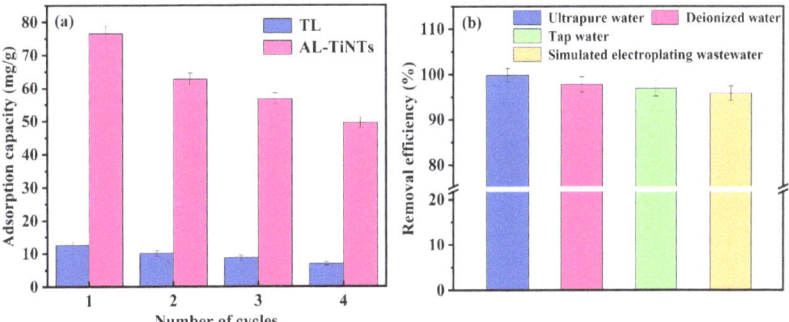

Figure 9. (a) The reusability of AL-TiNTs; (b) the applicability of AL-TiNTs in different water resources. (AL-TiNTs: 0.05 g; pH = 2; temperature: 25 °C.).

A comparison of the adsorption capacity of AL-TiNTs and other adsorbents for Cr(VI) is presented in Table 2. The adsorption capacity of AL-TiNTs after four cycles was still higher than that of the first-used iron oxide-coated cellulose/hydrotalcite (Fe$_3$O$_4$@CelHT) (29.00 mg/g) [62], multi-wall carbon nanotube (MWCNT) (11.41 mg/g) [63], mesoporous carbon nitride (MCN) (48.31 mg/g) [64], and fox nutshell activated carbon (FNAC) (46.21 mg/g) [65].

Table 2. Comparison of Cr(VI) adsorption capacity of AL-TiNTs with some reported adsorbents.

Adsorbent	Q_m (mg/g)	References
AL-TiNTs	77.15	This study
TNTs-RNH$_2$	69.10	[37]
Fe$_3$O$_4$@CelHT	29.00	[62]
MWCNT	11.41	[63]
MCN	48.31	[64]
FNAC	46.21	[65]

The applicability of AL-TiNTs in different water sources was further investigated (Figure 9b), and the detailed parameters of the simulated electroplating wastewater are described in Table S5. After 2.5 h of reaction, the removal efficiencies obtained by AL-TiNTs for Cr(VI) were 97.81%, 96.97%, and 95.86% in the deionized water, tap water, and simulated electroplating wastewater, respectively. Although the removal efficiencies were relatively lower compared to those in ultra-pure water (99.80%), they were all above 95%, indicating the potential of AL-TiNTs for practical applications.

3. Materials and Methods

3.1. Materials and Chemicals

The TL used was the same as we mentioned before [53]. TiO_2 (99.8%, 40 nm), potassium dichromate ($K_2Cr_2O_7$), formaldehyde (CH_2O), potassium ferricyanide ($K_3[Fe(CN)_6]$), triethylenetetramine (TETA), acetone (C_3H_6O), and phosphoric acid (H_3PO_4) were obtained from Damao Chemical Reagent Co., Ltd. (Tianjin, China). Further material information is provided in Text S7.

3.2. Preparation of AL-TiNTs

3.2.1. Amination of TL

Amine modification of TL by Mannich reaction [2]: First, 5 g of TL was dissolved in 0.4 mM sodium hydroxide (NaOH) solution (30 mL). Then, 4 mL of TETA and 3.6 mL of CH_2O were added and stirred (500 rpm) for 15 min. After that, the mixture was placed in water at 75 °C for 3 h. The mixed solution was filtered and collected. Then, an excess of $K_3[Fe(CN)_6]$ solution was added to the filtrate to allow the modified cationic lignin amine to precipitate out. To obtain aminated lignin (AL), the product was cleaned with ethanol and alternating ultra-pure water and freeze-dried for 24 h. The physical pictures of TL and AL samples are presented in Figure S9a,b.

3.2.2. Synthesis of AL-TiNTs

A total of 1.0 g of nano-titanium dioxide (nano-TiO_2) powder was dissolved into 10 M NaOH solution (50 mL) with magnetic stirring (500 rpm) for 30 min. After that, 1.0 g of AL was added and stirred magnetically (500 rpm) for 45 min. Then, AL and titanium dioxide nanoparticles were well dispersed in the NaOH solution. The suspension was transported to a 100 mL PTFE-lined reactor, and the reaction was performed continuously at 150 °C for 24 h [66]. Finally, we filtered the solution and washed the precipitate to pH 7. The product was freeze-dried for 24 h and named AL-TiNTs. AL-TiNTs with AL/nano-TiO_2 mass ratios of 2:1, 1:2, and 1:4 were obtained by varying the AL and nano-TiO_2 addition ratios, with other steps remaining the same. The physical image of the AL-TiNT sample is shown in Figure S9c.

3.3. Characterizations

The morphology of TL, AL, and AL-TiNTs was investigated by scanning electron microscopy (SEM) and energy-dispersive X-ray spectroscopy (EDS) (Czech TESCAN MIRA LMS). X-ray diffraction (XRD) was obtained with an Ultima IV X-ray diffractometer (Rigaku, Tokyo, Japan). The Fourier transform infrared spectrometer (FT-IR, Nicolet iS10 spectrometer) and the specific surface area (Micromeritics ASAP 2460 Version 3.01) were performed by the KBr particle technique and Brunauer–Emmett–Teller (BET) technique, respectively.

3.4. Experimental Design

3.4.1. Adsorption Experiment

A total of 0.05 g of AL-TiNTs was added to the conical flask, which contains Cr(VI) solution (50 mg/L), and shaken well with a constant temperature shaker (150 rpm). The pH was adjusted with HCl and NaOH. Using a disposable syringe, the sample solution was aspirated at a fixed time in 2 mL and filtered. Cr(VI) was measured with UV-Vis spectrophotometer (UV6100s, MAPADA, Shanghai, China). Details of the parameters influencing the adsorption performance of AL-TiNTs are described in Text S8.

3.4.2. Photocatalytic Degradation Experiments

Exploring Cr(VI) adsorption by AL-TiNTs through preliminary experiments, the adsorption–desorption balance was reached at around 40 min. Therefore, 0.05 g AL-TiNTs were incorporated into Cr(VI) (150 mg/L) solution (30 mL, pH = 2) at 25 °C and agitated magnetically for 40 min; then, the 300 W xenon lamp (wavelength > 420 nm) was switched on and illuminated continuously for 120 min. Sample solutions were collected at various times during the reaction and filtered with a 0.22 μm filter membrane. The total Cr concentration was detected by inductively coupled plasma mass spectrometry (iCAP6300). The *degradation rate* was calculated by

$$\text{Degradation rate } (\%) = \frac{(C_0 - C_t)}{C_0} \times 100\% \tag{4}$$

where C_0 (mg/L) and C_t (mg/L) represent the Cr(VI) concentrations at the beginning of photocatalysis with the light source on and at time t (min) during the photocatalytic process, respectively.

3.4.3. Reusability Study

The post-use AL-TiNTs were added to 0.1 M NaOH solution and stirred to desorb Cr(VI) to investigate the *reuse efficiency* of AL-TiNTs. The calculation of Cr(VI) desorption efficiency is provided in Text S6. The *reuse efficiency* of AL-TiNTs was determined as follows:

$$\text{Reuse efficiency } (\%) = \frac{q_n}{q_1} \times 100\% \tag{5}$$

where q_1 is the amount of adsorption of AL-TiNTs for the first time, and q_n is the amount adsorbed by recycling for the time n.

4. Conclusions

A novel layered nanotube material, AL-TiNT, was successfully prepared by a simple hydrothermal method. Under the synergistic effect of adsorption and photocatalysis, the AL-TiNTs provided the potential to adsorb and reduce Cr(VI) in one step. The adsorption capacity of AL-TiNTs for total Cr was 90.00 mg/g. The AL-TiNTs exhibited strong resistance to complex solutions and showed excellent selectivity for Cr(VI) in a mixed system of quaternary heavy metal ions at pH 2. In a ternary dye-mix system, the AL-TiNTs were also selective for the adsorption of MB at pH 7. In addition, AL-TiNTs displayed broad-spectrum adsorption of Zn^{2+}, Cd^{2+}, and Cu^{2+} in a neutral environment. The adsorption mechanisms of AL-TiNTs were mainly monolayer adsorption, electrostatic interaction, ion exchange, and complexation. AL-TiNTs performed well (95.86%) in simulated electroplating wastewater, which has the potential for practical application.

Supplementary Materials: The following supporting information can be downloaded at https://www.mdpi.com/article/10.3390/molecules28155789/s1. References [61,67–70] are cited in Supplementary Materials. Figure S1: Adsorption-desorption isotherms and pore distribution of AL; Figure S2: Effect of temperature on the adsorption of Cr (VI) by AL-TiNTs and kinetic analysis; Figure S3: Physical diagrams of light intensity corresponding to different xenon lamp source currents: (a) 5 A; (b) 8.5 A; (c) 12 A; Figure S4: Effect of different xenon lamp light source currents on the removal of Cr(VI) by AL-TiNTs: (a) 5 A; (b) 12 A (AL-TiNTs: 0.05 g; pH = 2; temperature: 25 °C); Figure S5: The effect of co-existing anions insolution on removal of Cr(VI) by AL-TiNTs: (a) cations and (b) anions of different ionic strengths (AL-TiNTs: 0.05 g; solution pH = 2; experimental temperature: 25 °C); Figure S6: Adsorption isotherms (a) Langmuir; (b) Freundlich; (c) Temkin; and (d) D-R isotherm models (AL-TiNTs: 0.05 g; pH = 2); Figure S7: Thermodynamics; Figure S8: Effect of pH on desorption efficiency of Cr(VI); Figure S9: (a) TL; (b) AL; (c) AL-TiNTs. Table S1: Pore structures and specific surface areas of samples; Table S2: Fitting results for different sorption models; Table S3: Thermodynamic parameters for Cr(VI) removal by AL-TiNTs; Table S4: Details of MB, CR and RhB and detection wavelengths; Table S5: The content of substances in simulated wastewater; Text S1: The reduction process of Cr(VI); Text S2: Kinetics models; Text S3: Isotherm models; Text S4: Thermodynamic; Text S5: Measurement methods of Zn^{2+}, Cd^{2+}, and Cu^{2+}; Text S6: Desorption of Cr(VI); Text S7: Materials; Text S8: Calculation equations of removal efficiency and capacity.

Author Contributions: Conceptualization, Q.W. and Y.S.; methodology, M.H.; software, Q.W.; formal analysis, F.Y.; data curation, Q.W.; writing—original draft preparation, Q.W.; writing—review and editing, Y.S.; visualization, M.H.; supervision, J.H.; funding acquisition, Y.S. All authors have read and agreed to the published version of the manuscript.

Funding: This research was funded by the Key Research and Development Program of Shaanxi (No. 2023-ZDLSF-61), the Shaanxi Union Research Center of University and Enterprise of Mine Eco-geological Environment Protection and Restoration in the Middle Reaches of Yellow River, and Huijin Technology Holding Group Corporation Limited.

Institutional Review Board Statement: Not applicable.

Informed Consent Statement: Not applicable.

Data Availability Statement: Not applicable.

Conflicts of Interest: The authors declare no conflict of interest.

References

1. Nakkeeran, E.; Patra, C.; Shahnaz, T.; Rangabhashiyam, S.; Selvaraju, N. Continuous biosorption assessment for the removal of hexavalent chromium from aqueous solutions using *Strychnos nux vomica* fruit shell. *Bioresour. Technol. Rep.* **2018**, *3*, 256–260. [CrossRef]
2. Pang, Y.; Chen, Z.; Zhao, R.; Yi, C.; Qiu, X.; Qian, Y.; Lou, H. Facile synthesis of easily separated and reusable silver nanoparticles/aminated alkaline lignin composite and its catalytic ability. *J. Colloid Interface Sci.* **2021**, *587*, 334–346. [CrossRef]
3. Jiang, B.; Liu, Y.; Zheng, J.; Tan, M.; Wang, Z.; Wu, M. Synergetic Transformations of Multiple Pollutants Driven by Cr(VI)-Sulfite Reactions. *Environ. Sci. Technol.* **2015**, *49*, 12363–12371. [CrossRef]
4. Yu, C.; Tang, X.; Li, L.S.; Chai, X.L.; Xiao, R.; Wu, D.; Tang, C.J.; Chai, L.Y. The long-term effects of hexavalent chromium on anaerobic ammonium oxidation process: Performance inhibition, hexavalent chromium reduction and unexpected nitrite oxidation. *Bioresour. Technol.* **2019**, *283*, 138–147. [CrossRef] [PubMed]
5. Yuan, P.; Liu, D.; Fan, M.; Yang, D.; Zhu, R.; Ge, F.; Zhu, J.; He, H. Removal of hexavalent chromium [Cr(VI)] from aqueous solutions by the diatomite-supported/unsupported magnetite nanoparticles. *J. Hazard. Mater.* **2010**, *173*, 614–621. [CrossRef] [PubMed]
6. Sheng, G.; Hu, J.; Li, H.; Li, J.; Huang, Y. Enhanced sequestration of Cr(VI) by nanoscale zero-valent iron supported on layered double hydroxide by batch and XAFS study. *Chemosphere* **2016**, *148*, 227–232. [CrossRef]
7. Jiang, B.; Guo, J.; Wang, Z.; Zheng, X.; Zheng, J.; Wu, W.; Wu, M.; Xue, Q. A green approach towards simultaneous remediations of chromium(VI) and arsenic(III) in aqueous solution. *Chem. Eng. J.* **2015**, *262*, 1144–1151. [CrossRef]
8. Peng, H.; Guo, J. Removal of chromium from wastewater by membrane filtration, chemical precipitation, ion exchange, adsorption electrocoagulation, electrochemical reduction, electrodialysis, electrodeionization, photocatalysis and nanotechnology: A review. *Environ. Chem. Lett.* **2020**, *18*, 2055–2068. [CrossRef]
9. Wang, Y.-L.; Zhao, P.-Y.; Liang, B.-L.; Chen, K.; Wang, G.-S. Carbon nanotubes decorated Co/C from ZIF-67/melamine as high efficient microwave absorbing material. *Carbon* **2023**, *202*, 66–75. [CrossRef]
10. GracePavithra, K.; Jaikumar, V.; Kumar, P.S.; SundarRajan, P. A review on cleaner strategies for chromium industrial wastewater: Present research and future perspective. *J. Clean. Prod.* **2019**, *228*, 580–593. [CrossRef]
11. Biswas, A.; Chandra, B.P.; Prathibha, C. Highly efficient and simultaneous remediation of heavy metal ions (Pb(II), Hg(II), As(V), As(III) and Cr(VI)) from water using Ce intercalated and ceria decorated titanate nanotubes. *Appl. Surf. Sci.* **2023**, *612*, 155841. [CrossRef]
12. Kim, I.J.; Zhao, W.; Park, J.G.; Meng, Z. Carbon nanotube filter for heavy metal ion adsorption. *Ceram. Int.* **2021**, *47*, 33280–33285. [CrossRef]
13. Rodrigues, E.; Almeida, O.; Brasil, H.; Moraes, D.; dos Reis, M.A.L. Adsorption of chromium (VI) on hydrotalcite-hydroxyapatite material doped with carbon nanotubes: Equilibrium, kinetic and thermodynamic study. *Appl. Clay Sci.* **2019**, *172*, 57–64. [CrossRef]
14. Guo, C.; Chen, W.; Yan, S.; Chen, Y.; Zhao, X.; Zhang, P.; Ma, W.; Yang, A. Novel zirconia-halloysite nanotube material for arsenite adsorption from water. *J. Environ. Chem. Eng.* **2023**, *11*, 109181. [CrossRef]
15. Lin, Y.-J.; Chen, J.-J.; Cao, W.-Z.; Persson, K.M.; Ouyang, T.; Zhang, L.; Xie, X.; Liu, F.; Li, J.; Chang, C.-T. Novel materials for Cr(VI) adsorption by magnetic titanium nanotubes coated phosphorene. *J. Mol. Liq.* **2019**, *287*, 110826. [CrossRef]
16. Wang, Y.; Liu, W.; Fu, H.; Yi, X.H.; Wang, P.; Zhao, C.; Wang, C.C.; Zheng, W. Simultaneous Cr(VI) reduction and Cr(III) removal of bifunctional MOF/Titanate nanotube composites. *Environ. Pollut.* **2019**, *249*, 502–511. [CrossRef]
17. Li, H.; Huang, Y.; Liu, J.; Duan, H. Hydrothermally synthesized titanate nanomaterials for the removal of heavy metals and radionuclides from water: A review. *Chemosphere* **2021**, *282*, 131046. [CrossRef]
18. Tu, C.; Luo, W.; Peng, Y.; Yu, P.; Shi, C.; Wu, Z.; Shao, L.; Zhan, P. Preparation of lignin-based carbon nanotubes using micelles as soft template. *Ind. Crops Prod.* **2023**, *191*, 116009. [CrossRef]
19. Osman, L.S.; Hamidon, T.S.; Latif, N.H.A.; Elias, N.H.H.; Saidin, M.; Shahidan, S.; Abdullah, S.H.A.; Ali, N.A.; Rusli, S.S.M.; Ibrahim, M.N.M.; et al. Rust conversion of archeological cannonball from Fort Cornwallis using oil palm frond lignin. *Ind. Crops Prod.* **2023**, *192*, 116107. [CrossRef]
20. Wang, B.; Sun, Y.-C.; Sun, R.-C. Fractionational and structural characterization of lignin and its modification as biosorbents for efficient removal of chromium from wastewater: A review. *J. Leather Sci. Eng.* **2019**, *1*, 5. [CrossRef]

21. Wang, C.; Kelley, S.S.; Venditti, R.A. Lignin-Based Thermoplastic Materials. *ChemSusChem* **2016**, *9*, 770–783. [CrossRef] [PubMed]
22. Huang, S.; Shuyi, S.; Gan, H.; Linjun, W.; Lin, C.; Danyuan, X.; Zhou, H.; Lin, X.; Qin, Y. Facile fabrication and characterization of highly stretchable lignin-based hydroxyethyl cellulose self-healing hydrogel. *Carbohydr. Polym.* **2019**, *223*, 115080. [CrossRef] [PubMed]
23. Bartczak, P.; Klapiszewski, L.; Wysokowski, M.; Majchrzak, I.; Czernicka, W.; Piasecki, A.; Ehrlich, H.; Jesionowski, T. Treatment of model solutions and wastewater containing selected hazardous metal ions using a chitin/lignin hybrid material as an effective sorbent. *J. Environ. Manag.* **2017**, *204*, 300–310. [CrossRef] [PubMed]
24. Li, Z.; Chen, J.; Ge, Y. Removal of lead ion and oil droplet from aqueous solution by lignin-grafted carbon nanotubes. *Chem. Eng. J.* **2017**, *308*, 809–817. [CrossRef]
25. Ying, W.; Shi, Z.; Yang, H.; Xu, G.; Zheng, Z.; Yang, J. Effect of alkaline lignin modification on cellulase-lignin interactions and enzymatic saccharification yield. *Biotechnol. Biofuels* **2018**, *11*, 214. [CrossRef]
26. He, T.; Jiang, Y.; Chang, S.; Zhou, X.; Ji, Y.; Fang, X.; Zhang, Y. Antibacterial and high-performance bioplastics derived from biodegradable PBST and lignin. *Ind. Crops Prod.* **2023**, *191*, 115930. [CrossRef]
27. Liu, Q.; Wang, F.; Zhou, H.; Li, Z.; Fu, Y.; Qin, M. Utilization of lignin separated from pre-hydrolysis liquor via horseradish peroxidase modification as an adsorbent for methylene blue removal from aqueous solution. *Ind. Crops Prod.* **2021**, *167*, 113535. [CrossRef]
28. Sun, Y.; Liu, X.; Lv, X.; Wang, T.; Xue, B. Synthesis of novel lignosulfonate-modified graphene hydrogel for ultrahigh adsorption capacity of Cr(VI) from wastewater. *J. Clean. Prod.* **2021**, *295*, 126406. [CrossRef]
29. Figueiredo, P.; Lintinen, K.; Hirvonen, J.T.; Kostiainen, M.A.; Santos, H.A. Properties and chemical modifications of lignin: Towards lignin-based nanomaterials for biomedical applications. *Prog. Mater Sci.* **2018**, *93*, 233–269. [CrossRef]
30. Ge, Y.; Song, Q.; Li, Z. A Mannich base biosorbent derived from alkaline lignin for lead removal from aqueous solution. *J. Ind. Eng. Chem.* **2015**, *23*, 228–234. [CrossRef]
31. Guo, Q.; Zhou, C.; Ma, Z.; Yang, X. Fundamentals of TiO_2 Photocatalysis: Concepts, Mechanisms, and Challenges. *Adv. Mater.* **2019**, *31*, 1901997. [CrossRef] [PubMed]
32. Shan, X.; Zhao, Y.; Bo, S.; Yang, L.; Xiao, Z.; An, Q.; Zhai, S. Magnetic aminated lignin/CeO_2/Fe_3O_4 composites with tailored interfacial chemistry and affinity for selective phosphate removal. *Sci. Total Environ.* **2021**, *796*, 148984. [CrossRef]
33. Liu, N.; Chen, X.; Zhang, J.; Schwank, J.W. A review on TiO_2-based nanotubes synthesized via hydrothermal method: Formation mechanism, structure modification, and photocatalytic applications. *Catal. Today* **2014**, *225*, 34–51. [CrossRef]
34. Fan, G.; Lin, R.; Su, Z.; Lin, X.; Xu, R.; Chen, W. Removal of Cr (VI) from aqueous solutions by titanate nanomaterials synthesized via hydrothermal method. *Can. J. Chem. Eng.* **2017**, *95*, 717–723. [CrossRef]
35. Li, K.; Yang, H.; Wang, L.; Chai, S.; Zhang, R.; Wu, J.; Liu, X. Facile integration of FeS and titanate nanotubes for efficient removal of total Cr from aqueous solution: Synergy in simultaneous reduction of Cr(VI) and adsorption of Cr(III). *J. Hazard. Mater.* **2020**, *398*, 122834. [CrossRef] [PubMed]
36. Liu, M.; Yin, W.; Qian, F.-J.; Zhao, T.-L.; Yao, Q.-Z.; Fu, S.-Q.; Zhou, G.-T. A novel synthesis of porous TiO_2 nanotubes and sequential application to dye contaminant removal and Cr(VI) visible light catalytic reduction. *J. Environ. Chem. Eng.* **2020**, *8*, 104061. [CrossRef]
37. Niu, G.; Liu, W.; Wang, T.; Ni, J. Absorption of Cr(VI) onto amino-modified titanate nanotubes using 2-bromoethylamine hydrobromide through SN_2 reaction. *J. Colloid Interface Sci.* **2013**, *401*, 133–140. [CrossRef] [PubMed]
38. Liu, W.; Sun, W.; Borthwick, A.G.L.; Wang, T.; Li, F.; Guan, Y. Simultaneous removal of Cr(VI) and 4-chlorophenol through photocatalysis by a novel anatase/titanate nanosheet composite: Synergetic promotion effect and autosynchronous doping. *J. Hazard. Mater.* **2016**, *317*, 385–393. [CrossRef] [PubMed]
39. Fu, Y.; Liu, X.; Chen, G. Adsorption of heavy metal sewage on nano-materials such as titanate/TiO_2 added lignin. *Results Phys.* **2019**, *12*, 405–411. [CrossRef]
40. Akbarzadeh, R.; Farhadian, N.; Asadi, A.; Hasani, T.; Salehi Morovat, S. Highly efficient visible-driven reduction of Cr(VI) by a novel black TiO_2 photocatalyst. *Environ. Sci. Pollut. Res. Int.* **2021**, *28*, 9417–9429. [CrossRef]
41. Liu, W.; Ni, J.; Yin, X. Synergy of photocatalysis and adsorption for simultaneous removal of Cr(VI) and Cr(III) with TiO_2 and titanate nanotubes. *Water Res.* **2014**, *53*, 12–25. [CrossRef] [PubMed]
42. Ge, Y.; Li, Z.; Kong, Y.; Song, Q.; Wang, K. Heavy metal ions retention by bi-functionalized lignin: Synthesis, applications, and adsorption mechanisms. *J. Ind. Eng. Chem.* **2014**, *20*, 4429–4436. [CrossRef]
43. Faix, O. Classification of Lignins from Different Botanical Origins by FT-IR Spectroscopy. *Holzforschung* **1991**, *45*, 21–27. [CrossRef]
44. Lu, Q.F.; Huang, Z.K.; Liu, B.; Cheng, X. Preparation and heavy metal ions biosorption of graft copolymers from enzymatic hydrolysis lignin and amino acids. *Bioresour. Technol.* **2012**, *104*, 111–118. [CrossRef]
45. Nosheen, S.; Galasso, F.S.; Suib, S.L. Role of Ti-O bonds in phase transitions of TiO_2. *Langmuir* **2009**, *25*, 7623–7630. [CrossRef]
46. Li, T.; Lu, S.; Wang, Z.; Huang, M.; Yan, J.; Liu, M. Lignin-based nanoparticles for recovery and separation of phosphate and reused as renewable magnetic fertilizers. *Sci. Total Environ.* **2021**, *765*, 142745. [CrossRef]
47. Ukhurebor, K.E.; Aigbe, U.O.; Onyancha, R.B.; Nwankwo, W.; Osibote, O.A.; Paumo, H.K.; Ama, O.M.; Adetunji, C.O.; Siloko, I.U. Effect of hexavalent chromium on the environment and removal techniques: A review. *J. Environ. Manag.* **2021**, *280*, 111809. [CrossRef]

48. Wang, L.; Liu, W.; Wang, T.; Ni, J. Highly efficient adsorption of Cr(VI) from aqueous solutions by amino-functionalized titanate nanotubes. *Chem. Eng. J.* **2013**, *225*, 153–163. [CrossRef]
49. Wang, X.; Cai, J.; Zhang, Y.; Li, L.; Jiang, L.; Wang, C. Heavy metal sorption properties of magnesium titanate mesoporous nanorods. *J. Mater. Chem. A* **2015**, *3*, 11796–11800. [CrossRef]
50. Chi, Z.; Hao, L.; Dong, H.; Yu, H.; Liu, H.; Wang, Z.; Yu, H. The innovative application of organosolv lignin for nanomaterial modification to boost its heavy metal detoxification performance in the aquatic environment. *Chem. Eng. J.* **2020**, *382*, 122789. [CrossRef]
51. Li, Y.; Bian, Y.; Qin, H.; Zhang, Y.; Bian, Z. Photocatalytic reduction behavior of hexavalent chromium on hydroxyl modified titanium dioxide. *Appl. Catal. B* **2017**, *206*, 293–299. [CrossRef]
52. Zafar, Z.; Fatima, R.; Kim, J.O. Effect of HCl treatment on physico-chemical properties and photocatalytic performance of Fe-TiO$_2$ nanotubes for hexavalent chromium reduction and dye degradation under visible light. *Chemosphere* **2021**, *284*, 131247. [CrossRef]
53. Sun, Y.; Wang, T.; Han, C.; Bai, L.; Sun, X. One-step preparation of lignin-based magnetic biochar as bifunctional material for the efficient removal of Cr(VI) and Congo red: Performance and practical application. *Bioresour. Technol.* **2022**, *369*, 128373. [CrossRef]
54. Hu, Z.; Ge, M.; Guo, C. Efficient removal of levofloxacin from different water matrices via simultaneous adsorption and photocatalysis using a magnetic Ag$_3$PO$_4$/rGO/CoFe$_2$O$_4$ catalyst. *Chemosphere* **2021**, *268*, 128834. [CrossRef] [PubMed]
55. Tang, D.; Zhang, G. Efficient removal of fluoride by hierarchical Ce–Fe bimetal oxides adsorbent: Thermodynamics, kinetics and mechanism. *Chem. Eng. J.* **2016**, *283*, 721–729. [CrossRef]
56. Sharaf El-Deen, S.E.A.; Ammar, N.S.; Jamil, T.S. Adsorption behavior of Co(II) and Ni(II) from aqueous solutions onto titanate nanotubes. *Fuller. Nanotub. Carbon Nanostructures* **2016**, *24*, 455–466. [CrossRef]
57. Alminderej, F.M.; Younis, A.M.; Albadri, A.E.A.E.; El-Sayed, W.A.; El-Ghoul, Y.; Ali, R.; Mohamed, A.M.A.; Saleh, S.M. The superior adsorption capacity of phenol from aqueous solution using Modified Date Palm Nanomaterials: A performance and kinetic study. *Arab. J. Chem.* **2022**, *15*, 104120. [CrossRef]
58. Tan, Y.; Lv, X.; Wang, W.; Cui, C.; Bao, Y.; Jiao, S. Surface-functionalization of hydrogen titanate nanowires for efficiently selective adsorption of methylene blue. *Appl. Surf. Sci.* **2023**, *615*, 156265. [CrossRef]
59. Heo, J.W.; An, L.; Chen, J.; Bae, J.H.; Kim, Y.S. Preparation of amine-functionalized lignins for the selective adsorption of Methylene blue and Congo red. *Chemosphere* **2022**, *295*, 133815. [CrossRef]
60. Charmas, B.; Ziȩzio, M.; Jedynak, K. Assessment of the Porous Structure and Surface Chemistry of Activated Biocarbons Used for Methylene Blue Adsorption. *Molecules* **2023**, *28*, 4922. [CrossRef]
61. Sun, Y.; Wang, T.; Han, C.; Lv, X.; Bai, L.; Sun, X.; Zhang, P. Facile synthesis of Fe-modified lignin-based biochar for ultra-fast adsorption of methylene blue: Selective adsorption and mechanism studies. *Bioresour. Technol.* **2022**, *344*, 126186. [CrossRef]
62. Periyasamy, S.; Gopalakannan, V.; Viswanathan, N. Fabrication of magnetic particles imprinted cellulose based biocomposites for chromium(VI) removal. *Carbohydr. Polym.* **2017**, *174*, 352–359. [CrossRef] [PubMed]
63. Huang, Z.-N.; Wang, X.-L.; Yang, D.-S. Adsorption of Cr(VI) in wastewater using magnetic multi-wall carbon nanotubes. *Water Sci. Eng.* **2015**, *8*, 226–232. [CrossRef]
64. Chen, H.; Yan, T.; Jiang, F. Adsorption of Cr(VI) from aqueous solution on mesoporous carbon nitride. *J. Taiwan Inst. Chem. Eng.* **2014**, *45*, 1842–1849. [CrossRef]
65. Kumar, A.; Jena, H.M. Adsorption of Cr(VI) from aqueous phase by high surface area activated carbon prepared by chemical activation with ZnCl$_2$. *Process Saf. Environ. Prot.* **2017**, *109*, 63–71. [CrossRef]
66. Liu, Y.; Liu, F.; Qi, Z.; Shen, C.; Li, F.; Ma, C.; Huang, M.; Wang, Z.; Li, J. Simultaneous oxidation and sorption of highly toxic Sb(III) using a dual-functional electroactive filter. *Environ. Pollut.* **2019**, *251*, 72–80. [CrossRef] [PubMed]
67. Liu, W.; Zhao, X.; Wang, T.; Fu, J.; Ni, J. Selective and irreversible adsorption of mercury(II) from aqueous solution by a flower-like titanate nanomaterial. *J. Mater. Chem. A* **2015**, *3*, 17676–17684. [CrossRef]
68. Xiong, Z.; Zheng, H.; Hu, Y.; Hu, X.; Ding, W.; Ma, J.; Li, Y. Selective adsorption of Congo red and Cu(II) from complex wastewater by core-shell structured magnetic carbon@zeolitic imidazolate frameworks-8 nanocomposites. *Sep. Purif. Technol.* **2021**, *277*, 119053. [CrossRef]
69. Wang, Z.; Guo, J.; Jia, J.; Liu, W.; Yao, X.; Feng, J.; Dong, S.; Sun, J. Magnetic Biochar Derived from Fenton Sludge/CMC for High-Efficiency Removal of Pb(II): Synthesis, Application, and Mechanism. *Molecules* **2023**, *28*, 4983. [CrossRef]
70. Yu, K.L.; Lee, X.J.; Ong, H.C.; Chen, W.H.; Chang, J.S.; Lin, C.S.; Show, P.L.; Ling, T.C. Adsorptive removal of cationic methylene blue and anionic Congo red dyes using wet-torrefied microalgal biochar: Equilibrium, kinetic and mechanism modeling. *Environ. Pollut.* **2021**, *272*, 115986. [CrossRef]

Disclaimer/Publisher's Note: The statements, opinions and data contained in all publications are solely those of the individual author(s) and contributor(s) and not of MDPI and/or the editor(s). MDPI and/or the editor(s) disclaim responsibility for any injury to people or property resulting from any ideas, methods, instructions or products referred to in the content.

Review

Effective Usage of Biochar and Microorganisms for the Removal of Heavy Metal Ions and Pesticides

Soumya K. Manikandan [1], Pratyasha Pallavi [1], Krishan Shetty [1], Debalina Bhattacharjee [2], Dimitrios A. Giannakoudakis [3,*], Ioannis A. Katsoyiannis [3] and Vaishakh Nair [1,*]

[1] Department of Chemical Engineering, National Institute of Technology Karnataka (NITK), Mangalore 575025, India
[2] Department of Physics, MVJ College of Engineering, Bangalore 560067, India
[3] Laboratory of Chemical and Environmental Technology, Department of Chemistry, Aristotle University of Thessaloniki, 54124 Thessaloniki, Greece
* Correspondence: dagchem@gmail.com (D.A.G.); vaishakhnair@nitk.edu.in (V.N.)

Citation: Manikandan, S.K.; Pallavi, P.; Shetty, K.; Bhattacharjee, D.; Giannakoudakis, D.A.; Katsoyiannis, I.A.; Nair, V. Effective Usage of Biochar and Microorganisms for the Removal of Heavy Metal Ions and Pesticides. *Molecules* **2023**, *28*, 719. https://doi.org/10.3390/molecules28020719

Academic Editor: Łukasz Chrzanowski

Received: 5 December 2022
Revised: 3 January 2023
Accepted: 6 January 2023
Published: 11 January 2023

Copyright: © 2023 by the authors. Licensee MDPI, Basel, Switzerland. This article is an open access article distributed under the terms and conditions of the Creative Commons Attribution (CC BY) license (https://creativecommons.org/licenses/by/4.0/).

Abstract: The bioremediation of heavy metal ions and pesticides is both cost-effective and environmentally friendly. Microbial remediation is considered superior to conventional abiotic remediation processes, due to its cost-effectiveness, decrement of biological and chemical sludge, selectivity toward specific metal ions, and high removal efficiency in dilute effluents. Immobilization technology using biochar as a carrier is one important approach for advancing microbial remediation. This article provides an overview of biochar-based materials, including their design and production strategies, physicochemical properties, and applications as adsorbents and support for microorganisms. Microorganisms that can cope with the various heavy metal ions and/or pesticides that enter the environment are also outlined in this review. Pesticide and heavy metal bioremediation can be influenced by microbial activity, pollutant bioavailability, and environmental factors, such as pH and temperature. Furthermore, by elucidating the interaction mechanisms, this paper summarizes the microbe-mediated remediation of heavy metals and pesticides. In this review, we also compile and discuss those works focusing on the study of various bioremediation strategies utilizing biochar and microorganisms and how the immobilized bacteria on biochar contribute to the improvement of bioremediation strategies. There is also a summary of the sources and harmful effects of pesticides and heavy metals. Finally, based on the research described above, this study outlines the future scope of this field.

Keywords: bioremediation; microbial cell; pollutant; immobilization

1. Introduction

The rapid expansion of industrialization has resulted in the depletion of natural resources and the production of vast volumes of hazardous waste that pollute water and soil, threatening the environment and human health [1]. The deterioration of soil and water quality due to releasing toxic pollutants has become a serious threat around the world. The release of these harmful wastes into the environment occurs in different forms; for example, atmospheric pollutants include noxious gases such as sulfur oxides and nitrogen oxides, while soil and water can be contaminated by organic pollutants (pesticides, hydrocarbons, phenols, etc.) and heavy metals (cadmium, arsenic, lead, chromium, mercury, etc.). Human health can be adversely affected by these environmental pollutants [2] through inhalation or ingestion (Figure 1). Additionally, some pollutants, such as heavy metal ions, can bioaccumulate in the food chain, and these persistent organic pollutants present significant risks to humans and other living creatures.

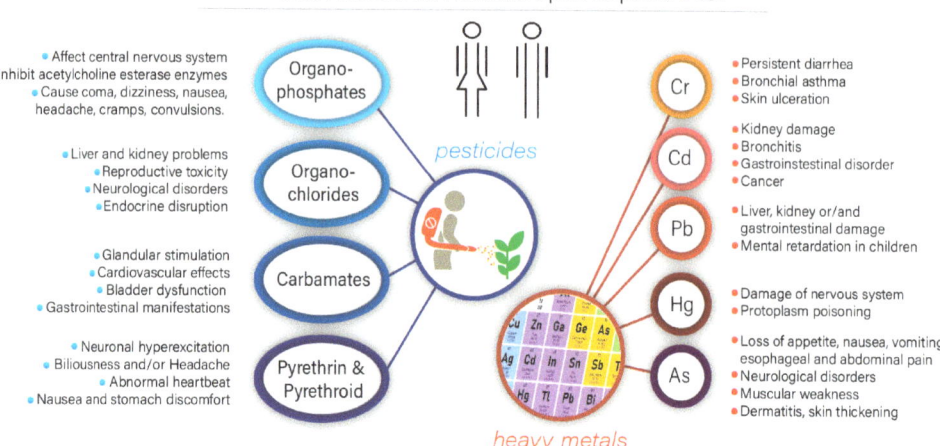

Figure 1. The effects of pesticides and heavy metal exposure on humans.

The accumulation of pesticides and their derivatives is becoming more prevalent, due to the rising population and rapid industrialization. As much as 80% to 90% of pesticides applied to crops in agricultural fields affect non-target life forms; they can relocate or volatilize from the treated area to pollute the air and soil and negatively affect non-target plants. The leaching of these accumulated pesticides leads to the contamination of groundwater and soil [3].

During the last few decades, the separation of pollutants from water systems and soil via several methods has been developed and successfully applied. Recently, technologies such as membrane filtration, ion exchange, and chemical precipitation have been utilized in real-life applications to remove pollutants such as metal ions from polluted areas. Chemical precipitation is a frequently used method for treating heavy metals because it is simple, inexpensive, and effective. However, chemical precipitation results in secondary pollution and eventually leads to additional difficulties in cleaning up the trace contaminants from large areas. Ion-exchange resin offers fast kinetics and is highly efficient for pollutant removal. However, the need for an acidic environment restricts their application in various contexts. Membrane filtration-based technologies can remove toxic substances with high efficiency, but the manufacture of membrane material is usually very complex and at a high cost. Conventional pollutant remediation methods are not eco-friendly and produce toxic chemical sludge. Therefore, there is a serious need to develop efficient and sustainable technologies for remediating toxic environmental pollutants.

Biochar is a carbonaceous material produced through the thermal treatment of different types of biomass, such as crop residues and biosolids [4,5]. Biochar production can be achieved via various processes, including slow or fast pyrolysis, flash carbonization, gasification, hydrothermal carbonization, torrefaction, etc. [6]. The key goal when designing the synthesis of the biochar is that the final material should possess high porosity, a large specific surface area, and elevated surface chemistry heterogeneity, as with oxygen-containing functional groups and minerals [7]. The physicochemical properties of the final obtained biochar can also be tuned by altering the microstructure. Such characteristics encourage biochar's rising application in (waste) water treatment, soil improvement, and its use in general air, water, and soil remediation [8]. Mechanisms such as physisorption, complexation, precipitation, ion exchange, and electrostatic interaction are involved in the removal of pollutants from aqueous solutions using biochar. Biochar with a high surface area and pore volume exhibits a higher metal–ion philicity because it can be physically entrapped within the pores on its surface [9]. The negatively charged surfaces of biochar can adsorb positively charged metal ions via electrostatic attraction. Compared to other adsorbents/microbial supports, biochar is a low-cost option and a promising candidate for pesticide and heavy metal treatment.

Bioremediation by microorganisms is considered a green technology that is acceptable to the general public. Microorganisms can bioadsorb, bioaccumulate, or biotransform the pollutants permanently at a low operating cost and without the generation of harmful secondary products [10]. Bioremediation can be effective even for contaminants in low concentrations that cannot otherwise be removed by chemical (e.g., incineration) or physical methods. According to some studies, microbial remediation has also been combined with other physical and chemical treatment methods. Hence, bioremediation reduces the health hazards to workers [11]. The microbial degradation of harmful and recalcitrant pesticides is efficient, cost-effective, and eco-friendly, with minimum hazards. The microbial consortia used for remediation have the additional advantage of promoting plant growth in the contaminated site.

Microorganisms can rapidly mutate and evolve in order to withstand environmental stress. The diversity and metabolic activity of the microorganisms are influenced by the presence of heavy metal ions and/or metalloids, which compels the microorganisms to develop resistance systems for overcoming this toxic metal ion stress. Furthermore, microorganisms convert toxic metal ions into inactive forms and can thus be utilized for bioremediation. Pesticide bioremediation involves biodegradation and biotransformation. In biodegradation, biological reactions modify the compound's chemical structure, decreasing its toxicity.

Microorganism immobilization on biochar is an efficient technology for treating wastewater and soil pollutants [12]. However, less information is available about the degradation of antibiotics, pesticides, heavy metals, PAHs, and other macromolecular organic pollutants that are immobilized on biochar by the microorganisms. Such pollutants are usually remediated via chemical methods or photocatalysis using biochar, which results in the production of free radicals that pose an ecotoxicological risk. As a result, biodegradation is becoming more important, and it is necessary to further investigate unexploited microorganisms for immobilization technology, based on biochar for pollutant biodegradation. However, biochar may cause toxicity to microorganisms according to their particle size [13]. To reduce this toxicity, the appropriate size of biochar must be chosen as a carrier for microorganism immobilization. Therefore, this work will evaluate the properties, influencing factors, strategies of immobilization, and removal efficiency of microbial cell-immobilized biochar (MCB) for the remediation of heavy metals and pesticides. The mechanisms involved in the bioremediation process will be explored. In the published research into the increasing environmental pollution caused by heavy metals and pesticides, much importance is given to remediation techniques. For biological remediation, there are many articles that discuss the role of microorganisms as an effective agent for the remediation of heavy metals and pesticides, as well as the role of biochar as an excellent adsorbent for the above pollutants. However, this review article focuses on the emerging role of biochar as an immobilization support for microbial cells.

2. Role of Biochar in the Removal of Metal Ions and Pesticides
2.1. Biochar Production, Properties, and Characterization

In general, biochar is a carbon-rich material derived from biomass (such as wood, manure, or leaves) upon thermal treatment at high temperatures in a closed container with minimal or in the absence of air [14]. Various processes, such as pyrolysis, gasification, and hydrothermal carbonization, are applied in biochar generation [15]. Biochar uses include (but are not limited to) carbon capture and storage, capacitive deionization, the Fenton process, microbial fuel cell electrodes, and electrochemical storage [16–18].

Biochar has been well established as a low-cost adsorbent that has adsorption capacities similar to carbon-based adsorbents, such as activated carbon, porous graphitic carbon nitride, graphene oxide, etc. The most crucial benefits are: (a) low cost of production,

(b) porous structure, (c) simple fabrication on a large scale, (d) eco-friendly nature promoting the cycle of the (bio)economy, (e) multiple surface-functional groups, especially oxygen-containing groups (thus enabling both hydrophobic and polar interactions), (f) ease of modification, etc. [19,20]. In addition, the preference for biochar as a catalyst support for photocatalysis and Fenton/photo-Fenton processes has become prevalent, due to its low cost and high surface area characteristics. The aromatic and other hetero-atom-containing functional groups that are present in biochar also provide moieties that are capable of electron transfer and facilitate the faster and more efficient degradation/reduction of pollutants because of electron delocalization and photo-induced e^-/h^+ pairs separation.

2.1.1. Biochar Production

Biochar production usually involves biomass collected from various plant/animal sources or wastewater sludge and thermal treatment using oxygen-deficient conditions, particularly pyrolysis. For instance, plant sources include olive pomace and rapeseed straw cereal waste, whereas animal sources include crustacean shells and animal manure [21–25]. Additionally, municipal wastewater sludge has also been used as biomass for biochar production [26]. The basic composition of biochar predominantly comprises amorphous phases and graphene sheets, as well as various aliphatic cyclic and aromatic groups as a matrix. The temperature of the treatment and the biomass source influence the final physicochemical features. For example, fibrous biomass sources such as wheat/rice straw generate tubular structures [27]. In contrast, the usage of sludge biochar prevents the formation of such structures in the biochar matrix [28].

Pyrolysis in oxygen-free conditions comprises the decomposition of lignocellulosic material, volatile matter release, and the reduction of carbonaceous material for plant biomass [29]. The types of pyrolysis include slow, fast, microwave-assisted, hydro- and co-pyrolysis. Slow pyrolysis operates for hours at lower temperature conditions (300–700 °C), resulting in a higher output percentage of biochar content compared to fast pyrolysis with lower residence time (<2–5 s), higher temperature conditions, and a lower output percentage of biochar. Increasing the temperature can lead to higher carbon content, alkalinity, and elevated specific surface area. In contrast, higher residence time can increase the specific surface area, due to prolonged temperature application.

Variations in high-temperature processes have also been tested in the context of biochar production. Microwave-assisted pyrolysis for biochar generation has also been demonstrated, with variations in absorbable power observed for biochar property analysis, with the demonstrated advantages of larger surface area and improved porosity characteristics [22]. Hydro-pyrolysis is usually conducted within a temperature range of 250–550 °C, with hydrogen gas application, ensuring the hydrocracking of the biomass [19]. Co-pyrolysis involves multiple biomass sources for biochar pyrolysis. The resultant physicochemical properties mainly depend on the biomass sources' blending ratios and pyrolysis temperature, improving the biochar sample's pore structure [30]. Gasification is another method of generating biochar in the presence of steam/oxygen at 750–900 °C, with the products being syngas and a low biochar yield. Torrefaction is conducted under oxygen-deficient conditions similar to those for biochar, apart from a temperature of 200–300 °C and a residence time of less than 30 min. Another method explored extensively for biochar production is hydrothermal carbonization, with an operating temperature range from 160 to 800 °C (preferably at lower temperatures) in the presence of water. The low-temperature environment results in higher O/C and H/C content, along with the creation of functional groups on the biochar surface; the process yields a low aromaticity level and low-porosity biochar (hydrochar). The conversion of the non-carbonized (amorphous) part of the biomass into a carbonized form can be enhanced by increasing the pyrolysis temperature, which also increases the aromaticity, π electron availability, etc. [30]. Both the negative effect of pore-size thermal shrinkage due to the collapse of micropore walls and the positive effect of pore-size increment due to the removal of volatile matter can be observed with increasing temperature conditions. Increasing the pyrolysis temperatures also decreases the biochar's stability in terms of chemical oxidation resistance [31].

2.1.2. Biochar: Physicochemical Properties and Characterization

Several characterization analyses can be conducted to elucidate biochar's physical and chemical properties. The proximate analysis involves the quantification of ash, fixed carbon, volatile matter, and moisture. High ash and fixed carbon contents are good indicators of high adsorbent capacity. Ultimate analysis, i.e., the quantification of C, H, N, and O composition in biochar samples, especially the H/C ratio and O/C and (O + N)/C ratios, is an indicator of the aromaticity and polarity of biochar [26].

The textural features, with an emphasis on the sizes and the volume of the pores and the specific surface area (S_{BET}), are usually estimated via N_2 sorption tests at 77 K, using the Barrett–Joyner–Halenda (BJH) or density functional theory (DFT) methods for the pore analysis and the Brunauer–Emmett–Teller (BET) theory for the S_{BET}. The definition of pore size category (micro-, meso-, and macropores) decides the interaction ability of biochar with the required moiety. For instance, biochar systems with microporous structures would show the lower adsorption capacity of higher molecular-weight pesticides, although a higher one is needed for metal cations [15].

Surface pH analysis, zeta potential, and electrical conductivity can define the range in which biochar–pesticide and biochar–metal ion interactions are maximized. The graphitization and alkalinity of the produced char increase at higher pyrolysis temperatures [32]. Surface functional group analyses, such as cation exchange capacity, Boehm titration, and humic substance analysis, are also used to evaluate the biochar's adsorption capacity and microbial support. Fourier transform infrared spectroscopic (FTIR) analysis also provides insight into the biochar matrix's multiple bond formation, with additional information on post-adsorption studies. The solid-state C-nuclear magnetic resonance technique can be used to study the relative abundance of the functional groups and the aliphatic and aromatic hydrocarbon contents [33].

The morphological and structural properties can be explored by scanning electron microscopy (SEM), transmission electron microscopy (TEM), X-ray diffraction (XRD), and atomic force microscopy (AFM). The surface chemistry can be analyzed by IR, Raman spectroscopy, X-ray photoelectron spectroscopy (XPS), potentiometric titration, and Boehm titration. It is always crucial to determine the surface pH and the point of zero charge, since they play a key role in the adsorption performance and activity when biochar is used as an adsorbent in aqueous phases.

The most important techniques for the physicochemical characterization of biochar are presented in Figure 2.

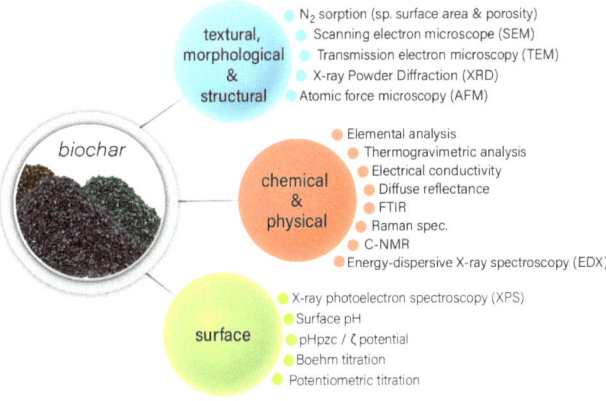

Figure 2. The predominant technique for the physicochemical characterization of biochar-based materials.

2.2. Biochar as Adsorbents

Heavy metals and pesticides can be directly adsorbed onto the biochar's surface. Modifying the outer surface of biochar via activation by tuning the chemical heterogeneity and/or by anchoring decorating different active species can lead to elevated and selective adsorption efficiency, exceptional stability, easy separation efficiency, and better recyclability [19]. Modification, including physical/thermal activation such as steam (for –OH functional group increment) and CO_2 activation, ball milling and sonication/ultrasonication, acid treatment (for deashing and demineralization) and base treatment, functional group activation, such as amine-functionalization, impregnation with metal oxides, doping, electrochemical treatment, plasma treatment, etc., can enhance the properties of biochar as an adsorbent [34].

Regarding the analysis of metal ions or pesticide removal using biochar-based materials, Langmuir and Freundlich's isotherm models are the most established ones. In general, a Langmuir versus Freundlich isotherm comparison explains monolayer adsorption vs. mono/multilayer adsorption, even though this approach is not absolutely correct in the case of studying adsorption in aqueous phases. Other isotherm models/approaches, such as the Jovanovich, Elovich, and Dubinin–Radushkevich (D–R) models, are also used in order to present an additional understanding of the role of adsorption conditions [35,36]. In addition, pseudo-first-order and pseudo-second-order models are the most widely applied models for kinetic studies for biochar–heavy metal/pesticide systems [37].

2.2.1. Removal of Metal Ions

The application of adsorbents for the removal of heavy metal ions involves physical and/or chemical adsorption via electrostatic interactions, ion exchange, complexation, reactions that have taken place on the material's surface, and/or precipitation [38]. When interacting with biochar, some metal ions undergo reduction and oxidation reactions, precipitation, and co-precipitation [39].

Multiple experimental condition parameters can affect the adsorption and removal capacity. The elevated adsorption of metal ions can be due to an increase in the specific surface area of the biochar as a result of optimizing the synthetic protocol, for instance, modifying the pyrolysis temperature [32]. A high pH directly affects the adsorbent's surface due to protonation, thus competing with metal ion adsorption [40]. Conversely, in alkaline pH conditions, hydroxy-complex formations can compete with other ions and impede adsorption [21]. Preferably, the point of zero-charge pH should be in the acidic region to efficiently adsorb metal ions and form complexes with a negative surface charge [41]. Cation-exchanging capacity also plays a crucial role in metal ion adsorption. For instance, Ma et al. [42] discovered that cation exchange significantly contributed to removing Cu^{2+} from lobster-shell-derived (HCl-treated) biochar, with 53–74% removal contributed by the cation exchange.

Biochar surface modifications are primarily conducted to improve the adsorption efficiency, and some of them are summarized in Table 1. A zirconium and iron composite with sludge biochar was generated to increase As^{5+} adsorption via complexation. The Zr-Fe biochar composite had a maximum adsorption capacity of 62.5 mg/g, compared to the pristine biochar capacity of 15.2 mg/g. The probable mechanism was suggested as the inner-sphere complexation of As^{5+} on the Zr-O-Fe surface [36]. Khan et al. [43] studied MoS_2-modified magnetic biochar with a maximum adsorption capacity of 139 mg/g, and hypothesized the presence of complexation, cation exchange, and Cd-π interactions. The deashing of biochar with acid solutions and potassium acetate improved lead adsorption, due to the pore size increment (unblocking SiO_2 particles out of biochar) and complexation of Pb^{2+} and C=C (π-electrons) [40].

Table 1. Biochar types involved in heavy metal removal.

Biomass Type	Pyrolysis Temperature (°C)	Modification	Metal Ion	System	Adsorption Capacity (mg/g)	Reference
Crab shell	350	Fe-La doped	Sb^{3+}	Water	498	[25]
Crab shell	350	Fe-La doped	SbO_6^{7-}	Water	337	[25]
Cattle manure	500	Fe-impregnated	Sb^{5+}	Water	58.3	[37]
Wood chip	600	Sulfurized	Hg^{2+}	Water	107.5	[37]
Sesbania bispinosa	450	MnO	AsO_4^{3-}	Water	7.35	[39]
Sesbania bispinosa	450	CuO	AsO_4^{3-}	Water	12.47	[39]
Rice straw	500	Thiol-modified	Cd^{2+}	Soil	45.1	[41]
Rice straw	500	Thiol-modified	Pb^{2+}	Soil	61.4	[41]
Lobster shell	600	HCl treatment	Cu^{2+}	Water	71.4	[42]
Lobster shell	600	HCl treatment	Cd^{2+}	Water	126	[42]
Peanut shell	600	MnO-embedded	Sb^{3+}	Water	248	[44]
Corn straw	600	Fe-impregnated	$HAsO_4^{2-}$	Water	6.80	[45]
Cornstalk	550	Mg-Al-LDH	As^{5+}	Soil	0.820	[46]
Cornstalk	550	Zn-Al-LDH	As^{5+}	Soil	0.916	[46]
Cornstalk	550	Cu-Al-LDH	As^{5+}	Soil	0.787	[46]
Canola straw	700	Steam activation	Pb^{2+}	Water	195	[47]
Rice husk	500	HA/Fe-Mn oxide-loaded	Cd^{2+}	Water	67.11	[48]
Rice husk	500	HA/Fe-Mn oxide-loaded	As^{5+}	Water	35.59	[48]
Rice husk	1 kW (microwave)	Fe_3O_4-magnetic	Cr^{6+}	Water	8.35	[49]
Pomelo peel	300	K_2FeO_4-promoted	Cr^{6+}	Water	209.64	[50]
Sawdust	180	Amino-functionalized (HNO_3, nicotinamide)	Sb^{5+}	Water	241.92	[51]
Sawdust	180	Amino-functionalized (HNO_3, nicotinamide)	Cr^{6+}	Water	132.74	[51]

2.2.2. Adsorption/Removal of Pesticides

Studies indicate that increased pesticide concentration and adsorption time has an asymptotic effect on adsorption capacity, whereas the adsorption capacity is enhanced by the increases in biochar concentration. The common mechanisms for pesticide adsorption onto biochar are the hydrophobic effect, π–π electron donor–acceptor interaction, pore filling, electrostatic interactions, ionic bonding, and H-bonding [26,52].

Several studies regarding the adsorption/removal of pesticides by biochar have been evaluated in Table 2, wherein the parameters of biochar pyrolysis temperature and surface modifications have been compiled, along with the adsorption capacity values. The pyrolysis temperature has a similar effect on biochar-pesticide adsorption as on biochar-heavy metal adsorption. The adsorption of carbendazim on dewatered sludge biochar was at a maximum at 700 °C, owing to the increased surface area and the increment in the partition coefficient [26]. Pore size governs the definitive adsorption capacity for pesticide–biochar interaction. Dichlorvos and pymetrozine had molecular sizes that were comparable to pore diameter; thus, adsorption was facile in both cases [53]. A decrease in the original biochar's H/C and O/C atomic ratios is expected to enhance the π–π electron donor-acceptor interactions, contributing to the sorption of certain pesticides, such as oxytetracycline and carbaryl [24]. Binh and Nguyen [52] concluded that a pH of 2 is a more favorable condition for the adsorption of 2,4 dichlorophenoxy acetic acid on corn-cob biochar based on the electrostatic interactions. In addition to the inherent functional groups and mechanisms involved in metolachlor adsorption onto biochar, Liu et al. [54] incorporated fulvic acid and citric acid into walnut-shell biochar that augmented the functional groups with oxygen, as shown in Figure 3. The removal capacity was also observed to decrease after 3 cycles in the metolachlor-simulated sewage biochar system.

Table 2. Biochars utilized in pesticide removal.

Biomass Type	Pyrolysis Temperature (°C)	Modification	Pesticide	System	Adsorption Capacity (mg/g)	Reference
Cow manure	600	HCl/HF	Carbaryl	Water	~55	[24]
Dewatered sludge	700	-	Carbendazim	Soil	0.144	[26]
Leonardite	550	-	Alachlor	Water	3.802	[35]
Corn cob	600	HF	2,4-dichloro-phenoxyacetic acid	Water		[52]
Coconut fiber	600	HCl	Dichlorvos	Water	90.9	[53]
Walnut shell powder	700	Fulvic acid	Metolachlor	Water	99.01	[54]
Walnut shell powder	700	Citric acid	Metolachlor	Water	74.07	[54]
Bagasse	500	-	Carbofuran	Water	18.9	[55]
Switchgrass	425	Fe^{3+}/Fe^{2+} magnetic	Metribuzin	Water	205	[56]
Switch grass	425	-	Metribuzin	Water	223	[56]

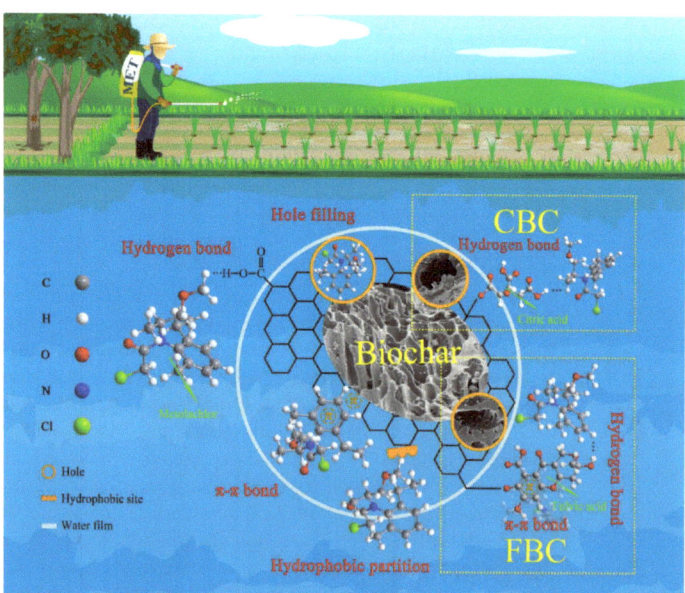

Figure 3. The fulvic acid- and citric acid-modified biochar adsorption mechanism for metolachlor in water Reprinted/adapted with permission from Ref. [54]. Copyright 2021, Elsevier.

2.3. Biochar as a Bioremediation Catalyst Support

The physicochemical properties of biochar that enable it to be an effective catalyst support include its large surface area, multi-scale porous structure, and surface functional group. Chen et al. [57] studied the volatilization of Hg^{2+} using the *Pseudomonas* strain, DC-B1, with biochar. The combined application of biochar and microbial strain resulted in the greatest Hg removal. Qiao et al. [58] demonstrated the stimulation of the microbial reduction of As^{5+} and Fe^{3+}, using oil palm fiber-derived biochar in synergy with soil microbes extracted from paddy for both studies. Biochar amended microcosm possessed a higher As^{5+} concentration than the control, indicating that biochar had an affinity to As^{5+} and Fe^{3+}. Both moieties were reduced in the biochar-amended microcosms since microbes drove the reduction reactions, and biochar behaved similarly to an electron shuttle. Qiao et al. [59] summarized the As^{5+} reduction with biochar and lactate. This reduction resulted in the identification of three ways: (i) Fe^{3+} reduction by microbial cells facilitated As^{5+} release; (ii) expression of As^{5+}-respiring gene transcripts in dissimilatory As^{5+}-reducing bacteria; (iii) the functioning electron transfer between the metal and As^{5+}-reducing bacteria.

Biochar is often employed as a good carrier in improving the photocatalytic activity of metal oxides. As a stable and inexpensive carbonaceous material, biochar effectively reduces the recombination rate of photogenerated electron-hole pairs, due to its excellent conductive property. An et al. [60] developed biochar-supported α-Fe_2O_3/MgO composites for photocatalytic degradation of organophosphorus pesticides and obtained a degradation efficiency of 90% in 80 min. Huang et al. [61] utilized pristine and manganese ferrite-modified biochar for Cu removal, confirming the role of biochar being principally an oxide carrier instead of an adsorbent. In addition, a preference for biochar as a carrier for photocatalysis and Fenton/photo-Fenton processes has been prevalent, due to its low cost and high surface area characteristics. The utilization of lignin-biochar as a catalyst support for $LaFeO_3$ in the catalytic photo-Fenton process had a positive effect on the degradation efficiency of pollutants, owing to enhanced adsorption capacity, a reduction in the charge transport resistance between $LaFeO_3$ and lignin-biochar, and the presence of oxygen-containing functional groups [62].

Several studies have been conducted for enzyme-immobilized biochar, particularly with laccase utilized as an enzyme to degrade pollutants [33]. The basic biochar–enzyme immobilization techniques are adsorption and covalent bonding. Comparatively, fewer instances of enzyme–biochar systems for the degradation of pesticides have been studied. Wang et al. [63] used laccase-immobilized biochar to degrade 2,4-dichlorophenol and obtained 64.6% degradation. The immobilized laccase improved the cation exchange capacity, organic matter content, stability, and catalytic degradation effect. A general outline for adsorption and the removal mechanisms for metals and pesticides via biochar systems are shown in Figure 4.

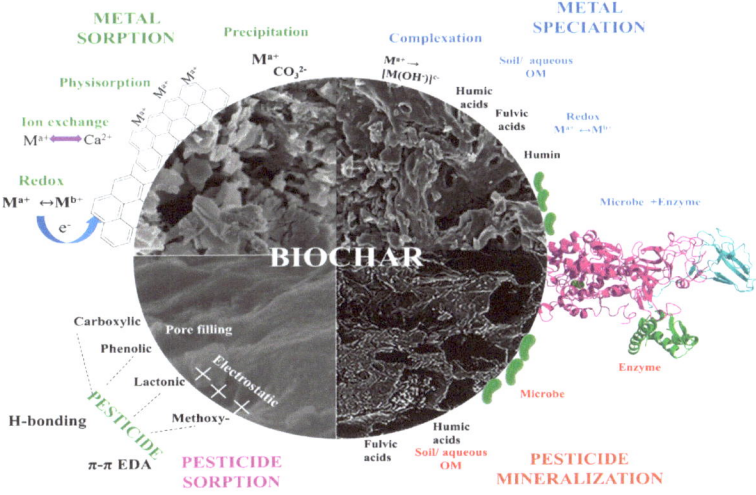

Figure 4. The overall involvement of biochar in heavy metal and pesticide remediation.

3. Role of Microorganisms in the Removal of Metal Ions and Pesticides

The surface area of microorganisms exhibits higher biological activity relative to their volume, resulting in greater interaction with their immediate environment. Thus, they can adapt and survive in polluted areas with the subsequent removal or detoxification of the pollutant [64]. The microorganisms use different strategies for their survival, including surface adsorption, micro-precipitation, extracellular or intracellular sequestration, reduction, enzymatic degradation, etc. Bioremediation is possible only when microbial activity and growth are allowed by environmental conditions. In certain situations, environmental factors can be altered to allow microbial population growth to eliminate contaminants [11]. As shown in Figure 5, various factors influence microbial degradation:

(i) Environmental factors

The pH can affect bioremediation by changing metal bioavailability; for instance, a decrease in soil pH value generally causes an increase in metal bioavailability [65]. This is because, at lower pH, the exchangeable capacity between metal cations and H^+ on the surface of soil particles is more prominent than at higher pH. Additionally, an optimum pH is essential for microbial growth, and some microbial degradation processes can be inhibited at an extreme pH. Temperature is another crucial factor influencing the bioremediation of metals and pesticides [65]. The solubility of these contaminants is increased at higher temperatures, which leads to their increased bioavailability. The physical nature and chemical composition of several organic pollutants and their adsorption-desorption mechanism are governed by temperature. Temperature also influences microbial growth, activity, and degradation potential. Furthermore, the soil moisture content is another parameter that affects the bioremediation process. A low soil moisture content limits the growth and metabolism of microorganisms, while high values can reduce soil aeration.

(ii) Type of microorganism and degradation capacity

The microorganism that is selected for biodegradation should be able to survive in a high-contamination environment and should be evaluated first for its degradation capacity before employing it for in situ remediation. The survival of these strains can be ensured by providing favorable growth conditions. It is also important to note that microbial strains selected for pollutant removal may need to meet certain ecological requirements. One such requirement is that the strains should be non-pathogenic. For instance, *Staphylococcus aureus*, as a typical pathogen, was resistant to many antibiotics and showed high bioremediation efficiency for heavy metals such as Cr and U through bioprecipitation [66]. However, certain metabolites that formed during the degradation of contaminants can be toxic. Therefore, deeper investigations of ecological security and the metabolic functions of microbial cells are indispensable before their possible application in environmental pollution control.

(iii) Bioavailability of the contaminants

The bioavailability of the contaminants can be defined as the fraction of a contaminant in a specific environment that is either adsorbed or degraded by the microbial cells within a given time. The control of bioavailability is dependent on the diffusion, uptake, and desorption of the contaminants. The slow mass transfer of contaminants into degrading microbes reduces their bioavailability. The significance of bioavailability depends very much on the properties of the pollutant, microorganism, and characteristics of the contaminated site [11].

(iv) Aerobic or anaerobic operating conditions

Depending on the type of organism and contaminant, bioremediation can be either aerobic or anaerobic. Most bioremediation systems work under aerobic conditions, but to effectively degrade the recalcitrant molecules, it is better to run the microbial degradation tests under anaerobic conditions. Apart from the abovementioned factors, the properties of the contaminated site (soil type, soil porosity, soil nutrients) and the properties of the contaminants (structure, hydrophobicity, recalcitrance, toxicity, solubility, and leaching ability) are also important in bioremediation.

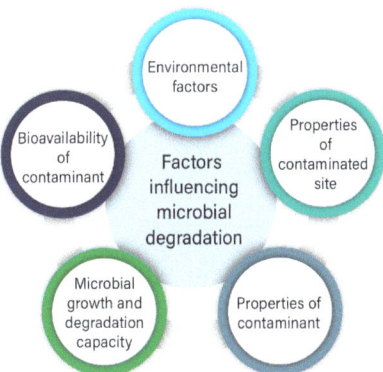

Figure 5. Factors influencing microbial degradation.

3.1. Removal of Heavy Metals Using Microorganisms

The removal of heavy metal ions by microorganisms is considered economical and sustainable. Any environmental stress can be withstood by microorganisms through rapid mutation and evolution, leading to toxic heavy metal resistance. They can sequester heavy metal ions, either intracellularly or extracellularly. Additionally, microorganisms can transform and reduce the metal ions to inactive forms. Table 3 summarizes the microorganisms used for various metal ion remediation conditions in recent years.

The factors influencing heavy metal remediation by microbes generally include pH, temperature, biomass concentration, the presence of other pollutants, etc. The inherent pH of the system defines the charges of the surface functional groups present on the microbial surfaces; pH in an unsuitable range may affect microbial growth. This shows pH to be an essential parameter in the degradation and removal of heavy metals by live biomass [81]. The pH also has an effect on the solubility of metal ions in the microbe-heavy metal system. A decrease in soil pH leads to an increase in the bioavailability of metals, thereby resulting in higher biosorption efficiency, as studied by Zhang et al. [73]. Another essential parameter in microbial growth and proliferation is the system's ambient temperature. With an increase in temperature, the solubility of metal ions increases; thus, the bioavailability of metals also increases [81]. High biomass or sorbent concentration will increase the overall biosorption efficiency, but any interference between binding sites reduces the specific metal ion uptake. The removal or adsorption of a particular heavy metal by microorganisms can also be positively or negatively affected by the co-existence of other metal ions.

The Mechanism of Heavy Metal Removal by Microorganisms

Microorganisms can adopt several mechanisms in order to survive in heavy-metal toxicity conditions. These mechanisms are depicted in Figure 6 and include biotransformation, extracellular polymeric substances secretion, metallothionein synthesis, etc. Heavy metal degradation by microorganisms can be described in two ways: biosorption and bioaccumulation.

Biosorption is the reversible physicochemical interaction of living (or dead) biomass or biomass-secreted products that act as biosorbents with sorbate molecules (e.g., metal ions). It was previously categorized as metabolism-dependent and metabolism-independent biosorption. Recently, the former has been widely accepted as bioaccumulation (also called active biosorption), and only the metabolism-independent processes are considered to be biosorption [82]. A metabolism-independent mechanism occurs passively on the dead or the living biomass cell surface. However, the biosorption of metal ions carried out by dead biomass is superior to that carried out by living cells. Cheng et al. [78] studied the biosorption of Cd^{2+} in the living and dead cells of the microalgae *Chlorella vulgaris*. The dead algal biomass removed 96.8% of the total cadmium, while the live algal biomass achieved

95.2% of cadmium adsorption. The steps involved in toxic heavy metal biosorption include binding the metal ions to various extracellular functional groups present on the microbial cell wall, via surface precipitation, chemical bonding (complexation/chelation), adsorption, or ion exchange. Physical adsorption depends on intermolecular or inter-ionic attraction forces. Complexation or chelation occurs due to the dative covalent bonds between metal ions, surface functional groups, and the ligands of biomass. When metal ion concentrations are higher than the solubility limit, surface precipitation or micro-precipitation has been observed. The exchange involves electrostatic interaction between the metal cations and the negatively charged functional groups on the cell surface; the interchange of the cations resulted in the metal ion being bound to the surface [82,83]. Surface-binding is found to be the principal phenomenon governing the biosorption of metal ions [84]. Physical modifications have been suggested to provide a cumulative effect on the biosorption capacity of the microorganisms by removing surface impurities or through the production of metal-binding sites. Li et al. [74] investigated the biosorption ability of a lactic acid bacterium, *Weissella viridescens* ZY-6, for Cd^{2+} removal from the aqueous solution, and achieved a 69.45–79.91% removal of Cd^{2+} from three kinds of juices: tomato, apple, and pear juices.

The extracellular sequestration of metal ions often occurs due to various biological structures produced by microbial cells, including extra-cellular polymeric substances, siderophores, glutathione, and biosurfactants. Under heavy metal stress, microorganisms often secrete extra-cellular polymeric substances or exopolysaccharides (EPS) as a protective response. EPS are constituted of proteins, lipids, complex carbohydrates, nucleic acids, uronic acid, humic acid, etc., which prevent the entrance of heavy metals into the cell [68,85]. Generally, EPS contain negatively charged functional groups and can interact electrostatically with heavy metals, resulting in the immobilization of the metal ions within the EPS. Some examples include the accumulation of Pb^{2+} and Zn^{2+} in the soluble EPS secreted by *Oceanobacillus profundus* KBZ 3-2 [68], and Pb^{2+} adsorption onto EPS of *Enterobacter* sp. FM-1 [69] and Cd^{2+} adsorption onto the EPS secreted by a living cyanobacteria, *Synechocystis* sp. PCC6803 [86]. Siderophores are secreted by microbes and act as metal chelators, with an extreme affinity for ferric iron. They can reduce the metal's bioavailability and toxicity by binding metal ions with variable affinities that have a similar chemistry to that of iron [87].

Biosurfactants are amphiphilic compounds that are produced extracellularly by microorganisms for the solubilization, desorption, complexation, and mobilization of pollutants in solutions. The induction of biosurfactants in microbe-heavy metal systems facilitates the extracellular sequestration and formation of biosurfactant–metal complex [88]. Rhamnolipids produced by *Pseudomonas aeruginosa* showed 53% As, 90% Cd, and 80% Zn extraction capacity from contaminated soil [89]. Ayangbenro and Babalola [90] observed that a lipopeptide biosurfactant generated by *Bacillus cereus* NWUABO1 could remove 69% of Pb, 54% of Cd, and 43% of Cr from the soil. Several microorganisms, including *Pseudomonas* sp., *Bacillus subtilis*, *Candida tropicalis*, *Candida* sp., *Burkholderia* sp., and *Citrobacter freundii* can produce biosurfactants, demonstrating heavy metal removal capacity [88].

Table 3. Microorganisms that are involved in heavy metal removal.

Heavy Metal	Microorganism	Initial Heavy Metal Concentration	Incubation Time	Degradation Efficiency (%)	Reference
		Bacteria			
Pb	*Bacillus cereus* BPS-9	-	48 h	77.57	[67]
	Oceanobacillus profundus KBZ 3-2	50 mg/L	24 h	97	[68]
	Enterobacter sp. FM-1	100 mg/L	24 h	93.85	[69]
Cr	*Bacillus subtilis* SZMC 6179J	55 mg/L	24 h	93.50	[70]
	Pseudomonas aeruginosa	20 ppm	21 days	89.67	[71]
	Pseudomonas stutzeri L1	100 mg/L	24 h	97	[72]
	Bacillus cohnii	100 mg/L	25 h	94	[73]
	Bacillus licheniformis	100 mg/L	25 h	95	[73]
Cd	*Weissella viridescens* ZY-6	NM	2 h	69.45–79.91	[74]
Zn	*Oceanobacillus profundus* KBZ 3-2	2 mg/L	24 h	54	[68]
Cu	*Pseudomonas aeruginosa*	15 ppm	14 days	90.89	[71]
As	*Bacillus* sp.	100 ppm	72 h	53.29	[75]
	Aneurinibacillus aneurinilyticus	100 ppm	72 h	50.37	[75]
		Fungi			
Pb	*Trichoderma brevicompactum* QYCD-6	50 mg/L	5 days	97.5	[76]
Cr	*Trichoderma brevicompactum* QYCD-6	100 mg/L	5 days	31.83	[76]
Cd	*Penicillium notatum*	10 ppm	14 days	77.67	[71]
	Trichoderma brevicompactum QYCD-6	30 mg/L	5 days	20.13	[76]
Cu	*Trichoderma brevicompactum* QYCD-6	50 mg/L	5 days	64.46	[76]
Ni	*Aspergillus niger*	20 ppm	28 days	81.07	[71]
		Microalgae			
Cd	*Desmodesmus* sp. MAS1	5 mg/L	7 days	>58%	[77]
	Heterochlorella sp. MAS3	5 mg/L	7 days	>58%	[77]
	Chlorella vulgaris	100 mg/L	5–15 min	Live cells—95.2 Dead cells—96.8	[78]
Zn	*Chlorophyceae* spp.	3 mg/L	3 h	91.9	[79]
Cu	*Chlorella vulgaris*	1.9–11.9 mg/L	12 days	39	[80]
	Chlorophyceae spp.	3 mg/L	10 min	88	[79]
As	*Scenedesmus almeriensis*	12 mg/L	3 h	40.7	[79]
Ni	*Chlorella vulgaris*	1.9–11.9 mg/L	12 days	32	[80]
Mn	*Scenedesmus almeriensis*	3 mg/L	3 h	99.4	[79]

Biosorption has been determined to be simple, fast, reversible, and inexpensive compared to bioaccumulation and can concentrate heavy metals, even from a very dilute aqueous solution. The advantageous properties of biosorption include the presence of multi-functional groups and the uniform distribution of binding sites on the cell surface, low operational cost, the absence of metal toxicity limitations, minimal preparatory steps, high efficiency and selectivity for metal ions, no production of secondary waste, and the possibility of the toxic heavy metal recovery and reusability of the biosorbent [84,91]. Several microbial strains have been identified to show multi-metal resistance and remediation abilities. Nokman et al. [92] isolated a *Pseudomonas putida* strain from effluent water generated from a tannery that exhibited resistance to Ag^{2+} and Co^{2+} and enhanced resistance to lead and chromium. Conversely, bioaccumulation is the metabolism-dependent active transportation of metal ions across the membrane into the living cell, as represented in Figure 6. The microorganisms selected for bioaccumulation should have specific properties, such as adaptation to the polluted environment, resistance to high loads of metal ions, and a mechanism of intracellular binding [93]. The mechanism consists of two steps; the first step is identical to biosorption and involves the attachment of heavy metals to charged functional groups on the cell surface. The second step is metabolism-dependent, relatively slow, and involves the penetration/transport of a metal-ligand complex into the cell membrane. The subsequent interaction of the complexes with intracellular metal-binding proteins (such as metallothionein and phytochelatins) occurs within the cell, leading to bioaccumulation [85]. Metallothioneins (MTs) help to regulate the intracellular metabolism of metals and protect against oxidative stress and toxic heavy metals [86,87]. Engineered recombinant *E. coli* expressed the *Corynebacterium glutamicum* metallothionein gene and achieved improved intracellular biosorption of Pb^{2+} and Zn^{2+}. Hu et al. [86] constructed a bio composite of immobilizing metallothionein, expressing *Pseudomonas putida* for the sorption of Cu^{2+}. Similarly, phytochelatins are metal-binding proteins that are analogous to the metallothioneins produced from microalgae, which can also chelate and detoxify heavy metal ions intracellularly.

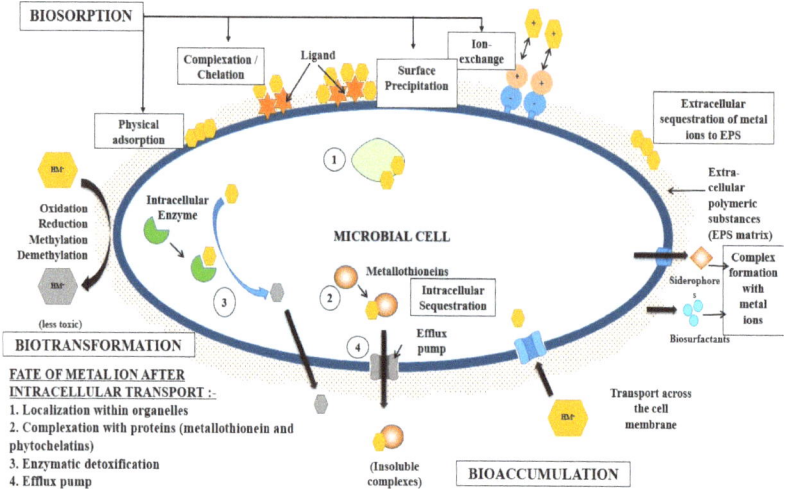

Figure 6. Mechanism of the microbial bioremediation of heavy metals.

3.2. Removal of Pesticides Using Microorganisms

The major types of pesticides and persistent organic pollutants include insecticides, herbicides, and fungicides. As with heavy metals, the microbial remediation of these persistent pesticides is economical and sustainable, compared to physical or chemical removal processes. It involves the degradation of complex pesticide molecules into simpler inorganic chemicals. Table 4 includes the commonly used microorganisms for the removal of pesticides. Indigenous soil microbial consortia have been more effective for the microbial degradation of pesticides than the non-indigenous strains, as non-indigenous strains are exposed to pesticide-contaminated regions exhibiting unfamiliar conditions. Several studies have reported on the degrading ability of indigenous microbes. Some of them show organophosphate degradation by indigenous *Kosakinia oryzae* [94], herbicide glyphosate degradation by *Providencia rettgeri* [95], and herbicide atrazine remediation by indigenous microbial consortia [96]. Individual or mixed microbial cultures can degrade the various sources of pesticides. Single microbial cells abide by their metabolic pathways for pesticide degradation, whereas mixed microbial cultures can achieve the same result through coupled metabolic pathways [97]. Thus, pesticides can rapidly be degraded by applying the combined microbial consortia isolated from indigenous sites.

Table 4. Microorganisms involved in pesticide removal.

Pesticide	Microorganism	Initial Pesticide Concentration	Incubation Time	Degradation Efficiency (%)	Reference
Bacteria					
Chlorpyrifos	*Pseudomonas nitroreducens* AR-3	100 mg/L	8 h	97	[98]
Chlorpyrifos	*Lactobacillus plantarum*	0.20–0.80 mg/kg	-	24.9–34.4	[99]
Malathion	*Escherichia coli* IES-02	50 ppm	4 h	99	[100]
Mesotrione	*Bacillus megaterium* Mes11	1 mM	5 h	99	[101]
Carbofuran	*Enterobacter* sp.	4 µg/ml	7 days	80	[102]
Fungi					
Chlorpyrifos Methyl parathion Profenos	*Aspergillus sydowii* CBMAI 935	50 mg/L	30 days	32 80 52	[103]
Pyrethroid mixture (cypermethrin, cyfluthrin, cyhalothrin)	*Aspergillus* sp.	500 mg/L	15 days	≈100	[104]
Microalgae					
Paraoxon, Malathion and Diazinon	*Coccomyxa subellipsoidea*	0.1 mg/ml	10 days	-	[105]
Atrazine	*Chlorella* sp.	40 µg/L 80 µg/L	8 days	83.0 64.3	[106]

However, certain recalcitrant pesticides have resilience against biodegradation by the indigenous microbial community. In such situations, bio-augmentation and bio-stimulation are considered promising approaches for the remediation of contaminated sites. Bio-augmentation involves the introduction of specific exogenous microbes to improve the degradative capacity of the contaminated sites. The two main strategies of bio-augmentation are autochthonous bio-augmentation, where the microbes are isolated from the same site and then re-injected, and allochthonous bio-augmentation, where the microbes are cultured from another site [107]. In one study, bio-augmentation with *Paenarthrobacter* sp. W11 significantly accelerated the degradation rate of atrazine in soil and dampened its toxic effect on wheat growth [108]. The success of bio-augmentation strategies depends on several factors, including the selection of appropriate microorganisms, the target pollutant's bioavailability, and the inoculum's survival capability in the toxic environment [11,109].

Bio-stimulation can be performed by providing the necessary nutrients or electron acceptors, such as oxygen or nitrate, to promote the proliferation of indigenous microbes. Aldas-Vargas et al. [110] investigated the biodegradation of herbicides, namely, mecoprop-p and 2,4-dichlorophenoxyacetic acid (2,4-D), in groundwater. They concluded that bio-stimulation with oxygen and dissolved organic carbon had the potential for field application. Raimondo et al. [111] bio-augmented lindane-contaminated soil with actinobacteria (mixed culture) and bio-stimulated it with sugarcane filter cake, further noticing enhanced lindane removal, along with microbial cell counts and enzyme activities.

The removal of pesticides depends, firstly, on the optimal conditions of the biomass, its survival and activity, and, secondly, on the pesticide's chemical structure, along with several biotic and abiotic factors, such as suitable microbial strains, nutrient availability, salinity, pH, temperature, etc. [112]. In the case of the in situ remediation of soil contaminated by the extensive use or overuse of pesticides for agricultural purposes, the growth of pesticide-degrading soil microbes depends on the soil characteristics [11].

Mechanisms Involved in Pesticide Removal by Microorganisms

There are several mechanisms by which microorganisms transform pesticides into their non-toxic forms in a contaminated site. Some include the surface adsorption, enzymatic degradation, or co-metabolism of the pesticide molecules, as depicted in Figure 7. Adsorption of the pesticide molecules is categorized as a passive process and involves the direct interaction of molecules with the microbial cell surface. As a result, the efficiency of pesticide adsorption by microorganisms is primarily determined by the available surface-active groups. The ultimate result of adsorption is the reduced mobility of the toxic pesticides. The extent of removal and the degradation efficiency are influenced by various components, such as the charge, polarity, solubility, volatility, and solubility of the pesticide molecules. Extra-cellular polymeric substances (EPS) and biosurfactants produced by the microorganisms also aid in the removal of pesticides. EPS can be produced by the microbial cell as a byproduct of pesticide degradation. This approach can have two benefits: (i) the reduction of excess toxic pesticides, and (ii) the production of EPS, which can have further environmental applications. Gupta et al. [113] observed 98% carbofuran degradation within 96 h by *Cupriavidus* sp. with simultaneous EPS production. Satapute and Jogaiah [114] reported that surfactin, a biosurfactant produced by a bacterial strain, could degrade 91% of difenoconazole.

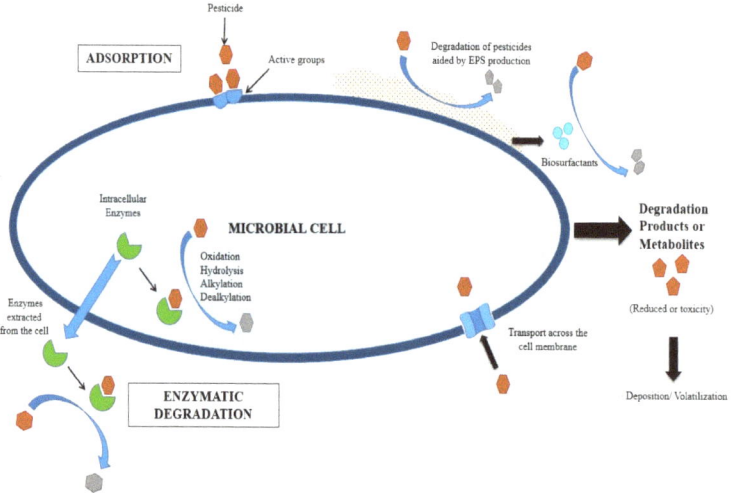

Figure 7. The mechanism of the microbial bioremediation of pesticide.

Microbial enzymes can catalyze the breakdown of pesticides. The enzymatic degradation processes may include an alteration in the structural components, the removal of undesirable pesticide properties, oxidation, and reduction [115]. Dash and Osborne [116] investigated monocrotophos degradation by *Bacillus aryabhattai* (VITNNDJ5) instead of the bacterial enzyme. The enzymatic degradation of pesticides can either be performed by intracellular enzymes that are present in the microbial cell or by extracting the enzymes capable of degradation from the cells. Sirajuddin et al. [100] isolated the *E. coli* IES-02 strain from a site contaminated with the organophosphate malathion, and the strain showed efficient degradation, utilizing it as the sole carbon source. They also purified carboxylesterase enzyme from the IES-02 strain and achieved 81% malathion degradation under optimized conditions within 20 min, whereas the IES-02 cell degradation was completed from 99.0% to 95.0% within 4 h. However, the extracted enzymes can be affected by solution properties, such as pH, temperature, etc. Depending on the environmental factors, enzymes may lose their degradation potential due to varied ambient conditions [117]. Oxidation, hydrolysis, alkylation, and dealkylation reactions have been predominantly observed in the microbial degradation process [118]. Some studies that have reported enzymatic degradation are on cypermethrin by esterase and laccase [119], carbendazim by carbendazim hydrolase [120], malathion by phosphotriesterase [121], and isoproturon, procymidone, chlorpyrifos, dichlorophos, and monocrotophos by laccase [122–124]. The enzymatic biodegradation mechanism of pesticides is often complex, and this diverse biodegradation pathway needs further investigation to understand enzyme involvement properly.

3.3. Challenges of Using Microorganisms as a Catalyst

The microbial degradation of metal ions and pesticides tends to be an appealing approach for bioremediation, even though certain challenges hinder their commercial application. These include: (i) the loss of microorganisms or reduced microbial survival because of the toxicity to microorganisms at a higher metal ion or pesticide concentration, (ii) reduced microbial proliferation, (iii) uneven microbial growth with high concentrations of the pollutant, (iv) the washing out of the microbial cells during the application, (v) the longer time required for the completion of the process, (vi) the presence of other co-existing metal ions and organics that can positively or negatively affect the remediation process.

Microbial immobilization on a support material can overcome the above drawbacks by fixing the free microbial cells to a specific carrier, either chemically or physically, and keeping them active for longer. An ideal carrier provides operational stability and cell protection from the toxic external environment, leading to efficient biodegradation. A support material retains the microbes and contributes to the sorption of the pollutants [125]. Hence, immobilizing the microorganism accelerates the pollutant's biodegradation capacity, enhances the robustness of the immobilized strains, and improves their tolerance to high pollutant concentrations.

4. Microbial Cell-Immobilized Biochar for the Removal of Metal Ions and Pesticides

Bioremediation with free microbial cells is generally inefficient, due to the lesser amount of microbes utilized for degradation, microbial loss, and the inhibition of growth and functioning from indigenous microorganisms [126]. Immobilizing the microorganisms creates a safe environment for microbial cells to perform specific functions, such as highly efficient physiochemical sorption and microbial metabolism. Pollutant adsorption/binding on the carrier material allows the degrading cells to outcompete indigenous microbes, overcoming the limitations of using free cells for bioremediation [127]. Biochar has been a prominent carrier for microbial cell immobilization, due to its minimal toxicity and abundant generation. Immobilized microbes have commonly been observed for better remediation efficiency than pristine biochar or free cell [128].

4.1. Immobilization Methods

Biochar-immobilized microorganisms are produced through the adsorption of microbes on biochar, entrapment with the help of crosslinking materials, or a combination of both methods. Adsorption is a simple and inexpensive method for immobilizing microorganisms [129,130]. Adsorbed cells colonize the biochar after being transferred from a bulk solution to its surface. The adsorption technique involves physical interactions, such as van der Waals forces, ionic interactions, and hydrogen bonding between the surface functional groups of microorganisms and functional groups on the surface of carriers, particularly the oxygen-containing groups, such as carboxylic, phenolic, and sulphonate groups. Microorganisms have a low affinity for carriers; there will thus be a high rate of desorption of cells from carriers [125]. As a result, appropriate carriers with high cell-binding characteristics are required for improved remediation. With a relatively weak interaction between the carrier and microbial cells, immobilization does not affect the intrinsic structure of the original microbes if the adsorption method is utilized. As a result, this method is better suited for immobilizing viable cells and biodegrading pollutants in the laboratory. Entrapment is a standard method of physical immobilization that is irreversible and provides better stability of microbes than adsorption [126]. Due to the improved stability of the thus-prepared immobilized cells, the entrapment method is preferred and is exercised in industrial applications for pollution abatement.

4.2. Factors that Influence Bioremediation Using Immobilized Microorganisms

The effective pollutant removal capacity of MCB is affected by pollutant concentration and its bioavailability, the incubation time of the cell, and various parameters, such as temperature, pH, etc. The biochar-immobilized microorganism technology requires a thorough understanding of the best conditions for maximum contamination removal.

Initial pollutant concentration influences the removal of pollutants, wherein setting the initial pollutant concentration until the saturation point increases the adsorption capacity of the biosorbed pollutants per unit weight of MCB [131]. The bioavailability of pollutants is defined as the total amount of a contaminant that is either available or that may be made available for uptake by microorganisms from its surroundings within a given period. The significance of bioavailability depends on the pollutant's physicochemical properties, microorganisms, and contaminated site characteristics [11]. Incubation time is another critical parameter affecting bioremediation because it has been observed to affect the growth pattern of microorganisms directly. *Proteus mirabilis* YC801, immobilized on biochar, achieved a 42.5% Cr bioreduction and adsorption capacity after 6 h of incubation [132]. The temporal requirement is high for microbial degradation, and the reaction time for complete degradation is higher than that for other removal processes. The time scale of the microbial degradation process can be reduced by selecting suitable microorganisms with quicker growth phases for pollutant degradation or removal. However, choosing biochar with a high adsorption potential for pollutants is critical for reducing the bacterial adaptation time.

The pH value also influences microbial metabolic processes, particularly growth, cell membrane transport, the zeta potential of sorbate, and changes in the sorbent surface characteristics [133]. Huang et al. [132] observed an increase in Cr^{6+} reduction with a pH increment from 6.0 to 7.0, showing a maximum removal of 83.7% at pH 7.0. However, alkalifying the Cr^{6+}-MCB system from pH = 8.0 to pH = 10.0 inhibited the removal capacity of MCB for Cr^{6+}. Similarly, the highest Cr bioreduction was found at 30 °C, similar to the optimal culture temperature for the strain. Bioreduction significantly decreased with a further increase in temperature above 30 °C, which might be attributed to the loss of cell viability and the inhibition of the essential enzymes and proteins responsible for microbial growth and biodegradation at elevated temperatures [12,132].

Similarly, temperature and pH significantly influenced tebuconazole degradation by *Alcaligenes faecalis* WZ2, and degradation efficiency was strongly correlated with bacterial growth [125]. Tebuconazole degradation efficiency reached 88.5% under ideal conditions (a temperature of 30–35 °C and a pH of 6–8). Because of bacterial growth inhibition and a decrease in the catalytic activities of microbial enzymes involved in tebuconazole degradation, the efficiency was significantly reduced below the ideal temperature and pH.

4.3. Heavy Metal Ions and Pesticide Removal Using MCB

The advantages of immobilizing cell systems onto carriers in the bioremediation of metal ions and pesticides are far superior to those of using biochar or free cells alone [134]. Pollutant transfer into the microbial community from the contaminated sites can be enhanced by immobilizing the microbial strains onto biochar. Biochar can enhance the biological community composition of the soil through physisorption; in return, these microorganisms, adsorbed on the biochar surface, have a metabolizing capability for the pollutants present in the soil [135]. The porous structure of biochar enhances the growth and reproduction of the microorganisms and can also act as a source of nutrients for the microorganisms [136]. The immobilization also ensures that the microorganisms are assimilated for degradation to form biofilms around the porous structure complex of the biochar microbes [136]. Biochar can alleviate the contaminant concentration and reduce the inhibitory effect of these contaminants on the growth of microorganisms via the adsorption and subsequent decrease in contaminant concentration in soil/aqueous medium [137].

Using biochar and bioremediation in tandem with functional microbial strains is a viable and emerging strategy for the long-term remediation of contaminated water and soil. Numerous microbial strains with strong metal tolerance or adsorption capability have been isolated and used for bioremediation, either as free-living cells or by immobilizing a microbial cell with a specific carrier substance. Metal-tolerant microorganisms immobilized on biochar have been used as a bio-augmentation method to improve heavy metal phytoremediation, indirectly reducing heavy metal contamination in soil. Incorporating bacteria immobilized on biochar into the soil may indirectly improve Cd removal by promoting plant growth and the phytoremediation effect [138]. Cd-resistant bacteria immobilized on biochar improved the phytoextraction efficiency by *Chlorophytum laxum* R. Br. via cadmium phytoaccumulation in the shoots and roots, and Cd translocation from the roots to the shoots. Insoluble phosphate solubilization can be achieved via microbial phosphate solubilizers (PSB). Teng et al. [134] observed that combining PSB and biochar improved Pb^{2+} immobilization by forming a stable crystal texture on its surface. Zhang et al. [139] used the PSB bacteria *Pseudomonas chlororaphis* for lead removal. However, the organism could not proliferate in indigenous bacteria, whereas the addition of PSB-immobilized biochar (PIB) improved bacterial growth and reduced Pb concentrations to less than 1 mg/kg. As a result, soil inoculation with PIB can be used as a substitute for Pb immobilization, avoiding the secondary pollution caused by phosphorus toxicity.

The microorganism immobilization with biochar carrier was also influential in remediating soil polluted with a combination of heavy metals. Tu et al. [140] introduced *Pseudomonas* sp. NT-2, loaded onto maize straw biochar, into Cd-Cu mixed soil. The application of *Pseudomonas* sp. NT-2-loaded biochar effectively reduced the bioavailability of Cd and Cu and increased the soil enzymatic activities in the soil system. Qi et al. [135] used three strains of mixed bacteria, *Bacillus subtilis*, *Bacillus cereus*, and *Citrobacter* sp.-loaded biochar for U and Cd removal. They discovered that MCB promoted growth in celery and reduced the U and Cd phytoaccumulation, compared to free cell and biochar treatments. Research on Cr^{6+} removal by immobilized microorganisms with biochar has attracted increased interest recently. The metal ion-resistant bacterium, *Proteus mirabilis* YC80, was immobilized using biochar derived from the bloom-forming cyanobacterium, *D. flos-aquae* [132]. The ability of biochar-immobilized *Proteus mirabilis* PC801 to remove Cr^{6+} was superior, compared to a free cell. The removal efficiency of Cr^{6+} by PC801-immobilized biochar was 100%, with 87.7% total Cr immobilized on the carrier and only 12.3% Cr^{3+} remaining in the solution. Table 5 includes microbial cell immobilized biochar reported for heavy metal and pesticide abatement.

Table 5. Microbial cell immobilized biochar for heavy metal and pesticide abatement.

Microorganism	Catalyst Support	Pollutant Type	Mechanism	System Water/Soil	Quantification of Heavy Metal Removal	Reference
Bacillus sp.TZ5	Coconut shell	Cd^{2+}	Adsorption	Soil	48.49%	[141]
Delftia sp B9	Cornstalk	Cd^{2+}	Adsorption	soil	0.33 mg/kg reduced to 0.06–0.13 mg/kg	[142]
Chlorella sp.	Water hyacinth	Cd^{2+}	Adsorption	water	92.5%	[143]
Leclercia adecarboxylata	Rice hull	Pb^{2+}	Entrapment	water	93%	[144]
Bacillus subtilis	Pig manure	Hg^{2+}, Pb^{2+} co-contamination	Adsorption	water	69 mg/g Hg 112.3 mg/g Pb	[145]
Bacillus subtilis	Corn straw	Hg^{2+}, Pb^{2+} co-contamination	Adsorption	water	53.7 mg/g Hg; 83.0 mg/g Pb	[145]
Enterobacter sp.	Rice husk BC	Pb^{2+}	Adsorption	-	24.1%	[146]
Enterobacter sp.	Sludge BC	Pb^{2+}	Adsorption	-	60.9%	[146]

Physical adsorption, ion exchange, surface complexation, precipitation, and biotransformation are some of the mechanisms involved in MCB-mediated heavy metals removal (Figure 7). Biochar containing oxygen functional groups, mineral components such as carbonates and phosphates, and microbial surface functional groups contribute to the removal of metal cations. Shen et al. [143] investigated the mechanism of cadmium removal using biochar-immobilized microalgae. They discovered that electrostatic attraction, surface complexation, and ion exchange were responsible for cadmium removal (maximum adsorption 217.41 mg/g) from wastewater. Similarly, Tu et al. [140] noted that surface complexation with different functional groups on cells, cation exchange, and surface complexation on biochar contributed to the enhanced stability of Cd^{2+} and Cu^{2+} in the contaminated soil. Microorganisms secrete enzymes that mediate redox reactions and surface complexation. These are the mechanisms involved in removing As^{3+}, As^{5+}, Cr^{6+}, U^{6+}, and Mn^{2+}. Youngwilai et al. [147] examined the mechanism of Mn^{2+} removal by the *Streptomyces violarus* strain, immobilized on biochar. They found that the two processes, namely, biological oxidation by the immobilized strain and adsorption by biochar, work together.

The presence of multiple contaminants at a particular contaminated site is a widespread phenomenon that could severely affect the microorganisms' remediation potential [148]. This limitation can be addressed by the associative effect of the benefits of biochar and the microorganisms via the immobilization of functional bacteria (such as organic contaminants–degraders) on biochar, as this could potentially remediate various types of contaminants. The application of biochar-microbial complex also increased the soil microbial and enzymatic activity, along with conducting the simultaneous bioremediation of multiple contaminants in several studies [126,148]. Several studies report that the degradation efficiency by biochar-immobilized bacterial consortia in co-contaminated sites is significantly enhanced compared to the free bacteria, due to their bioaugmentation abilities. For instance, Li et al. [149] immobilized PAH-degrading bacteria (*Citrobacter* sp.) into biochar, increased the degradation rate of PAH and reduced the toxicity of Ni by bio-transforming the available Ni into a stable form.

Pesticide degradation can be enhanced by introducing exogenous free cells to polluted soil. However, this method has several drawbacks, including the growth and survival of microbial cells, inadequate nutrients, lesser adaptability to surroundings, and competition with native microorganisms [139,150]. Immobilizing the exogenous pollutant-degrading bacteria on a support material can be an alternative strategy. This can be an ideal environment for their survival in different soil conditions [151]. Microorganisms immobilized in biochar have the potential to directly or indirectly reduce environmental pollution, while also allowing for the long-term maintenance of catalytic activity. Due to its superior porosity, ample surface area, and functional groups, biochar is an ideal medium and a rich nutrient composition for immobilizing and reproducing microbial cells [125].

Biochar can improve the soil's pollutant adsorption capacity while providing the nutrients for microbial growth and function [152]. Adsorption and covalent-binding methods were used to immobilize *Pseudomonas putida* onto coconut fiber-derived biochar. The efficacy of MCB in paraquat removal from contaminated water was studied by Ha et al. [129]. After 48 h of incubation, MCB could convert paraquat to 4,4-bipyridyl and malic acid. According to Wahla et al. [148], the immobilization of the MB3R consortium was achieved on biochar-remediated soil contaminated with metribuzin. The immobilization of a microbial consortium on biochar increased the rate of cypermethrin degradation and removal efficiency while lowering the cypermethrin's bioavailability to indigenous organisms [153]. Sun et al. [125] isolated and identified *Alcaligenes faecalis* WZ-2 as a tebuconazole-degrading strain and supported it on wheat straw biochar as a carrier. The biochar-immobilized WZ-2 reduced the half-life of tebuconazole in soil from 40.8 to 13.3 days and affected the microbial population and enzyme activities in polluted soil.

5. Conclusions and Future Prospective

Recent research on removing heavy metal ions and/or pesticides using biochar and microorganisms has revealed their enormous potential. Biomass-derived materials, such as biochar, have gradually been established as a viable platform for advancing the design and development of carbon-based materials and their suitability for various uses, such as, for instance, environmental remediation applications. A plethora of biochar production and activation approaches can be used, depending on the final application. In this review, the role of microorganisms and biochar in bioremediation is thoroughly discussed. Along with its usage as an adsorbent for heavy metal ions and pesticides, biochar can also be utilized as an immobilization support for microorganisms. Carbonaceous materials have been frequently used as carrier materials for bacterial immobilization, to enhance the bioremediation efficiency of organic pollutants. Compared with expensive carbon materials, biochar is more competitive as a carrier material, as it is cheaper but has acceptably high porosity, which could provide shelter and nutrients for microbial cells, facilitating the colonization of microbial cells and the formation of microbial hot spots on the surface and in the pores of biochar. According to previous research, adsorption and entrapment are the most common methods for preparing the MCB. Toxic metal ions and pesticides have been successfully removed using immobilized cells. The key factors influencing the removal efficiencies are the pollutant's concentration, incubation time, temperature, and pH.

The physical and chemical properties of biochar make it a suitable carrier/platform for microbial cell immobilization; however, this research area is still in its initial stages. The limitations related to the loss of activity of MCB and mass transfer potential have not been studied widely. Even though the immobilization of metal ions and pesticide-degrading microorganisms are cost-effective, stable, and environmentally friendly approaches, research can be conducted to enhance the treatment efficiency and improve the stability of microbial cells. The regeneration of the immobilized cells and recovery of the adsorbed pollutant can be improved. Most of the research focusing on immobilized microbes on biochar is mainly laboratory-based and involves the remediation of soil or an aqueous environment. The practical application of this in situ method is restricted, as the actual contamination sites are usually complicated. Research can be conducted to elucidate the heavy metal and pesticide degradation ability of a particular MCB from the soil and aqueous environment. The practical use of MCB can be further improved by increasing the efficacy and viability of the immobilized microbial cells and exploring approaches that would make the usage of MCB easier in contaminated sites. Moreover, the microbe-immobilized biochar can be employed in co-contaminated sites with heavy metals and pesticides for remediation. Genetically modified microorganisms are of increasing interest for the treatment of targeted pollutants. Therefore, further studies can be performed to genetically modify the microorganism for the targeted remediation of metal ions and pesticides, as well as to study the immobilization characteristics of these microbes on biochar.

Author Contributions: Conceptualization, V.N.; writing—original draft preparation, S.K.M., P.P. and K.S.; writing—review and editing, V.N., D.B., I.A.K. and D.A.G.; visualization, S.K.M. and D.A.G.; supervision, V.N. All authors have read and agreed to the published version of the manuscript.

Funding: This research received no external funding.

Institutional Review Board Statement: Not applicable.

Informed Consent Statement: Not applicable.

Data Availability Statement: Not applicable.

Acknowledgments: The authors acknowledge the National Institute of Technology, Karnataka, India, for providing the facility and financial support.

Conflicts of Interest: The authors declare no conflict of interest.

References

1. Priyadarshanee, M.; Das, S. Biosorption and removal of toxic heavy metals by metal tolerating bacteria for bioremediation of metal contamination: A comprehensive review. *J. Environ. Chem. Eng.* **2021**, *9*, 104686. [CrossRef]
2. Amin, M.; Chetpattananondh, P. Biochar from extracted marine *Chlorella* sp. residue for high efficiency adsorption with ultrasonication to remove Cr(VI), Zn(II) and Ni(II). *Bioresour. Technol.* **2019**, *289*, 121578. [CrossRef]
3. Sun, S.; Sidhu, V.; Rong, Y.; Zheng, Y. Pesticide pollution in agricultural soils and sustainable remediation methods: A review. *Curr. Pollut. Rep.* **2018**, *4*, 240–250. [CrossRef]
4. Enders, A.; Hanley, K.; Whitman, T.; Joseph, S.; Lehmann, J. Characterization of biochars to evaluate recalcitrance and agronomic performance. *Bioresour. Technol.* **2012**, *114*, 644–653. [CrossRef] [PubMed]
5. Li, L.; Lai, C.; Huang, F.; Cheng, M.; Zeng, G.; Huang, D.; Li, B.; Liu, S.; Zhang, M.M.; Qin, L.; et al. Degradation of naphthalene with magnetic biochar activate hydrogen peroxide: Synergism of bio-char and Fe–Mn binary oxides. *Water Res.* **2019**, *160*, 238–248. [CrossRef]
6. Kung, C.C.; Mu, J.E. Prospect of China's renewable energy development from pyrolysis and biochar applications under climate change. *Renew. Sustain. Energy Rev.* **2019**, *114*, 109343. [CrossRef]
7. Tan, X.; Liu, Y.; Zeng, G.; Wang, X.; Hu, X.; Gu, Y.; Yang, Z. Application of biochar for the removal of pollutants from aqueous solutions. *Chemosphere* **2015**, *125*, 70–85. [CrossRef]
8. Ahmad, M.; Rajapaksha, A.U.; Lim, J.E.; Zhang, M.; Bolan, N.; Mohan, D.; Vithanage, M.; Lee, S.S.; Ok, Y.S. Biochar as a sorbent for contaminant management in soil and water: A review. *Chemosphere* **2014**, *99*, 19–33. [CrossRef]
9. Kumar, S.; Loganathan, V.A.; Gupta, R.B.; Barnett, M.O. An Assessment of U(VI) removal from groundwater using biochar produced from hydrothermal carbonization. *J. Environ. Manag.* **2011**, *92*, 2504–2512. [CrossRef]
10. Jiang, Y.; Yang, F.; Dai, M.; Ali, I.; Shen, X.; Hou, X.; Alhewairini, S.S.; Peng, C.; Naz, I. Application of microbial immobilization technology for remediation of Cr(VI) contamination: A review. *Chemosphere* **2022**, *286*, 131721. [CrossRef]
11. Karimi, H.; Mahdavi, S.; Asgari Lajayer, B.; Moghiseh, E.; Rajput, V.D.; Minkina, T.; Astatkie, T. Insights on the bioremediation technologies for pesticide-contaminated soils. *Environ. Geochem. Health* **2022**, *44*, 1329–1354. [CrossRef]
12. Abu Talha, M.; Goswami, M.; Giri, B.S.; Sharma, A.; Rai, B.N.; Singh, R.S. Bioremediation of Congo red dye in immobilized batch and continuous packed bed bioreactor by *Brevibacillus parabrevis* using coconut shell bio-char. *Bioresour. Technol.* **2018**, *252*, 37–43. [CrossRef] [PubMed]
13. Liang, L.; Xi, F.; Tan, W.; Meng, X.; Hu, B.; Wang, X. Review of organic and inorganic pollutants removal by biochar and biochar-based composites. *Biochar* **2021**, *3*, 255–281. [CrossRef]
14. Noronha, F.R.; Manikandan, S.K.; Nair, V. Role of coconut shell biochar and earthworm (*Eudrilus euginea*) in bioremediation and palak spinach (*Spinacia oleracea* L.) growth in cadmium-contaminated soil. *J. Environ. Manag.* **2022**, *302*, 114057. [CrossRef] [PubMed]
15. Yaashikaa, P.R.; Kumar, P.S.; Varjani, S.; Saravanan, A. A critical review on the biochar production techniques, characterization, stability and applications for circular bioeconomy. *Biotechnol. Rep.* **2020**, *28*, e00570. [CrossRef]
16. Lee, J.; Kim, K.H.; Kwon, E.E. Biochar as a catalyst. *Renew. Sustain. Energy Rev.* **2017**, *77*, 70–79. [CrossRef]
17. Dehkhoda, A.M.; Gyenge, E.; Ellis, N. A novel method to tailor the porous structure of KOH-activated biochar and its application in capacitive deionization and energy storage. *Biomass Bioenergy* **2016**, *87*, 107–121. [CrossRef]
18. Liu, S.; Lai, C.; Zhou, X.; Zhang, C.; Chen, L.; Yan, H.; Qin, L.; Huang, D.; Ye, H.; Chen, W.; et al. Peroxydisulfate activation by sulfur-doped ordered mesoporous carbon: Insight into the intrinsic relationship between defects and $^1O_2$ generation. *Water Res.* **2022**, *221*, 118797. [CrossRef]
19. Akhil, D.; Lakshmi, D.; Kartik, A.; Vo, D.-V.N.; Arun, J.; Gopinath, K.P. Production, characterization, activation and environmental applications of engineered biochar: A review. *Environ. Chem. Lett.* **2021**, *19*, 2261–2297. [CrossRef]
20. Jindo, K.; Mizumoto, H.; Sawada, Y.; Sanchez-Monedero, M.A.; Sonoki, T. Physical and chemical characterization of biochars derived from different agricultural residues. *Biogeosciences* **2014**, *11*, 6613–6621. [CrossRef]
21. Park, S.J.; Lee, Y.J.; Kang, J.K.; Lee, J.; Lee, C.G. Application of Fe-impregnated biochar from cattle manure for removing pentavalent antimony from aqueous solution. *Appl. Sci.* **2021**, *11*, 9257. [CrossRef]
22. Kostas, E.T.; Durán-Jiménez, G.; Shepherd, B.J.; Meredith, W.; Stevens, L.A.; Williams, O.S.A.; Lye, G.J.; Robinson, J.P. Microwave pyrolysis of olive pomace for bio-oil and bio-char production. *Chem. Eng. J.* **2020**, *387*, 123404. [CrossRef]
23. Wang, J.; Wang, S. Preparation, modification and environmental application of biochar: A review. *J. Clean. Prod.* **2019**, *227*, 1002–1022. [CrossRef]
24. Li, M.; Zhao, Z.; Wu, X.; Zhou, W.; Zhu, L. Impact of mineral components in cow manure biochars on the adsorption and competitive adsorption of oxytetracycline and carbaryl. *RSC Adv.* **2017**, *7*, 2127–2136. [CrossRef]
25. Zhang, X.; Wang, Y.; Ju, N.; Ai, Y.; Liu, Y.; Liang, J.; Hu, Z.N.; Guo, R.; Xu, W.; Zhang, W.; et al. Ultimate resourcization of waste: Crab shell-derived biochar for antimony removal and sequential utilization as an anode for a Li-Ion Battery. *ACS Sustain. Chem. Eng.* **2021**, *9*, 8813–8823. [CrossRef]
26. Ding, T.; Huang, T.; Wu, Z.; Li, W.; Guo, K.; Li, J. Adsorption-desorption behavior of carbendazim by sewage sludge-derived biochar and its possible mechanism. *RSC Adv.* **2019**, *9*, 35209–35216. [CrossRef]

27. Igalavithana, A.D.; Mandal, S.; Niazi, N.K.; Vithanage, M.; Parikh, S.J.; Mukome, F.N.D.; Rizwan, M.; Oleszczuk, P.; Al-Wabel, M.; Bolan, N.; et al. Advances and future directions of biochar characterization methods and applications. *Crit. Rev. Environ. Sci. Technol.* **2017**, *47*, 2275–2330. [CrossRef]
28. Leng, L.; Xiong, Q.; Yang, L.; Li, H.; Zhou, Y.; Zhang, W.; Jiang, S.; Li, H.; Huang, H. An overview on engineering the surface area and porosity of biochar. *Sci. Total Environ.* **2021**, *763*, 144204. [CrossRef] [PubMed]
29. Ogura, A.P.; Lima, J.Z.; Marques, J.P.; Massaro Sousa, L.; Rodrigues, V.G.S.; Espíndola, E.L.G. A review of pesticides sorption in biochar from maize, rice, and wheat residues: Current status and challenges for soil application. *J. Environ. Manag.* **2021**, *300*, 113753. [CrossRef]
30. Ahmed, M.J.; Hameed, B.H. Insight into the co-pyrolysis of different blended feedstocks to biochar for the adsorption of organic and inorganic pollutants: A review. *J. Clean. Prod.* **2020**, *265*, 121762. [CrossRef]
31. Xu, Z.; He, M.; Xu, X.; Cao, X.; Tsang, D.C.W. Impacts of different activation processes on the carbon stability of biochar for oxidation resistance. *Bioresour. Technol.* **2021**, *338*, 125555. [CrossRef] [PubMed]
32. Yan, C.; Xu, Y.; Wang, L.; Liang, X.; Sun, Y.; Jia, H. Effect of different pyrolysis temperatures on physico-chemical characteristics and lead(II) removal of biochar derived from chicken manure. *RSC Adv.* **2020**, *10*, 3667–3674.
33. Pandey, D.; Daverey, A.; Arunachalam, K. Biochar: Production, properties and emerging role as a support for enzyme immobilization. *J. Clean. Prod.* **2020**, *255*, 120267. [CrossRef]
34. Vijayaraghavan, K. Recent advancements in biochar preparation, feedstocks, modification, characterization and future applications. *Environ. Technol. Rev.* **2019**, *8*, 47–64. [CrossRef]
35. Sakulthaew, C.; Watcharenwong, A.; Chokejaroenrat, C.; Rittirat, A. Leonardite-derived biochar suitability for effective sorption of herbicides. *Water Air Soil Pollut.* **2021**, *232*, 36. [CrossRef]
36. Rahman, M.A.; Lamb, D.; Rahman, M.M.; Bahar, M.M.; Sanderson, P.; Abbasi, S.; Bari, A.S.M.F.; Naidu, R. Removal of arsenate from contaminated waters by novel zirconium and zirconium-iron modified biochar. *J. Hazard. Mater.* **2021**, *409*, 124488. [CrossRef] [PubMed]
37. Park, J.H.; Wang, J.J.; Kim, S.H.; Kang, S.W.; Jeong, C.Y.; Jeon, J.R.; Park, K.H.; Cho, J.S.; Delaune, R.D.; Seo, D.C. Cadmium adsorption characteristics of biochars derived using various pine tree residues and pyrolysis temperatures. *J. Colloid Interface Sci.* **2019**, *553*, 298–307. [CrossRef]
38. Inyang, M.I.; Gao, B.; Yao, Y.; Xue, Y.; Zimmerman, A.; Mosa, A.; Pullammanappallil, P.; Ok, Y.S.; Cao, X. A review of biochar as a low-cost adsorbent for aqueous heavy metal removal. *Crit. Rev. Environ. Sci. Technol.* **2016**, *46*, 406–433. [CrossRef]
39. Imran, M.; Iqbal, M.M.; Iqbal, J.; Shah, N.S.; Khan, Z.U.H.; Murtaza, B.; Amjad, M.; Ali, S.; Rizwan, M. Synthesis, characterization and application of novel MnO and CuO impregnated biochar composites to sequester arsenic (As) from water: Modeling, thermodynamics and reusability. *J. Hazard. Mater.* **2021**, *401*, 123338. [CrossRef]
40. Zhang, J.; Shao, J.; Jin, Q.; Zhang, X.; Yang, H.; Chen, Y.; Zhang, S.; Chen, H. Effect of deashing on activation process and lead adsorption capacities of sludge-based biochar. *Sci. Total Environ.* **2020**, *716*, 137016. [CrossRef]
41. Fan, J.; Cai, C.; Chi, H.; Reid, B.J.; Coulon, F.; Zhang, Y.; Hou, Y. Remediation of cadmium and lead polluted soil using thiol-modified biochar. *J. Hazard. Mater.* **2020**, *388*, 122037. [CrossRef]
42. Ma, J.; Huang, W.; Zhang, X.; Li, Y.; Wang, N. The utilization of lobster shell to prepare low-cost biochar for high-efficient removal of copper and cadmium from aqueous: Sorption properties and mechanisms. *J. Environ. Chem. Eng.* **2021**, *9*, 104703. [CrossRef]
43. Khan, Z.H.; Gao, M.; Qiu, W.; Song, Z. Properties and adsorption mechanism of magnetic biochar modified with molybdenum disulfide for cadmium in aqueous solution. *Chemosphere* **2020**, *255*, 126995. [CrossRef] [PubMed]
44. Wan, S.; Qiu, L.; Li, Y.; Sun, J.; Gao, B.; He, F.; Wan, W. Accelerated antimony and copper removal by manganese oxide embedded in biochar with enlarged pore structure. *Chem. Eng. J.* **2020**, *402*, 126021. [CrossRef]
45. He, R.; Peng, Z.; Lyu, H.; Huang, H.; Nan, Q.; Tang, J. Synthesis and characterization of an iron-impregnated biochar for aqueous arsenic removal. *Sci. Total Environ.* **2018**, *612*, 1177–1186. [CrossRef] [PubMed]
46. Gao, X.; Peng, Y.; Guo, L.; Wang, Q.; Guan, C.Y.; Yang, F.; Chen, Q. Arsenic adsorption on layered double hydroxides biochars and their amended red and calcareous soils. *J. Environ. Manag.* **2020**, *271*, 111045. [CrossRef]
47. Kwak, J.H.; Islam, M.S.; Wang, S.; Messele, S.A.; Naeth, M.A.; El-Din, M.G.; Chang, S.X. Biochar properties and lead(II) adsorption capacity depend on feedstock type, pyrolysis temperature, and steam activation. *Chemosphere* **2019**, *231*, 393–404. [CrossRef] [PubMed]
48. Guo, J.; Yan, C.; Luo, Z.; Fang, H.; Hu, S.; Cao, Y. Synthesis of a novel ternary HA/Fe-Mn oxides-loaded biochar composite and its application in cadmium(II) and arsenic(V) adsorption. *J. Environ. Sci.* **2019**, *85*, 168–176. [CrossRef]
49. Liu, X.; Wang, Y.; Gui, C.; Li, P.; Zhang, J.; Zhong, H.; Wei, Y. Chemical forms and risk assessment of heavy metals in sludge-biochar produced by microwave-induced low temperature pyrolysis. *RSC Adv.* **2016**, *6*, 101960–101967. [CrossRef]
50. Yin, Z.; Xu, S.; Liu, S.; Xu, S.; Li, J.; Zhang, Y. A novel magnetic biochar prepared by K_2FeO_4-promoted oxidative pyrolysis of pomelo peel for adsorption of hexavalent chromium. *Bioresour. Technol.* **2020**, *300*, 122680. [CrossRef]
51. Deng, J.; Li, X.; Wei, X.; Liu, Y.; Liang, J.; Shao, Y.; Huang, W.; Cheng, X. Different adsorption behaviors and mechanisms of a novel amino-functionalized hydrothermal biochar for hexavalent chromium and pentavalent antimony. *Bioresour. Technol.* **2020**, *310*, 123438. [CrossRef]
52. Binh, Q.A.; Nguyen, H.H. Investigation the isotherm and kinetics of adsorption mechanism of herbicide 2,4-dichlorophenoxyacetic acid (2,4-D) on corn cob biochar. *Bioresour. Technol. Rep.* **2020**, *11*, 100520. [CrossRef]

53. Binh, Q.A.; Tungtakanpoung, D.; Kajitvichyanukul, P. Similarities and differences in adsorption mechanism of dichlorvos and pymetrozine insecticides with coconut fiber biowaste sorbent. *J. Environ. Sci. Health-Part B Pestic. Food Contam. Agric. Wastes* **2020**, *55*, 103–114. [CrossRef]
54. Liu, L.; Fang, W.; Yuan, M.; Li, X.; Wang, X.; Dai, Y. Metolachlor-adsorption on the walnut shell biochar modified by the fulvic acid and citric acid in water. *J. Environ. Chem. Eng.* **2021**, *9*, 106238. [CrossRef]
55. Vimal, V.; Patel, M.; Mohan, D. Aqueous carbofuran removal using slow pyrolyzed sugarcane bagasse biochar: Equilibrium and fixed-bed studies. *RSC Adv.* **2019**, *9*, 26338–26350. [CrossRef] [PubMed]
56. Essandoh, M.; Wolgemuth, D.; Pittman, C.U.; Mohan, D.; Mlsna, T. Adsorption of metribuzin from aqueous solution using magnetic and nonmagnetic sustainable low-cost biochar adsorbents. *Environ. Sci. Pollut. Res.* **2017**, *24*, 4577–4590. [CrossRef] [PubMed]
57. Chen, J.; Dong, J.; Chang, J.; Guo, T.; Yang, Q.; Jia, W.; Shen, S. Characterization of an Hg(II)-volatilizing *Pseudomonas* sp. strain, DC-B1, and its potential for soil remediation when combined with biochar amendment. *Ecotoxicol. Environ. Saf.* **2018**, *163*, 172–179. [CrossRef]
58. Qiao, J.; Li, X.; Li, F. Roles of different active metal-reducing bacteria in arsenic release from arsenic-contaminated paddy soil amended with biochar. *J. Hazard. Mater.* **2018**, *344*, 958–967. [CrossRef]
59. Qiao, J.T.; Li, X.M.; Hu, M.; Li, F.B.; Young, L.Y.; Sun, W.M.; Huang, W.; Cui, J.H. Transcriptional activity of arsenic-reducing bacteria and genes regulated by lactate and biochar during arsenic transformation in flooded paddy soil. *Environ. Sci. Technol.* **2018**, *52*, 61–70. [CrossRef]
60. An, X.; Chen, Y.; Ao, M.; Jin, Y.; Zhan, L.; Yu, B.; Wu, Z.; Jiang, P. Sequential photocatalytic degradation of organophosphorus pesticides and recovery of orthophosphate by biochar/α-Fe_2O_3/MgO composite: A new enhanced strategy for reducing the impacts of organophosphorus from wastewater. *Chem. Eng. J.* **2022**, *435*, 135087. [CrossRef]
61. Huang, W.H.; Wu, R.M.; Chang, J.S.; Juang, S.Y.; Lee, D.J. Pristine and manganese ferrite modified biochars for copper ion adsorption: Type-wide comparison. *Bioresour. Technol.* **2022**, *360*, 127529. [CrossRef] [PubMed]
62. Chen, X.; Zhang, M.; Qin, H.; Zhou, J.; Shen, Q.; Wang, K.; Chen, W.; Liu, M.; Li, N. Synergy effect between adsorption and heterogeneous photo-Fenton-like catalysis on $LaFeO_3$/lignin-biochar composites for high efficiency degradation of ofloxacin under visible light. *Sep. Purif. Technol.* **2022**, *280*, 119751. [CrossRef]
63. Wang, Z.; Ren, D.; Wu, J.; Jiang, S.; Yu, H.; Cheng, Y.; Zhang, S.; Zhang, X. Study on adsorption-degradation of 2,4-dichlorophenol by modified biochar immobilized laccase. *Int. J. Environ. Sci. Technol.* **2022**, *19*, 1393–1406. [CrossRef]
64. Castro, C.; Urbieta, M.S.; Plaza Cazón, J.; Donati, E.R. Metal biorecovery and bioremediation: Whether or not thermophilic are better than mesophilic microorganisms. *Bioresour. Technol.* **2019**, *279*, 317–326. [CrossRef]
65. Zhang, H.; Yuan, X.; Xiong, T.; Wang, H.; Jiang, L. Bioremediation of co-contaminated soil with heavy metals and pesticides: Influence factors, mechanisms and evaluation methods. *Chem. Eng. J.* **2020**, *398*, 125657. [CrossRef]
66. Tariq, M.; Waseem, M.; Rasool, M.H.; Zahoor, M.A.; Hussain, I. Isolation and molecular characterization of the indigenous *Staphylococcus aureus* strain K1 with the ability to reduce hexavalent chromium for its application in bioremediation of metal-contaminated sites. *PeerJ* **2019**, *2019*, 1–20. [CrossRef] [PubMed]
67. Sharma, B.; Shukla, P. Lead bioaccumulation mediated by *Bacillus cereus* BPS-9 from an industrial waste contaminated site encoding heavy metal resistant genes and their transporters. *J. Hazard. Mater.* **2021**, *401*, 123285. [CrossRef]
68. Mwandira, W.; Nakashima, K.; Kawasaki, S.; Arabelo, A.; Banda, K.; Nyambe, I.; Chirwa, M.; Ito, M.; Sato, T.; Igarashi, T.; et al. Biosorption of Pb (II) and Zn (II) from aqueous solution by *Oceanobacillus profundus* isolated from an abandoned mine. *Sci. Rep.* **2020**, *10*, 21189. [CrossRef]
69. Li, Y.; Xin, M.; Xie, D.; Fan, S.; Ma, J.; Liu, K.; Yu, F. Variation in extracellular polymeric substances from *Enterobacter* sp. and their Pb^{2+} adsorption behaviors. *ACS Omega* **2021**, *6*, 9617–9628. [CrossRef]
70. Liu, J.; Xue, J.; Wei, X.; Su, H.; Xu, R. Optimization of Cr^{6+} removal by *Bacillus subtilis* strain SZMC 6179J from chromium-containing soil. *Indian J. Microbiol.* **2020**, *60*, 430–435. [CrossRef]
71. Oyewole, O.A.; Zobeashia, S.S.L.T.; Oladoja, E.O.; Raji, R.O.; Odiniya, E.E.; Musa, A.M. Biosorption of heavy metal polluted soil using bacteria and fungi isolated from soil. *SN Appl. Sci.* **2019**, *1*, 857. [CrossRef]
72. Sathishkumar, K.; Murugan, K.; Benelli, G.; Higuchi, A.; Rajasekar, A. Bioreduction of hexavalent chromium by *Pseudomonas stutzeri* L1 and *Acinetobacter baumannii* L2. *Ann. Microbiol.* **2017**, *67*, 91–98. [CrossRef]
73. Kumaresan Sarankumar, R.; Arulprakash, A.; Devanesan, S.; Selvi, A.; AlSalhi, M.S.; Rajasekar, A.; Ahamed, A. Bioreduction of hexavalent chromium by chromium resistant alkalophilic bacteria isolated from tannery effluent. *J. King Saud Univ.-Sci.* **2020**, *32*, 1969–1977. [CrossRef]
74. Li, W.; Chen, Y.; Wang, T. Cadmium biosorption by lactic acid bacteria *Weissella viridescens* ZY-6. *Food Control* **2021**, *123*, 107747. [CrossRef]
75. Dey, U.; Chatterjee, S.; Mondal, N.K. Isolation and characterization of arsenic-resistant bacteria and possible application in bioremediation. *Biotechnol. Rep.* **2016**, *10*, 1–7. [CrossRef] [PubMed]
76. Zhang, D.; Yin, C.; Abbas, N.; Mao, Z.; Zhang, Y. Multiple heavy metal tolerance and removal by an earthworm gut fungus *Trichoderma brevicompactum* QYCD-6. *Sci. Rep.* **2020**, *10*, 6940. [CrossRef] [PubMed]

77. Abinandan, S.; Subashchandrabose, S.R.; Venkateswarlu, K.; Perera, I.A.; Megharaj, M. Acid-tolerant microalgae can withstand higher concentrations of invasive cadmium and produce sustainable biomass and biodiesel at pH 3.5. *Bioresour. Technol.* **2019**, *281*, 469–473. [CrossRef]
78. Cheng, J.; Yin, W.; Chang, Z.; Lundholm, N.; Jiang, Z. Biosorption capacity and kinetics of cadmium(II) on live and dead *Chlorella vulgaris*. *J. Appl. Phycol.* **2017**, *29*, 211–221. [CrossRef]
79. Saavedra, R.; Muñoz, R.; Taboada, M.E.; Vega, M.; Bolado, S. Comparative uptake study of arsenic, boron, copper, manganese and zinc from water by different green microalgae. *Bioresour. Technol.* **2018**, *263*, 49–57. [CrossRef]
80. Rugnini, L.; Costa, G.; Congestri, R.; Bruno, L. Testing of two different strains of green microalgae for Cu and Ni removal from aqueous media. *Sci. Total Environ.* **2017**, *601–602*, 959–967. [CrossRef]
81. Kumar, V.; Dwivedi, S.K. Mycoremediation of heavy metals: Processes, mechanisms, and affecting factors. *Environ. Sci. Pollut. Res.* **2021**, *28*, 10375–10412. [CrossRef] [PubMed]
82. Ilyas, S.; Srivastava, R.R.; Ilyas, N. Biosorption of strontium from aqueous solutions. *Handb. Environ. Chem.* **2020**, *88*, 65–83. [CrossRef]
83. Escudero, L.B.; Quintas, P.Y.; Wuilloud, R.G.; Dotto, G.L. Recent advances on elemental biosorption. *Environ. Chem. Lett.* **2019**, *17*, 409–427. [CrossRef]
84. Hansda, A.; Kumar, V.; Anshumali. A comparative review towards potential of microbial cells for heavy metal removal with emphasis on biosorption and bioaccumulation. *World J. Microbiol. Biotechnol.* **2016**, *32*, 170. [CrossRef]
85. Gupta, P.; Diwan, B. Bacterial exopolysaccharide mediated heavy metal removal: A review on biosynthesis, mechanism and remediation strategies. *Biotechnol. Rep.* **2017**, *13*, 58–71. [CrossRef]
86. Shen, L.; Li, Z.; Wang, J.; Liu, A.; Li, Z.; Yu, R.; Wu, X.; Liu, Y.; Li, J.; Zeng, W. Characterization of extracellular polysaccharide/protein contents during the adsorption of Cd(II) by *Synechocystis* sp. PCC6803. *Environ. Sci. Pollut. Res.* **2018**, *25*, 20713–20722. [CrossRef]
87. Xie, Y.; He, N.; Wei, M.; Wen, T.; Wang, X.; Liu, H.; Zhong, S.; Xu, H. Cadmium biosorption and mechanism investigation using a novel *Bacillus subtilis* KC6 isolated from pyrite mine. *J. Clean. Prod.* **2021**, *312*, 127749. [CrossRef]
88. Mishra, S.; Lin, Z.; Pang, S.; Zhang, Y.; Bhatt, P.; Chen, S. Biosurfactant is a powerful tool for the bioremediation of heavy metals from contaminated soils. *J. Hazard. Mater.* **2021**, *418*, 126253. [CrossRef]
89. Lopes, C.S.C.; Teixeira, D.B.; Braz, B.F.; Santelli, R.E.; de Castilho, L.V.A.; Gomez, J.G.C.; Castro, R.P.V.; Seldin, L.; Freire, D.M.G. Application of rhamnolipid surfactant for remediation of toxic metals of long- and short-term contamination sites. *Int. J. Environ. Sci. Technol.* **2021**, *18*, 575–588. [CrossRef]
90. Ayangbenro, A.S.; Babalola, O.O. Genomic analysis of *Bacillus cereus* NWUAB01 and its heavy metal removal from polluted soil. *Sci. Rep.* **2020**, *10*, 19660. [CrossRef]
91. Elgarahy, A.M.; Elwakeel, K.Z.; Mohammad, S.H.; Elshoubaky, G.A. A critical review of biosorption of dyes, heavy metals and metalloids from wastewater as an efficient and green process. *Clean. Eng. Technol.* **2021**, *4*, 100209. [CrossRef]
92. Nokman, W.; Benluvankar, V.; Maria Packiam, S.; Vincent, S. Screening and molecular identification of heavy metal resistant *Pseudomonas putida* S4 in tannery effluent wastewater. *Biocatal. Agric. Biotechnol.* **2019**, *18*, 101052. [CrossRef]
93. Bose, S.; Kumar, P.S.; Vo, D.V.N.; Rajamohan, N.; Saravanan, R. Microbial degradation of recalcitrant pesticides: A review. *Environ. Chem. Lett.* **2021**, *19*, 3209–3228. [CrossRef]
94. Dash, D.M.; Osborne, W.J. Rapid biodegradation and biofilm-mediated bioremoval of organophosphorus pesticides using an indigenous *Kosakonia oryzae* strain -VITPSCQ3 in a Vertical-flow Packed Bed Biofilm Bioreactor. *Ecotoxicol. Environ. Saf.* **2020**, *192*, 110290. [CrossRef]
95. Xu, B.; Sun, Q.J.; Lan, J.C.W.; Chen, W.M.; Hsueh, C.C.; Chen, B.Y. Exploring the glyphosate-degrading characteristics of a newly isolated, highly adapted indigenous bacterial strain, *Providencia rettgeri* GDB 1. *J. Biosci. Bioeng.* **2019**, *128*, 80–87. [CrossRef] [PubMed]
96. Kolekar, P.D.; Patil, S.M.; Suryavanshi, M.V.; Suryawanshi, S.S.; Khandare, R.V.; Govindwar, S.P.; Jadhav, J.P. Microcosm study of atrazine bioremediation by indigenous microorganisms and cytotoxicity of biodegraded metabolites. *J. Hazard. Mater.* **2019**, *374*, 66–73. [CrossRef] [PubMed]
97. Bhatt, P.; Zhou, X.; Huang, Y.; Zhang, W.; Chen, S. Characterization of the role of esterases in the biodegradation of organophosphate, carbamate, and pyrethroid pesticides. *J. Hazard. Mater.* **2021**, *411*, 125026. [CrossRef]
98. Aswathi, A.; Pandey, A.; Sukumaran, R.K. Rapid degradation of the organophosphate pesticide—Chlorpyrifos by a novel strain of *Pseudomonas nitroreducens* AR-3. *Bioresour. Technol.* **2019**, *292*, 122025. [CrossRef]
99. Zhang, Y.H.; Xu, D.; Zhao, X.H.; Song, Y.; Liu, Y.L.; Li, H.N. Biodegradation of two organophosphorus pesticides in whole corn silage as affected by the cultured *Lactobacillus plantarum*. *3 Biotech* **2016**, *6*, 73. [CrossRef] [PubMed]
100. Sirajuddin, S.; Khan, M.A.; Qader, S.A.U.; Iqbal, S.; Sattar, H.; Ansari, A. A comparative study on degradation of complex malathion organophosphate using of *Escherichia coli* IES-02 and a novel carboxylesterase. *Int. J. Biol. Macromol.* **2020**, *145*, 445–455. [CrossRef]
101. Carles, L.; Joly, M.; Bonnemoy, F.; Leremboure, M.; Donnadieu, F.; Batisson, I.; Besse-Hoggan, P. Biodegradation and toxicity of a maize herbicide mixture: Mesotrione, nicosulfuron and S-metolachlor. *J. Hazard. Mater.* **2018**, *354*, 42–53. [CrossRef]

102. Ekram, M.A.E.; Sarker, I.; Rahi, M.S.; Rahman, M.A.; Saha, A.K.; Reza, M.A. Efficacy of soil-borne *Enterobacter* sp. for carbofuran degradation: HPLC quantitation of degradation rate. *J. Basic Microbiol.* **2020**, *60*, 390–399. [CrossRef]
103. Soares, P.R.S.; Birolli, W.G.; Ferreira, I.M.; Porto, A.L.M. Biodegradation pathway of the organophosphate pesticides chlorpyrifos, methyl parathion and profenofos by the marine-derived fungus *Aspergillus sydowii* CBMAI 935 and its potential for methylation reactions of phenolic compounds. *Mar. Pollut. Bull.* **2021**, *166*, 112185. [CrossRef]
104. Kaur, P.; Balomajumder, C. Effective mycoremediation coupled with bioaugmentation studies: An advanced study on newly isolated *Aspergillus* sp. in Type-II pyrethroid-contaminated soil. *Environ. Pollut.* **2020**, *261*, 114073. [CrossRef] [PubMed]
105. Nicodemus, T.J.; DiRusso, C.C.; Wilson, M.; Black, P.N. Reactive Oxygen Species (ROS) mediated degradation of organophosphate pesticides by the green microalgae *Coccomyxa subellipsoidea*. *Bioresour. Technol. Rep.* **2020**, *11*, 100461. [CrossRef]
106. Hu, N.; Xu, Y.; Sun, C.; Zhu, L.; Sun, S.; Zhao, Y.; Hu, C. Removal of atrazine in catalytic degradation solutions by microalgae *Chlorella* sp. and evaluation of toxicity of degradation products via algal growth and photosynthetic activity. *Ecotoxicol. Environ. Saf.* **2021**, *207*, 111546. [CrossRef] [PubMed]
107. Cycoń, M.; Mrozik, A.; Piotrowska-Seget, Z. Bioaugmentation as a strategy for the remediation of pesticide-polluted soil: A review. *Chemosphere* **2017**, *172*, 52–71. [CrossRef]
108. Chen, S.; Li, Y.; Fan, Z.; Liu, F.; Liu, H.; Wang, L.; Wu, H. Soil bacterial community dynamics following bioaugmentation with *Paenarthrobacter* sp. W11 in atrazine-contaminated soil. *Chemosphere* **2021**, *282*, 130976. [CrossRef]
109. Papadopoulou, E.S.; Genitsaris, S.; Omirou, M.; Perruchon, C.; Stamatopoulou, A.; Ioannides, I.; Karpouzas, D.G. Bioaugmentation of thiabendazole-contaminated soils from a wastewater disposal site: Factors driving the efficacy of this strategy and the diversity of the indigenous soil bacterial community. *Environ. Pollut.* **2018**, *233*, 16–25. [CrossRef]
110. Aldas-Vargas, A.; van der Vooren, T.; Rijnaarts, H.H.M.; Sutton, N.B. Biostimulation is a valuable tool to assess pesticide biodegradation capacity of groundwater microorganisms. *Chemosphere* **2021**, *280*, 130793. [CrossRef]
111. Raimondo, E.E.; Aparicio, J.D.; Bigliardo, A.L.; Fuentes, M.S.; Benimeli, C.S. Enhanced bioremediation of lindane-contaminated soils through microbial bioaugmentation assisted by biostimulation with sugarcane filter cake. *Ecotoxicol. Environ. Saf.* **2020**, *190*, 110143. [CrossRef]
112. Nie, J.; Sun, Y.; Zhou, Y.; Kumar, M.; Usman, M.; Li, J.; Shao, J.; Wang, L.; Tsang, D.C.W. Bioremediation of water containing pesticides by microalgae: Mechanisms, methods, and prospects for future research. *Sci. Total Environ.* **2020**, *707*, 136080. [CrossRef] [PubMed]
113. Gupta, J.; Rathour, R.; Singh, R.; Thakur, I.S. Production and characterization of extracellular polymeric substances (EPS) generated by a carbofuran degrading strain *Cupriavidus* sp. ISTL7. *Bioresour. Technol.* **2019**, *282*, 417–424. [CrossRef]
114. Satapute, P.; Jogaiah, S. A biogenic microbial biosurfactin that degrades difenoconazole fungicide with potential antimicrobial and oil displacement properties. *Chemosphere* **2022**, *286*, 131694. [CrossRef]
115. Kumar, M.; Yadav, A.N.; Saxena, R.; Paul, D.; Tomar, R.S. Biodiversity of pesticides degrading microbial communities and their environmental impact. *Biocatal. Agric. Biotechnol.* **2021**, *31*, 101883. [CrossRef]
116. Dash, D.M.; Osborne, J.W. Biodegradation of monocrotophos by a plant growth promoting *Bacillus aryabhattai* (VITNNDJ5) strain in artificially contaminated soil. *Int. J. Environ. Sci. Technol.* **2020**, *17*, 1475–1490. [CrossRef]
117. Marican, A.; Durán-Lara, E.F. A review on pesticide removal through different processes. *Environ. Sci. Pollut. Res.* **2018**, *25*, 2051–2064. [CrossRef]
118. Kumar, S.; Kaushik, G.; Dar, M.A.; Nimesh, S.; López-chuken, U.J.; Villarreal-chiu, J.F. Microbial Degradation of Organophosphate Pesticides: A Review. *Pedosphere* **2018**, *28*, 190–208. [CrossRef]
119. Gangola, S.; Sharma, A.; Bhatt, P.; Khati, P.; Chaudhary, P. Presence of esterase and laccase in *Bacillus subtilis* facilitates biodegradation and detoxification of cypermethrin. *Sci. Rep.* **2018**, *8*, 12755. [CrossRef] [PubMed]
120. Zhang, Y.; Wang, H.; Wang, X.; Hu, B.; Zhang, C.; Jin, W.; Zhu, S.; Hu, G.; Hong, Q. Identification of the key amino acid sites of the carbendazim hydrolase (MheI) from a novel carbendazim-degrading strain *Mycobacterium* sp. SD-4. *J. Hazard. Mater.* **2017**, *331*, 55–62. [CrossRef]
121. Bigley, A.N.; Desormeaux, E.; Xiang, D.F.; Bae, S.Y.; Harvey, S.P.; Raushel, F.M. Overcoming the challenges of enzyme evolution to adapt phosphotriesterase for V-Agent decontamination. *Biochemistry* **2019**, *58*, 2039–2053. [CrossRef]
122. Chauhan, P.S.; Jha, B. Pilot scale production of extracellular thermo-alkali stable laccase from *Pseudomonas* sp. S2 using agro waste and its application in organophosphorous pesticides degradation. *J. Chem. Technol. Biotechnol.* **2018**, *93*, 1022–1030. [CrossRef]
123. Sarker, A.; Lee, S.H.; Kwak, S.Y.; Nandi, R.; Kim, J.E. Comparative catalytic degradation of a metabolite 3,5-dichloroaniline derived from dicarboximide fungicide by laccase and MnO_2 mediators. *Ecotoxicol. Environ. Saf.* **2020**, *196*, 110561. [CrossRef] [PubMed]
124. Zeng, S.; Qin, X.; Xia, L. Degradation of the herbicide isoproturon by laccase-mediator systems. *Biochem. Eng. J.* **2017**, *119*, 92–100. [CrossRef]
125. Sun, T.; Miao, J.; Saleem, M.; Zhang, H.; Yang, Y.; Zhang, Q. Bacterial compatibility and immobilization with biochar improved tebuconazole degradation, soil microbiome composition and functioning. *J. Hazard. Mater.* **2020**, *398*, 122941. [CrossRef] [PubMed]

126. Wang, C.; Gu, L.; Ge, S.; Liu, X.; Zhang, X.; Chen, X. Remediation potential of immobilized bacterial consortium with biochar as carrier in pyrene-Cr(VI) co-contaminated soil. *Environ. Technol.* **2019**, *40*, 2345–2353. [CrossRef]
127. Yu, T.; Wang, L.; Ma, F.; Wang, Y.; Bai, S. A bio-functions integration microcosm: Self-immobilized biochar-pellets combined with two strains of bacteria to remove atrazine in water and mechanisms. *J. Hazard. Mater.* **2020**, *384*, 121326. [CrossRef]
128. Zhang, S.; Wang, J. Removal of chlortetracycline from water by *Bacillus cereus* immobilized on Chinese medicine residues biochar. *Environ. Technol. Innov.* **2021**, *24*, 101930. [CrossRef]
129. Ha, N.T.H.; Toan, N.C.; Kajitvichyanukul, P. Enhanced paraquat removal from contaminated water using cell-immobilized biochar. *Clean Technol. Environ. Policy* **2022**, *24*, 1073–1085. [CrossRef]
130. Manikandan, S.K.; Giannakoudakis, D.A.; Prekodravac, J.R.; Carlos, J. Role of catalyst supports in biocatalysis. *J. Chem. Technol. Biotechnol.* **2022**, *98*, 7–21. [CrossRef]
131. Guo, Q.; Bandala, E.R.; Goonetilleke, A.; Hong, N.; Li, Y.; Liu, A. Application of *Chlorella pyrenoidosa* embedded biochar beads for water treatment. *J. Water Process Eng.* **2021**, *40*, 1073–1085. [CrossRef]
132. Huang, S.W.; Chen, X.; Wang, D.D.; Jia, H.L.; Wu, L. Bio-reduction and synchronous removal of hexavalent chromium from aqueous solutions using novel microbial cell/algal-derived biochar particles: Turning an environmental problem into an opportunity. *Bioresour. Technol.* **2020**, *309*, 123304. [CrossRef] [PubMed]
133. Chojnacka, K. Biosorption and bioaccumulation—The prospects for practical applications. *Environ. Int.* **2010**, *36*, 299–307. [CrossRef]
134. Manikandan, S.K.; Nair, V. *Pseudomonas stutzeri* immobilized sawdust biochar for nickel ion removal. *Catalysts* **2022**, *12*, 1495. [CrossRef]
135. Qi, X.; Gou, J.; Chen, X.; Xiao, S.; Ali, I.; Shang, R.; Wang, D.; Wu, Y.; Han, M.; Luo, X. Application of mixed bacteria-loaded biochar to enhance uranium and cadmium immobilization in a co-contaminated soil. *J. Hazard. Mater.* **2021**, *401*, 123823. [CrossRef] [PubMed]
136. Xing, Y.; Luo, X.; Liu, S.; Wan, W.; Huang, Q.; Chen, W. A novel eco-friendly recycling of food waste for preparing biofilm-attached biochar to remove Cd and Pb in wastewater. *J. Clean. Prod.* **2021**, *311*, 127514. [CrossRef]
137. Osman, A.I.; Fawzy, S.; Farghali, M.; El-Azazy, M.; Elgarahy, A.M.; Fahim, R.A.; Maksoud, M.I.A.A.; Ajlan, A.A.; Yousry, M.; Saleem, Y.; et al. Biochar for agronomy, animal farming, anaerobic digestion, composting, water treatment, soil remediation, construction, energy storage, and carbon sequestration: A review. *Environ. Chem. Lett.* **2022**, *20*, 2385–2485. [CrossRef]
138. Chuaphasuk, C.; Prapagdee, B. Effects of biochar-immobilized bacteria on phytoremediation of cadmium-polluted soil. *Environ. Sci. Pollut. Res.* **2019**, *26*, 23679–23688. [CrossRef]
139. Zhang, X.; Li, Y.; Li, H. Enhanced bio-immobilization of Pb contaminated soil by immobilized bacteria with biochar as carrier. *Pol. J. Environ. Stud.* **2017**, *26*, 413–418. [CrossRef]
140. Tu, C.; Wei, J.; Guan, F.; Liu, Y.; Sun, Y.; Luo, Y. Biochar and bacteria inoculated biochar enhanced Cd and Cu immobilization and enzymatic activity in a polluted soil. *Environ. Int.* **2020**, *137*, 105576. [CrossRef]
141. Ma, H.; Wei, M.; Wang, Z.; Hou, S.; Li, X.; Xu, H. Bioremediation of cadmium polluted soil using a novel cadmium immobilizing plant growth promotion strain *Bacillus* sp. TZ5 loaded on biochar. *J. Hazard. Mater.* **2020**, *388*, 122065. [CrossRef] [PubMed]
142. Liu, Y.; Tie, B.; Peng, O.; Luo, H.; Li, D.; Liu, S.; Lei, M.; Wei, X.; Liu, X.; Du, H. Inoculation of Cd-contaminated paddy soil with biochar-supported microbial cell composite: A novel approach to reducing cadmium accumulation in rice grains. *Chemosphere* **2020**, *247*, 125850. [CrossRef]
143. Shen, Y.; Li, H.; Zhu, W.; Ho, S.H.; Yuan, W.; Chen, J.; Xie, Y. Microalgal-biochar immobilized complex: A novel efficient biosorbent for cadmium removal from aqueous solution. *Bioresour. Technol.* **2017**, *244*, 1031–1038. [CrossRef] [PubMed]
144. Teng, Z.; Shao, W.; Zhang, K.; Yu, F.; Huo, Y.; Li, M. Enhanced passivation of lead with immobilized phosphate solubilizing bacteria beads loaded with biochar/ nanoscale zero valent iron composite. *J. Hazard. Mater.* **2020**, *384*, 121505. [CrossRef]
145. Wang, T.; Sun, H.; Ren, X.; Li, B.; Mao, H. Adsorption of heavy metals from aqueous solution by UV-mutant *Bacillus subtilis* loaded on biochars derived from different stock materials. *Ecotoxicol. Environ. Saf.* **2018**, *148*, 285–292. [CrossRef] [PubMed]
146. Chen, H.; Zhang, J.; Tang, L.; Su, M.; Tian, D.; Zhang, L.; Li, Z.; Hu, S. Enhanced Pb immobilization via the combination of biochar and phosphate solubilizing bacteria. *Environ. Int.* **2019**, *127*, 395–401. [CrossRef]
147. Youngwilai, A.; Kidkhunthod, P.; Jearanaikoon, N.; Chaiprapa, J.; Supanchaiyamat, N.; Hunt, A.J.; Ngernyen, Y.; Ratpukdi, T.; Khan, E.; Siripattanakul-Ratpukdi, S. Simultaneous manganese adsorption and biotransformation by *Streptomyces violarus* strain SBP1 cell-immobilized biochar. *Sci. Total Environ.* **2020**, *713*, 136708. [CrossRef] [PubMed]
148. Wahla, A.Q.; Anwar, S.; Mueller, J.A.; Arslan, M.; Iqbal, S. Immobilization of metribuzin degrading bacterial consortium MB3R on biochar enhances bioremediation of potato vegetated soil and restores bacterial community structure. *J. Hazard. Mater.* **2020**, *390*, 121493. [CrossRef]
149. Li, X.; Wang, Y.; Luo, T.; Ma, Y.; Wang, B.; Huang, Q. Remediation potential of immobilized bacterial strain with biochar as carrier in petroleum hydrocarbon and Ni co-contaminated soil. *Environ. Technol.* **2022**, *43*, 1068–1081. [CrossRef]
150. de-Bashan, L.E.; Bashan, Y. Immobilized microalgae for removing pollutants: Review of practical aspects. *Bioresour. Technol.* **2010**, *101*, 1611–1627. [CrossRef]
151. Chen, B.; Yuan, M.; Qian, L. Enhanced bioremediation of PAH-contaminated soil by immobilized bacteria with plant residue and biochar as carriers. *J. Soils Sediments* **2012**, *12*, 1350–1359. [CrossRef]

152. Zheng, Y.; Han, X.; Li, Y.; Yang, J.; Li, N.; An, N. Effects of biochar and straw application on the physicochemical and biological properties of paddy soils in Northeast China. *Sci. Rep.* **2019**, *9*, 16531. [CrossRef] [PubMed]
153. Liu, J.; Ding, Y.; Ma, L.; Gao, G.; Wang, Y. Combination of biochar and immobilized bacteria in cypermethrin-contaminated soil remediation. *Int. Biodeterior. Biodegrad.* **2017**, *120*, 15–20. [CrossRef]

Disclaimer/Publisher's Note: The statements, opinions and data contained in all publications are solely those of the individual author(s) and contributor(s) and not of MDPI and/or the editor(s). MDPI and/or the editor(s) disclaim responsibility for any injury to people or property resulting from any ideas, methods, instructions or products referred to in the content.

MDPI AG
Grosspeteranlage 5
4052 Basel
Switzerland
Tel.: +41 61 683 77 34

Molecules Editorial Office
E-mail: molecules@mdpi.com
www.mdpi.com/journal/molecules

Disclaimer/Publisher's Note: The title and front matter of this reprint are at the discretion of the Guest Editors. The publisher is not responsible for their content or any associated concerns. The statements, opinions and data contained in all individual articles are solely those of the individual Editors and contributors and not of MDPI. MDPI disclaims responsibility for any injury to people or property resulting from any ideas, methods, instructions or products referred to in the content.

www.ingramcontent.com/pod-product-compliance
Lightning Source LLC
LaVergne TN
LVHW072318090526
838202LV00019B/2307